MOLECULES AND RADIATION

An Introduction to Modern Molecular Spectroscopy

SECOND EDITION

JEFFREY I. STEINFELD
Massachusetts Institute of Technology

DOVER PUBLICATIONS, INC.
Mineola, New York

Copyright

Bibliographical Note

This Dover edition, first published in 2005, is an unabridged and slightly corrected republication of the second edition of the work, originally published by The MIT Press, Cambridge, Massachusetts, in 1985 [first edition: 1974].

Library of Congress Cataloging-in-Publication Data

Steinfeld, Jeffrey I.
Molecules and radiation : an introduction to modern molecular spectroscopy / Jeffrey I. Steinfeld.—2nd ed.
p. cm.
Originally published: 2nd ed. Cambridge, Mass. : MIT Press, c1985.
Includes bibliographical references and index.
ISBN 0-486-44152-0 (pbk.)
1. Molecular spectroscopy. I. Title.

QC454.M6S83 2005
543'.54—dc22

2005041347

Manufactured in the United States of America
Dover Publications, Inc., 31 East 2nd Street, Mineola, N.Y. 11501

To my family

Contents

Preface to the Second Edition xiii
Preface to the First Edition xv

1 Review of the Quantum Mechanical Background 1

1 State Vector Notation 1
2 Wave Function and Matrix Representations 4
3 Basis Functions for the Energy 7
4 Perturbation Theory of Energy Corrections 8
5 Time-Dependent Perturbations: Absorption and Emission of Electromagnetic Radiation 14
6 The Electric Dipole Interaction 19
7 Optical Constants of Atoms and Molecules 25
8 Homogeneous and Inhomogeneous Line Shapes 31
9 The Road Ahead 36
Problems 36
References 38

2 Atomic Spectra 39

1 The Hydrogen Atom—a One-Electron System 39
2 Two- and Many-Electron Atoms 40
3 Spin-Orbit Coupling in a One-Electron Atom 48
4 Vector Precession Diagrams: *LS* and *jj* Coupling 51
5 Examples of Energy Levels and Spectra 52
6 Nuclear Hyperfine Interactions in Atoms 60
7 The Stark Effect 64
8 The Zeeman Effect 67
Problems 74
References 75

3 Diatomic Molecules 77

1 The Born-Oppenheimer Approximation 77
2 Electronic Energy Levels for Molecules with Stationary Nuclei 79
3 Determination of Molecular States from Separated Atom States 85

4 Selection Rules for Electronic Transitions 87
5 Simple Molecular Orbital Theory and Molecular Potential Curves 92
Problems 111
References 112

4 Rotation and Vibration of Diatomic Molecules 113

1 Rigid Rotor Hamiltonian, Eigenfunctions, and Spectra 113
2 Harmonic Oscillator Hamiltonian, Eigenfunctions, and Spectra 120
3 Combination of Rotational and Vibrational Motions 124
4 The Nonrigid Rotor: Centrifugal Distortions 127
5 The Anharmonic Oscillator 128
6 Rotational and Vibrational Raman Spectra 134
Problems 142
References 144

5 Electronic Spectra of Diatomic Molecules 145

1 "Ro-vibronic" Energy Levels 145
2 The Franck-Condon Principle 145
3 Vibrational Band Analysis 150
4 Rotational Fine Structure of Bands 157
5 Angular Momentum Coupling Cases 161
6 Selection Rules in Electronic Transitions 167
7 Perturbations and Predissociation 171
8 Zeeman Effect in Molecules 179
9 Stark Effect in Molecules 183
10 Magnetooptic Rotation (Faraday Effect) in Molecules 184
11 Photofragment Spectroscopy of Continuum States 187
Problems 192
References 196

6 Rudimentary Group Theory 198

1 Symmetry Elements E, C_n, σ, and i 199
2 Point Groups for Molecules 202
3 Group Properties and Multiplication Tables 202

4 Representations and Character Theory 205
5 Direct Products of Representations 207
6 Selected Applications 209
Problems 211
References 212

7 **Rotational Spectra of Polyatomic Molecules** 214

1 Rigid Body Hamiltonian, Eigenfunctions, and Spectra 214
2 Microwave Spectroscopy of Rotational Levels 219
3 Additional Topics in Rotational Spectroscopy 223
Problems 226
References 228

8 **Vibrational Spectra of Polyatomic Molecules** 230

1 Lagrangian Mechanics 230
2 Normal Coordinate Analysis of a Linear Triatomic Molecule 232
3 "FG" Matrix Method and Symmetry of Normal Vibrations 241
4 Selection Rules for Vibrational Transitions 248
5 Rotational Structure of Vibrational Bands 252
Problems 256
References 261

9 **Electronic Spectra of Polyatomic Molecules** 262

1 Energy Levels and Spectra 262
2 Electronic States of Polyatomic Molecules: Walsh's Rules 263
3 Electronic Spectroscopy of Formaldehyde 268
4 The 2,600-Å System of Benzene 272
5 Molecular Photoelectron Spectroscopy 280
6 Radiationless Transitions 285
Problems 290
References 292

10 **From Molecular Beams to Masers to Lasers** 293

1 Hyperfine Structure and Line Width 293
2 Optical Pumping 296

3 Optical Level-Crossing Spectroscopy 300
4 Molecular Beam Methods 301
5 Supersonically Cooled Molecular Beam Spectroscopy 307
6 Molecular Beam Masers 308
7 Optical Masers 311
References 325

11 Optical Resonance Spectroscopy 327

1 The Idealized Two-Level System 327
2 The Rabi Solution for a Two-Level System 329
3 Saturation in a Two-Level System 334
4 Feynman-Vernon-Hellwarth Theorem 337
5 Optical Bloch Equations 340
6 Magnetic Resonance Analogues of Coherent Optical Spectroscopy 347
Problems 353
References 355

12 Coherent Transient Spectroscopy 356

1 Transient Nutation 356
2 Photon Echo 364
3 Fourier Transform Spectroscopy 373
4 Self-Induced Transparency 374
5 Saturation Revisited 388
6 Saturated Absorption Spectroscopy: The "Lamb Dip" 391
7 Double-Resonance Spectroscopy 394
Problems 402
References 402

13 Multiple-Photon Spectroscopy 404

1 Nonlinear Polarizability and Second-Harmonic Generation 404
2 Optical Parametric Oscillation 413
3 Multiphoton Spectra of Molecules 415
4 Infrared Multiple-Photon Absorption 426

5 Resonant Multiphoton Ionization 429
6 Stimulated Raman Scattering 429
7 Coherent Anti-Stokes Raman Scattering 434
8 Production and Detection of Picosecond Light Pulses 438
Problem 442
References 443

14 Spectroscopy beyond Molecular Constants 444

1 Molecules at High Excitation Levels 444
2 Spectra from Molecular Dynamics 448
References 453

Appendix A Direct Product Tables 454

Appendix B Lagrangian Mechanics 459

Appendix C Density Matrix Methods 463

Appendix D Dipole Correlation and Spectral Density Functions 471

Appendix E The Literature of Spectroscopy 475

Appendix F IUPAC/IUPAP Energy Conversion Constants (Revised 1973) 483

Index 485

Preface to the Second Edition

There are two principal objectives in this revised edition. First, I have attempted to correct as many as possible of the errors and misstatements that appeared in the first edition. Second, I have sought to bring up to date the discussion of new topics, particularly in chapters 10–14. Tremendous advances in spectroscopy, particularly in laser spectroscopy, have taken place in the past ten years, and any treatment of modern spectroscopy must reflect at least some of these developments. At the same time, I have tried not to lose sight of the lore and tradition of classical spectroscopy, which forms the basis for many of these advances.

Many individuals have contributed suggestions, corrections, and new material for this edition; in particular, I would like to thank Professors W. Chupka and R. Hochstrasser for their detailed critiques. Professor R. Beaudet made it possible for me to spend a semester at the University of Southern California, during which the material appearing in chapters 11 and 12 was prepared for an advanced topics course. Doctor J. Francisco and Professors W. Klemperer and K. Innes read the manuscript of the revised edition and made many valuable suggestions.

A note on references and cross references may be helpful to the reader: A cross reference to a section within a chapter is by section number; a cross reference to a section from one chapter to another is by chapter number *and* section number. For example, in chapter 1 a cross reference to the first section in that chapter is *section 1*, whereas in chapter 7 a cross reference to this same section is *chapter 1.1*. Also, a list of commonly used abbreviations for the journals included in the chapter references may be found in appendix E.4.

Preface to the First Edition

The fact that molecules possess quantized internal energy levels and that this enables them to absorb and emit electromagnetic radiation at discrete frequencies is one of the basic principles of chemistry. As such, the concepts of molecular spectroscopy appear in the curriculum at several different points. A descriptive treatment of atomic and molecular energy levels, without recourse to solutions of the Schrödinger equation, is usually included in first-year college chemistry or in advanced high school courses. These appear again, this time with the solutions for the particle-in-a-box, the harmonic oscillator, and the one-electron atom, in the typical undergraduate physical chemistry course. In graduate courses in molecular spectroscopy, one finds that these simple solutions are only first approximations to actual molecular eigenstates and that spectroscopy, in practice, is really an exercise in finding the best basis set for use in a perturbation expansion.

It is from a course of this last type that this book has developed. A number of satisfactory texts exist at the elementary and advanced undergraduate level; this is not equally so when one looks for a single book by means of which graduate students can gain entry to current research literature as well as to the detailed manuals of Herzberg, Townes and Schawlow, and Condon and Shortley,[1] works by which practicing spectroscopists regulate their professional lives. There are several reasons for this. One is that Herzberg's volumes themselves are basically phenomenological, and discussions of the problem of finding the representation in which the Hamiltonian operator is most nearly diagonal must be culled from the literature. Another is that much of the research effort in spectroscopy today is directed not at the traditional field of determining structural parameters, but at using quantitative spectroscopy, laser devices, and coherent optical phenomena to probe molecular dynamics. Books and monographs exist on many of these topics, but at the present time there seems to be no unified treatment suitable for use as a textbook.

This, then, is the origin of the present book. It is the outgrowth of several iterations of a one-semester graduate course in molecular spectroscopy at the Massachusetts Institute of Technology, with supplementary material added. The emphasis of the course was on introducing students to the concepts and the methods of modern molecular spectroscopy so that the language would be familiar when the course proceeded to discuss quantum electronics, lasers, and related coherent and nonlinear optical phenomena. The course, and thus

1. See appendix E for a compilation of basic reference and data sources in spectroscopy.

this book, also reflect the author's prejudice that the most interesting areas of current spectroscopic research deal with the dynamics of the interaction of radiation and molecular matter, rather than with the more traditional subject matter of the tabulation and assignment of spectral lines and their interpretation in terms of molecular structures.

There is surely more material in the present text than can be covered in a single one-semester course, but a good survey can be obtained by selecting portions of the text to cover in detail and using the rest for independent study. Clearly, the book has been designed for use in a graduate-level course with chemical physics orientation. However, it should also be appropriate for a small number of senior undergraduates who have already had the usual treatment of quantum chemistry found in the undergraduate physical chemistry course; they can use it either as a supplementary text or for self-study. No quantum mechanical background is assumed other than that found in such an undergraduate course; the mathematical background required is at a corresponding level. Some background in the elements of electromagnetic theory (Maxwell's equations, etc.) will be helpful. When advanced concepts, such as a relativistically invariant Hamiltonian, group-theoretical manipulations, or the density matrix, are required, they are introduced in a heuristic way. This means that they are treated without complete mathematical rigor in most cases. Although the orientation of the book is to provide the foundation necessary for modern spectroscopic research in considerable detail, it does not get involved in the fundamental issues of radiation theory and quantum mechanics that lie behind it all. At the same time, we have attempted to avoid focusing solely on current research topics, appropriate for a review monograph that is expected to be obsolete in a few years, but certainly not for a textbook.

There are many people who have helped bring this book to fruition, but a few should be singled out in particular. My thanks go to the students enrolled in these courses for their forbearance during the process of finding out the best way of expressing certain concepts, and to Professors R. C. Lord and S. G. Kukolich, who shared in the teaching of the course. I have also made free use of the problems originally prepared by Professor Lord and Professor Ian Mills. There are several places in the text at which the best way of presenting a particular topic seemed to be to make use of notes that had been prepared by other lecturers for other courses; these are indicated at appropriate points in the text, and my thanks go to the various individuals for permission to use this material. Special thanks are due Professor Kent R. Wilson of the University of

California at San Diego, who made possible a pleasant stay at his campus, during which a large portion of the first draft of the book was written, and also to Professor C. Bradley Moore of the University of California at Berkeley for similar arrangements during which the final manuscript was prepared. Thanks are also due Ms. Sandra V. Sutton for her expert typing of the manuscript, to Professor Victor Laurie for a valuable first reading of the manuscript, and to B. Garetz and B. D. Green for additional proofreading and comments.

1974

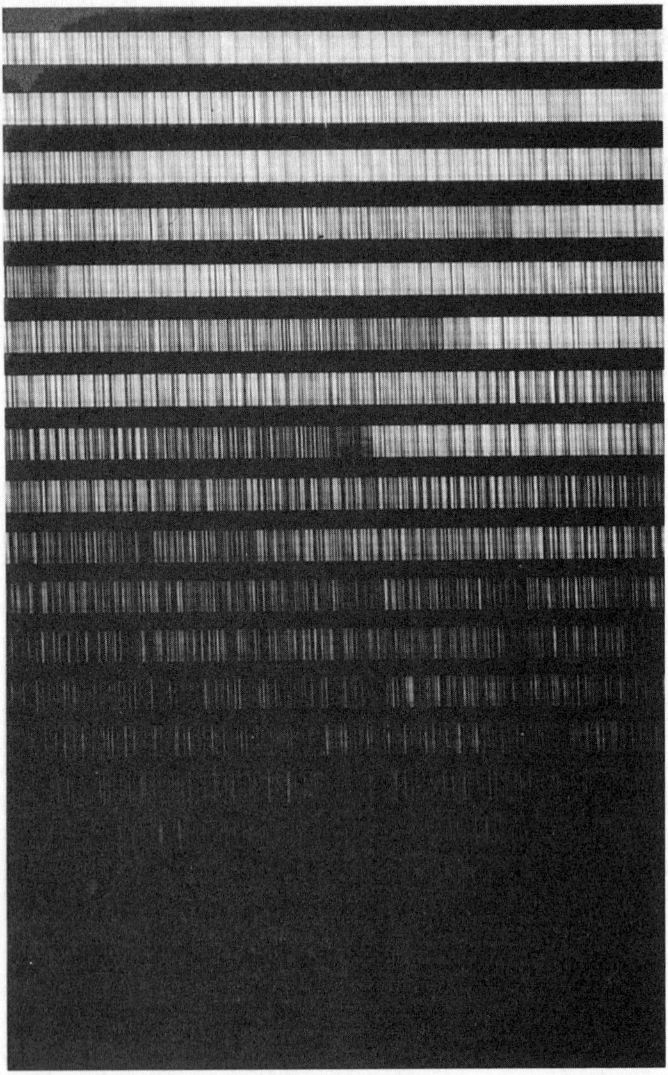

A portion of the visible absorption spectrum of the I_2 molecule, taken on an echelle spectrograph at MIT. Every line in the discrete part of the spectrum can be assigned and accounted for by the methods to be discussed in this book. [From J. D. Campbell, S.B. Thesis, MIT, June 1967. Reproduced with permission.]

1 Review of the Quantum Mechanical Background

1 State Vector Notation

If one were to ask for a definition of the basic operation in spectroscopy, the answer would probably be in terms of carrying out a *measurement* on a molecule, using electromagnetic radiation as a measuring tool. The essence of the quantum mechanical behavior of systems as small as atoms and molecules is that carrying out a measurement on such a system forces it to take on a sharp value of the one or more observables being measured. In spectroscopic measurements, this observable is almost always the total energy of the system. The measured quantity is the set of differences between the possible energy levels, which is related to an observed set of resonances in the electromagnetic radiation spectrum by the Bohr-Einstein law,

$$\Delta E = h\nu = \frac{hc}{\lambda} = hc\bar{\nu},$$

where ν is the frequency of the radiation, in hertz (the term cycles per second, or $\sec^{-1}$, is no longer officially used);
λ is the wavelength of the radiation in centimeters;
$\bar{\nu}$ is the wave number of the radiation in reciprocal centimeters, or cm^{-1}.

Energy and frequency are related by Planck's constant $h = 6.62617_4 \times 10^{-27}$ erg sec, while frequency and wavelength are related by the speed of light in vacuum, $c = 2.99792458 \times 10^{10}$ cm/sec.[1]

Since the concept of measurement is a key aspect of the whole, it seems appropriate to introduce the use of a quantum mechanical notation known as *measurement algebra* at this point. The primary advantage of this notation, as we shall see, is that it provides a very compact and handy way of expressing the *matrix elements*, and relations between them, in terms of which spectroscopic theory is formulated. It is well to remember that merely using a new notation, while it may be convenient, does not introduce any more content than was expressed by the more familiar language of wave functions and eigenvalues. Several classic quantum mechanics textbooks employ this notation, and the reader is encouraged to look at these as an aid in becoming familiar with this algebra. Especially recommended are the texts by Messiah, Gottfried, and Feynman, Leighton, and Sands (references 1–3).

1. Values of these and other fundamental constants are periodically reevaluated by overall least-squares adjustments; see, for example, B. N. Taylor and E. R. Cohen, *J. Phys. Chem. Ref. Data* **2**, 663 (1973).

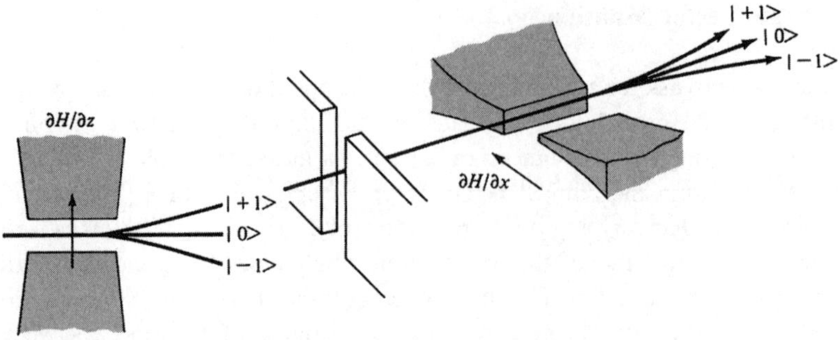

Figure 1.1
Schematic representation of a Stern-Gerlach experiment. A beam in a $j = 1$ state enters from the left through a magnetic field inhomogeneous in the z direction. It is split into three j_z components; the $+1$ component is selected and sent through a magnetic field inhomogeneous in the x direction, whereupon it splits up into three j_x components. If one of these components is then sent through another z magnet, it will again split up into three j_z components, despite the earlier z measurement.

The basic object of this algebra is the *state* of a system, which is given the symbol $|n\rangle$. This state may correspond to an energy level E_n, or to an angular momentum state (in which the angular momentum will have the value $\sqrt{j(j+1)}\hbar$, the quantum number j being conventionally used for the angular momentum), or perhaps to a composite state in which several observables have sharp values. One way of visualizing this state is as the actual physical particle in that state; that is particularly appropriate, for example, in the Stern-Gerlach experiment.

The Stern-Gerlach apparatus, shown in figure 1.1, is a molecular beam in which particles travel from a source, through various inhomogeneous magnetic fields, to a detector that records the presence or absence of molecules at a particular position in space. Let us suppose we have chosen a beam source such that the particles coming out all have exactly one unit of angular momentum (a ^{1}P or ^{3}S atom, for example; see chapter 2.6). If this beam passes through a magnet shaped to have a field gradient in a particular direction—say the z direction—then we know that the beam will split up into three subbeams, each moving in a slighty different direction. Each of these beams is composed of a *pure state*, $|jj_z\rangle$, having a definite value of the total angular momentum j and the projection of the angular momentum in the z direction, j_z. Since all the particles entering the magnetic field must leave it in one of the three beams, we have an example of a *completeness relation,*

$$\sum_{j_z} |j j_z\rangle = |j\rangle. \tag{1.1}$$

In the second part of figure 1.1, we have taken one of the beams—the $|1\ {+1_z}\rangle$ state—and passed it through a magnet having a field gradient in the orthogonal x direction. The result of doing this, as we know, is that we again obtain three beams, this time spread out in a plane perpendicular to that in which the beams spread out after passing through the z magnet. This entire experimental situation can be described simply by letting the x magnet be represented by an operator J_x, and saying that a state with a quantized component of angular momentum in the z direction is not an eigenstate of the operator J_x, since an eigenstate of a particular operator is not split into further substates by application of that operator.

Now suppose we take one of the states that have passed through the x magnet, *after* having passed through the z magnet—say, the $|1\ {-1_x}\rangle$ state obtained by operating J_x on $|1\ {+1_z}\rangle$. The result of a second J_z operation is again to produce three beams. In other words, the application of the J_x operator has destroyed the previous quantization of J_z; we say that J_x and J_z are not simultaneously measurable observables.

Since each of the subbeams has one-third the intensity of the parent beam, the effect of operating J_x and J_z on $|1\ 1_z\rangle$ is to produce three beams dispersed in the z direction, each with one-ninth the initial intensity. Suppose the order had been reversed, and J_z had been applied first, followed by J_x. Since $|1\ 1_z\rangle$ is an eigenstate of J_z, a single beam would have emerged from the z magnet this time; application of J_x to this beam would have produced three beams dispersed in the x direction, each with one-third the initial intensity. Clearly, the results of the two possible orders of applying these operators are inequivalent, so we can write

$$J_x J_z |1\ 1_z\rangle \neq J_z J_x |1\ 1_z\rangle,$$

or, since this result is true for any initial choice of state,

$$J_x J_z - J_z J_x = [J_x, J_z] \neq 0, \tag{1.2}$$

where the expression $[J_x, J_z]$ is the *commutator* of J_x and J_z. Saying that two operators do not commute is equivalent to saying that the physical quantities they represent cannot be determined simultaneously in the same particle.

2 Wave Function and Matrix Representations

A connection with the forms of quantum mechanics with which we may be more familiar can be obtained by considering the wave function $\psi_n(\mathbf{r})$ corresponding to the state $|n\rangle$. In measurement algebra notation, we write it as $\psi_n(\mathbf{r}) = \langle\mathbf{r}|n\rangle$. The square of the magnitude of the wave function (which may be a complex number) is a function of $\mathbf{r}$, which gives the probability density for finding a particle in state n at the position $\mathbf{r}$:

$$|\psi_n(\mathbf{r})|^2 = |\langle\mathbf{r}|n\rangle|^2 = \langle n|\mathbf{r}\rangle\langle\mathbf{r}|n\rangle = P_n(\mathbf{r}).$$

The integral of $P_n(\mathbf{r})$ over all space is, of course, unity:

$$\int P_n(\mathbf{r})\, d\mathbf{r} = \int \langle n|\mathbf{r}\rangle\, d\mathbf{r}\, \langle\mathbf{r}|n\rangle = 1.$$

In general, a combination of the form $\langle m|n\rangle$ is a number (called the projection or transformation matrix element), the square of which gives the probability that if the system is in state n, it is also in state m. When m is a set of coordinates, $\mathbf{r}$, this is just the ordinary coordinate space wave function. If m and n are two eigenstates of the same operator, then $\langle m|n\rangle = \delta_{mn}$ ($= 0$ if $m \neq n$, and $= 1$ if $m = n$); if a system has just been measured as being in some particular state, then it is in that state, and no other.

A problem, which is often encountered, is that of not knowing the set of eigenstates for a particular physical situation, although knowing the set for a closely related, usually simpler one. Fortunately, there exist mathematical procedures for solving this problem. Any arbitrary state (or function) can always be expanded in terms of a suitable complete set of states (or functions); a familiar example is Fourier's series:

$$f(x) = \sum_{k=0}^{\infty} (a_k \sin kx + b_k \cos kx),$$

or, in integral form,

$$f(x) = (2\pi)^{-1/2} \int_{-\infty}^{\infty} g(k) e^{ikx}\, dk.$$

Such an expansion can be carried out with any set, so long as it is complete. We may use, for example, Hermite polynomials (as occur in harmonic oscillator wave functions), spherical harmonics (rigid rotor wave functions), or the hydrogenlike orbitals for a particle in a Coulomb field equally as well as sines and cosines, which themselves happen to be particle-in-a-box wave functions. Suppose we do this and express our desired exact wave function for some system as

$$\psi_n^{\text{exact}}(\mathbf{r}) = \sum_k C_k^n \phi_k^0(\mathbf{r}). \tag{1.3}$$

If we write this familiar expression in terms of measurement algebra, it becomes

$$\langle \mathbf{r}|n\rangle = \sum_k \langle \mathbf{r}|k^0\rangle\langle k^0|n\rangle,$$

where the transformation number $\langle k^0|n\rangle$ is seen to be just the expansion coefficient. A convenient formal abbreviation immediately suggests itself, namely,

$$\sum_k |k\rangle\langle k| = 1, \tag{1.4}$$

where the individual objects $|k\rangle\langle k|$ are known as *projection operators.* Equation (1.4) is a more general form of the completeness relation, an example of which we have seen. The corresponding expression for a continuous variable, such as position $\mathbf{r}$ or momentum $\mathbf{p}$, would be just

$$\int |k\rangle\, dk\, \langle k| = 1.$$

We can use this last form to show the equivalence of the Schrödinger and Heisenberg representations of quantum mechanics. The matrix element of an operator $\mathbf{A}$ is defined as

$$A_{mn} = \int \psi_m^*(\mathbf{r}) A_{\text{op}} \psi_n(\mathbf{r})\, d\mathbf{r},$$

where A_{op} represents a mathematical operation on the Schrödinger wave function ψ_n (multiplication by a constant, differentiation, and so on). Using the state vector notation gives

$$\begin{aligned} A_{mn} &= \int \langle m|\mathbf{r}\rangle\, \mathbf{A}\, \langle \mathbf{r}|n\rangle\, d\mathbf{r} \\ &= \langle m|[\textstyle\int |\mathbf{r}\rangle\, d\mathbf{r}\, \langle \mathbf{r}|]\mathbf{A}|n\rangle = \langle m|\mathbf{A}|n\rangle. \end{aligned}$$

Since this is true for any continuous variable, not only $\mathbf{r}$, we see that the matrix elements of the Heisenberg representation are indeed independent of

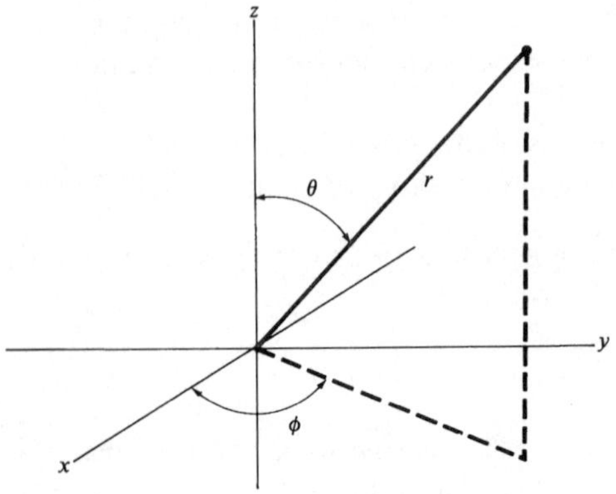

Figure 1.2
Definition of spherical polar coordinates, r, θ, and ϕ.

any coordinate system. From here on, we shall use the general form $\langle m|\mathbf{A}|n\rangle$ for a matrix element.

In order to make the general principles more concrete, let us consider once more the example we used in section 1, namely, the angular momentum in a $j=1$ system. The operator for angular momentum is obtained from the classical $\mathbf{r} \times \mathbf{p}$ form by substituting $\mathbf{p} = (\hbar/i)\nabla$; in the spherical (r, θ, ϕ) coordinate system shown in figure 1.2, this becomes

$$J_x = \frac{\hbar}{i}\left(-\sin\phi\frac{\partial}{\partial\theta} - \cot\theta\cos\phi\frac{\partial}{\partial\phi}\right),$$

$$J_z = \frac{\hbar}{i}\frac{\partial}{\partial\phi},$$

$$J^2 = -\hbar^2\left[\frac{1}{\sin\theta}\frac{\partial}{\partial\theta}\left(\sin\theta\frac{\partial}{\partial\theta}\right) + \frac{1}{\sin^2\theta}\left(\frac{\partial^2}{\partial\phi^2}\right)\right].$$

Consider the wave function

$$\langle\theta\,\phi|1\,1_z\rangle = Y_{11}(\theta,\phi) = \frac{\sqrt{3}}{8\pi}\sin\theta\, e^{i\phi},$$

which is just one of the spherical harmonic functions. When J^2 is operated on this function, we find by explicit algebraic substitution that

$$J^2 Y_{11}(\theta,\phi) = 2\hbar^2 Y_{11}(\theta,\phi)$$

and, similarly,

$$J_z Y_{11}(\theta, \phi) = \hbar Y_{11}(\theta, \phi).$$

Y_{11} is, thus, an eigenfunction of these two operators, and the coefficient that appears when the indicated operation is performed is the eigenvalue of the operator for that eigenfunction.

If we try to carry out the operation $J_x Y_{11}(\theta, \phi)$, however, we shall find that it cannot be reduced to a form (constant) $\cdot\, Y_{11}(\theta, \phi)$. What we obtain, in fact, are three terms, which turn out to be Y_{11}, Y_{10}, and Y_{1-1}, corresponding to the three physical beams produced in figure 1.1. An explicit verification of the commutation rule, equation (1.2), can be carried out in the same way.

If we evaluate the matrix element $\langle 1\, 1_z | J_x | 1\, 1_z \rangle$, we find its value to be zero. This is because we are averaging over equal contributions from the $+1$, 0, and -1 components. The corresponding matrix element of J_x^2, though, is *not* equal to zero, so that

$$\langle J_x^2 \rangle - \langle J_x \rangle^2 \neq 0; \tag{1.5}$$

we say that there is a *dispersion* in the operator spectrum. More commonly, one says that J_x is not a "good quantum number" in the $|1\, 1_z\rangle$ representation. This statement implies the dispersion of equation (1.5) and the noncommutation of equation (1.2). We shall see, later on, that the key step in interpreting complex spectra is to establish what the "good quantum numbers" are for the system we are examining.

3 Basis Functions for the Energy

The quantum mechanical operator with which we shall be most concerned in molecular spectroscopy is the Hamiltonian operator, of which the eigenvalue is the energy:

$$\mathscr{H}|n\rangle = E_n|n\rangle, \tag{1.6}$$

which is the Schrödinger equation. Experimental spectroscopy makes use of atoms and molecules themselves as a sort of miniature analog computer to solve this equation for the E_n; the theoretical interpretation consists essentially of finding the basis in which we can most easily effect the same solution, that is, in which the Hamiltonian operator is most nearly diagonal. It is possible to say the same thing in a variety of ways: That the Hamiltonian is diagonal, that we have found the good quantum numbers for the system, that certain dynamical variables are constants of the motion, or that the operators for these dynamical variables all commute with one another are all equivalent statements.

There are a few simple systems for which the differential equation represented by equation (1.6) can be solved by elementary methods. These systems will be familiar to anyone who has taken an elementary course in quantum chemistry, and so we shall simply tabulate the results of their solution (see table 1.1). Of these four systems, the first is not of much interest to the spectroscopy of bound systems. The last three, however, form the framework within which nearly all atomic and molecular problems are discussed.[2] Only a few simple systems are actually describable by these elementary solutions; more often, we have a system that is close to being so described, but is not exactly equivalent. In such cases, we make use of *perturbation theory*, which is described in the following section.

4 Perturbation Theory of Energy Corrections

Suppose that we are interested in a system for which the Hamiltonian is given by

$$\mathscr{H} = \mathscr{H}^0 + \lambda\mathscr{H}', \tag{1.7}$$

where $\mathscr{H}^0$ is one of the elementary systems listed in table 1.1 and $\lambda\mathscr{H}'$ is the additional part that makes our particular system different from these and harder to solve. The multiplicative factor λ is included as a mathematical convenience, in order to keep track of orders of magnitude during the derivation to follow. We shall seek an expansion for the eigenstates and energy eigenvalues as follows:

$$|n\rangle = |n^0\rangle + \lambda|n^{(1)}\rangle + \lambda^2|n^{(2)}\rangle + \cdots, \tag{1.8}$$

$$E_n = E_n^0 + \lambda E_n^{(1)} + \lambda^2 E_n^{(2)} + \cdots. \tag{1.9}$$

Rewriting the Schrödinger equation (1.6) and keeping together all terms in the same power of λ, we have

2. A notable addition is that of the Zeeman Hamiltonian for spin systems in a magnetic field; this will be introduced in chapters 2.8 and 5.8.

Table 1.1[a]

System	Hamiltonian $\mathscr{H}^0$	Energy eigenvalue E_n^0
Free particle (one dimension)	$T_1 = -\frac{\hbar^2}{2m}\frac{d^2}{dx^2}$	$\frac{p^2}{2m}$ (continuous)
Rigid rotor (three dimensions)	$T_3 = -\frac{\hbar^2}{2\mu}\nabla^2(R, \theta, \phi)$	$\begin{cases} \frac{\hbar^2}{2I_A} j(j+1) + \frac{\hbar^2}{2}\left(\frac{1}{I_C} - \frac{1}{I_A}\right)k^2 & \text{(general symmetric top)} \\ \frac{\hbar^2}{2\mu R^2} j(j+1) & \text{(linear rotor)} \end{cases}$
Harmonic oscillator (one dimension)	$T_1 + \frac{1}{2}k_F x^2$	$(n + \frac{1}{2})\hbar\omega,\ \omega = 2\pi\sqrt{\frac{k_F}{m}}$
Hydrogen atom	$T_3 - \frac{Ze^2}{R}$	$-\frac{\mu e^4 Z^2}{2\hbar^2} \cdot \frac{1}{n^2}$

a. Key:
m = mass;
μ = reduced mass;
R = interparticle separation;
I_A, I_C = moments of inertia;
k_F = force constant of harmonic oscillator;
Z = nuclear charge;
n, j, k = integers ≥ 0 ($n \geq 1$ for the hydrogen atom; $k \leq j$ for the symmetric top).

$$
\begin{aligned}
(\mathscr{H} - E_n)|n\rangle = (\mathscr{H}^0 - E_n^0)|n\rangle \\
&+ \lambda[(\mathscr{H}^0 - E_n^0)|n^{(1)}\rangle + (\mathscr{H}' - E_n^{(1)})|n^0\rangle] \\
&+ \lambda^2[(\mathscr{H}^0 - E_n^0)|n^{(2)}\rangle + (\mathscr{H}' - E_n^{(1)})|n^{(1)}\rangle] \\
&+ \cdots = 0.
\end{aligned}
\tag{1.10}
$$

The first term, to zero order in λ, is just the Schrödinger equation for $\mathscr{H}^0$ and is equal to zero. We shall also set terms smaller than $\lambda\mathscr{H}'$ in the Hamiltonian operator itself equal to zero. Then, to find the first-order corrections to the states and energies, we expand the state correct to first order in terms of the zero-order basis set; thus,

$$|n^{(1)}\rangle = \sum_k |k^0\rangle\langle k^0|n^{(1)}\rangle.$$

This is substituted into equation (1.10). $\mathscr{H}^0$, operating on this state, yields

$$
\begin{aligned}
\mathscr{H}^0|n^{(1)}\rangle &= \sum_k \mathscr{H}^0|k^0\rangle\langle k^0|n^{(1)}\rangle \\
&= \sum_k E_k^0|k^0\rangle\langle k^0|n^{(1)}\rangle.
\end{aligned}
$$

Collecting all the terms to the first power in λ gives

$$\sum_k |k^0\rangle\langle k^0|n^{(1)}\rangle(E_k^0 - E_n^0) = (E_n^{(1)} - \mathscr{H}')|n^0\rangle. \tag{1.11}$$

In order to isolate the desired quantity $E_n^{(1)}$, we form the matrix element of the preceding equation by premultiplying both sides by the complex conjugate state $\langle n^0|$; in wave function language, this corresponds to multiplying both sides of the equation by ψ_n^{0*} and integrating over all space. The result of doing this is

$$\langle n^0|\left\{\sum_k |k^0\rangle\langle k^0|n^{(1)}\rangle(E_k^0 - E_n^0)\right\} = \langle n^0|(E_n^{(1)} - \mathscr{H}')|n^0\rangle,$$

or, using the fact that both $|n^0\rangle$ and $|k^0\rangle$ are part of the same orthonormal set,

$$\sum_k \langle k^0|n^{(1)}\rangle(E_k^0 - E_n^0)\,\delta_{nk} = E_n^{(1)} - \langle n^0|\mathscr{H}'|n^0\rangle.$$

But the left-hand side clearly equals zero, for the only term in the sum for which $\delta_{nk} \neq 0$ is when $n = k$, and then $E_k^0 - E_n^0 = E_n^0 - E_n^0 = 0$. This gives us the simple result

$$E_n^{(1)} = \langle n^0|\mathscr{H}'|n^0\rangle = \int \psi_n^{0*}\mathscr{H}'\psi_n^0\, d\mathbf{r}. \tag{1.12}$$

In other words, the first-order correction to the energy, is simply the diagonal matrix element of the perturbing term in the Hamiltonian, in the zero-order basis set.

In many cases, the first-order correction is not sufficiently accurate, and we have to go one step further and find the second-order correction. In order to do this, we shall need to know the wave functions to first-order.[3] We can find them by taking an off-diagonal matrix element of equation (1.11):

$$\langle l^0| \left\{\sum_k |k^0\rangle\langle k^0|n^{(1)}\rangle (E_k^0 - E_n^0)\right\} = \langle l^0|E_n^{(1)} - \mathscr{H}'|n^0\rangle, \qquad l \neq n.$$

This simplifies to

$$\sum_k \langle k^0|n^{(1)}\rangle\, \delta_{lk}(E_k^0 - E_n^0) = \langle l^0|n^{(1)}\rangle (E_l^0 - E_n^0) = -\langle l^0|\mathscr{H}'|n^0\rangle.$$

This gives us an expression for $\langle l^0|n^{(1)}\rangle$, which, when substituted into the definition of $|n^{(1)}\rangle$, namely,

$$|n^{(1)}\rangle = \sum_l |l^0\rangle\langle l^0|n^{(1)}\rangle,$$

gives

$$|n^{(1)}\rangle = \sum_{l\neq n} \frac{|l^0\rangle\langle l^0|\mathscr{H}'|n^0\rangle}{E_n^0 - E_l^0}.$$

We can now proceed to find $E_n^{(2)}$. The terms to zero and first orders in λ have already been made to equal zero; so we simply collect the terms to

3. This is a particular case of a more general principle (the Brillouin-Wigner theorem), which states that if we know the states to order m, we can find the energy to order $2m+1$. Thus, the zero-order states suffice to determine the energy to first order, since

$$E_n^{(1)} = \langle n^0|\mathscr{H}'|n^0\rangle.$$

The first-order states enable us to find both the second- and third-order corrections to the energy, to wit,

$$E_n^{(2)} = \langle n^0|\mathscr{H}'|n^{(1)}\rangle$$

and

$$E_n^{(3)} = \langle n^{(1)}|\mathscr{H}'|n^{(1)}\rangle - E_n^{(1)}\langle n^{(1)}|n^{(1)}\rangle.$$

See P.-O. Löwdin, *J. Mol. Spectroscopy* **13**, 326 (1964); J. O. Hirschfelder and W. J. Meath, *Adv. Chem. Phys.* **12**, 29ff. (1967).

second order in λ, giving

$$(\mathscr{H}^0 - E_n^0)|n^{(2)}\rangle + (\mathscr{H}' - E_n^{(1)})|n^1\rangle - E_n^{(2)}|n^0\rangle = 0. \tag{1.13}$$

We let

$$|n^{(2)}\rangle = \sum_k |k^0\rangle\langle k^0|n^{(2)}\rangle,$$

and we shall need to evaluate

$$\mathscr{H}^0|n^{(2)}\rangle = \sum_k E_k^0|k^0\rangle\langle k^0|n^{(2)}\rangle$$

and

$$\mathscr{H}'|n^{(1)}\rangle = \sum_{l\neq n} \frac{\mathscr{H}'|l^0\rangle\langle l^0|\mathscr{H}'|n^0\rangle}{E_n^0 - E_l^0}$$
$$= \sum_m \sum_{l\neq n} \frac{|m^0\rangle\langle m^0|\mathscr{H}'|l^0\rangle\langle l^0|\mathscr{H}'|n^0\rangle}{E_n^0 - E_l^0}.$$

Making all these substitutions into equation (1.13) gives

$$\sum_k (E_k^0 - E_n^0)|k^0\rangle\langle k^0|n^{(2)}\rangle = E_n^{(2)}|n^0\rangle + \sum_{l\neq n} \frac{|l^0\rangle\langle l^0|\mathscr{H}'|n^0\rangle\langle n^0|\mathscr{H}'|n^0\rangle}{E_n^0 - E_l^0}$$
$$- \sum_m \sum_{l\neq n} \frac{|m^0\rangle\langle m^0|\mathscr{H}'|l^0\rangle\langle l^0|\mathscr{H}'|n^0\rangle}{E_n^0 - E_l^0}.$$

If we now form the matrix elements of this complicated equation by premultiplying by $\langle n^0|$, and make liberal use of the orthonormality of the basis set, we obtain as a result

$$E_n^{(2)} = \sum_{l\neq n} \frac{\langle n^0|\mathscr{H}'|l^0\rangle\langle l^0|\mathscr{H}'|n^0\rangle}{E_n^0 - E_l^0}. \tag{1.14}$$

Equation (1.14) can be interpreted in terms of a "mixing" of certain characteristics of other members of the basis states into the perturbed states, with the strength of mixing inversely proportional to the difference in energy between the two states. Such interpretations can easily be overdone, though; equation (1.14) really just reflects the fact that we have used the basis states for the unperturbed system as a convenient complete set for expansion of the perturbed states.

There may be several distinct states with the same energy eigenvalue in the unperturbed system we are considering; these are termed degenerate states. The effect of a perturbation in such a case is often to remove the degeneracy, or, as it is often called, "split" the levels. The origin of the term "splitting" may be seen from the effects of electric and magnetic fields on actual spectroscopic lines (see figures 2.14 and 2.18). A different, somewhat more elaborate mathematical procedure is called for in this case.

Let us suppose there are α different degenerate states belonging to the energy eigenvalue E_n^0, so that

$$\mathscr{H}^0|n^0 i\rangle = E_n^0|n^0 i\rangle, \qquad i = 1, \ldots, \alpha.$$

The first effect of the perturbation will be to produce new linear combinations of these basis states, given by

$$|\psi_{nl}^0\rangle = \sum_{i=1}^{\alpha} |n^0 i\rangle\langle n^0 i|\psi_{nl}^0\rangle = \sum_{i=1}^{\alpha} |n^0 i\rangle\, c_{inl}^0,$$

where we have used the notation c_{inl}^0 for the mixing coefficient, just for compactness. The $|\psi_{nl}^0\rangle$ are known as the "correct zero-order states." The states in the presence of the perturbation can, as usual, be expanded in a power series in λ,

$$|\psi_{nl}\rangle = |\psi_{nl}^0\rangle + \lambda|\psi_{nl}^{(1)}\rangle + \cdots,$$

but, as before, we need to know only the correct zero-order states to determine the energy to first order. Substituting into equation (1.10) and picking out the terms to first order in λ give

$$(\mathscr{H}^0 - E_n^0)|\psi_{nl}^{(1)}\rangle = (E_{nl}^{(1)} - \mathscr{H}')|\psi_{nl}^0\rangle.$$

If we let

$$|\psi_{nl}^{(1)}\rangle = \sum_k \sum_{i=1}^{\alpha} |k^0 i\rangle\langle k^0 i|\psi_{nl}^{(1)}\rangle = \sum_k \sum_{i=1}^{\alpha} |k^0 i\rangle\, c_{ikl}^{(1)},$$

we have

$$\mathscr{H}^0|\psi_{nl}^{(1)}\rangle = \sum_k \sum_{i=1}^{\alpha} |k^0 i\rangle\, c_{ikl}^{(1)} E_k^0,$$

so that

$$\sum_k \sum_{i=1}^{\alpha} |k^0 i\rangle\, c_{ikl}^{(1)}(E_k^0 - E_n^0) = \sum_{i=1}^{\alpha} (E_{nl}^{(1)} - \mathscr{H}')|n^0 i\rangle\, c_{inl}^0.$$

We again form the matrix element of both sides of the equation by premultiplying by $\langle n^0 j|$ and using orthonormality; this time, we get a set of α simultaneous linear equations,

$$\sum_{i=1}^{\alpha} c^0_{inl}(\langle n^0 j|\mathscr{H}'|n^0 i\rangle - \langle n^0 i\rangle E^{(1)}_{nl}) = 0$$

for $j = 1, \ldots, \alpha$. The method for the solution of such a set of simultaneous equations requires that the determinant of the coefficients be equal to zero, or

$$\begin{vmatrix} H'_{11} - E^{(1)} & H'_{12} & \cdots & H'_{1\alpha} \\ H'_{21} & H'_{22} - E^{(1)} & \cdots & H'_{2\alpha} \\ \vdots & \vdots & & \vdots \\ H'_{\alpha 1} & \cdots & & H_{\alpha\alpha} - E^{(1)} \end{vmatrix} = 0. \tag{1.15}$$

The left-hand side of equation (1.15) is known as a *secular determinant* for a particular n. We have used the additional orthonormality relation $\langle n^0 j|n^0 i\rangle = \delta_{ij}$ in constructing the determinant. The solution to equation (1.15) is found, first of all, as the α algebraic roots of the equation, which gives us up to α different values for $E^{(1)}$. (Some of the roots may be the same numerically —it is conceivable that a particular perturbation may not remove all the degeneracy present in a system.) This operation is equivalent to finding a transformation of the basis set that makes the Hamiltonian matrix diagonal.

A simple example of the foregoing involves an initially doubly degenerate level, for which the secular determinantal equation is particularly easy to solve. The correct zero-order states are just

$$|\psi^0_n\rangle = \frac{1}{\sqrt{2}}[|n^0 1\rangle \pm |n^0 2\rangle],$$

and the energies of the states, to first order, are

$$E_n = E^0_n + H'_{nn} \pm H'_{12}.$$

5 Time-Dependent Perturbations: Absorption and Emission of Electromagnetic Radiation

Thus far, we have been considering stationary states of atoms and molecules that are generated by time-independent Hamiltonian operators. What spectroscopy is primarily concerned with, however, is transitions between different energy levels; so we must turn our attention to those time-varying operators that can induce transitions.

The Schrödinger equation including the time is just

$$\mathscr{H}^0|n,t\rangle = -\frac{\hbar}{i}\frac{\partial}{\partial t}|n,t\rangle.$$

If $\mathscr{H}^0$ is time independent, as we have been considering it until this point, it is easy to show that if the time-dependent state has the form

$$|n,t\rangle = e^{-iE_nt/\hbar}|n\rangle,$$

then we regain the time-independent equation we have been using,

$$\mathscr{H}^0|n\rangle = E_n|n\rangle.$$

Suppose we now add a perturbation which is time dependent, so that

$$(\mathscr{H}^0 + \mathscr{H}'(t))|\Psi\rangle = -\frac{\hbar}{i}\frac{\partial}{\partial t}|\Psi\rangle. \tag{1.16}$$

We take our time-dependent state as a linear combination of the basis states of the same problem with $\mathscr{H}'(t) = 0$, with the appropriate energy phase factor associated with each term, that is,

$$|\Psi\rangle = \sum_n c_n(t)e^{-iE_nt/\hbar}|n\rangle, \qquad c_n = \langle n,t|\Psi\rangle, \qquad \sum_n |c_n(t)|^2 = 1.$$

If we substitute this into equation (1.16), we obtain

$$\sum_n c_n(t)e^{-iE_nt/\hbar}(\mathscr{H}^0 + \mathscr{H}')|n\rangle = -\frac{\hbar}{i}\left[\sum_n \frac{d}{dt}c_n(t)e^{-iE_nt/\hbar}\right]|n\rangle$$
$$-\frac{\hbar}{i}\sum_n c_n(t)e^{-iE_nt/\hbar}\frac{\partial}{\partial t}|n\rangle.$$

Because $|n\rangle$ represents a time-independent state, the terms involving $\partial|n\rangle/\partial t$ in this equation are all zero. This gives the simplified equation

$$\sum_n c_n(t)e^{-iE_nt/\hbar}(\mathscr{H}_0 + \mathscr{H}'(t))|n\rangle = -\frac{\hbar}{i}\left[\sum_n \frac{d}{dt}c_n(t)e^{-iE_nt/\hbar}\right]|n\rangle.$$

Premultiplying both sides of this equation by the complex conjugate state vector $\langle m|$ gives

$$\sum_n c_n(t)e^{-iE_nt/\hbar}[\langle m|\mathscr{H}_0|n\rangle + \langle m|\mathscr{H}'(t)|n\rangle] = -\frac{\hbar}{i}\sum_n \frac{d}{dt}c_n(t)e^{-iE_nt/\hbar}\langle m|n\rangle.$$

We make use of the orthonormality of the basis functions to eliminate all the terms on the right-hand side of the above equation except the one in which $m = n$. This leaves us with the following set of equations for the time development of the coefficients $c_m(t)$:

$$-\frac{\hbar}{i}\frac{d}{dt}c_m(t)e^{-iE_mt/\hbar} = c_m(t)e^{-iE_mt/\hbar}E_m + \sum_n c_n(t)e^{-iE_nt/\hbar}\langle m|\mathscr{H}'|n\rangle.$$

Multiplying both sides of this equation by $e^{+iE_mt/\hbar}$ gives

$$\frac{d}{dt}c_m(t) = -\frac{i}{\hbar}\sum_{n=0}^{\infty} c_n(t)e^{-i(E_n-E_m)t/\hbar}\langle m|\mathscr{H}'(t)|n\rangle. \tag{1.17}$$

By defining $(E_n - E_m)/\hbar = \omega_{nm}$ and $\langle m|\mathscr{H}'(t)|n\rangle = V_{mn}$, this becomes

$$\frac{d}{dt}c_m(t) = -\frac{i}{\hbar}\sum_{n=0}^{\infty} c_n(t)e^{-i\omega_{nm}t}V_{mn}. \tag{1.18}$$

It should be emphasized that (1.17) or (1.18) is an *exact* result; no approximations have yet been introduced.

The first-order perturbation result describes transitions between two isolated levels under the influence of a weak time-dependent perturbation connecting the levels. To obtain this result, we let all the $c_n(t)$ be zero except for one, say, $c_i(t) \equiv a(t)$, which initially has the value

$$c_i(0) = a(0) = 1.$$

Then the value of the amplitude of any other state m is given by

$$c_m(t) = -\frac{i}{\hbar}\int_0^t a(t')e^{-i\omega_{am}t'}V_{ma}(t')\,dt', \qquad m \neq i.$$

If we further assume that the interaction is sufficiently weak so that $c_m(t)$ never becomes large, we can replace $a(t')$ by 1; then

$$c_m(t) = -\frac{i}{\hbar}\int_0^t e^{-i\omega_{am}t'}V_{ma}(t')\,dt'.$$

We now need to specify $V_{ma}(t)$. Let us assume that the perturbation is an *oscillating* or *harmonic* function such as an electromagnetic field (see following section), so that

$$V_{ma}(t) = V_{ma}^0 \cos \omega t = \frac{V_{ma}^0}{2}\left[e^{i\omega t} + e^{-i\omega t}\right].$$

This gives

$$\begin{aligned} c_m(t) &= -\frac{i}{2\hbar} V_{ma}^0 \int_0^t e^{-i\omega_{am}t'}\left[e^{+i\omega t'} + e^{-i\omega t'}\right] dt' \\ &= -\frac{i}{2\hbar} V_{ma}^0 \left\{ \frac{e^{i(\omega+\omega_{am})t} - 1}{\omega + \omega_{am}} - \frac{e^{-i(\omega-\omega_{am})t} - 1}{\omega - \omega_{am}} \right\}. \end{aligned} \tag{1.19}$$

In a typical physical situation, there will be a whole set of transitions from state a to other states m, with a frequency $\omega_{am} = (E_a - E_m)/\hbar$ associated with each transition. The resonant denominator in (1.19) picks out the transition having $\omega_{am} \approx \omega$; the amplitudes for all the other states, for which this condition is not met, will remain close to zero. Let us call this near coincidence the *resonant transition*, with amplitude $c_b(t) \equiv b(t)$. Then, since $\omega \approx \omega_{ab}$, the first "antiresonant" term in (1.19) can be neglected in comparison with the second term, since it will be rapidly oscillating and thus will average to zero over any significant time interval. We shall encounter this removal of the antiresonant term later, as the *rotating wave approximation* (RWA); it is equivalent to replacing an oscillating perturbation $V^0 \cos \omega t$ with a rotating perturbation $V^0 e^{i\omega t}$. We can then write, in the RWA,

$$b(t) = -\frac{1}{2\hbar} V_{ba}^0 \frac{e^{-i(\omega-\omega_{ab})t} - 1}{\omega - \omega_{ab}}.$$

The *probability* of a transition from state $|a\rangle$ to state $|b\rangle$, given by the square of this amplitude, is then

$$P_{a\to b}(t) = |b(t)|^2 = \left|\frac{V_{ba}^0}{\hbar}\right|^2 \frac{\sin^2(\frac{1}{2}(\omega - \omega_{ab})t)}{(\omega - \omega_{ab})^2}. \tag{1.20}$$

For further discussion, the reader is referred to any of several standard texts on quantum mechanics (reference 4).

If the radiation source is monochromatic and exactly on resonance ($\Delta = \omega - \omega_{ab} = 0$), then (1.20) reduces to

$$P_{a\to b}(t) = \frac{|V_{ba}^0|^2}{4\hbar^2} t^2.$$

This is a first-order perturbation result valid only for $V_{ba}^0 t \ll \hbar$—that is, for short time *and* for weak coupling between the molecule and the radiation field. Ordinarily, the oscillating frequency dependence of (1.20) is not seen, since the interaction must be averaged over some distribution of frequencies $g(\omega)$:

$$P_{a\to b}(t) = \left|\frac{V_{ba}^0}{\hbar}\right|^2 \int_{-\infty}^{\infty} \frac{\sin^2(\frac{1}{2}(\omega - \omega_{ab})t)}{(\omega - \omega_{ab})^2} g(\omega)\, d\omega$$

$$= 2\pi\left|\frac{V_{ba}^0}{\hbar}\right|^2 g(\omega_{ab})t.$$

The rate of transitions from state $|a\rangle$ to state $|b\rangle$, which determines the optical absorption coefficient, is then

$$R_{a\to b} = 2\pi\left|\frac{V_{ba}^0}{\hbar}\right|^2 g(\omega_{ab}). \tag{1.21}$$

In other words, for transitions to occur, there must be a nonvanishing spectral density at the resonant frequency ω_{ab}; this is just a restatement of the Bohr-Einstein frequency rule introduced at the beginning of this chapter. The frequency dependence can be displayed, however, under the proper conditions, such as in a molecular beam experiment. (See chapter 10.4 for a discussion of this method.) In this experiment, a beam of molecules, all at a preselected velocity, was sent through a radio frequency field of finite spatial extent. The measurement made was of what fraction of the molecules had undergone a transition at ω_{ab} after passing through this field; clearly, all the molecules were subjected to the influence of the field for the same time

$$t = \frac{\text{length of field region}}{\text{molecular velocity}}.$$

The frequency distribution $g(\omega)$ was essentially monochromatic, $\delta(\omega - \omega_0)$, in which ω_0 was continuously varied. The resulting spectrum, shown in figure 1.3, clearly displays the oscillatory dependence on ω.

The questions that remain to be addressed at this point can be stated as follows:

1. What is the physical form of V_{ba}^0, and what does this form imply about which molecular states can couple with each other?
2. What is the relation of (1.20) and (1.21) to bulk optical properties of materials, such as the absorption coefficients?
3. How can we deal with situations in which the perturbation approximation breaks down as the strength of the coupling becomes large?

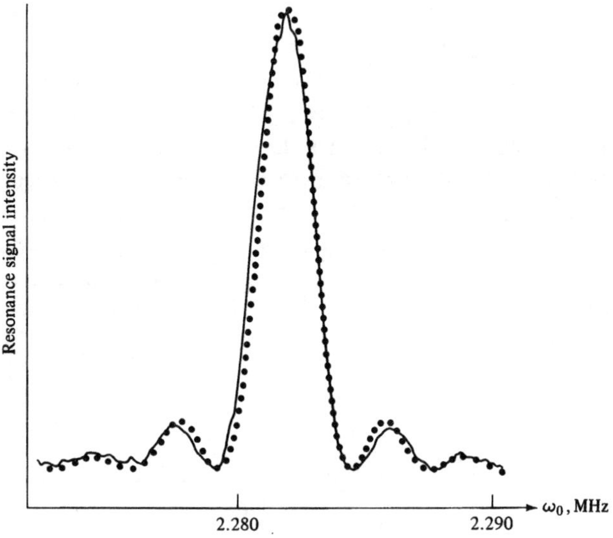

Figure 1.3
Frequency-dependent transition probability for a molecular beam resonance experiment: ———, experimental curve for HCN molecule, $J = 1$, $\Delta M_J = 1$, $\Delta M_F = 0$; . . . , calculated $[\sin^2\{(\omega - \omega_{ab})t/2\}/(\omega - \omega_{ab})^2]$ behavior for molecular velocity $v_0 = (8 \times 10^4 \text{ cm sec}^{-1}) \pm 10\%$. [From T. R. Dyke, G. R. Tomasevich, W. Klemperer, and W. Falconer, *J. Chem. Phys.* **57**, 2277 (1972). Reproduced with permission.]

The first two of these questions will be dealt with in the following sections; discussion of the last point will be deferred until the Rabi solution for two-level systems is taken up in chapter 11.

6 The Electric Dipole Interaction

The interaction of atoms and molecules with electromagnetic radiation is usually treated by writing the electric field associated with the light wave as

$$\mathscr{E}(\mathbf{r}, t) = \mathbf{E}_0 \cos(\omega t - \mathbf{k}\cdot\mathbf{r}),$$

where the electric field $\mathbf{E}_0 = \hat{\mathbf{i}}E_{0x} + \hat{\mathbf{j}}E_{0y} + \hat{\mathbf{k}}E_{0z}$; $\mathbf{k}$ is the propagation vector of the light, and the interaction $\mathscr{H}'(t)$ is simply

$$\mathscr{H}'(t) = \mathscr{E}\cdot\boldsymbol{\mu} = e\sum_k E_k \sum_j x_{kj}, \tag{1.22}$$

where the sum over k includes the x, y, and z components of $\mathbf{E}_0$, and the sum over j includes all the charges (nuclei and electrons) in the molecule. Equation (1.22) simply expresses the interaction of the electric field with the dipole moment operator. This treatment will not be adequate for our purposes, however, because we whall want to consider higher-order interactions.

The more complete treatment of these processes begins with obtaining the Hamiltonian operator in the correct relativistically invariant form, in which all coordinate systems moving at constant velocity are equivalent. This is necessary because we are dealing with electromagnetic radiation, which is a relativistic phenomenon.[4] The conventional Hamiltonian has kinetic and potential energy terms of the general form

$$\mathscr{H} = \frac{p^2}{2m} + V(\mathbf{r}),$$

with the operator for the momentum $\mathbf{p}$ being $(\hbar/i)\nabla$. The relativistically invariant form is

$$\mathscr{H} = \frac{1}{2m}\left(\mathbf{p} - \frac{e}{c}\mathbf{A}\right)^2 + [V(\mathbf{r}) + e\phi],$$

where $\mathbf{A}$ is the vector potential and ϕ is the scalar potential associated with the electromagnetic field. Writing out the square of the momentum in detail, we obtain the form of the Schrödinger equation as

$$\begin{aligned}\mathscr{H}|\Psi\rangle &= \left\{\frac{1}{2m}\left(-\hbar^2\nabla^2 + i\hbar\frac{e}{c}\nabla\cdot\mathbf{A} + 2i\hbar\frac{e}{c}\mathbf{A}\cdot\nabla + \frac{e^2}{c^2}|\mathbf{A}|^2\right)\right.\\ &\quad \left. + V(\mathbf{r}) + e\phi\right\}|\Psi\rangle\\ &= -\frac{\hbar}{i}\frac{\partial}{\partial t}|\Psi\rangle.\end{aligned}$$

The time dependence of the radiation field is now that of the vector potential:

$$\mathbf{A}(\mathbf{r}, t) = \mathbf{A}_0 \cos(\omega t - \mathbf{k}\cdot\mathbf{r}). \tag{1.23}$$

4. This treatment follows closely that of D. Bohm in chapter 18 of his book *Quantum Theory* (reference 5). Many points that are stated without proof here can be found more fully discussed in this reference. For a discussion of vector potential, gauge invariance, and other aspects of electromagnetic theory involved in the derivation, the reader is referred to the text by Jackson (reference 6).

The electric field is given by

$$\mathscr{E}(\mathbf{r}, t) = -\frac{1}{c}\frac{\partial \mathbf{A}}{\partial t} = k\mathbf{A}_0 \sin(\omega t - \mathbf{k}\cdot\mathbf{r}), \tag{1.24}$$

and the associated magnetic field is

$$\mathbf{H}(\mathbf{r}, t) = \nabla \times \mathbf{A} = \mathbf{k} \times \mathbf{A}_0 \sin(\omega t - \mathbf{k}\cdot\mathbf{r}), \tag{1.25}$$

always oriented perpendicularly to the electric field.

We are always free to choose a gauge (reference 6) that makes the problem more tractable mathematically; in this case, the best choice is the Coulomb gauge, which provides that $\nabla \cdot \mathbf{A} = 0$ and the scalar potential $\phi = 0$. The Schrödinger equation can be written as

$$\begin{aligned}\mathscr{H}|\Psi\rangle &= [\mathscr{H}^0 + \mathscr{H}'(t)]|\Psi\rangle \\ &= \left\{\left[-\frac{\hbar^2}{2m}\nabla^2 + V(\mathbf{r})\right] + \left[\frac{ie\hbar}{mc}\mathbf{A}\cdot\nabla + \frac{e^2}{2mc^2}|\mathbf{A}|^2\right]\right\}|\Psi\rangle.\end{aligned}$$

We usually neglect the term in $|\mathbf{A}|^2$, unless the fields are very high (as in a focused laser beam, for example—see chapter 13).

The time dependence of **A** has already been treated in the preceding section; we turn our attention now to evaluation of the matrix elements of $\mathbf{A}\cdot\nabla$, which will tell us which molecular states can be coupled by radiation and with what intrinsic strength; in other words, these quantities provide the *selection rules* for radiative transitions. The matrix elements are complex vector quantities, which are best evaluated one component at a time. A typical component might be

$$(V_{ba}^0)_x = \frac{e\hbar}{mc} A_{0x}\langle b|e^{ikz}\frac{\partial}{\partial x}|a\rangle.$$

We shall use the Taylor's series expansion of e^{ikz},

$$e^{ikz} = 1 + ik(z - z_0) - \frac{k^2}{2}(z - z_0)^2 + \cdots.$$

If $kz = 2\pi z/\lambda$ is much less than 1, we can make the *dipole approximation*, which consists simply of replacing e^{ikz} by the first term in the expansion, namely, 1. This is certainly valid for any wavelength of radiation greater than, say, 1,000 Å, which lies deep in the vacuum ultraviolet, when we are talking about molecules with dimensions of the order of angstroms (10^{-8} cm). In this approximation,

$$(V^0_{ba})_x = \frac{e\hbar}{mc} A_{0x} \langle b|\frac{\partial}{\partial x}|a\rangle, \tag{1.26}$$

the operator $\partial/\partial x$ is the *dipole velocity operator*. It is worth stopping a moment at this point and noting this fact well: The leading term in the interaction of a radiation field with a molecule is the dipole velocity operator, not the dipole moment operator introduced in equation (1.22). There is no hint that this is the case when we begin our derivation by assuming that the only interaction of note is that between the electric field of the radiation and the charge distribution in the molecule. The reason that we can get the correct answer by either approach—apart from saying that this is simply another beneficent manifestation of the correspondence principle at work—is that there is a relation between the dipole velocity and dipole moment operators, namely,[5]

$$\langle b|\frac{\partial}{\partial x}|a\rangle = \frac{-m}{\hbar^2}(E_b - E_a)\langle b|x|a\rangle. \tag{1.27}$$

5. The derivation of this relation is both simple and not often given in elementary texts, and so it is given here. This is one case in which it is easier to use the wave function notation than state vector notation; note also that this theorem is valid only when the ψ are the *exact* wave functions. We rearrange a one-dimensional Schrödinger equation to give

$$\frac{d^2}{dx^2}\psi^*_m + \frac{2m}{\hbar^2}[E_m - V(x)]\psi^*_m = 0$$

and multiply this equation by $x\psi_k$; similarly,

$$\frac{d^2}{dx^2}\psi_k + \frac{2m}{\hbar^2}[E_k - V(x)]\psi_k = 0,$$

and this is to be multiplied by $x\psi^*_m$. We subtract the second equation from the first and integrate over all space to give

$$\int_{-\infty}^{\infty} x\psi_k \frac{d^2}{dx^2} 2\psi^*_m - x\psi^*_m \frac{d^2}{dx^2}\psi_k \, dx = \frac{2m}{\hbar^2}(E_k - E_m)\int_{-\infty}^{\infty} \psi^*_m \psi_k x \, dx.$$

When this is integrated by parts, we obtain

$$\frac{2m}{\hbar^2}(E_k - E_m)\langle m|x|k\rangle = -\int_{-\infty}^{\infty}\left[\frac{d}{dx}(x\psi_k)\frac{d}{dx}\psi^*_m - \frac{d}{dx}(x\psi^*_m)\frac{d}{dx}\psi_k\right]dx$$

$$= -\int_{-\infty}^{\infty}\left[\psi_k\frac{d}{dx}\psi^*_m - \psi^*_m\frac{d}{dx}\psi_k\right]dx = 2\langle m|\frac{d}{dx}|k\rangle,$$

since d/dx is an anti-Hermitean operator, that is, $\int\psi^*_m \, d/dx \, \psi_k \, dx = -\int\psi_k \, d/dx \, \psi^*_m \, dx$. Q.E.D.

Using equation (1.27) in equation (1.26) gives us an expression for the x component of V_{ba}^0,

$$
\begin{aligned}
(V_{ba}^0)_x &= -\frac{e}{\hbar c}(E_b - E_a)A_{0x}\langle b|x|a\rangle \\
&= -\frac{e}{c}\omega_{ba}A_{0x}\langle b|x|a\rangle,
\end{aligned}
\tag{1.28}
$$

with similar expressions for the y and z polarizations. Substituting this into the frequency-averaged form of equation (1.20) gives a final expression for the probability of a radiative transition:

$$
P_{a\to b}(t) = \frac{2\pi}{\hbar^2}\left(\frac{e}{c}\right)^2 \omega_0^2 |A_0(\omega_0)|^2 |\langle b|\mathbf{r}|a\rangle|^2 t. \tag{1.29}
$$

We can make the substitutions [see (1.24), and recall that $k = 2\pi/\lambda = \omega/c$]

$$\omega_0^2|A_0|^2 = c^2|E_0|^2$$

and

$$e\mathbf{r} = \boldsymbol{\mu},$$

the dipole moment operator. This gives

$$P_{a\to b}(t) = \frac{2\pi}{\hbar^2}|E_0|^2|\langle b|\boldsymbol{\mu}|a\rangle|^2 t,$$

so that the rate of transitions from $|a\rangle$ to $|b\rangle$ is probability per unit time,

$$
\begin{aligned}
R_{a\to b} &= \frac{2\pi}{\hbar^2}|E_0|^2|\langle b|\boldsymbol{\mu}|a\rangle|^2 \\
&= \frac{2\pi}{\hbar^2}|\langle b|\boldsymbol{\mu}\cdot\mathbf{E}_0|a\rangle|^2,
\end{aligned}
\tag{1.30}
$$

which is identical to the result that would have been obtained by starting with the electric dipole formulation of equation (1.22). Since the result we have obtained is just the same as if we had started with the simple interaction postulated in equation (1.22), one may well ask what the point was of going through this much more complicated derivation in order to reach the same result. The proper reply is that the latter route provides us with the framework to obtain the higher-order terms in the interaction that are important when the electric dipole matrix elements vanish. These are found by retaining sucessively higher terms in the expansion of e^{ikz}. For example, if we include the term $ik(z - z_0)$, the x component of V_{ba}^0 becomes

$$(V_{ba}^0)_x = -\frac{e\hbar k}{mc} A_{0x} \langle b | z \frac{\partial}{\partial x} | a \rangle,$$

and so on. To evaluate this term, we divide it in two parts and add and subtract terms involving $x\,\partial/\partial z$; thus

$$(V_{ba}^0)_x = -\frac{e\hbar k}{2mc} A_{0x} \left[\langle b | z \frac{\partial}{\partial x} + x \frac{\partial}{\partial z} | a \rangle + \langle b | z \frac{\partial}{\partial x} - x \frac{\partial}{\partial z} | a \rangle \right].$$

The reason for doing this is that the two resulting terms can each be identified with a term in the multipole expansion of the molecular charge distribution. The first term is

$$\frac{e}{mc} A_{0x} \frac{\hbar k}{2} \langle b | z \frac{\partial}{\partial x} + x \frac{\partial}{\partial z} | a \rangle = \frac{e}{mc} A_{0x} im\omega_{ba} \langle b | xz | a \rangle,$$

which is the *ba*th element of the xz component of the quadrupole moment tensor of the molecule. The second term has the form of an angular momentum operator, so that

$$\frac{e}{mc} A_{0x} \frac{\hbar k}{2} \langle b | z \frac{\partial}{\partial x} - x \frac{\partial}{\partial z} | a \rangle = \frac{ie}{mc} A_{0x} \frac{k}{2} \langle b | J_y | a \rangle$$

$$= iA_{0x} k \langle b | \mu_y | a \rangle,$$

where $\boldsymbol{\mu}$ is the magnetic dipole moment. Since $\mathbf{H} = \nabla \times \mathbf{A}$, it is evident that the y component of $\boldsymbol{\mu}$ interacts with the y component of $\mathbf{H}$, and so on, to give an effective Hamiltonian $\mathscr{H}'_{\mathrm{MD}}(t) = \boldsymbol{\mu} \cdot \mathbf{H}(t)$.

We can summarize the results of this section as follows. The strongest transitions in absorption and emission spectra will follow electric dipole selection rules; that is, the intensity of the transition will be proportional to the square of the matrix element $\langle b | \mathbf{r} | a \rangle$. When these matrix elements are zero, the second-order (and thus much weaker) interactions go as the square of the matrix elements of either the electric quadrupole moment, $\langle b | \mathbf{rr} | a \rangle$, with typical terms x^2, y^2, z^2, xy, and so on, or as the magnetic dipole moment, $\langle b | \mathbf{J} | a \rangle$, where $\mathbf{J}$ is the vector angular momentum operator. In later chapters, we shall consider a different type of high-order interaction with the electromagnetic field, namely, with fields of such a high intensity

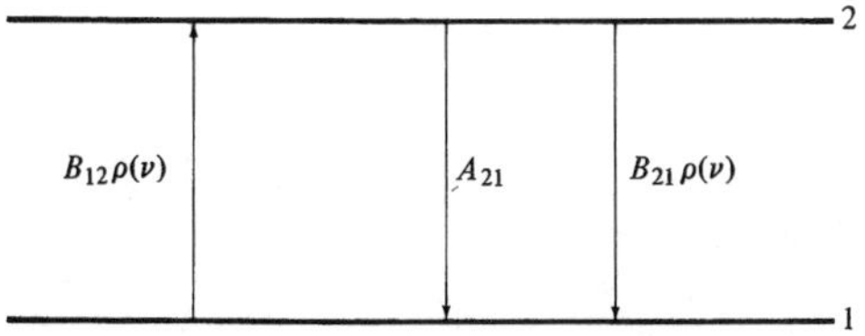

Figure 1.4
Einstein coefficients B_{12} for absorption, A_{21} for spontaneous emission, and B_{21} for stimulated emission.

that a higher-order perturbation in time must be carried out in order to obtain correct results.

7 Optical Constants of Atoms and Molecules

In treating the interaction of radiation with matter, as occurs in optical pumping and selective laser excitation, we must be able to calculate the rate of radiative transitions induced by a field. This requires knowing the relations between such quantities as the transition dipole moment, the absorption coefficient, the Einstein A and B coefficients, and so on. For this development, we shall switch from angular frequency units (rad/sec, or Avis—see reference 7, footnote 14) to ordinary frequency units (cycles/sec, or Hertz) because, while most of the formal theoretical developments are written in terms of the former units, it appears that most of the practical literature dealing with concrete problems use the latter units.

We begin by considering two levels, coupled by an electric dipole allowed transition, in a radiation field having a distribution of frequencies following Planck's blackbody law,

$$\rho(\nu) = \frac{8\pi h\nu^3}{c^3}[e^{h\nu/kT} - 1]^{-1}, \tag{1.31}$$

where $\rho(\nu)$, the energy density *per unit frequency interval*, is in the units $(\text{erg}\,\text{cm}^{-3})/\text{sec}^{-1}$. The rate of transition from the lower level to the upper level (conventionally labeled 1 and 2, respectively) is seen from figure 1.4 to be just $N_1 B_{12}\rho(\nu)$; the rate of transitions from the upper level to the lower level is $N_2[B_{21}\rho(\nu) + A_{21}]$, where A_{21} is the spontaneous radiative decay rate, which is the inverse of the natural radiative lifetime. At equilibrium, $dN_1/dt = dN_2/dt = 0$, so that these two competing rates are equal, and

$$\frac{N_2}{N_1} = \frac{B_{12}\rho(\nu)}{B_{21}\rho(\nu) + A_{21}}.$$

But we also must have an equilibrium, that is, a Boltzmann distribution of the populations between the two levels,

$$\frac{N_2}{N_1} = \frac{g_2}{g_1} e^{-h\nu/kT},$$

where g_i is the degeneracy of the ith state ($i = 1$ or 2); the temperature T in this distribution law is the same as that appearing in the blackbody distribution law for the radiation. If we solve for $\rho(\nu)$, we find

$$\rho(\nu) = \frac{A(g_2/g_1)e^{-h\nu/kT}}{B_{12} - B_{21}(g_2/g_1)e^{-h\nu/kT}} = \frac{8\pi h\nu^3}{c^3}\frac{1}{e^{h\nu/kT} - 1}.$$

This can be true only if two conditions are fulfilled:

1. $B_{12} = B_{21}(g_2/g_1)$ and
2. $A_{21} = (8\pi h\nu^3/c^3)B_{21} = (8\pi h\nu^3/c^3)(g_1/g_2)B_{12}$.

These A and B coefficients were introduced by Einstein in 1917 (reference 8).

This phenomenological approach is a convenient way of describing stimulated and spontaneous emission without having to go to the more rigorous and difficult treatments of quantum electrodynamics. Under normal circumstances, that is, weak radiation fields, N_2 remains much less than N_1, and we shall be able to recover linear absorption. In systems far from equilibrium, such as a laser, there may be population inversion, with $g_1N_2 > g_2N_1$, and the system can act as an amplifier, rather than an attenuator, of radiation. This situation may be described by a negative temperature in (1.32); if such a system is coupled to a second system at an ordinary, positive temperature, the former can do work on the latter—separating isotopes, perhaps, or some other chemical work.

The relation between these coefficients and the transition dipole strength can be obtained by observing that the upward pumping rate per molecule, $B_{12}\rho(\nu)$, is the same as the rate at which probability accumulates in the upper state, $(2\pi/\hbar^2)\mu_{ab}^2E_0^2$. Furthermore, the energy density $\rho(\nu)$ in cgs units is $(\varepsilon/4\pi)|E_0(\omega_0)|^2$, where $\omega_0 = 2\pi\nu$ and ε is the dielectric constant ($\varepsilon = 1$ for a vacuum and $\varepsilon > 1$ for a liquid or solid medium). Combining these relations gives

$$B_{12}\rho(\nu) = \frac{8\pi^2}{\hbar^2}\left(\frac{1}{4\pi}E_0^2(\omega_0)\right)\mu_{ab}^2 = \frac{8\pi^2}{\hbar^2}\mu_{ab}^2\rho(\nu), \tag{1.32}$$

so that

$$B_{12} = \frac{8\pi^2}{\hbar^2}\mu_{ab}^2. \tag{1.33}$$

Now we wish to develop the relation between these optical constants and the phenomenologically measured absorption coefficient.[6] What one observes in an absorption experiment is that the intensity is attenuated according to the law

$$I(\nu) = I_0(\nu)e^{-k_\nu l} \tag{1.34}$$

for sufficiently weak $I_0(\nu)$. Here we are using the intensity distribution per unit frequency interval, in the units W/cm^2 sec^{-1}, just as the energy density $\rho(\nu)$ was in the units erg cm^{-3}/sec^{-1}. I represents a flux of energy across unit area per unit time, while ρ represents a density of energy per unit volume. The two quantities are related by the velocity of light c:

$$I = c\rho.$$

We can define A and B coefficients for light intensity instead of energy density simply by dividing by c, so that

$$B^I = \frac{1}{c}B^\rho.$$

It is worth examining the units of these B coefficients at this point; this will save much confusion in the future. The A coefficient, being the reciprocal of a lifetime, must be in the unit time^{-1}; by the same token, $B^\rho\rho$ or $B^I I$ must also be in the unit time^{-1}. The energy density ρ is in erg cm^{-3}/sec^{-1}, or mass length^{-1} time^{-1}; hence B^ρ must be in the units mass^{-1} length, or cm/g in cgs units. Similarly, I is in erg/cm^2 sec sec^{-1}, or mass time^{-2}, and B^I must have the units mass^{-1} time, or sec/g in cgs units. In the following development, we shall use B^I exclusively.

Equation (1.34) arises from the differential absorption law

$$-dI = Ik_\nu\, dl.$$

6. The development here follows that of Mitchell and Zemansky (reference 9).

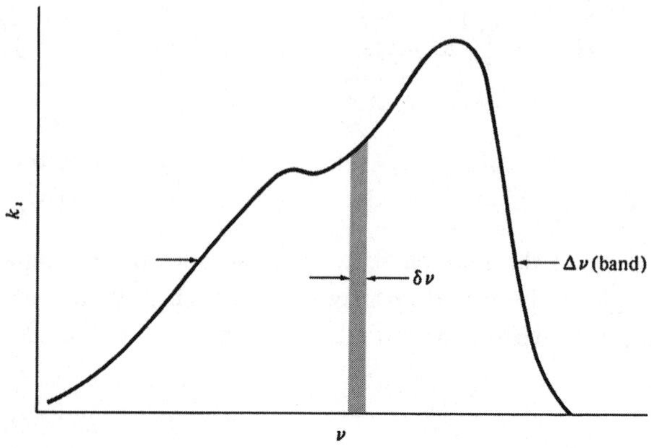

Figure 1.5
Definition of the parameter β as the ratio of the homogeneous packet width $\delta\nu$ to the total inhomogeneous band width $\Delta\nu$.

If we set the rate of energy removal from the light beam equal to the net rate of upward transitions times the energy per quantum $h\nu$, this becomes

$$-d[I(\nu)\,\delta\nu] = N_1 h\nu B_{12} I(\nu)\,dl - N_2 h\nu B_{21} I(\nu)\,dl, \tag{1.35}$$

where $\delta\nu$ is the frequency range over which the molecules in levels 1 and 2 can absorb and emit, respectively. If the absorption band is inhomogenously broadened, then N_1 and N_2 are small subsets of the entire ground and excited state populations, respectively. If we call this fraction β,

$$\beta = \frac{N_1}{N_{\text{ground}}} \approx \frac{N_2}{N_{\text{excited}}},$$

then $\delta\nu$ will be or the order of β times the full inhomogenous band width $\Delta\nu$, as figure 1.5 shows. Rearrangement of equation (1.35) gives

$$\frac{1}{I(\nu)}\frac{dI(\nu)}{dl}\delta\nu = h\nu(B_{12}N_1 - B_{21}N_2) = k_\nu\delta\nu. \tag{1.36}$$

For the case in which $N_2 \ll N_1$ [$h\nu \gg kT$ and $I(\nu)$ sufficiently small so that the equilibrium populations are not appreciably perturbed], we have $N_{\text{ground}} \approx N_{\text{total}}$, and we can neglect populations in N_2 or N_{excited}. There are two ways of solving for B_{12}, either directly from equation (1.36),

$$B_{12} = \frac{k_\nu\,\delta\nu}{h\nu N_1},$$

or by integration over the entire absorption band,

$$B_{12} = \frac{\int k_\nu \, d\nu}{h\nu_0 N_{\text{total}}}.$$

Either expression gives the same value for B_{12}, since there is a common factor of β in the numerator and denominator. Also, note that both expressions give the correct units for B^I.

Sometimes an optical absorption cross section is defined, by analogy with scattering experiments, as

$$\frac{I}{I_0} = e^{-\sigma_\nu N l} = e^{-k_\nu l},$$

where N is the total density of molecules per cubic centimeter. By substituting σN for k, we have an expression for B_{12}:

$$B_{12} = \int \frac{\sigma_\nu \, d\nu}{h\nu_0}.$$

One additional useful measure of optical absorption is the oscillator strength, defined as

$$f_{21} = \frac{mc^3}{8\pi^2 \nu_0^2 e^2} \frac{g_2}{g_1} A_{21}. \tag{1.37}$$

If we substitute, successively, B_{12} for A_{21} and the integrated absorption coefficient for B_{12}, this becomes

$$\int k_\nu \, d\nu = \frac{\pi e^2}{mc} N f.$$

The usefulness of this quantity is that for transitions in which a single electron changes its quantum state in the molecule or atom,

$$\sum_j f_{jl} = 1;$$

that is, the sum of the oscillator strengths of all the transitions arising from any one electronic state to all other levels is unity. If we substitute $1/\tau$ for A_{21} in equation (1.37), we can obtain

$$f\tau = \frac{mc}{8\pi^2 e^2} \frac{g_2}{g_1} \lambda_0^2 = 1.51 \frac{g_2}{g_1} \lambda_0^2,$$

where λ_0 is in centimeters; this gives a handy relation between oscillator strength and radiative lifetime for an electronic transition at a given wavelength. For example, a transition with $f = 1$ in the visible ($\lambda_0 = 5{,}000$ Å) will have an excited state with a lifetime of approximately 4×10^{-9} sec. The accompanying set of "radiation relations"[7] summarizes the numerical relations among a number of these optical constants.

Radiation Relations The rate of accumulation of probability in the upper level of a two-level system is

$$\frac{dN_2}{dt} = -AN_2 + BN_1 I - BN_2\left(\frac{g_1}{g_2}\right) I$$

molecules per cubic centimeter per second. Here I is light flux in quanta per square centimeter per second per unit frequency interval. The blackbody flux per solid angle is

$$I_{BB} = \frac{2}{\lambda^2}\frac{1}{e^{hc/k\lambda T} - 1}.$$

For numerical evaluation, take λ in centimeters, $g = g_{el}(2J + 1)$. The degeneracies and oscillator strengths are related by

$$(gf) = g_1 f_{abs} = -g_2 f_{em}.$$

Also take $|\mu_{ab}|^2 = |R|^2$ in the atomic units $a_0^2 e^2 = 6.459 \times 10^{-36}$ esu^2 cm^2. We then have the following expressions equivalent to the Einstein A coefficient:

$$\frac{g_2 A}{g_1} = \frac{8\pi}{\lambda^2}B = \frac{8\pi^2 e^2}{mc\lambda^2}\frac{(gf)}{g_1} = \frac{8\pi c}{\lambda^2}\int k\,d\left(\frac{1}{\lambda}\right) = \frac{4\pi^{3/2}c}{\lambda^2\sqrt{\ln 2}}k_0\,\Delta\left(\frac{1}{\lambda_D}\right)$$

$$= \frac{64\pi^4}{3h\lambda^3}\frac{|R|^2}{g_1} = 25.133\lambda^{-2}B = 0.6670\lambda^{-2}\frac{(gf)}{g_1}$$

$$= 7.5345 \times 10^{11}\lambda^{-2}\int k\,d\lambda^{-1} = 8.020 \times 10^{11}\lambda^{-2}k_0\,\Delta\lambda_D^{-1}$$

$$= 2.026 \times 10^{-6}\lambda^{-3}\frac{|R|^2}{g_1}.$$

Expressed in terms of the oscillator strength f, these are

7. Based on a compilation by L. Brewer, the University of California at Berkeley. (Reprinted with permission).

$$\frac{(gf)}{g_1} = \frac{mc\lambda^2}{8\pi^2 e^2}\frac{g_2 A}{g_1} = \frac{mc}{\pi e^2}B = \frac{mc^2}{\pi e^2}\int k\,d\left(\frac{1}{\lambda}\right) = \frac{mc^2 k_0 \Delta(1/\lambda_D)}{2e^2\sqrt{\pi \ln 2}} = \frac{8\pi^2 mc}{3he^2\lambda}\frac{|R|^2}{g_1}$$

$$= 1.499\lambda^2\frac{g_2 A}{g_1} = 37.68B = 1.1296 \times 10^{12}\int k\,d\lambda^{-1}$$

$$= 1.202 \times 10^{12} k_0\,\Delta\lambda_D^{-1} = 3.038 \times 10^{-6}\lambda^{-1}\frac{|R|^2}{g_1}.$$

Expressed in terms of the Einstein B coefficient, these are

$$B = \frac{\lambda^2}{8\pi}\frac{g_2 A}{g_1} = \frac{\pi e^2}{mc}\frac{(gf)}{g_1} = c\int k\,d\left(\frac{1}{\lambda}\right) = \frac{c}{2}\sqrt{\frac{\pi}{\ln 2}}\,k_0\,\Delta\left(\frac{1}{\lambda_D}\right) = \frac{8\pi^3}{3h\lambda}\frac{|R|^2}{g_1}$$

$$= 0.03979\lambda^2\frac{g_2 A}{g_1} = 2.654 \times 10^{-2}\frac{(gf)}{g_1} = 2.998 \times 10^{10}\int k\,d\lambda^{-1}$$

$$= 3.191 \times 10^{10} k_0\,\Delta\lambda_D^{-1} = 8.061 \times 10^{-8}\lambda^{-1}\frac{|R|^2}{g_1}.$$

Other useful relations are the follows:

1. gas constant:

$$R = 8.3147 \times 10^7 \text{ erg deg}^{-1} = 82.06 \text{ cm}^3 \text{ atm deg}^{-1};$$

2. molecular density:

$$N = \frac{N_0 P_{\text{atm}}}{RT} = 7.339 \times 10^{21}\frac{P_{\text{atm}}}{T} = \frac{N_0 P_{\text{torr}}}{760RT}$$

$$= 9.657 \times 10^{18}\frac{P_{\text{torr}}}{T} \text{ molecules/cm}^3;$$

3. $\ln\left(\frac{I}{I^0}\right) = -klN_1 = \ln(1-\alpha), \qquad \frac{N}{C} = \frac{N_0}{1{,}000};$

4. $\log\left(\frac{I}{I^0}\right) = -\varepsilon/C$, C in moles per liter, $\qquad \frac{\varepsilon}{k} = \frac{N_0}{2{,}303};$

5. $g_2 A = 2.880 \times 10^{-9}\lambda^{-2}\int \varepsilon\,d\lambda^{-1} = 2.880 \times 10^{-9}\lambda^{-3}\int \varepsilon\,d\ln\lambda^{-1}.$

8 Homogeneous and Inhomogenous Line Shapes

Another important feature of the interaction of radiation with molecules is the *line shape*, that is, the frequency dependence of the transition probability in the vicinity of a resonance. There are two basic types of line-broadening mechanisms to consider. First, consider that the upper, or excited level of an isolated transition (for example, as in figure 1.4) decays spontaneously in time with a decay rate γ_b. In the absence of any other decay precesses, γ_b will be the *radiative* width,

$$\gamma_b^{\text{rad}} = A_{21} = 1/\tau_{\text{rad}}.$$

If the excited atom or molecule can be affected by collisions, then there is an additional density-dependent component to the line width,

$$\gamma_b^{\text{coll}} = (\text{collision rate coefficient}) \times (\text{density}).$$

Other *nonradiative* decay mechanisms may also contribute to this line width; an example is discussed in chapter 9.6.

The effect of such a decay term can be readily represented by adding an imaginary term to the energy of the state,

$$E_b = \hbar\left(\omega_{ab} - \frac{i}{2}\gamma_b\right),$$

so that

$$e^{-iE_b t/\hbar} = e^{-i\omega_{ab}t - \gamma_b t/2}.$$

The unperturbed time evolution of state $|b\rangle$ is then given by

$$n_b(t) = |c_b(t)|^2 = |c_b(0)|^2 |e^{-i\omega_{ab}t}|^2 |e^{-\gamma_b t/2}|^2 = n_b(0)e^{-\gamma_b(t)}.$$

The oscillatory time dependence $e^{i\omega_{ab}t}$ is multiplied by its complex conjugate and therefore gives a constant amplitude; the decay terms add, however, and give exponential decay of the population n_b.

In the presence of a field at frequency ω, however, things get a bit more complicated. From (1.19), we write the coefficient $b(t)$ as

$$b(t) = -\frac{V_{ba}^0}{2\hbar}\frac{e^{-i(\omega-\omega_{ab})t} - 1}{\omega - \omega_{ab}}.$$

If we let $\omega_{ab} \to \omega_{ab} - i\gamma_b/2$, this becomes

$$\begin{aligned} b(t) &= -\frac{V_{ba}^0}{2\hbar(\omega - \omega_{ab} + i\gamma_b/2)} e^{-i(\omega-\omega_{ab}+i\gamma_b/2)t} - 1 \\ &= \frac{-iV_{ba}^0}{2\hbar[i(\omega - \omega_{ab}) + \gamma_b/2]} e^{-i\omega t}e^{+i\omega_{ab}t}e^{-\gamma_b t/2} - 1. \end{aligned}$$

The steady-state line shape is found as the amplitude, or square of the coefficient,

$$g(\omega) = \frac{|V_{ba}^{0}|^2}{4\hbar^2[(\omega - \omega_{ab})^2 + \gamma_b^2/4]}. \tag{1.38}$$

From this we find a generalized *Lorentzian* line shape,

$$k(\nu) = \frac{k_0\gamma}{(\nu - \nu_0)^2 + \gamma^2/4}, \tag{1.39}$$

where we have switched from angular to circular frequency units, and k_0 is given by the integral over the line shape,

$$\int k(\nu)\,d\nu = k_0.$$

This type of line shape, in which each molecule may absorb or emit radiation over the entire line width, is known as a *homogeneous* form of line broadening.[8] A homogeneous line nearly always has a Lorentzian shape.[9] This type of broadening is really a reflection of the Heisenberg uncertainty principle: Some decay process prevents the molecule from remaining in a specified energy state for longer than Δt, on the average, and the line width $\gamma = 1/\Delta t$. An alternative derivation of (1.39) will be given in appendix D.

A different sort of line broadening is one in which different molecules undergoing the same nominal transition absorb radiation corresponding to this transition at slightly different frequencies, the shifts arising from small differences in the environments of the molecules. This is known as inhomogeneous broadening.[8] The most extreme example of this type of broadening is a molecular band, in which molecules in different rotational states absorb at different frequencies (See chapters 4 and 5). A more typical example of this is Doppler broadening, which results from a Maxwellian distribution of velocities of gas molecules relative to a spectroscopist observing the molecules. The Gaussian distribution of each component of velocity is reflected in the Gaussian form of the Doppler line shape,

8. The terms homogeneous and inhomogeneous broadening were first introduced by A. M. Portis, *Phys. Rev.* **91**, 1071 (1953), for the case of paramagnetic resonance line shapes, but they are equally applicable to any spectroscopic transition.
9. Exceptions to this rule occur when the system decays as $\exp(-t/\tau)^2$, which leads to a non-Lorentzian line shape—see Behmenburg, *Progress in Atomic Spectroscopy* (Plenum, New York, 1979); MacQuarrie, *Statistical Mechanics* (Harper and Row, New York, 1976), p. 499.

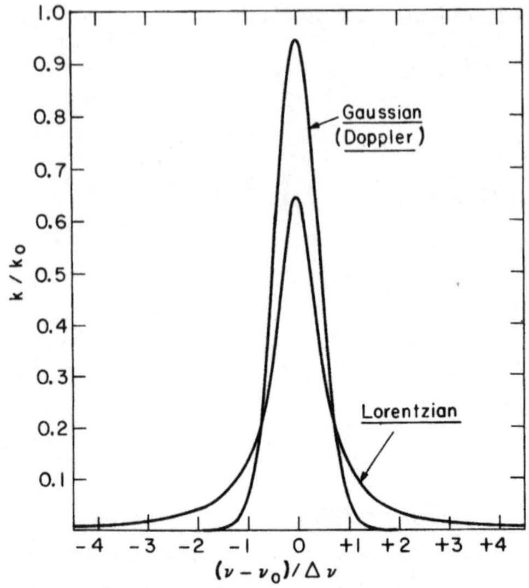

Figure 1.6
Comparison between Gaussian (Doppler) and Lorentzian line shapes having the same full width at half-height and the same integrated areas ($\int k\,d\nu = 1$). Notice that the Lorentzian line has much wider "wings," that is, far off-resonance absorption components. [From J. I. Steinfeld, *Laser and Coherence Spectroscopy*, Plenum, New York, 1978. Reproduced with permission.]

$$k(\nu) = k_0 \frac{2\sqrt{\ln 2}}{\sqrt{\pi}(\Delta\nu_0)} \exp\left[-4\ln 2\frac{(\nu-\nu_0)^2}{(\Delta\nu_0)^2}\right]. \tag{1.40}$$

A comparison between homogeneous and inhomogeneous, that is, Lorentzian and Gaussian line shapes, is shown in figure 1.6. Many sources of inhomogeneous broadening have a similar distribution, so any inhomogeneous line shape is generally taken to be Gaussian. In a more general sense, though, the irregular absorption and emission spectrum contours of large polyatomic molecules, which are composed of a large number of overlapping vibrational and rotational subfeatures, can equally well be considered a case of an inhomogeneously broadened band.

In gases, the frequency shift due to line-of-sight velocity is

$$\nu = \nu_0(1 - v/c),$$

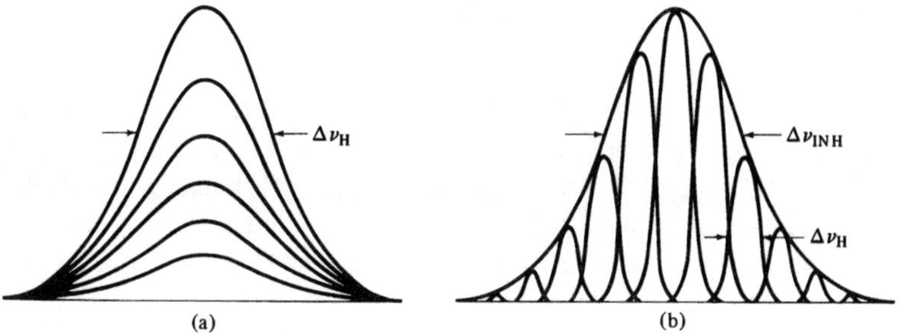

Figure 1.7
Homogeneous (a) and inhomogeneous (b) band shapes. The Gaussian inhomogeneous (INH) line, of total width $\Delta\nu_{INH}$, is made up of a superposition of Lorentzian homogeneous (H) lines each of width $\Delta\nu_H$. [From A. M. Stoneham, *Rev. Mod. Phys.* **41**, 82 (1969). Reproduced with permission.]

and the velocity is distributed according to a one-dimensional Maxwell-Boltzmann law,

$$f(v) = e^{-Mv^2/2RT},$$

where M is the molecular weight of the absorbing or emitting species. From these, we may find an expression for the Doppler width,

$$\Delta\nu_0 = \frac{2}{c}\sqrt{2R\ln 2}\,\nu_0\sqrt{\frac{T}{M}} \simeq 7.2\times 10^{-7}\nu_0\sqrt{\frac{T}{M}}. \tag{1.41}$$

For example, in atomic sodium ($M = 23$ amu) at 500°K, the Doppler width of the sodium D line at $\nu_0 = 17{,}000\ \mathrm{cm}^{-1}$ is $\Delta\nu_D = 0.057\ \mathrm{cm}^{-1} = 1{,}700$ MHz. The radiative width, however, is $\Delta\nu_{rad} = 1/2\pi\tau_{rad} \simeq 1/(2\pi\times 10^{-8}\ \mathrm{sec}) = 16\ \mathrm{MHz} = 0.0005\ \mathrm{cm}^{-1}$. Thus the Doppler width is typically much larger than the homogeneous radiative width, and is also larger than many spectroscopic features, such as hyperfine splittings, width spectroscopists want to be able to observe. A number of techniques now exist for achieving *sub-Doppler resolution* by eliminating all or part of the inhomogeneous line width; some of these will be discussed in detail in chapter 10.

On closer examination, an inhomogeneously broadened band is made up of a number of closely spaced components, each of which has its own associated homogeneous line shape, as shown in figure 1.7. If we wish to calculate the composite line shape resulting from the two broadening mechanisms acting simultaneously, we must carry out the convolution integral

$$\phi(\omega) = \int_{-\infty}^{\infty} I_{\mathrm{INH}}(\omega - \omega') I_{\mathrm{H}}(\omega' - \omega_0)\, d\omega',$$

Where inhomogeneous and homogeneous are denoted INH and H, respectively. The commonest form of this integral, that for combined Doppler and Lorentz broadening (called the Voigt line shape), unfortunately cannot be evaluated in closed form. Detailed numerical tables and approximate forms for this problem are given in the work of Mitchell and Zemansky (reference 9), to which the reader seriously concerned about line shape problems is referred.

9 The Road Ahead

Now that we have dealt with these essential preliminaries, there are two distinct directions in which we may proceed. One is to explore the radiative dynamics of abstract, idealized two-level systems. This is the approach taken in many recent textbooks, and it will be taken here following chapter 10. First, however, it is really necessary to have a thorough grounding in the "classical" spectroscopy of the atomic and molecular systems one actually uses to carry out experiments. This material is presented in chapters 2–9 and should be thoroughly understood before proceeding to the study of the nonclassical phenomena.

Problems

1. Carry out the angular momentum algebra indicated in section 2; that is, show explicitly that $\langle 1\, 1_z | j_x | 1\, 1_z \rangle = 0$ and that $\langle J_x^2 \rangle - \langle J_x \rangle^2 \neq 0$.

2.

(a) Given the matrix elements of the coordinate x for a harmonic oscillator:

$$\langle v | x | v' \rangle = \int \psi_v^* x \psi_v' \, dx = 0 \qquad \text{unless } v' = v \pm 1$$

and

$$\langle v + 1 | x | v \rangle = (2\beta)^{-1/2}(v + 1)^{1/2},$$

$$\langle v - 1 | x | v \rangle = (2\beta)^{-1/2}(v)^{1/2},$$

where $\beta = 4\pi^2 m\nu/h$ and v is the vibrational quantum number.

Evaluate the nonzero matrix elements of x^2, x^3, and x^4; that is, evaluate the integrals

$$\langle v|x^r|v'\rangle = \int \psi_v x^r \psi_{v'}\, dx$$

for $r = 2, 3$, and 4 (without actually doing the explicit integrals, of course!).
(b) From the results of (a), evaluate the average values of x, x^2, x^3, and x^4 in the vth vibrational state. Is it true that $\overline{x^2} = (\bar{x})^2$, or that $\overline{x^4} = (\overline{x^2})^2$? What conclusions can you draw about the results of a measurement of x in the vth vibrational state?

3. Show that A_{10} for the electric dipole transition of an electron bound to an origin by a one-dimensional harmonic potential is just

$$A_{10} = \frac{2\omega^2 e^2}{3mc^3}$$

if the dipole moment $\mu = ex$ and $\omega = \sqrt{k/m}$. The oscillator strength of an electronic transition is defined as

$$f_{ml} = A_{ml}\frac{mc^3}{2\omega_{ml}^2 e^2}.$$

The Thomas-Kuhn sum rule states that the sum of the oscillator strengths of all the transitions from one given level equals the number of electrons N in the system, or

$$\sum_m f_{ml} = N.$$

That, of course, is the reason for the definition. Can you prove this sum rule?

4. Determine the peak absorption coefficient $k(\nu_0)$ due to a transition at $\nu_0 = 3 \times 10^{14}$ Hz, where $N_2 \simeq 0$, $N_1 = 10^{18}$ cm^{-3}; the full width of the Gaussian absorption curve is 400 cm^{-1}, and $f = 10^{-3}$. Define the optical density as

$$\log_{10}\left(\frac{I_{\text{in}}}{I_{\text{out}}}\right),$$

where I denotes intensity. What is the optical density at ν_0 for a 1-cm path length of material? What is the spontaneous lifetime of the upper state?

At what temperature will the rate for the transition induced by blackbody radiation equal the spontaneous emission rate?

5. Consider that the broadening of a spectral line of a radiating atomic gas of mass m in thermodynamic equilibrium at temperature T is due only to the "Doppler effect,"

$$\lambda = \lambda_0\left(1 + \frac{v_x}{c}\right),$$

where v_x is the velocity component in the line of sight of the observer. Derive an expression for the observed intensity I as a function of λ.

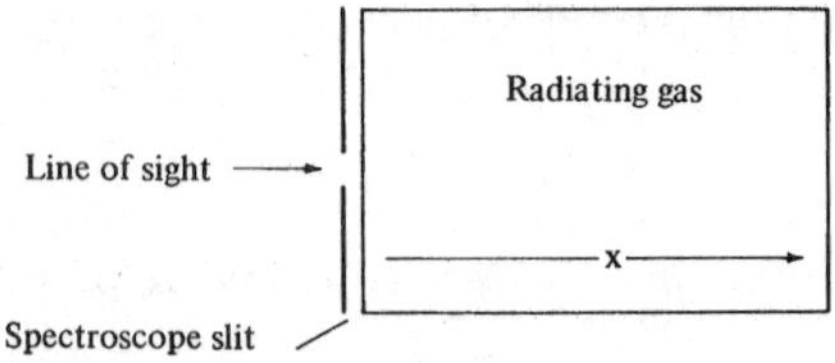

References

1. A. Messiah, *Quantum Mechanics* (North-Holland, Amsterdam, 1961).
2. K. Gottfried, *Quantum Mechanics* (Benjamin, Menlo Park, CA, 1966).
3. R. P. Feyman, R. B. Leighton, and M. Sands, *Lectures on Physics*, Vol. 3 (Addison-Wesley, Reading, MA, 1965).
4. L. Pauling and E. B. Wilson, Jr., *Introduction to Quantum Mechanics* (McGraw-Hill, New York, 1935).
5. D. Bohm, *Quantum Theory* (Prentice-Hall, Englewood Cliffs, NJ, 1951).
6. J. D. Jackson, *Classical Electrodynamics* (Wiley, New York, 1962).
7. J. S. Waugh, C. H. Wang, L. M. Huber, and R. L. Vold, *J. Chem: Phys.* **48**, 662 (1968).
8. A. Einstein, *Physik. Z.* **18**, 121 (1917).
9. A. C. G. Mitchell and M. W. Zemansky, *Resonance Radiation and Excited Atoms* (Cambridge University Press, Cambridge, 1961).

2 Atomic Spectra

1 The Hydrogen Atom—a One-Electron System

The nonrelativistic Hamiltonian operator for a hydrogenlike atom consisting of one electron and one positively charged nucleus is

$$\mathscr{H} = -\frac{\hbar^2}{2\mu}\nabla^2 - \frac{Ze^2}{r}, \tag{2.1}$$

where $Z = 1$ for hydrogen, 2 for He^+, 3 for Li^{2+}, and so on, and $\mu = m_p m_e/(m_p + m_e)$, the reduced mass for the orbiting electron. Since $m_p/m_e = 1{,}837$, μ is very nearly equal to the electron mass itself. In spherical polar coordinates (see figure 1.2), which are suitable ones for the separation and solution of the differential equation, the Hamiltonian becomes

$$\mathscr{H} = -\frac{\hbar^2}{2\mu}\left\{\frac{1}{r^2}\frac{\partial}{\partial r}\left(r^2\frac{\partial}{\partial r}\right) + \frac{1}{r^2\sin^2\theta}\left[\sin\theta\frac{\partial}{\partial\theta}\left(\sin\theta\frac{\partial}{\partial\theta}\right) + \frac{\partial^2}{\partial\phi^2}\right]\right\} - \frac{Ze^2}{r}.$$

We are familiar with the solutions to the Schrödinger equation involving this Hamiltonian. The energy eigenvalues are

$$E_{nlm} = -\frac{\mu e^4 Z^2}{2\hbar^2}\cdot\frac{1}{n^2}. \tag{2.2}$$

The natural unit of energy in (2.2) is the *Rydberg* $R = m_e e^4/4\pi\hbar^3 c = 109737.318\ \text{cm}^{-1}$ for infinite nuclear mass. For the electron-proton system, $\mu = 0.9994557 m_e$ and $R_H = 0.9994557R = 109{,}677.59\ \text{cm}^{-1}$. Another frequently used energy unit in atomic calculations is the *Hartree*, equal to two Rydberg units. In addition to the principal quantum number n, which determines the energy, the system is characterized by the angular momentum quantum number l and the angular momentum orientation quantum number m. $l = 0$ states are denoted s, $l = 1$ states p, $l = 2$ states d, and so on, for historical reasons. The notation is associated with the sharp, principal, diffuse, and so on, series in atomic spectroscopy. The energy is independent of l and m.[1]

We are also familiar with the form of $\langle r\theta\phi|nlm\rangle = \psi_{nlm}(r, \theta, \phi)$; it is

1. The $(2l + 1)$-fold degeneracy associated with the m quantum numbers is a characteristic of any ordinary spherically symmetrical system, for which the Hamiltonian is invariant under any rotations, reflections, and inversions in three-dimensional space. The fact that the different l states are also all degenerate for the hydrogen atom means that the special form of the $1/r$ potential is also invariant under these operations in a suitably defined four-dimensional coordinate system. See R. Aldrovandi, J. A. C. Alcaras, and P. L. Ferreira, *Am. J. Phys.* **35**, 520 (1967).

$$\psi_{nlm}(r,\theta,\phi) = N_r\left(\frac{Z}{a_0}\right)^{3/2}\left(\frac{2\rho}{n}\right)^l e^{-\rho/n}L_{n+l}^{2l+1}\left(\frac{2\rho}{n}\right)N_a P_l^{|m|}(\cos\theta)\frac{1}{\sqrt{2\pi}}e^{im\phi}, \quad (2.3)$$

where N_r and N_a are the radial and angular normalization factors, respectively; $\rho = 2Zr/na_0$ (a_0 = the Bohr radius $= h^2/4\pi^2me^2 = 0.52917$ Å); L_{n+l}^{2l+l} is the Laguerre polynomial; and P_l^m is the associated Legendre polynomial. The reason for writing this wave function explicitly is that it displays the basic form of the "hydrogenlike atomic orbitals," in which almost all discussions of approximate solutions for complex many-electron atoms are cast. The "atomic orbital pictures," with which we have all been familiar since high school or college chemistry courses, are just representations of the real parts of the angular terms in this wave function.

The wave function [equation (2.3)] is still not complete, however, without the addition of a factor describing the orientation of the spin of the electron. This is simply added a posteriori to indicate whether the spin is aligned parallel or antiparallel to some external field direction. In the purely electrostatic Hamiltonian of equation (2.1), there is no coupling between the spin and the spatial coordinates of the electron; there will be none until we begin to consider relativistic effects and magnetic interactions. Some of these are considered in sections 3 and 8; for a complete discussion of these effects, the reader is referred to the comprehensive text by Bethe and Salpeter (reference 1).

The spectrum arising from the energy levels in equation (2.2) consists of an extremely simple set of series of lines, which are indicated in the Grotrian diagram in figure 2.1.[2]

2 Two- and Many-Electron Atoms

The nonrelativistic Hamiltonian for a many-electron atom is similar to that of equation (2.1), with the addition of terms representing the Coulomb repulsion between the several electrons:

2. The term "Grotrian diagram" is often used and may seem a little mysterious in its origin. It derives from the name of Walter Grotrian, who, in his book *Graphische Darstellung der Spektren von Atomen und Ionen mit Ein, Zwei, und Drei Valenzelektronen* (Springer-Verlag, Berlin, 1928) first used energy level and spectrum diagrams of this type. A recent revision of this classic work is S. Bashkin and J. O. Stoner, Jr., *Atomic Energy Levels and Grotrian Diagrams*, Vols. 1–4 (North-Holland, Amsterdam, 1975–1982).

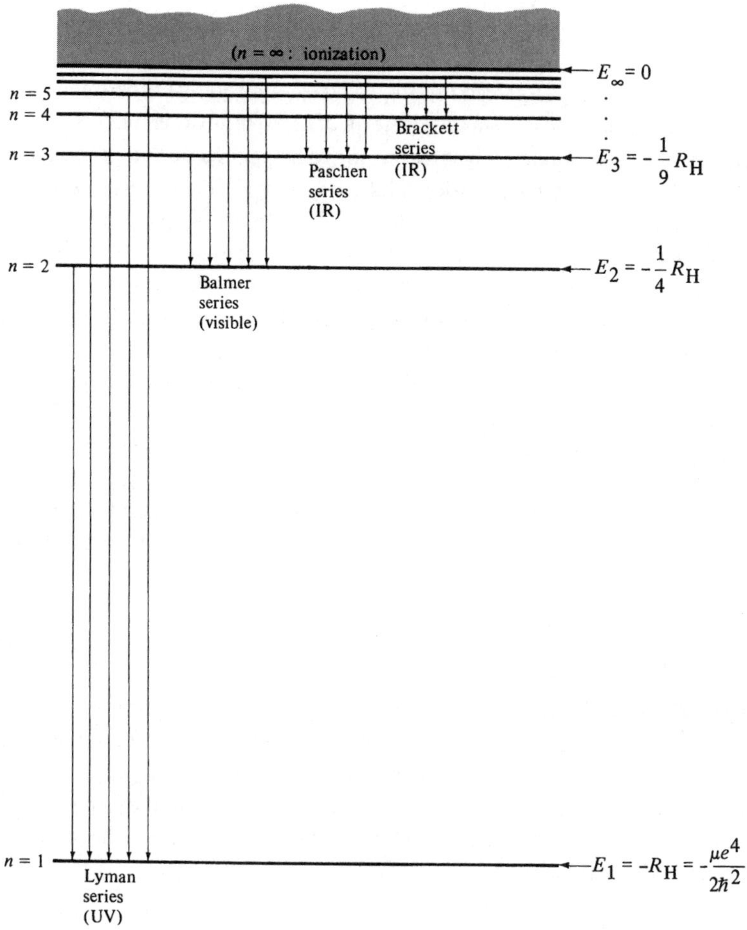

Figure 2.1
Energy levels and observed spectra of the hydrogen atom, arranged in a Grotrian diagram.

$$\mathscr{H} = -\frac{\hbar^2}{2m}\sum_i \nabla_i^2 - \sum_i \frac{Ze^2}{r_i} + \sum_{i>j}\frac{e^2}{r_{ij}}. \tag{2.4}$$

The terms involving $1/r_{ij}$ describe the electrostatic repulsion between electrons; the specification $i > j$ in the last term of (2.4) simply precludes counting each repulsion term twice. These terms make an exact solution of $\mathscr{H}|\psi\rangle = E|\psi\rangle$, with $\mathscr{H}$ given by equation (2.4), impossible. These terms also lift the l degeneracy present in the simple one-electron problem.

Observed atomic spectra can be described quite adequately by treating the correlation terms by means of perturbation theory (chapter 1.4). This is so even though the magnitudes of the perturbations may be as large as the zero-order energies; this means that we may not expect quantitative prediction of the locations of all the energy levels, but that at least we can establish the "good quantum numbers" for the many-electron system, which will provide the framework within which to carry out more sophisticated calculations, such as self-consistent field treatments. The zero-order Hamiltonian is taken as a sum of one-electron terms of the type of equation (2.1). The necessary matrix elements of $1/r_{ij}$ are most easily evaluated by using the expansion in spherical polar coordinates:

$$\frac{1}{r_{ij}} = \sum_l \sum_m \frac{(l-|m|)!}{(l+|m|)!}\frac{r_<^l}{r_>^{l+1}} P_l^{|m|}(\cos\theta_i) P_l^{|m|}(\cos\theta_j) e^{im(\phi_i-\phi_j)}. \tag{2.5}$$

Consider two electrons in an atom, with quantum numbers n_1, l_1 and n_2, l_2. All the m_{l_1} and m_{l_2} states are degenerate in the absence of a magnetic field. From results of degenerate perturbation theory (chapter 1.4), the correct zero-order states and first-order energies can be found by diagonalizing the matrix $\mathscr{H} - E\mathbf{1}$, where $\mathbf{1}$ is the diagonal unit matrix. The required off-diagonal matrix elements of $\mathscr{H}$ are

$$\langle n_1 n_2 l_1 m_{l_1} l_2 m_{l_2} | \frac{1}{r_{12}} | n_1' n_2' l_1' m_{l_1}' l_2' m_{l_2}' \rangle.$$

The operation indicated by this expression is to multiply a product of two wave functions of the type of equation (2.3), with quantum numbers $n_1 l_1 m_{l_1}$ and $n_2 l_2 m_{l_2}$, by the expansion of $1/r_{12}$ given by equation (2.5), and then by another product of two wave functions, with quantum numbers $n_1' l_1' m_{l_1}'$ and $n_2' l_2' m_{l_2}'$, and integrate over the coordinates $r_1\theta_1\phi_1 r_2\theta_2\phi_2$. As in so many seemingly formidable problems of this type, much of the calculation can be done away with by examining one key part of the integral in which the symmetry of the problem is embodied. In this case, this is the part involving ϕ. This part of the matrix element involves a sum of terms over m, each one of which has the form

$$\iint_0^{2\pi} e^{-im_{l_1}\phi_1}e^{-im_{l_2}\phi_2}e^{im(\phi_1-\phi_2)}e^{im'_{l_1}\phi_1}e^{im'_{l_2}\phi_2}\,d\phi_1\,d\phi_2$$

$$= \int_0^{2\pi} e^{i(m'_{l_1}-m_{l_1}+m)\phi_1}\,d\phi_1 \int_0^{2\pi} e^{i(m'_{l_2}-m_{l_2}-m)\phi_2}\,d\phi_2.$$

This integral will be zero unless

$$m'_{l_1} - m_{l_1} + m = 0$$

and

$$m'_{l_2} - m_{l_2} - m = 0,$$

that is,

$$m'_{l_1} + m'_{l_2} = m_{l_1} + m_{l_2} = M_L, \tag{2.6}$$

where L denotes the total angular momentum (and S is the total spin angular momentum). Physically, equation (2.6) means that the total z component of the orbital angular momentum is a good quantum number of the system, even though l_1 and l_2 individually are not. No other restrictions can be placed on this integral, so the matrix element takes the form

$$\delta(m_{l_1} + m_{l_2}, m'_{l_1} + m'_{l_2})\,\langle n_1 n_2 l_1 l_2 M_L|\frac{1}{r_{12}}|n'_1 n'_2 l'_1 l'_2 M'_L\rangle.$$

A similar restriction holds for the coupling of the individual electron spins. Both the total spin angular momentum (S^2) and one component (S_z) commute with the total atomic Hamiltonian, as well as with the total orbital angular momentum (L^2) and one component (L_z); for a detailed proof, see Bethe and Jackiw (reference 2). The matrix elements of $\mathscr{H}$ thus have the form

$$\langle n_1 n_2 l_1 l_2 m_{l_1} m_{l_2} s_1 s_2|\mathscr{H}|n'_1 n'_2 l'_1 l'_2 m'_{l_1} m'_{l_2} s'_1 s'_2\rangle$$

$$= \delta(L, L')\delta(M_L, M'_L)\delta(S, S')\delta(M_S, M'_S)\langle n_1 \cdots|\mathscr{H}|n'_1 \cdots\rangle.$$

There is a dipolar magnetic interaction coupling the individual electron spins having the form

$$\mathscr{H}'_{ss} = \left(\frac{e}{mc}\right)^2 \frac{(\mathbf{S}_1 \cdot \mathbf{S}_2)r_{12}^2 - 3\mathbf{S}_1 \cdot (\mathbf{r}_2 - \mathbf{r}_1)\mathbf{S}_2 \cdot (\mathbf{r}_2 - \mathbf{r}_1)}{r_{12}^5},$$

but this makes a negligible contribution to the total energy.

The most important effect of the spin can only be introduced on an ad hoc basis in this sort of treatment, and that is the *Pauli exclusion principle*. This can be stated in the following form: The total wave function of a system of electrons (or, indeed, of all half-integral spin particles—reference 3) must be *antisymmetric* with respect to exchange of any two of the particles. The important consequence for the construction of atomic states is that no two electrons in the system (in our case, the atom) can have all the same quantum numbers. If they did, then exchanging these two electrons would leave the total state of the system completely unchanged, which is contrary to the form of the Pauli principle just stated—namely, that exchanging two electrons must change the sign of the wave function for the state. A second implication is that electrons are indistinguishable particles, and that when counting up possible many-electron states, one must take care not to count as distinct states those that differ only by the labeling of individual electrons. The operation of these principles will become evident as we now proceed to construct the states for a typical two-electron system.

Let us consider the problem of finding the total orbital angular momentum L and total spin angular momentum S of the states arising from a configuration of two equivalent p electrons, each with $l = 1$ and $s = \frac{1}{2}$. This is the situation, for example, for the atoms in the fourth column of the periodic table, carbon through lead (table 2.1). The first step is to make a table of all the possible combinations of the m_l and m_s values.

The first state in the table is Pauli prohibited because the two electrons have the same quantum numbers. The third state is the same as the second because they differ just in the labeling of the electrons as 1 and 2. Similarly, the fifth and fourth lines are equivalent. From the sixth line down, only the Pauli-allowed states are tabulated; it should be obvious which ones have been eliminated. All together, there are 15 distinct states arising from the p^2 configuration.

Finding the terms is simply a question of picking out those multiplets that can give rise to the tabulated states. There are several methods of doing this; a convenient one is to use the diagrams introduced by Slater (reference 4), as shown in figure 2.2. The states are arranged in an $M_L \times M_S$ array. It is clear that this array is composed of one vertical column of five M states corresponding to a ^{1}D multiplet ($L = 2, S = 0$), and a 3×3 array corre-

Table 2.1

m_{l_1}	m_{l_2}	m_{s_1}	m_{s_2}	M_L	M_S
1	1	+	+	No good (Pauli principle)	
1	1	+	−	2	0
1	1	−	+	Same as preceding state (Pauli principle)	
1	0	+	+	1	1
0	1	+	+	Same as preceding state (Pauli principle)	
1	0	+	−	1	0
1	0	−	+	1	0
1	0	−	−	1	−1
1	−1	+	+	0	1
1	−1	+	−	0	0
1	−1	−	+	0	0
0	0	+	−	0	0
1	−1	−	−	0	−1
0	−1	+	+	−1	1
0	−1	+	−	−1	0
0	−1	−	+	−1	0
0	−1	−	−	−1	−1
−1	−1	+	−	−2	0

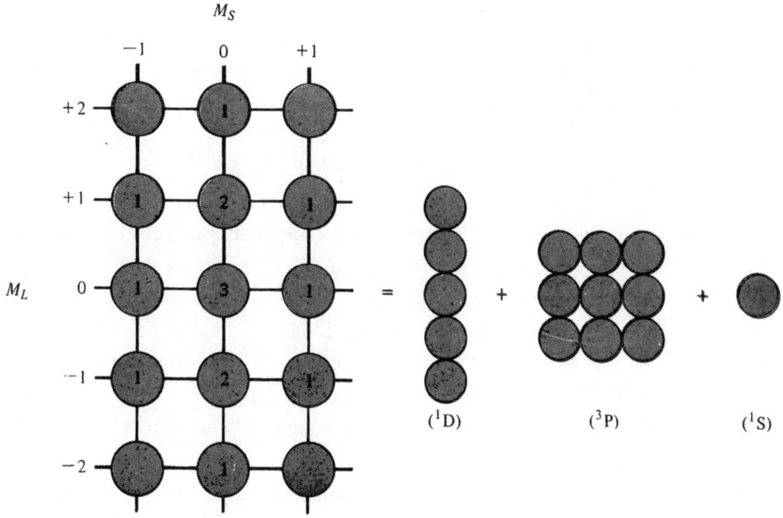

Figure 2.2
Slater diagram for a p^2 electron configuration and its resolution into ^{1}D, ^{3}P, and ^{1}S terms.

sponding to a ^{3}P multiplet ($L = 1, S = 1$). This accounts for 14 states, and the fifteenth is a ^{1}S ($L = 0, S = 0$). The multiplets arising from a p^2 configuration are thus ^{1}D, ^{3}P, and ^{1}S. Other atomic configurations can be analyzed in this same way. In a very real sense, however, there is no need anymore to do so—all configurations of practical interest have already been worked out and tabulated in the authoritative text by Condon and Shortley (reference 5).

We now consider, in a schematic way, the method of finding the energies of these multiplet states. Previously, we have had to calculate the sum of one-electron energies; in this approximation, all configurations with the same set of principal quantum numbers (nn) will be degenerate. Then, for our p^2 configuration, we write down the matrix of the perturbation term, $1/r_{12}$, in the $|m_{l_1} m_{l_2} m_{s_1} m_{s_2}\rangle$ representation; let us say that the ordering of the columns and rows of this matrix, shown in figure 2.3, is the same as that in table 2.1. We can immediately use the restriction of equation (2.6) to say that all the matrix elements are identically zero except for the shaded ones shown in the figure—one 3 × 3, two 2 × 2s, and eight 1 × 1s along the diagonal. Our rather formidable 15 × 15 matrix diagonalization problem is thus reduced to, at worst, the solution of a cubic equation. But we need not even do that much work.

The diagonal matrix element in the upper left-hand corner, belonging to the $|1\,1 + -\rangle$ state, clearly transforms into a member of the ^{1}D state upon diagonalization. The energy correction for this state is just

$$\langle n n 1\,1 + 1 + 1 + \tfrac{1}{2} - \tfrac{1}{2} | \frac{e^2}{r_{12}} | n n 1\,1 + 1 + 1 + \tfrac{1}{2} - \tfrac{1}{2}\rangle.$$

This integral can most easily be done numerically, using a digital computer. There will be 5 eigenvalues of the matrix having this same value. Similarly, the next one down the diagonal belongs to the ^{3}P state, and there will be a total of 9 eigenvalues of this magnitude. This accounts for 14 of the 15 eigenvalues of the matrix, and the fifteenth, corresponding to the energy correction for the ^{1}S level, can be found by invoking the principle of trace invariance. This principle is important in spectroscopic theory, so it will be stated and proved here.

The trace of a matrix[3] is defined as the sum of its diagonal elements,

3. For a good introduction to matrix algebra, see F. B. Hildebrand, *Methods of Applied Mathematics*, 2nd ed., (Prentice-Hall, Englewood Cliffs, NJ, 1965) pp. 1–118; C. L. Perrin, *Mathematics for Chemists*, (Interscience-Wiley, New York, 1970), chapter 8.

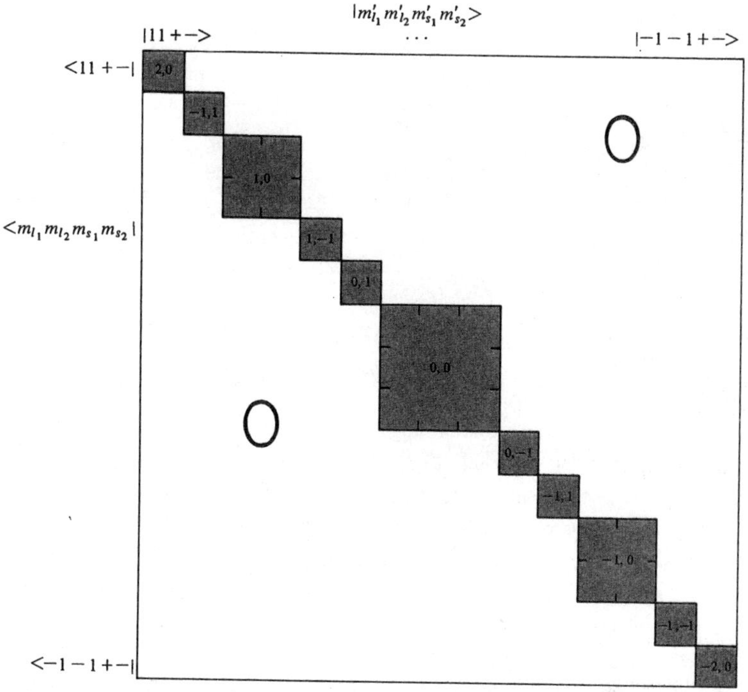

Figure 2.3
Matrix of $1/r_{12}$ for the p^2 electron configuration in the $(m_{l_1} m_{l_2} m_{s_1} m_{s_2})$ representation. Each submatrix is labeled by the appropriate value of (M_L, M_S). Elements in the fields labeled 0 are all zero.

$$\text{Tr}\,\mathbf{M} = \sum_j M_{jj}.$$

One of the properties of a trace is that the trace of the product of two matrices is the same, no matter which way the product is formed; that is,

$$\text{Tr}(\mathbf{AB}) = \text{Tr}(\mathbf{BA}).$$

Proof The proof is simple. Using the definition of a matrix product gives

$$\text{Tr}(\mathbf{AB}) = \sum_j (AB)_{jj} = \sum_j \sum_k A_{jk} B_{kj}$$

and

$$\mathrm{Tr}(\mathbf{BA}) = \sum_j (BA)_{jj} = \sum_j \sum_k B_{jk} A_{kj}.$$

Reversing the order of summation yields

$$\mathrm{Tr}(\mathbf{BA}) = \sum_j \sum_k A_{kj} B_{jk} = \sum_k (AB)_{kk} = \mathrm{Tr}(\mathbf{AB}). \qquad \text{Q.E.D.}$$

A Hermitian matrix can be diagonalized by a suitable unitary transformation:

$$\mathbf{UMU}^{-1} = \lambda \mathbf{1}.$$

If we form the trace of this transformed matrix,

$$\mathrm{Tr}(\mathbf{UMU}^{-1}) = \mathrm{Tr}((\mathbf{MU}^{-1})\mathbf{U}) = \mathrm{Tr}(\mathbf{M}(\mathbf{U}^{-1}\mathbf{U})) = \mathrm{Tr}(\mathbf{M}).$$

Thus the trace of a matrix is invariant under a unitary transformation that diagonalizes it, or, in practical terms, the sum of the eigenvalues of the diagonal matrix is equal to the sum of the diagonal elements of the original matrix from which it was derived. Thus we can find the energy correction for the ^{1}S state simply by adding up the 15 diagonal elements in our original $1/r_{12}$ matrix and subtracting five times the ^{1}D energy and nine times the ^{3}P energy; the entire problem is solved by carrying out two one-center integrals.

This discussion was intended merely to indicate the procedure whereby one can calculate, at least to first order, the energies of atomic multiplets. The entire procedure is described in detail in the texts by Slater (reference 4) and Condon and Shortley (reference 5). Much more detailed procedures, such as Hartree-Fock or configuration interaction calculations, designed for large digital computers, are now used to calculate much more accurate atomic energy levels than can be obtained by this simple approach.

Examples of atomic multiplet energies and the spectra arising from them will be deferred until after the discussion of spin-orbit coupling in the following section.

3 Spin-Orbit Coupling in a One-Electron Atom

The Hamiltonian we have been considering thus far, given in equation (2.1), takes no account of any relativistic effects. The velocity distribution of an electron in a hydrogen atom is in the range 10^8 cm/sec, which is less than 1% of the speed of light. Thus any relativistic mass corrections are small for this atom, but they can become appreciable for the inner electrons of heavier atoms. More significant is the magnetic interaction between the spin of the electron and the internal magnetic field generated by this orbital velocity.[4] The energy of a magnetic moment moving in an electromagnetic field is given[5] by

$$W = \boldsymbol{\mu} \cdot \mathbf{B}_{\text{eff}} = \boldsymbol{\mu} \cdot \left(\mathbf{B}_0 - \frac{\mathbf{v} \times \mathbf{E}}{c^2} \right).$$

For an electron in hydrogen, with no external fields, $\mathbf{B}_0 = 0$,

$$\boldsymbol{\mu} = \boldsymbol{\mu}_s = \frac{e\hbar}{2m}\mathbf{S},$$

and the orbital velocity $\mathbf{v} = \mathbf{p}/m$. The spin-orbit Hamiltonian is

$$\mathscr{H}'_{\text{so}} = \boldsymbol{\mu}_s \cdot \frac{\mathbf{E} \times \mathbf{v}}{c^2} = \frac{e\hbar}{2mc^2}\mathbf{S} \cdot (\mathbf{E} \times \mathbf{v}) = \frac{e\hbar}{2m^2c^2}\mathbf{S} \cdot (\mathbf{E} \times \mathbf{p}).$$

For a central field, such as exists in a one-electron atom,

$$\mathbf{E} = \frac{|E|}{r}\mathbf{r} = \frac{1}{r}\frac{\partial U}{\partial r}\mathbf{r},$$

so that

$$\mathscr{H}'_{\text{so}} = \frac{e\hbar}{2m^2c^2}\frac{1}{r}\frac{\partial U}{\partial r}\mathbf{S} \cdot (\mathbf{r} \times \mathbf{p}).$$

But $\mathbf{r} \times \mathbf{p}$ is just the orbital angular momentum $\mathbf{L}$, and so

$$\mathscr{H}'_{\text{so}} = \frac{e\hbar}{2m^2c^2}\frac{1}{r}\frac{\partial U}{\partial r}(\mathbf{L} \cdot \mathbf{S}) = \mathsf{A}\mathbf{L} \cdot \mathbf{S}.$$

We wish to estimate the magnitude of this spin-orbit perturbation on the energy of an atom; in order to do this, we must evaluate matrix elements of the form

$$\langle n\, l\, s\, m_l\, m_s | \mathscr{H}'_{\text{so}} | n'\, l'\, s'\, m'_l\, m'_s \rangle.$$

We can find the representation in which $\mathscr{H}'_{\text{so}}$ is diagonal by using the appropriate vector expression for $\mathbf{L} \cdot \mathbf{S}$. If we note that

4. A number of other relativistic effects are cataloged by Bethe and Salpeter (reference 1).
5. See reference 6 in chapter 1.

$$(\mathbf{L}+\mathbf{S})^2 = L^2 + S^2 + 2\mathbf{L}\cdot\mathbf{S},$$

we can write

$$\mathbf{L}\cdot\mathbf{S} = \tfrac{1}{2}[(\mathbf{L}+\mathbf{S})^2 - L^2 - S^2].$$

The quantum numbers l and s, thus, remain good for the system, but the z components of the spin and orbital angular momenta are mixed. If we define $\mathbf{L}+\mathbf{S}=\mathbf{J}$, then j and $m_j = m_l + m_s$ will be good quantum numbers, replacing m_l and m_s by themselves. As is the usual rule for addition of angular momenta, j can take on values from $l+s$ to $l-s$, in integral steps between these limits.

The spin-orbit perturbation is thus given by

$$E_{\text{so}}^{(1)} = \langle nlsjm_j|\frac{e\hbar}{2m^2c^2}\frac{1}{r}\frac{\partial U}{\partial r}\mathbf{L}\cdot\mathbf{S}|nlsjm_j\rangle.$$

The angular momentum part is simply

$$\begin{aligned}\langle lsjm_j|\mathbf{L}\cdot\mathbf{S}|lsjm_j\rangle &= \tfrac{1}{2}\langle lsjm_j|J^2 - L^2 - S^2|lsjm_j\rangle \\ &= \tfrac{1}{2}[j(j+1) - l(l+1) - s(s+1)].\end{aligned}$$

The radial part involves the expectation value of $\partial U/\partial r$. For a one-electron central Coulomb potential, $U(r) = -Ze^2/r$, so that $\partial U/\partial r = Ze^2/r^2$, and thus

$$\begin{aligned}E_{\text{so}}^{(1)} &= [j(j+1) - l(l+1) - s(s+1)]\frac{mZ^3e^8}{2c^2\hbar^4}\int_0^\infty R_{nl}^2(r)r^{-3}\,dr \\ &= \frac{mc^2\alpha^4Z^4}{n^3}\,\frac{j(j+1) - l(l+1) - s(s+1)}{2l(l+\frac{1}{2})(l+1)},\end{aligned} \tag{2.7}$$

where α is the dimensionless fine structure constant, $e^2/\hbar c = 1/137.0388$ according to the most recent determinations. The main point to notice from this development is that the spin-orbit energy goes as the fourth power of Z, the charge on the nucleus, and thus can become very large for the heavier atoms.[6]

6. For example, in sodium, which is a relatively light atom, the splitting between the $J=\frac{1}{2}$ and the $J=\frac{3}{2}$ levels is approximately 20 cm^{-1}. The equivalent internal magnetic field required to produce this splitting (section 8) is of the order of 400,000 G (gauss); fields of this magnitude are indeed produced by electronic orbital motion in atoms and molecules.

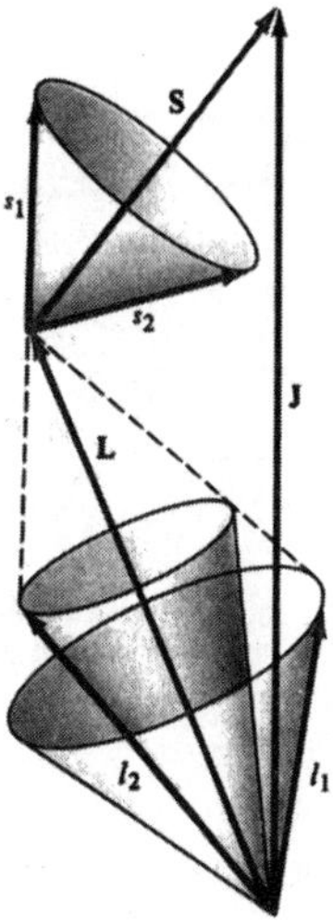

(a)

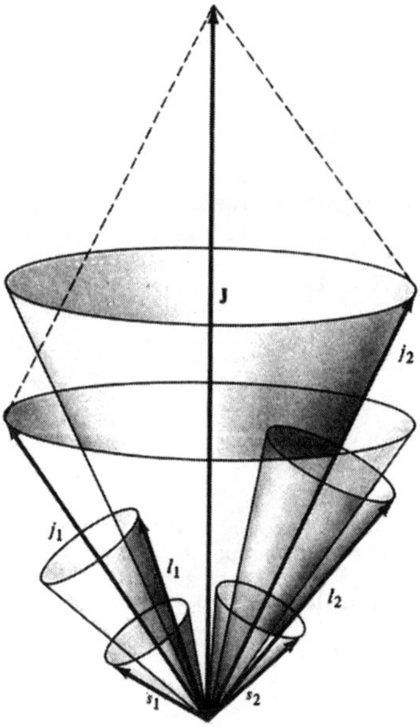

(b)

Figure 2.4
(a) Vector precession diagram in the *LS* coupling limit. (b) Vector precession diagram in the *jj* coupling limit.

4 Vector Precession Diagrams: *LS* and *jj* Coupling

The foregoing mathematical results can be embodied in a simple intuitive picture known as the "vector coupling model," in which an interaction such as electron correlation or spin-orbit coupling is represented by the *precession* of the two angular momentum vectors involved around their resultant sum. For cases in which the electron repulsion is larger than the spin-orbit coupling (that is, light atoms), the l_1 and l_2 angular momenta are first coupled to form their resultant **L**, and the s_1 and s_2 are coupled to form **S**; **L** and **S** are coupled to form **J**, as shown in figure 2.4a. This scheme, naturally enough, is known as *LS* coupling, and also as Russel-Saunders coupling, after the proposers of the model. For cases in which the spin-orbit coupling is larger than the electron repulsion (that is, heavy atoms), l_1 and s_1 are

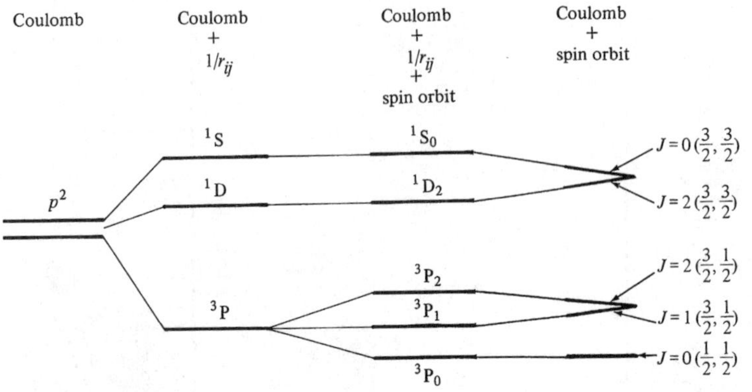

Figure 2.5
Correlation diagram for the p^2 configuration. From the left: The first effect shown is that of electron-electron repulsion, which splits the configuration into multiplets; spin-orbit coupling then splits nonsinglet states into different J levels. As spin-orbit dominates electron repulson energy, with higher Z, levels labeled by equal j values tend to cluster together.

coupled to form j_1; l_2 and s_2 are coupled to form j_2; and finally j_1 and j_2 are coupled to form **J**. This is known as the *jj* coupling scheme, shown in figure 2.4b. The usefulness of these vector models, aside from providing a physical picture, is that the geometrical properties of the coupled vectors can be used in further calculations, such as Zeeman (magnetic field) splittings, as is done in section 8.

Neither the *LS* nor the *jj* coupling scheme is, in itself, a completely adequate description of any atom; rather, they are two limits that blend smoothly from one to the other as a function of the atomic number Z. Such a connection of two limiting approximate cases is known as a *correlation diagram* and is very extensively used in spectroscopy. The correlation diagram for the vector coupling model is shown in figure 2.5 for the p^2 configuration previously considered.

5 Examples of Energy Levels and Spectra

It is now appropriate to ask how well real atoms reflect the quantization scheme we have developed thus far; for example, how far along the Z scale in figure 2.5 do the various atoms lie that possess p^2 ground configurations? The answer to this question may be seen in figure 2.6, which represents the observed atomic energy levels in the series C I through Pb I. (In atomic

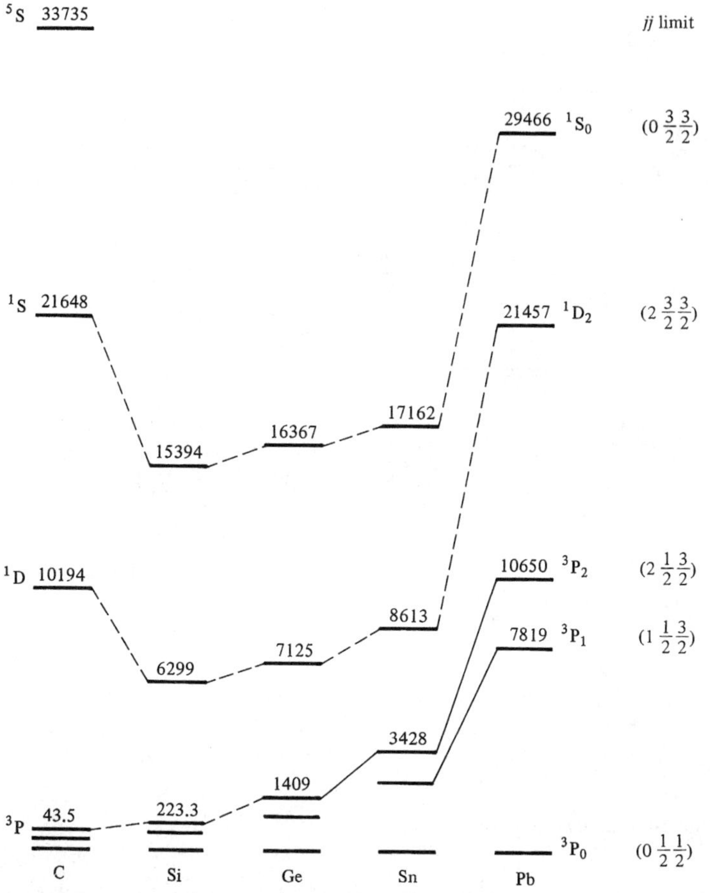

Figure 2.6
Energy levels for neutral atoms of the fourth column of the periodic table. All states arise from the p^2 configuration except for the ^{5}S, which arises from sp^3. Energies and splittings in this figure and in figures 2.7 and 2.8 are all in cm^{-1}. The right-hand (high Z) terms are labeled with the jj limit quantum numbers.

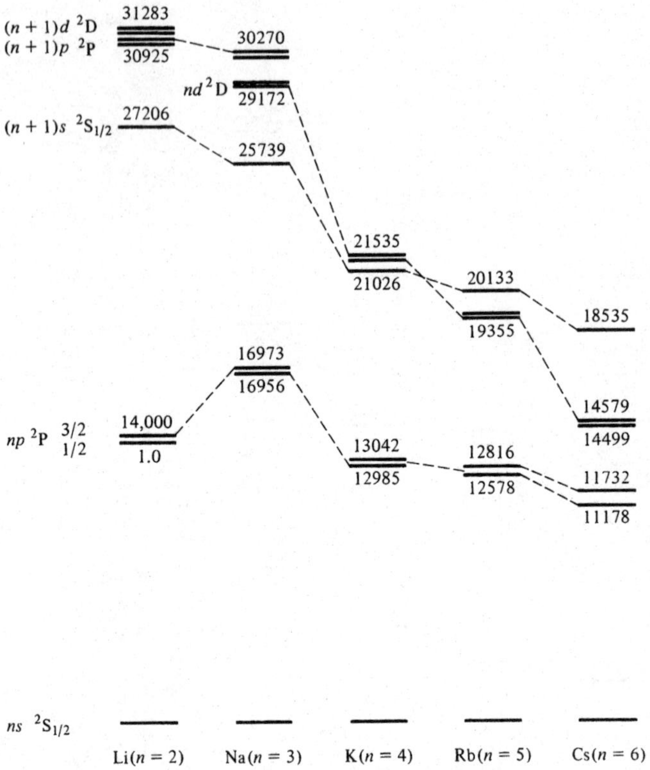

Figure 2.7
Energy levels for neutral alkali atoms.

spectroscopy, the neutral atom is designated M I, the singly ionized M^+ is M II, and so on; this once more reflects the activity of spectroscopists who cataloged lines, early in the twentieth century, without a real understanding of their origin.) These data, as well as those in figures 2.7 and 2.8, are all taken from Moore's tables of atomic energy levels (reference 6). We see that in this series, at least, even the heaviest member, lead, is far from being at the pure *jj* coupling limit; the members of the ^{3}P multiplet are still grouped together, distinct from the ^{1}S and ^{1}D.

A simpler example is shown in figure 2.7, which depicts the energy levels for the alkali metal atoms. Again, even for cesium, the spin-orbit splitting

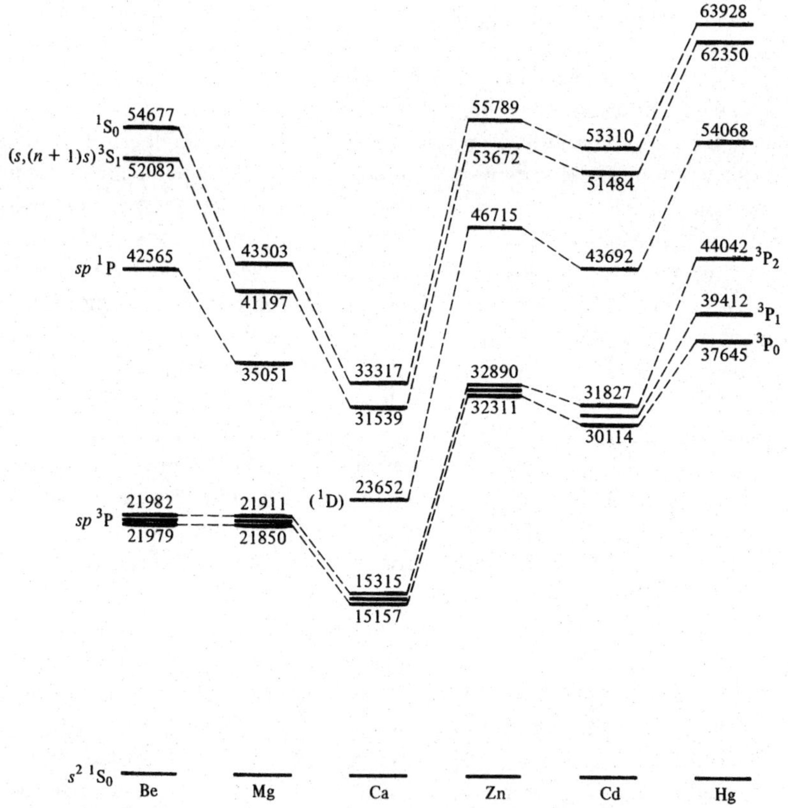

Figure 2.8
Energy levels for neutral group II atoms, having two outer electrons.

is appreciably less than the intermultiplet splitting. Finally, a somewhat more complex series is shown in figure 2.8, that of the configuration having s^2 outer electrons, which includes Be I through Hg I.

One generalization to be gleaned from an inspection of these and a number of other atomic energy level spectra is that among the multiplet levels arising from a given configuration, it is almost always the one of highest spin multiplicity that has the lowest energy. That is, triplet states are lower in energy than singlets, quartet states are lower than doublets, and so on. Among states of the same multiplicity, the state with the larger L value is often the one of lower energy. This generalization (and it is no more than that—not a hard-and-fast rule, especially for complex configurations) is known as *Hund's rule*. The basis for this rule lies in the antisymmetry of the electronic wave functions, which is embodied in the Pauli exclusion principle: The states of highest multiplicity have totally symmetric spin wave functions, which means that the electronic spatial wave functions must be antisymmetric. This eliminates orbital functions in which two electrons have the same coordinates, and thus reduces the electron repulsion energy. For a fuller discussion, see the treatment of the helium atom in the text by Heitler (reference 7) or Coulson (reference 8).

We should now turn our attention to the kinds of spectral lines that can originate from these energy levels, and this brings us to consider the question of specific selection rules. In order to find the strongest lines in the spectrum, we should consider electric dipole allowed transitions, that is, those for which the matrix element $\langle m|\mathbf{r}|k\rangle$ in equations (1.22)–(1.26) is nonzero. If we substitute hydrogenlike wave functions for $\langle m|$ and $|k\rangle$ in the indicated integral, and write the components of $\mathbf{r}$ in spherical polar coordinates, we have an integral of the form

$$\iiint \psi^*_{nlm}(r,\theta,\phi)\begin{pmatrix} r\sin\theta\cos\phi \\ r\sin\theta\sin\phi \\ r\cos\theta \end{pmatrix}\psi_{n'l'm'}(r,\theta,\phi)r^2\,dr\sin\theta\,d\theta\,d\phi.$$

From the properties of the associated Legendre polynomials, this integral will be zero unless $l - l' = 0$ or ± 1. Since the spin coordinates do not appear at all in this integral, we must have $s - s' = 0$. When spin-orbit coupling is appreciable, we find that the operative restriction is that $j - j' = 0$ or ± 1. One additional restriction appears: If both l and l', or j and j', are zero, then the integral vanishes, and the two levels are not connected by electromagnetic radiation, not only in the electric dipole approximation, but to all multipole orders.

We should note here that an approximation implicit in this entire treatment is that we can treat changes in the state of an atom as single-electron transitions between one orbital and another, with no attendant change in the orbitals of all the other electrons. (This is sometimes called the "virtual orbital approximation.") We know that this cannot be true, but empirically it is found to be a pretty good model, not only for description of spectra, but even for the quantitative calculation of oscillator strengths.

To recapitulate: We have the following selection rules for atomic spectra in the *LS* coupling limit:

$\Delta L = 0$ or ± 1 and $\Delta S = 0$

or

$\Delta J = 0$ or ± 1

with

$L = 0 \nleftrightarrow L = 0$ and $J = 0 \nleftrightarrow J = 0.$

In figure 2.9 we can see some of the allowed transitions subject to these selection rules for some typical cases. For sodium (figure 2.9a), the strongest feature in the spectrum is the ${}^2\mathrm{P}$–${}^2\mathrm{S}$ transition, which produces the characteristic yellow D lines. In mercury (figure 2.9b), the strongest transition is the ${}^1\mathrm{P}_1$–${}^1\mathrm{S}_0$ line at the edge of the vacuum ultraviolet. Spin-orbit coupling is sufficiently great, however, to give a great deal of oscillator strength to the ${}^3\mathrm{P}_1$–${}^1\mathrm{S}_0$ line at 2,537 Å, which is a $\Delta J = 1$ transition. The familiar visible lines of the mercury arc—4,358 Å blue, 5,460 Å green, and 5,770/5,790 Å yellow—all correspond to transitions between excited atomic states. Note that the strongest allowed transitions are always between states arising from different electron configurations; Pauli-allowed states that can be constructed from a given configuration are generally such that so-called intramultiplet transitions are forbidden by one of the selection rules.

The operation of these selection rules leads to one other noteworthy phenomenon, namely, the existence of metastable atomic states, which are levels lying at higher energy than the ground state, but which cannot get rid of their energy by radiative emission because the transition required to do so is forbidden. An example is the $6\,{}^3\mathrm{P}_0$ state of mercury [figure 2.9b], which cannot decay to the ${}^1\mathrm{S}_0$ ground state because of the $J = 0 \nleftrightarrow J = 0$

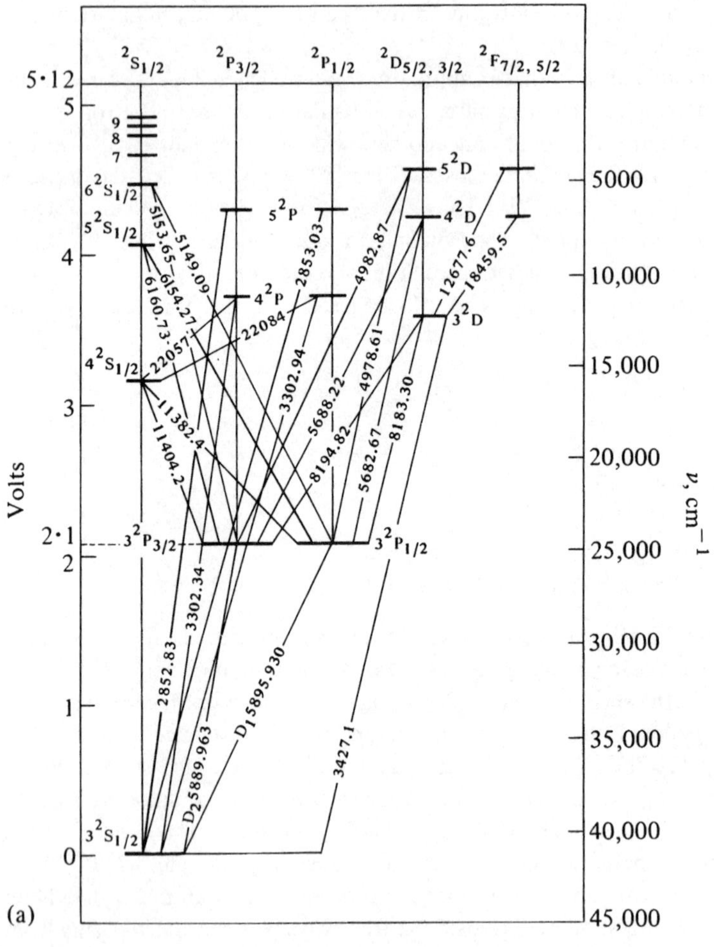

Figure 2.9
Allowed transitions in the spectrum of (a) sodium, (b) mercury. [From A. C. G. Mitchell and M. W. Zemansky, *Resonance Radiation and Excited Atoms*, Cambridge University Press, Cambridge, 1961. Reproduced with permission.]

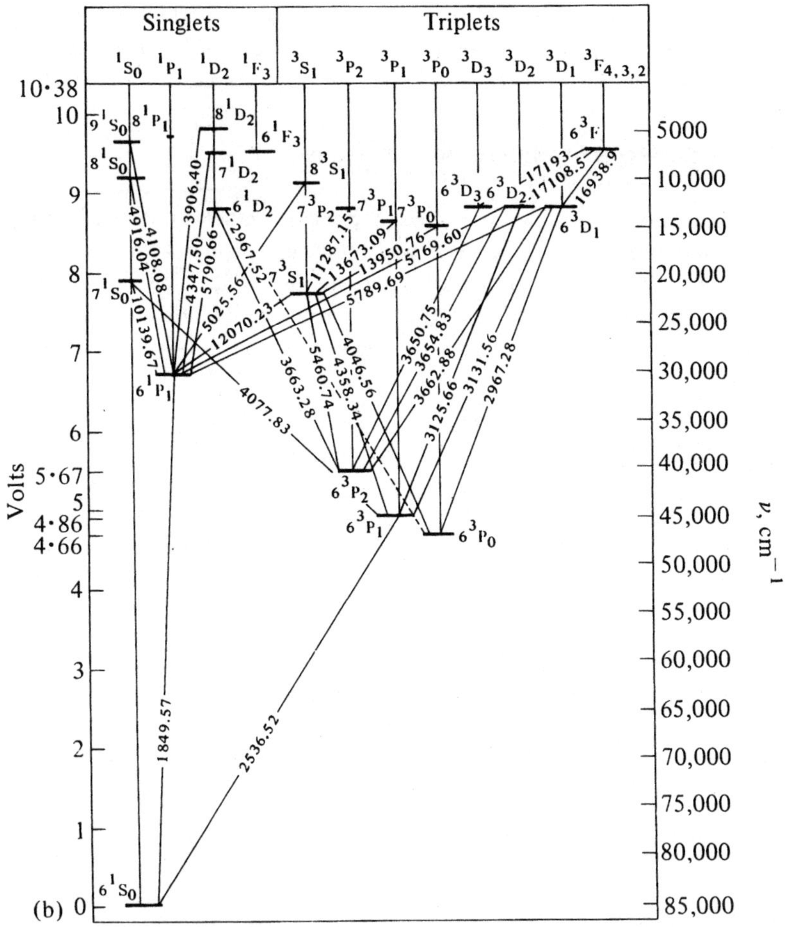
Singlets
Triplets
Volts
ν, cm⁻¹
(b)
1849.57
2536.52
4916.04
4108.08
10139.67
4077.83
3906.40
4347.50
5790.66
5025.56
12070.23
2967.52
3663.28
5460.74
4358.34
4046.56
11287.15
13673.09
13950.76
5769.60
5789.69
3650.75
3654.83
3662.88
3125.66
3131.56
2967.28
17193
17108.5
16938.9

prohibition. Other examples are the ^{1}D state in oxygen and the 2^1S and 2^3S states of helium. These metastable atoms can be very useful energy carriers in gas phase photochemistry or in molecular beam studies.

6 Nuclear Hyperfine Interactions in Atoms

As the quality and resolution of spectroscopic instruments improved during the development of the science, it was found that many atomic lines possessed a structure even more detailed than multiplet and spin-orbit splitting. These splittings came to be known as "hyperfine structure," and eventually it was realized that their origin lay in the interaction of the orbital electron with the atomic nucleus.

The Hamiltonian that gives rise to these splittings can be written as a sum of three terms:[7]

$$\mathscr{H}_{\text{hfs}} = T(\{\mathbf{p}\}) + V(\{\mathbf{r}\}) + \mathscr{H}'(\text{spin}). \tag{2.8}$$

The first term embodies the effects of the finite nuclear mass. Let us write the kinetic energy of the atom as

$$T = \tfrac{1}{2}m_e \sum_i \dot{\mathbf{r}}_i^2 + \tfrac{1}{2}M\dot{\mathbf{r}}^2,$$

where m_e is the electron mass and M the nuclear mass. If we introduce

$$\mathbf{R} = \frac{1}{M + Nm_e}\left(M\mathbf{r} + m_e \sum_i \mathbf{r}_i\right)$$

and

$$\mathbf{S}_i = \mathbf{r}_i - \mathbf{r},$$

then we can write the conjugate momenta as

$$\dot{\mathbf{P}} = \frac{\partial \mathscr{H}}{\partial \mathbf{R}}, \qquad \dot{\mathbf{p}}_i = \frac{\partial \mathscr{H}}{\partial \mathbf{S}_i}$$

and express the total kinetic energy as

$$T = \frac{1}{2\mu}\sum_i \mathbf{p}_i^2 + \frac{1}{M}\sum_{i \neq j} \mathbf{p}_i \cdot \mathbf{p}_j + \frac{1}{2(M + Nm_e)} \cdot \mathbf{P}^2.$$

7. A more thorough treatment of these factors may be found in chapter 18 of Condon and Shortley (reference 5).

Let us examine these three terms individually. The first involves the electron-reduced mass, $\mu = m_e M/(m_e + M)$, and yields the dominant contribution to the isotope Rydberg shift of a spectroscopic line. The second, involving cross terms, appears in many-electron atoms alone, and, because of the nuclear mass factor out front, is several orders of magnitude smaller than the first term. The last term takes account of photon recoil in the center-of-mass system. This term is completely negligible; the frequency shift from this effect is typically of the order of 100 Hz in a typical line frequency of 5×10^{14} Hz. The photon recoil effect can be observed in molecular beam experiments, however, in which a laser is used to "push" absorbing atoms out of the flight path of the beam.

The second term in equation (2.8) takes account of the non-Coulombic potential experienced by the electron when it is within the nucleus. The potential can be written formally as

$$V = \sum_{i \neq j} \frac{e^2}{r_{ij}} - \sum_{i} \frac{e^2}{r_i}(r_i > r_0) + V'(r_i < r_0),$$

where r_0 is the radius of the nucleus, determined, for example, from nuclear scattering experiments.

The third term in equation (2.8) takes account of the magnetic-dipolar coupling between the angular momentum **J** of the electron and the spin **I** of the nucleus. It is by far the most important contributor to hyperfine structure. The energy terms that give rise to the hyperfine splittings have the form

$$\langle n l s j m_j i m_i | \mathbf{I} \cdot \mathbf{J} | n' l' s' j' m_j' i' m_i' \rangle.$$

Since these matrix elements vanish unless $m_j + m_i = m_j' + m_i'$, we can use the same vector coupling model as we did for the coupling of several electrons to form multiplet states or for spin-orbit coupling. That is, the operator $\mathbf{I} \cdot \mathbf{J}$ is diagonalized in a new $|n l s j i f m_f\rangle$ basis, where $\mathbf{F} = \mathbf{I} + \mathbf{J}$; the vector coupling diagram is shown in figure 2.10. By the same geometric arguments we used before, the first-order hyperfine energy is found to be

$$E_{\text{hfs}}^{(1)} = \frac{a}{2}[F(F+1) - J(J+1) - I(I+1)], \tag{2.9}$$

where a is the hyperfine coupling constant (see reference 4 for a further discussion of this quantity). We may also note that the selection rule for dipole-allowed transitions is just

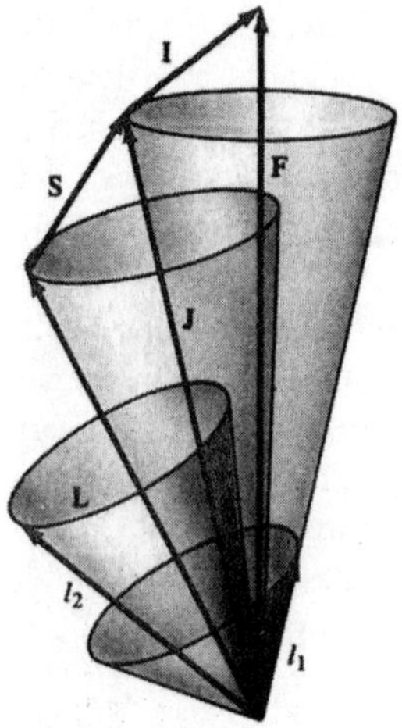

Figure 2.10
Vector coupling diagram, in the *LS* coupling limit, including nuclear spin.

$$\Delta F = 0 \text{ or } \pm 1$$

except

$$F = 0 \not\leftrightarrow F = 0.$$

As an example of nuclear hyperfine structure, let us consider a particular line of the mercury (Hg I) emission spectrum, the green line at 5,460.74 Å. The assignment is $7^3S_1 \rightarrow 6^3P_2$. There are six naturally occurring isotopes of mercury:

$$^{198}\text{Hg}, {}^{200}\text{Hg}, {}^{202}\text{Hg}, {}^{204}\text{Hg}: \quad I = 0;$$
$$^{199}\text{Hg}: \quad I = \tfrac{1}{2};$$
$$^{201}\text{Hg}: \quad I = \tfrac{3}{2}.$$

The nuclear spin of the first four isotopes is zero, in accordance with the so-called even-even rule of nuclear physics, namely, that all nuclei with even atomic numbers ($Z = 80$ for mercury) and even nucleon numbers possess no nuclear spin. These isotopes each give rise to a single line, slightly displaced from one another by virtue of the kinetic energy and non-Coulomb potential terms in equation (2.8). (These are labeled α, β, γ, and δ in figure 2.12.)

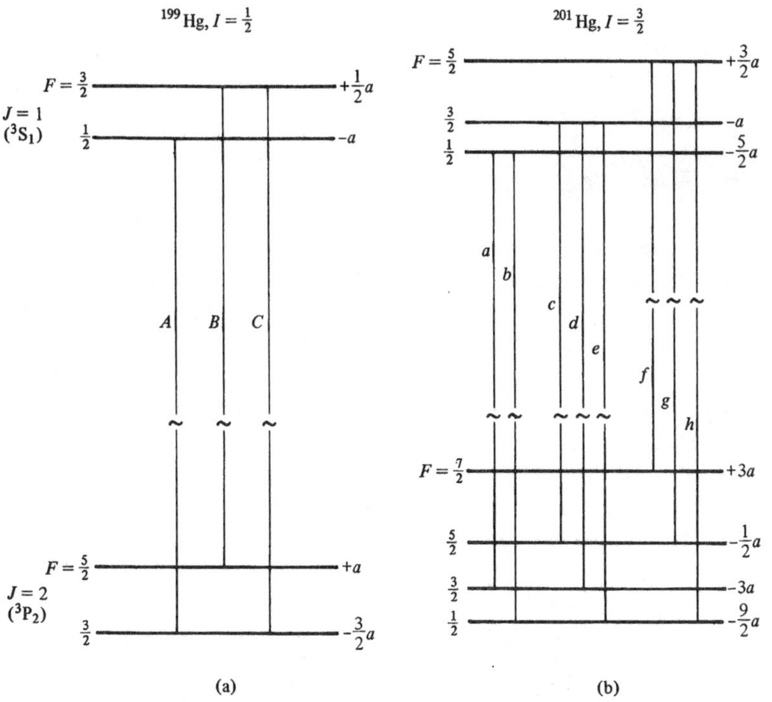

Figure 2.11
Nuclear hyperfine splitting for (a) mercury 199 (^{199}Hg) and (b) mercury 201 (^{201}Hg).

To find the splittings due to the dipolar spin coupling term, we first find the levels for each energy level of each isotope. Figure 2.11a shows the construction for the ^{199}Hg isotope with $I = \frac{1}{2}$. For the upper level, $^3S_1$, $J = 1$, and we can have $F = \frac{3}{2}$ or $\frac{1}{2}$. Using the energy expression [equation (2.9)], we find the splittings $+\frac{1}{2}$ and -1, in units of the hyperfine coupling constant a. Similarly, for the lower $^3P_2$ level, $J = 2$, $F = \frac{5}{2}$ or $\frac{3}{2}$, and the splitting is $+1$, $-\frac{3}{2}$. Using our selection rule for this quantization, we can have the lines labeled A, B, and C in figures 2.11a and 2.12. The analogous construction for ^{201}Hg, with $I = \frac{3}{2}$, is shown in figure 2.11b; we obtain lines a–h. Notice, too, that when the splittings are weighted by the degeneracy of each sublevel ($g_F = 2F + 1$) and added together, the trace invariance rule discussed in section 2 is obeyed.

When all of these hyperfine components are put together, the predicted spectrum shown in figure 2.12 is obtained. An experimental trace of this emission line is displayed above; the structure is faithfully reproduced to within the limit of resolution of the spectrometer. The entire hyperfine pattern occupies less than 0.5 Å of the spectrum.

7 The Stark Effect

This discussion of atomic spectra will conclude with a consideration of the effects of externally imposed fields on the energy levels and spectra. Like a great deal of the rest of atomic spectroscopy, this is not an area of active research at the moment; rather, the study of these effects can serve as an introduction to the concepts involved, and as a simple example for later application to molecular spectroscopy.

The Hamiltonian for the interaction of an atom with a static electric field [called the Stark effect after its discoverer, Johannes Stark (1874–1957); also called the electrochromic effect by other spectroscopists who did not like Stark] is just the electric dipole interaction:

$$\mathscr{H}'_{\text{Stark}} = -\mathbf{E}\cdot\sum_i e\mathbf{r}_i. \tag{2.10}$$

Therefore the states coupled in by the field are the same as those in dipole selection rules for radiative transitions. If we let z be the field axis, the first-order Stark shift is given by

$$E^{(1)}_{\text{Stark}} = eE_z\langle n\,l\,s\,j\,m_j|\sum_i z_i|n\,l\,s\,j\,m_j\rangle.$$

This is zero for all atoms except hydrogen, to be discussed shortly. Since the first-order perturbation vanishes, we go to second order, which has the form

$$E^{(2)}_{\text{Stark}} = e^2E_z^2\sum_k \frac{\langle 0|\sum_i z_i|k\rangle\langle k|\sum_i z_i|0\rangle}{E_0 - E_k}.$$

After some lengthy computations, this is found to have the form

$$E^{(2)}_{\text{Stark}} = A - BM_J^2.$$

The Stark coefficients A and B are both functions of J and are proportional to E^2. The field dependence of the energy is shown in figure 2.13. Experimentally, the splittings can be observed by photographing an atomic emission line in a discharge between two electrodes shaped to set up an inhomogeneous electric field, as diagrammed in figure 2.14.

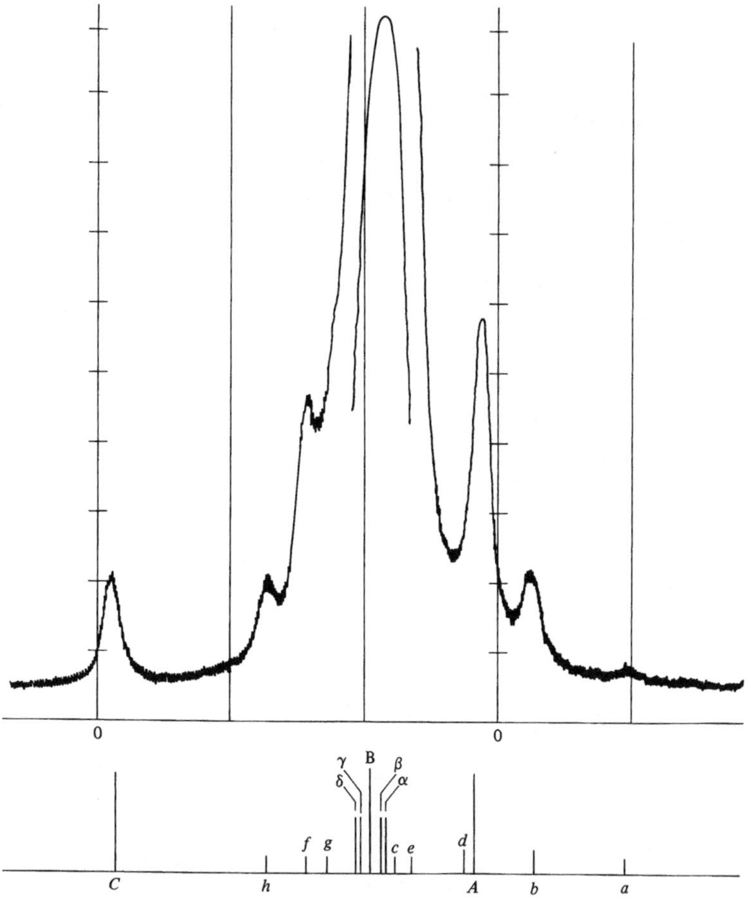

Figure 2.12
The green line of mercury at 5,460.75 Å, showing resolved nuclear hyperfine structure. The experimental spectrum is a photoelectric tracing using the tenth order of a Harrison grating mounted in a 35-foot Czerny-Turner system at the Spectroscopy Laboratory of MIT. The central portion of the line was attenuated by one-third as it was being recorded in order to keep the peak on scale. [Reproduced with the permission of A. V. Nowak, MIT.]

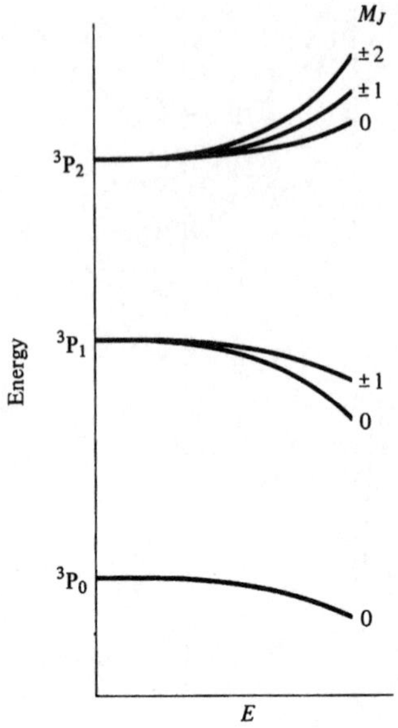

Figure 2.13
Stark splittings in a ^3P level.

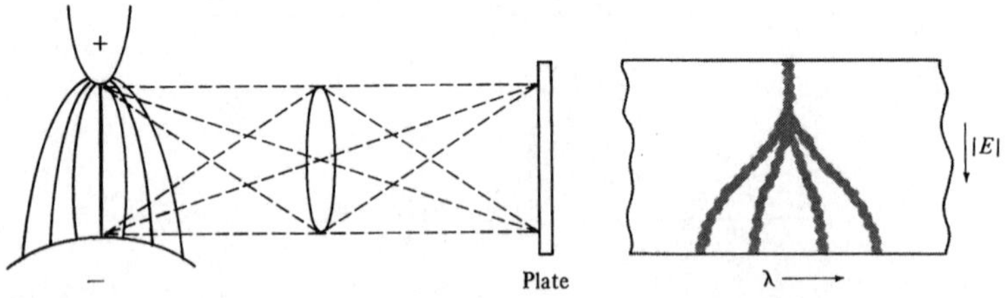

Figure 2.14
Experimental arrangement for observing Stark splittings (dispersing elements not shown) and appearance of line on a spectrum plate (recall that the focusing optics reverse the image), also known as a LoSurdo photograph.

This scheme applies for all neutral atoms except hydrogen. In hydrogen, the different l levels are degenerate, so that we must use the results of first-order degenerate perturbation theory, which gives the energy as

$$E_n = E_n^0 + E_{nn}^{(1)} \pm E_{12}^{(1)}.$$

The diagonal first-order term is still zero, but now the splitting for a state such as the $2s2p\,^2\mathrm{P}$ is given by

$$E_{12}^{(1)} = eE\langle 2s|z|2p\rangle;$$

thus the splitting is linear in electric field strength.

8 The Zeeman Effect

In contrast with the Stark effect, which has had only limited usefulness in the elucidation of atomic energy levels, the interaction of an atom with a static magnetic field [called the Zeeman effect, after Dutch physicist Pieter Zeeman (1865–1943)] has been widely used in the determination of level multiplicities and nuclear spins. The Hamiltonian is just the magnetic dipole interaction,

$$\mathscr{H}'_{\mathbf{Zeeman}} = -\boldsymbol{\mu}\cdot\mathbf{B}, \tag{2.11}$$

and since $\boldsymbol{\mu} = (e/2mc)g(J)\mathbf{J}$, where $g(J)$ is the gyromagnetic ratio, which will be derived shortly, there will always be one component of $\boldsymbol{\mu}$ that commutes with J_z (unless $J = 0$), and thus m_j remains a good quantum number in the presence of a magnetic field. The Zeeman splitting thus appears in first order, and is easily calculated as

$$\begin{aligned} E_{\mathbf{Zeeman}}^{(1)} &= \langle n\,l\,s\,j\,m_j|\mathscr{H}'_{\mathbf{Zeeman}}|n'\,l'\,s'\,j'\,m_j'\rangle \\ &= -\mu_z B_z\,\delta(l,l')\,\delta(s,s')\,\delta(j,j')\,\delta(m_j,m_j') \\ &= -\frac{e\hbar}{2mc}B_z g(J)m_j. \end{aligned}$$

The problem remaining for quantitative interpretation is that of finding $g(J)$. To do this, we return to the geometric arguments of the classical vector coupling model, as shown in figure 2.15. Because of the dipole interaction, the angular momentum vector $\mathbf{J}$ precesses around the magnetic field vector $\mathbf{B}$. The coupling is weak, however, so that $\mathbf{L}$ and $\mathbf{S}$ are still coupled to form

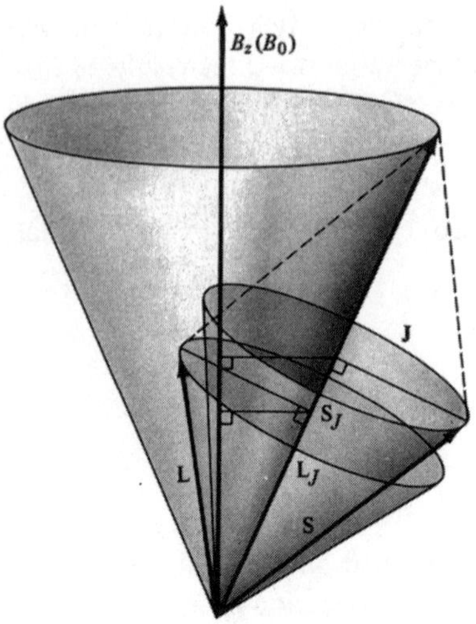

Figure 2.15
Construction, using vector coupling model, to find the gyromagnetic ration $g(J)$. The projections of **L** and **S** on **J** are first found; the projections of these projections on the magnetic field vector then give the magnitude of the interaction with the field.

J; this is equivalent to requiring that the Zeeman energy be much less than the spin-orbit interaction energy (we shall shortly see what happens when this requirement is not fulfilled).

The first step is to find the individual projections of **L** and **S** on **J**:

$$\mathbf{L}_J = \frac{\mathbf{J}(\mathbf{L}\cdot\mathbf{J})}{|\mathbf{J}|^2} = \frac{\mathbf{J}(\mathbf{L}\cdot\mathbf{J})}{J(J+1)},$$

$$\mathbf{S}_J = \frac{\mathbf{J}(\mathbf{S}\cdot\mathbf{J})}{|\mathbf{J}|^2} = \frac{\mathbf{J}(\mathbf{S}\cdot\mathbf{J})}{J(J+1)}.$$

We then take the Zeeman energy as the sum of contributions from the electron orbital and electron spin angular momenta, projected onto the field vector as shown in figure 2.15. This gives

$$\begin{aligned} E^{(1)}_{\text{Zeeman}} &= -\frac{e\hbar}{2mc} B_0 \cdot (\text{orbital magnetic moment and spin magnetic moment}) \\ &= -\frac{e\hbar}{2mc} B_0 \left[M_J g_L \frac{\mathbf{L}\cdot\mathbf{J}}{J(J+1)} + M_J g_S \frac{\mathbf{S}\cdot\mathbf{J}}{J(J+1)} \right]. \end{aligned}$$

The orbital g factor, g_L, is unity, and the electron spin g factor, g_S, is very nearly equal to two.[8] Thus we have

$$E^{(1)}_{\text{Zeeman}} = -\frac{e\hbar}{2mc} B_0 M_J \frac{(\mathbf{L} + 2\mathbf{S}) \cdot \mathbf{J}}{J(J+1)}$$

$$= -\frac{e\hbar}{2mc} B_0 M_J \frac{(\mathbf{J} + \mathbf{S}) \cdot \mathbf{J}}{J(J+1)},$$

since $\mathbf{J} = \mathbf{L} + \mathbf{S}$. We know what the matrix element of J^2 will be; to find the matrix element of $\mathbf{S} \cdot \mathbf{J}$, we make use of $\mathbf{L} = \mathbf{J} - \mathbf{S}$, so that $L^2 = J^2 + S^2 - 2\mathbf{J} \cdot \mathbf{S}$, and so $\mathbf{J} \cdot \mathbf{S} = \frac{1}{2}(J^2 + S^2 - L^2)$. If we extract the part of the expression that corresponds to $g(J)$ and substitute for $\mathbf{J} \cdot \mathbf{S}$, we obtain

$$g(J) = \langle nsjm | \frac{\mathbf{J}^2 + \frac{1}{2}(\mathbf{J}^2 + \mathbf{S}^2 - \mathbf{L}^2)}{J(J+1)} | nsjm \rangle$$

$$= 1 + \frac{J(J+1) + S(S+1) - L(L+1)}{2J(J+1)}.$$

In summary, the magnetic field lifts the degeneracy of the M_J levels (by destroying the spherical symmetry of the total Hamiltonian), and produces a splitting linear the field strength.

Paschen-Back Limit When the magnetic field strength is turned up sufficiently high, so that the Zeeman energy becomes appreciably larger than the spin-orbit coupling energy, we no longer have $\mathbf{J}$ as a good quantum number. $\mathbf{L}$ and $\mathbf{S}$ individually are decoupled (that is, are good quantum numbers), and the Zeeman energy is just

$$E^{(1)}_{\text{Zeeman}} = E(M_L, M_S) = -\frac{e\hbar}{2mc} B_0 (M_L + 2M_S) + \mathbf{A} M_L M_S,$$

where $\mathbf{A}$ is the spin-orbit couping parameter previously introduced.

We can summarize each of these limiting cases by considering the graph shown in figure 2.16 for the Zeeman splitting of a ^{2}P term. In the low-field

8. Quantum considerations predict that the electron, with spin angular momentum equal to $\frac{1}{2}\hbar$, should have a magnetic moment equivalent to one unit of angular momentum; thus $g_S = 2$. This was originally termed the "anomalous" g factor for the electron, to distinguish it from the "normal" $g = 1$. Careful measurements have shown, however, that actually $g_S = 2.00232277$. One of the triumphs of modern quantum electrodynamic theory is the prediction that $g_S = 2(1 + e^2/2\pi\hbar c)$, extremely close to the measured value. The ratio $e^2/\hbar c$ is the dimensionless *fine structure* constant α, which has the value 1/137.0388.

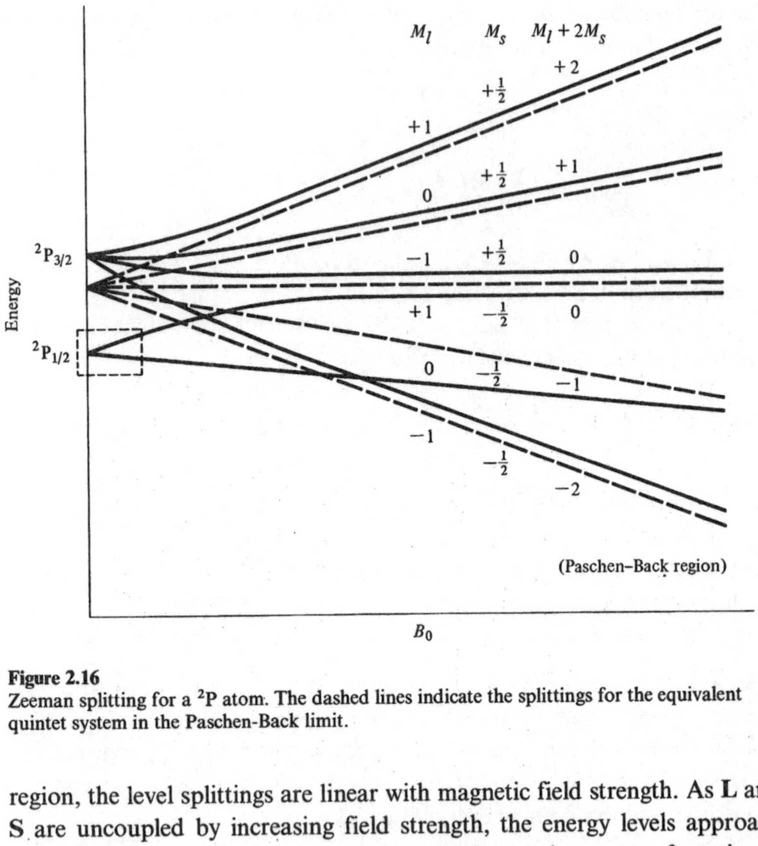

Figure 2.16
Zeeman splitting for a ^2P atom. The dashed lines indicate the splittings for the equivalent quintet system in the Paschen-Back limit.

region, the level splittings are linear with magnetic field strength. As **L** and **S** are uncoupled by increasing field strength, the energy levels approach asymptotically to those of a quintet originating at the center of gravity of the $^2P_{3/2}$ and $^2P_{1/2}$ levels; this is because $M_L + 2M_S$ can range over the values $+2, +1, 0, -1, -2$.

Zeeman Splittings with Nuclear Hyperfine Interactions The Hamiltonian for magnetic field interaction with an atom possessing both electronic and nuclear spin angular momentum is just

$$\mathscr{H}' = a\mathbf{I}\cdot\mathbf{J} + (g_J\mu_0\mathbf{J} + g_I\mu_N\mathbf{I})\cdot\mathbf{B}. \tag{2.12}$$

The nuclear magneton μ_N is defined as $e\hbar/2M_{\text{proton}}c$, by analogy with the Bohr magneton, $\mu_0 = e\hbar/2m_ec$; we have also included the $\mathbf{I}\cdot\mathbf{J}$ dipolar hyperfine coupling in equation (2.12). By geometric construction and manipulation similar to the foregoing for g_J, we find for the energy

$$E^{(1)}_{\text{Zeeman}}(F, M_F) = \frac{a}{2}[F(F+1) - I(I+1) - J(J+1)]$$

$$= -M_F B_0\left[g_J\mu_0\frac{F(F+1) + J(J+1) - I(I+1)}{2F(F+1)}\right. \qquad (2.13)$$

$$\left. + g_I\mu_N\frac{F(F+1) + I(I+1) - J(J+1)}{2F(F+1)}\right].$$

The detailed derivation of this expression is omitted here, but left for one of the exercises at the end of this chapter. The first term is just the hyperfine splitting in the absence of a field, usually referred to as zero-field splitting. This and the second term have wave number values of a few cm^{-1}, that is, lie in the microwave region of the spectrum. Since μ_N, which involves the nuclear mass in the denominator, is several orders of magnitude smaller than μ_0, the third term (nuclear Zeeman splitting) involves energies in the radio frequency portion of the spectrum.

When the Zeeman energy is considerably greater than the zero-field splitting, but still less than the spin-orbit splitting, **I** and **J** become uncoupled, and the appropriate energy expression for this intermediate region is

$$E(M_I, M_J) = aM_IM_J - B_0[g_J\mu_0M_J + g_I\mu_NM_I].$$

The energy expression for the complete Paschen-Back limit for an atom possessing nuclear spin is just

$$E(M_I, M_L, M_S) = aM_IM_J + \mathbf{A}M_LM_S + B_0[\mu_0(M_L + 2M_S) + \mu_N g_I\mu_I].$$

Figure 2.17 shows the Zeeman effect for a $^2P_{1/2}$ atom with nuclear spin $I = \frac{3}{2}$, for which F can equal 2 or 1.

In order to see what spectra will actually be observed upon application of a magnetic field, the selection rules between the different M levels must be derived. The M quantum numbers are associated only with the $e^{\pm iM\phi}$ dependence of the wave functions in the integral in section 5, so that these rules are quite simply arrived at. The three components of the electric dipole moment can be associated with three polarizations of the light wave, for this purpose.

If the incoming radiation is linearly polarized, with the electric vector coincident with the z axis defined by the field used to split the M levels, then there is no ϕ dependence in the $\mathbf{E}\cdot\mathbf{r}$ part of the integral, and we must

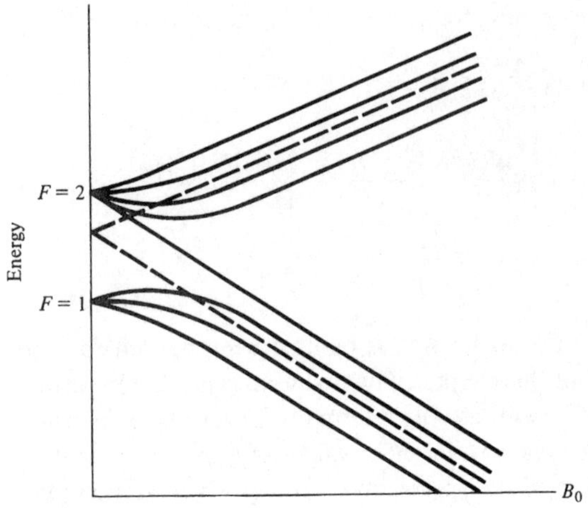

Figure 2.17
Zeeman splitting for a $^2P_{1/2}$ atom with nuclear spin. The area of this graph corresponds to the area enclosed by the little dashed box in figure 2.16.

have $M' = M$ in order for the integral to be nonzero. This defines a $\Delta M = 0$ selection rule for linearly polarized light; this is termed a "π-component" of the transition (see problem 4). For the other polarizations, it is advantageous to use circularly polarized light to get the simplest result. Right circularly polarized (RCP) light can be represented as $x + iy \propto \cos\phi + i\sin\phi = e^{+i\phi}$, and left circularly polarized (LCP) light similarly as $e^{-i\phi}$. For RCP light, the integral

$$\int e^{-iM'\phi}e^{+i\phi}e^{iM\phi}\,d\phi$$

is nonzero only if $M' - M = +1$, so that $\Delta M = +1$; similarly, for LCP light, $\Delta M = -1$. These transitions are termed σ^+ and σ^- components, respectively. Some typical Zeeman split atomic line spectra are shown in figure 2.18.

The treatment thus far should enable the reader to go directly to chapter 10, which deals with optical pumping experiments on just these atomic sublevels. The intervening chapters deal with the spectroscopy of diatomic and polyatomic molecules, and familiarity with the concepts appropriate to these systems will be assumed in later discussions of molecular laser systems and the like.

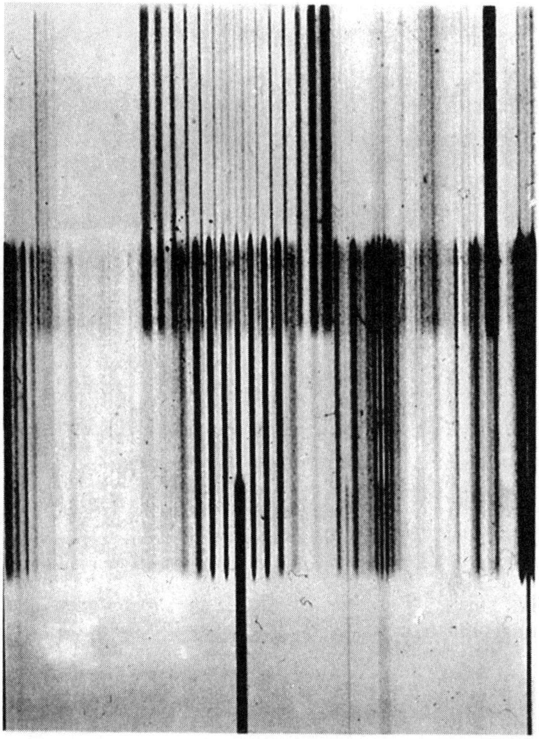

Figure 2.18
Zeeman splitting in an atomic emission line of praseodymium. Bottom panel: Zero magnetic field. Middle panel: Field on, parallel (π) components. Top panel: Field on, perpendicular (σ) components. [Spectrum is reproduced with the permission of the late G. R. Harrison, MIT Spectroscopy Laboratory.]

Problems

1. In the spectrum of rubidium, an alkali metal, the short-wavelength limit of the diffuse series is 4,775 Å. The lines of the first doublet in the principal series ($5^2P_{3/2} \to 5^2S_{1/2}$ and $5^2P_{1/2} \to 5^2S_{1/2}$) have wavelengths of 7,800 Å and 7,947 Å, respectively.

(a) By means of term symbols, write a general expression for the doublets of the sharp series, giving explicitly the possible values for n, the principal quantum number.
(b) What is the spacing in cm^{-1} of the first doublet in the sharp series?
(c) Compute the first ionization potential of rubidium in cm^{-1} and electron volts (eV).

2. In the first transition row of the periodic table there is a regular trend in ground state multiplicities from calcium (singlet) to manganese (sextet) to zinc (singlet), with one exception.

(a) Why does the multiplicity rise to a maximum and then fall?
(b) Explain the discontinuity shown by chromium (atomic number 24.)
(c) Niobium, the element under vanadium in the second transition row, also shows a discontinuity in multiplicity, though vanadium does not. Explain.

3. The number of possible spin eigenfunctions for a single particle of spin I is $2I + 1$.

(a) How many linearly independent spin eigenfunctions are possible for two equivalent particles of spin I?
(b) For a particle with $I = 1$, denote the three spin eigenfunctions by α, β, and γ, corresponding to the eigenvalues $M_z = +, 0, -$. How many linearly independent symmetric and how many linearly independent antisymmetric spin states are there for two equivalent particles with $I = 1$?

4. Atomic eigenfunctions contain a factor $\exp(iM\phi)$. When the atom is in a magnetic field B, the quantum number M represents the projection of the J vector on $B(-J \leq M \leq +J)$. The usual selection rules for L, S, and J still hold for moderate B, and in addition a selection rule governing the values of ΔM becomes important. The dipole moment operators for transitions involving M are $ce\varepsilon_{\parallel}$ and $c'e\varepsilon_{\perp} \cos \phi$. The coefficients c and c' are nonzero constants (for the purposes of this problem): e is the charge on the electron; and $\varepsilon_{\parallel}$ and $\varepsilon_{\perp}$ are the components of the electric field of the radiation

parallel and perpendicular to B. Derive the selection rules for ΔM for radiation polarized (a) parallel and (b) perpendicular to B.

5. Carry out the full calculation leading from equation (2.12) to equation (2.13). Use the classical vector coupling model, by analogy with the calculation carried out for the Landé g factor neglecting hyperfine structure.

6. Calculate the Zeeman pattern to be expected for the sodium D lines at 10,000 gauss (G). You may neglect nuclear hyperfine interactions. Indicate the polarization of each Zeeman line, that is, whether the electric vector of the emitted radiation is parallel to the applied magnetic field (π component) or perpendicular to it (σ compenent).

(a) Show qualitatively the Stark effect to be expected for the sodium D lines. The splittings are proportional to what power of the electric field strength?
(b) What do you think might happen to a beam of ground state sodium atoms passing through a strong inhomogeneous magnetic field? a strong inhomogeneous electric field?

7. Evaluate the transition dipole moment matrix element between the $(n = 1, l = 0, m = 0[1^2\text{S}])$ and the $(n = 2, l = 1, m = 1[2^2\text{P}])$ states of atomic hydrogen. The wave functions are

$$\psi_{100} = \frac{1}{\pi^{1/2} a_0^{3/2}} e^{-r/a_0} Y_{00}(\theta, \phi),$$

$$\psi_{211} = \frac{1}{4(2\pi)^{1/2} a_0^{5/2}} r e^{-r/2a_0} Y_{11}(\theta, \phi),$$

neglecting electron and nuclear spin. Remember that the dipole operator is a 3-vector,

$$\boldsymbol{\mu} = e_0 \mathbf{r} = e_0(\hat{\mathbf{i}} \sin\theta\cos\phi + \hat{\mathbf{j}} \sin\theta\sin\phi + \hat{\mathbf{k}} \cos\theta).$$

References

1. H. A. Bethe and E. E. Salpeter, *Quantum Mechanics of One- and Two-Electron Atoms* (Springer-Verlag, Berlin, 1957).

2. H. A. Bethe and R. W. Jackiw, *Intermediate Quantum Mechanics*, 2nd ed. (Benjamin/Cummings, Reading, MA, 1968).

3. R. F. Streater and A. S. Wightman, *PCT, Spin, Statistics, and All That* (Benjamin, Menlo Park, CA, 1964).

4. J. C. Slater, *Quantum Theory of Atomic Structure* (McGraw-Hill, New York, 1960).

5. E. U. Condon and G. H. Shortley, *The Theory of Atomic Spectra* (Cambridge University Press, Cambridge, 1963).

6. C. E. Moore, *Atomic Energy Levels*, N.B.S. Circular 467 (US Government Printing Office, Washington, DC, 1949).

7. W. Heitler, *Elementary Wave Mechanics* (Clarendon Press, Oxford, 1956).

8. C. A. Coulson, *Valence* (Oxford University Press, Oxford, 1961).

3 Diatomic Molecules

1 The Born-Oppenheimer Approximation

As with so many other aspects of human experience, the spectroscopy of molecules, while more interesting than that of atoms, is at the same time much more complicated. This can be seen immediately by examining the Hamiltonian[1] for a system of nuclei and electrons, even when we eliminate spin and keep only the one- and two-electron terms:

$$\mathscr{H} = \underbrace{-\frac{\hbar^2}{2m}\sum_i \nabla_i^2}_{\text{(electron kinetic energy)}} - \underbrace{\sum_A \frac{\hbar^2}{2M_A}\nabla_A^2}_{\text{(nuclear kinetic energy)}} - \underbrace{\sum_{A,i}\frac{Z_A e^2}{r_{Ai}}}_{\text{(electron-nuclear attractions)}} + \underbrace{\sum_{A>B}\frac{Z_A Z_B e^2}{R_{AB}}}_{\text{(nuclear-nuclear repulsions)}} + \underbrace{\sum_{i>j}\frac{e^2}{r_{ij}}}_{\text{(electron-electron repulsions)}}. \tag{3.1}$$

The standard, universally applied approach to solving the Schrödinger equation with this Hamiltonian in order to find the energy levels of the molecule is to assume with Born and Oppenheimer (reference 1) that the wave function is separable into nuclear and electronic parts; that is,

$$\Psi_{\text{molecular}}(\mathbf{r}, \mathbf{R}) = \psi_e(\mathbf{r}; \mathbf{R})\chi_N(\mathbf{R}). \tag{3.2}$$

If we use this product wave function in the Schrödinger equation for the molecule, we find that it separates into two equations, one for ψ_e and one for χ_N:

$$\mathscr{H}_e\psi_e = \left\{-\frac{\hbar^2}{2m}\sum_i \nabla_i^2 - \sum_{A,i}\frac{Z_A e^2}{|R_A - r_i|} + \sum_{i>j}\frac{e^2}{r_{ij}}\right\}\psi_e = E_e(R_A)\psi_e(R_A), \tag{3.3a}$$

$$\mathscr{H}_N\chi_N = \left\{-\sum_A \frac{\hbar^2}{2M_A}\nabla_A^2 + E_e(R_A) + \sum_{A>B}\frac{Z_A Z_B e^2}{|R_A - R_B|}\right\}\chi_N = E_{\text{total}}\chi_N. \tag{3.3b}$$

The physical content of this separation is based on the relative velocities of the nuclei and the electrons, which, in turn, reflect their relative masses. The Born-Oppenheimer separation assumes that the set of nuclear coordinates **R** is constant, that is, that the nuclei are stationary, and sets up a potential that is a sum of Coulomb attractions in which the electrons move and interact with one another. The energy eigenvalue for the electrons, $E_e(R_A)$, is a function of **R**, as is the form of the electronic wave function itself, $\psi_e(R_A)$. The energy E_e then acts as a term in the potential in which the nuclei themselves move, at much slower speeds. There are some terms that are

1. It may be worth recalling Dirac's comment at this point, to the effect that all of chemistry is implicit in the diagonalization of this Hamiltonian, if we could but do the arithmetic.

neglected in this approximation, however, and before proceeding any further, we should investigate their magnitude.

If we insert the product wave function [equation (3.2)] directly into the Schrödinger equation using the Hamiltonian [equation (3.1)], we obtain

$$-\frac{\hbar^2}{2m}\sum_i \nabla_i^2 \psi_e \chi_N - \sum_A \frac{\hbar^2}{2M_A}\nabla_A^2 \psi_e \chi_N + V(\mathbf{R}, \mathbf{r})\psi_e \chi_N = E_{\text{total}}\psi_e \chi_N.$$

Now

$$\nabla_i^2 \psi_e \chi_N = \chi_N \nabla_i^2 \psi_e$$

since χ_N does not depend on $\mathbf{r}$ at all. But in the second term,

$$\nabla_A^2 \psi_e \chi_N = \psi_e \nabla_A^2 \chi_N + 2(\nabla_A \psi_e)(\nabla_A \chi_N) + \chi_N \nabla_A^2 \psi_e.$$

We need not worry further about the potential energy terms. If we write out all the individual terms and collect those that belong to a particular eigenvalue equation, we obtain

$$\begin{aligned}
&\psi_e\left(-\sum_A \frac{\hbar^2}{2M_A}\nabla_A^2 \chi_N\right) - \chi_N\left(E_e \psi_e + \sum_{A>B} \frac{Z_A Z_B e^2}{|R_A - R_B|}\psi_e\right) \\
&+ \left\{-\sum_A \frac{\hbar^2}{2M_A} 2(\nabla_A \psi_e)(\nabla_A \chi_N) + \chi_N \nabla_A^2 \psi_e\right\} = E_{\text{total}}\psi_e \chi_N.
\end{aligned} \tag{3.4}$$

Equation (3.4) differs from equation (3.3b) multiplied through by ψ_e in just the last terms in brackets. We can estimate their magnitude as follows: A typical contribution has the form $(\hbar^2/2M_A)\nabla_A^2\psi_e$. Now $\hbar\nabla_A\psi_e$ is of the same order of magnitude as $\hbar\nabla_i\psi_e$, since the derivatives operate over approximately the same molecular dimensions. But this is just the momentum of the electron. Therefore, we can estimate $(\hbar^2/2M_A)\nabla_A^2\psi_e$ as $p_e^2/2M_A = (m/M_A)p_e^2/2m = (m/M_A)E_e$. Thus the energy contribution from terms neglected in the Born-Oppenheimer approximation is of the magnitude of the electronic energy itself times the ratio of electron mass to nuclear mass, which is typically of the order of 1:10,000. Since typical electronic energies in molecules are of the order of 10^4–10^5 cm^{-1}, this energy difference may be typically of the order of a few cm^{-1}.

Another, perhaps more physical way of looking at this is to recognize that the quantity $\nabla_A\psi_e$ just measures the variation of the electronic wave function with nuclear coordinates. Since this variation is usually small, the energy corrections involved in the Born-Oppenheimer separation are also

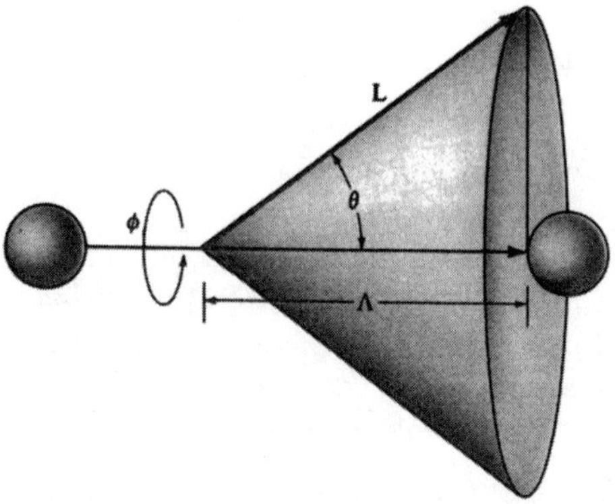

Figure 3.1
The electronic orbital angular momentum **L** precesses around the internuclear axis; the projection of the angular momentum on the axis is Λ.

small. One important situation in which this is not so, however, is when the electronic wave function is degenerate by virtue of some element of symmetry in the molecule. (See the following section, and also chapter 6, for a discussion of this point.) Then, any small displacement of nuclear coordinates that breaks this element of symmetry removes the associated degeneracy of the wave function and can cause a large energy splitting between the two no longer degenerate levels.

2 Electronic Energy Levels for Molecules with Stationary Nuclei

A model with which to begin thinking about the electronic eigenstates of a diatomic molecule is shown in figure 3.1. The orbital angular momentum of the electrons in the molecule precesses around the internuclear axis, just as if it were in a very strong Stark field set up by the positively charged nuclei. In this model, $L_z = \hbar/i(\partial/\partial\phi)$ is a constant of the motion; in the cylindrical coordinates appropriate to the symmetry of the problem, $L_z\psi_{el} = \pm\Lambda\hbar\psi_{el}$. This eigenvalue equation is readily solved for the ϕ dependence of ψ_{el}, which is just $(2\pi)^{-1/2}e^{\pm i\Lambda\phi}$. The quantum number Λ can take on any integral value starting with zero, as usual, and the molecular wave functions are classified according to the value of Λ appearing in $\psi_{el}(\phi)$.

The most useful way of discussing these wave functions is in terms of the appropriate molecular symmetry elements. The use of symmetry considerations in molecular spectroscopy is a very useful and powerful tool, primarily for the following reasons:

1. The squared amplitude of the wave function must have the symmetry of the total Hamiltonian.
2. Selection rules may be determined simply by inspection of symmetry relations, thus obviating the need to evaluate complex integral expressions.
3. Unbroken symmetry in a molecule indicates the presence of degeneracy of energy levels. We have already seen examples of this, such as the spherical symmetry of an atom leading to the $(2J + 1)$-fold degeneracy of each J level.

Diatomic molecules, which possess a simple cylindrical symmetry, form a good model system with which one can begin to use symmetry considerations because there is a direct correlation with the analytical mathematical expressions embodying the symmetry. Formal group theory (such as is discussed in chapter 6) becomes essential for polyatomic molecules, in which the mathematics is much more elaborate.

The basic procedure in symmetry, or group theoretical, arguments is the identification of those operations in space that leave the appearance of the object under study unchanged—that is, if an observer closed his eyes, and someone else was given the opportunity either to carry out the operation on the object or not, as he pleased, then the observer could not tell whether the operation had indeed been carried out, or the object simply left alone. For example, an equilateral triangle can be rotated through 120°, pivoting around a point defined by the intersection of, say, the angle bisectors, with no change in its appearance. This operation can be repeated three times before the triangle is actually back where it started from; this is a threefold rotation axis, denoted C_3. A square possesses fourfold axis, C_4; a rectangle possesses a C_2, while a trapezoid has only a C_1, corresponding to a rotation through a full 360°. But a trapezoid can be reflected across its centerline to produce a figure indistinguishable from the original. The other figures we have been considering also possess reflection symmetry. The square, in addition, has a point at the center, which has the property that if all the coordinates are inverted through it, we again obtain a duplicate of the original figure; this is known as inversion symmetry. Rotation, reflection, and inversion are the basic symmetry elements out of which the group theoretical study of molecules is constructed, as shown in figure 3.2.

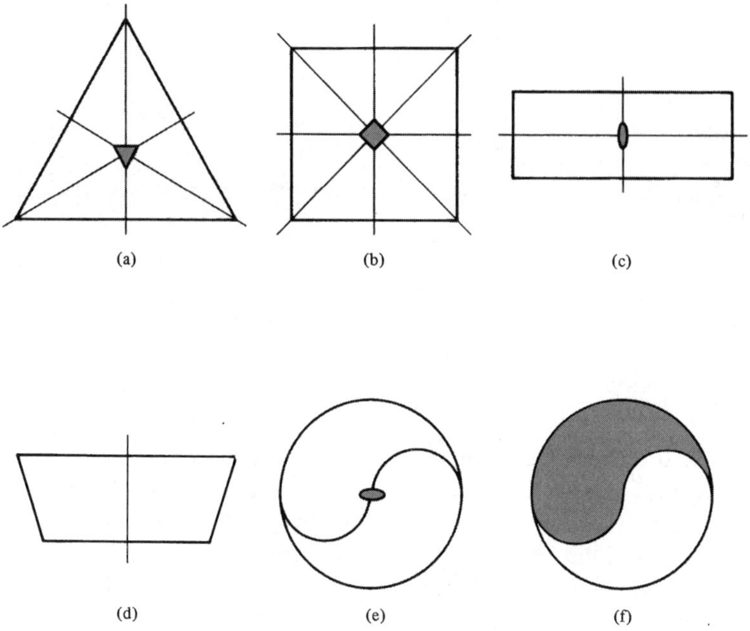

Figure 3.2
Symmetry elements of plane figures. In (a) is drawn an equilateral triangle with a threefold axis of symmetry C_3 and three reflections; in (b), a square with a C_4, four reflections, and a center of inversion; in (c), a rectangle with a C_2 and two reflections; and in (d), a trapezoid with only a single reflection symmetry. Sometimes the number of symmetry elements one sees depends on how closely one looks at the object under study. The "yin-yang" figure in outline in (e) appears to have a C_2; but in (f), when one distinguishes between the two opposing influences making up the figure, there is no element of symmetry aside from the identity operation C_1.

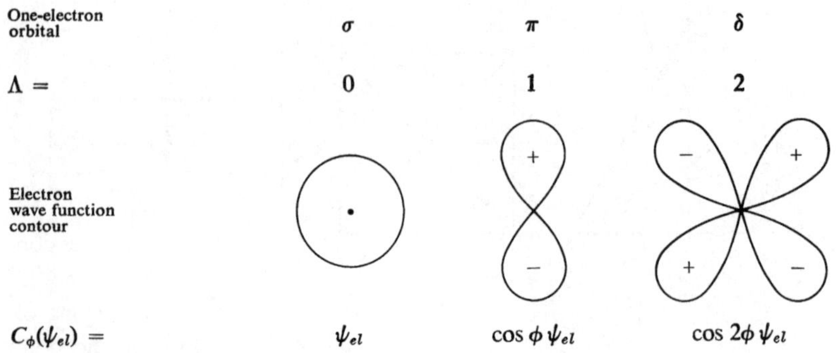

Figure 3.3
Cylindrical symmetry of one-electron orbitals.

Let us begin by considering the operation of rotation through an angle ϕ, denoted C_ϕ, on the real part of the electronic wave function of a diatomic molecule. Let us look at orbitals in which $\Lambda = 0$, 1, 2, and so on,[2] each projected down the internuclear axis. The mathematical effect of carrying out this operation is indicated below each diagram in figure 3.3. The general expression for the result of the operation of rotation is

$$C_\phi(\psi_{el}) = \cos(\Lambda\phi)\,\psi_{el}.$$

There are two other symmetry elements, and thus two other operations, appropriate to diatomic molecules. First, note that corresponding to the infinitefold axis of symmetry (C_∞), there exist an infinite number of mirror planes containing the axis. The operation appropriate to these planes is reflection, denoted σ (which must not be confused with the notation for a one-electron orbital with $\Lambda = 0$). The expression for the result of performing this operation can be simply deduced. Let us write

$$\sigma|\psi\rangle = C|\psi\rangle, \qquad \sigma^2|\psi\rangle = C^2|\psi\rangle = |\psi\rangle$$

2. These orbitals are denoted σ, π, δ, and so on, by analogy with the $s, p, d, \ldots$ notation for atomic orbitals; the general prescription in going from atomic to molecular notation is to replace the Latin letter by the corresponding Greek one, retaining upper or lower case as appropriate. The whole matter of spectroscopic notation for molecules is discussed in an authoritative article in *J. Chem. Phys.* **23**, 1997 (1955); although the author of this article is given as an International Commission, it was actually written by R. S. Mulliken.

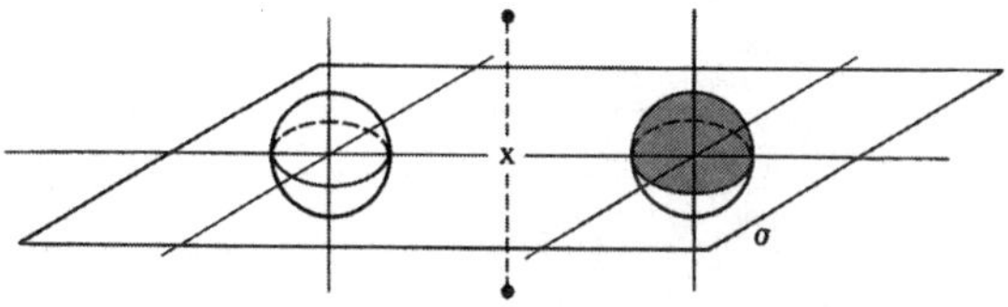

because it is plain that repeating the reflection operation twice in succession brings you right back to where you started from. Thus $C^2 = 1$, and so $C = \pm 1$. Therefore the effect of the σ operation on $|\psi\rangle$ can have one of only two possible results: to leave $|\psi\rangle$ unchanged or to change the sign. The wave functions having these reflection properties are denoted $+$ and $-$, respectively, and we write

$$\sigma|+\rangle = |+\rangle \qquad \text{and} \qquad \sigma|-\rangle = -|-\rangle.$$

The C_∞ rotation and reflection symmetries are possessed by all diatomic molecules. When, in addition, the molecule is homonuclear—that is, when its chemical formula can be written as X_2—an additional symmetry element appears, that of inversion through the point corresponding to the center of mass, denoted *i*. By virtue of the same sort of arguments we applied to reflection symmetry, this operation has eigenvalues $+1$ and -1. The states are labeled g and u, respectively, from the German *gerade* (even) and *ungerade* (odd), so that

$$i|g\rangle = |g\rangle, \qquad i|u\rangle = -|u\rangle.$$

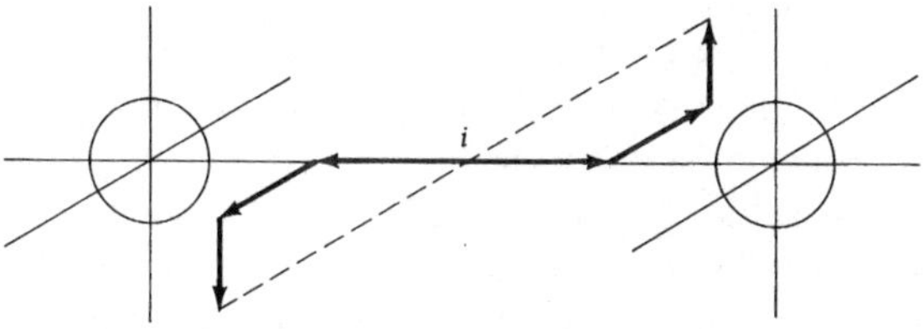

We can proceed to make up tables that give the results of carrying out the rotation, reflection, and inversion operations on the various possible symmetry species for diatomic molecules. Heteronuclear molecules having the rotation C_∞ and the reflection σ form the group $C_{\infty v}$, and homonuclear molecules possessing in addition the inversion operation i form the group

Table 3.1

Group $C_{\infty v}$				Group $D_{\infty h}$				
	E	$2C_\phi$	σ		E	$2C_\phi$	σ	i
Σ^+	1	1	+	Σ_g^+	1	1	+1	+1
Σ^-	1	1	−	Σ_u^+	1	1	+1	−1
Π	2	$2\cos\phi$	0	Σ_g^-	1	1	−1	+1
Δ	2	$2\cos 2\phi$	0	Σ_u^-	1	1	−1	−1
.				Π_g	2	$2\cos\phi$	0	+1
.				Π_u	2	$2\cos\phi$	0	−1
.				Δ_g	2	$2\cos 2\phi$	0	+1
				Δ_u	2	$2\cos 2\phi$	0	−1
				.				
				.				
				.				

$D_{\infty h}$. The various representations of these groups are labeled according to the orbital electronic angular momentum along the internuclear axis, and so are denoted Σ, Π, Δ, and so on. See table 3.1. (All of these terms will be more rigorously defined in chapter 6.)

The E, or identity, operation can be considered as equivalent to a C_1; it simply counts the number of degenerate species belonging to a particular representation. In Σ states ($\Lambda = 0$), states are distinguished which are either + or − to reflection; these are nondegenerate. In states with $\Lambda > 0$, we no longer have an infinite number of reflection planes containing the axis; as may be seen from the diagrams in figure 3.3, there exist only a small number of reflection planes at specific orientations. + and − reflection states can no longer be distinguished, and the states are doubly degenerate, so that the character, as it is called, of E is equal to two.

What we have done thus far is to characterize the possible electronic states of a diatomic molecule, in the same way as we enumerated S, P, D, . . . states of atoms. We shall come back and make use of these symmetry properties when we work out the selection rules for allowed transitions between these various states. To conclude this section, we consider the question of angular momentum coupling in molecules, when spin must be taken into account; that is, we wish, as usual, to find the good quantum numbers for a diatomic molecule.

The Hamiltonian for which we want to find the constants of the motion is that of equation (3.3a), including electron kinetic energy, electron–nuclear attraction, and electron–electron repulsion. The algebraic details get complicated, but it is possible to evaluate the commutators in detail and show

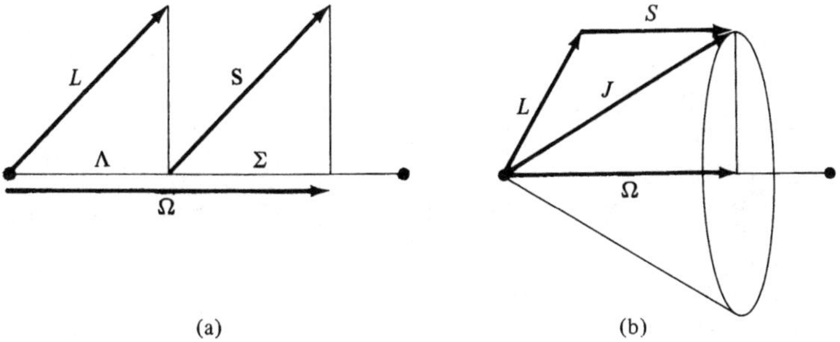

Figure 3.4
Angular-momentum vector coupling in (a) Russell-Saunders and (b) heavy-atom limit.

that L^2 does *not* commute with this Hamiltonian, essentially because the strong θ dependence of the potential destroys any possible spherical symmetry, which is required in order for the total angular momentum to be a constant of the motion. $L_z = (\hbar/i)(\partial/\partial\phi)$ does commute with $\mathscr{H}_e$, and thus is a conserved quantity. (The z coordinate is defined as the internuclear axis.)

When spin and orbital angular momenta are weakly coupled, that is, in molecules with light nuclei, we have an analogue of Russell-Saunders coupling in atoms. The z component of L is denoted Λ; the z component of the spin S is denoted Σ (once again, not to be confused with the molecular state in which $\Lambda = 0$); and their sum $\Lambda + \Sigma = \Omega$ (figure 3.4a). According to whether $\Sigma = 0, \frac{1}{2}, 1$, and so on, we can have singlet, doublet, triplet, and so on, states. In the heavy atom limit, when **L** and **S** are spin-orbit coupled to form **J**, the total electronic angular momentum **J** is coupled to the internuclear axis to give Ω, and we distinguish the various molecular states solely on the basis of the value of Ω (figure 3.4b). The entire matter of angular momentum coupling in diatomic molecules, including coupling with rotational angular momentum, will be dealt with in much greater detail in chapter 5.5.

3 Determination of Molecular States from Separated Atom States

A diatomic molecule is formed by bringing two atoms together. Each of these atoms is in a particular electronic state, and the first question we encounter in

Table 3.2

$^2P + {}^2P$					States including spin and spin-orbit coupling
m_{l_A}	m_{l_B}	M_L	Λ	State	
+1	+1	+2	2	Δ	$^1\Delta_2;\ {}^3\Delta_3,\ {}^3\Delta_2,\ {}^3\Delta_1$
−1	−1	−2			
+1	0	+1	1	Π	$^1\Pi_1;\ {}^3\Pi_2,\ {}^3\Pi_1,\ {}^3\Pi_0$
−1	0	−1			
0	+1	+1	1	Π	$^1\Pi_1;\ {}^3\Pi_2,\ {}^3\Pi_1,\ {}^3\Pi_0$
0	−1	−1			
+1	−1	0	0	Σ^+, Σ^-	$\{{}^1\Sigma^+,\ {}^3\Sigma_0^+,\ {}^3\Sigma_1^+$
−1	+1	0			$\{{}^1\Sigma^-,\ {}^3\Sigma_0^-,\ {}^3\Sigma_1^-$
0	0	0	0	Σ^+	$^1\Sigma^+,\ {}^3\Sigma_0^+,\ {}^3\Sigma_1^+$

the study of diatomic molecule spectroscopy is that of which molecular states can be formed from a given pair of atoms. Some of these cases will be worked out in detail in this section—not because this is anything a practicing spectroscopist does any more, since the results for all conceivable combinations of atoms have been worked out and are completely tabulated in Herzberg's text (reference 2), but because doing so can lead to a fuller understanding of the nature of the molecular states under discussion.

As an example, let us work out the molecular states arising from the interaction of two ^{2}P atoms. A heteronuclear species of this type would be a group III diatomic such as AlB, while a homonuclear species would be B_2, Al_2, and so on. Since each P state is triply degenerate, and the spin multiplicity is two, when we are finished, we should end up with $3 \times 3 \times 2 \times 2 = 36$ states altogether.

We proceed in exactly the same way as we did when deriving the atomic states arising from a given configuration in chapter 2.2; that is, we tabulate all possible combinations of m_{l_A} and m_{l_B}, remembering that m_l can take on only the values $\pm\Lambda$. The spin states are treated separately; in this case, two doublets ($S = \frac{1}{2}$) can combine to give either a singlet or a triplet ($\Sigma = 0$ or 1).

Since the two nuclei in the heteronuclear molecule are different, no states are excluded by the Pauli principle, and we obtain our full complement of 36 states (table 3.2). To count up the number of states correctly, we recall that each state with $\Omega > 0$ is doubly degenerate. There are 13 of these, namely, $^1\Delta_2$, $^3\Delta_3$, $^3\Delta_2$, $^3\Delta_1$, $^1\Pi_1$, $^3\Pi_2$, $^3\Pi_1$, $^1\Pi_1$, $^3\Pi_2$, $^3\Pi_1$, $^3\Sigma_1^+$, $^3\Sigma_1^-$, $^3\Sigma_1^+$. Each $^3\Pi_0$ has a + and a − component, and each $\Omega = 0$ state is singly degenerate. There are six of the latter, namely, $^1\Sigma^+$, $^3\Sigma_0^+$, $^1\Sigma^-$, $^3\Sigma_0^-$, $^1\Sigma^+$, and $^3\Sigma_0^+$, for a grand total of 36 states. The $\Omega = 0$ states occur in pairs, with + and − parity, except for the last one arising from two $m = 0$ atomic states. The parity of this state is determined by inspecting the sum $L_A + L_B + \Sigma l_A + \Sigma l_B$, where the sum is over the individual electrons in the constituent atoms. When the total sum is even, the parity is $+$, and when it is odd, the parity is $-$. In the case of the group III diatomics, we have $1 + 1 + 1 + 1 = 4$, which is even, and so the last state tabulated is a Σ^+.

In the case of homonuclear diatomic molecules, two different possibilities can be distinguished.

1. If the two like atoms are in different electronic states, then each state is doubled according to the u–g inversion symmetry. Thus, our ${}^{1,3}\Delta$ states give ${}^1\Delta_g$, ${}^3\Delta_g$, ${}^1\Delta_u$, ${}^3\Delta_u$, and so on, for a total of 72 molecular states. An example of this would be the states of excited Na_2^* arising from $Na^*\,{}^2P(2p)$ and $Na^*\,{}^2P(3p)$.
2. If the two atoms, in addition to being alike, are in the same electronic states, then the Pauli exclusion principle comes into play and gives us the same 36 states we had for the heteronuclear case, each with an appropriate u or g designation. The inversion symmetry of each state is given by the Wigner-Witmer rules (reference 4); since these are somewhat complicated to derive and use, and the results are completely tabulated by Herzberg (reference 2), we shall not pursue the details here.

As a final example, let us consider the molecular analogue of the jj coupling limit in atoms. This coupling scheme applies when l and s are no longer individually quantized in the separated atoms, but, rather, j is the only good quantum number. The procedure then is to couple m_{j_1} and m_{j_2} and identify the resulting value of Ω. For example, the molecular states of the halogens, formed from two ${}^2P_{3/2}$ atoms, are shown in figure 3.5.

4 Selection Rules for Electronic Transitions

When we come to consider the question of which pairs of molecular states will be coupled by electromagnetic radiation, and thus which molecular spectra will actually be observable, we employ the same principle as that previously used for atomic spectra in chapter 2.5. That is, we inspect the matrix elements of various quantum mechanical operators between the two states of interest, to see whether they have zero or nonzero values. As was shown in chapter 1.6, the strongest transitions are electric dipole transitions, for which the operator is the dipole moment μ; weaker, but still observable transitions arise from the

m_{j_1}	m_{j_2}	M_J	Ω	Russell-Saunders correlation
$+\frac{3}{2}$	$+\frac{3}{2}$	3		
$-\frac{3}{2}$	$-\frac{3}{2}$	-3	$3u$	
$+\frac{3}{2}$	$+\frac{1}{2}$	2		
$-\frac{3}{2}$	$-\frac{1}{2}$	-2		
$+\frac{1}{2}$	$+\frac{3}{2}$	2	$2u, 2g$	
$-\frac{1}{2}$	$-\frac{3}{2}$	-2		
$+\frac{3}{2}$	$-\frac{1}{2}$	1		
$-\frac{3}{2}$	$+\frac{1}{2}$	-1	$1u$	
$-\frac{1}{2}$	$+\frac{3}{2}$	1		
$+\frac{1}{2}$	$-\frac{3}{2}$	-1	$1g$	
$+\frac{1}{2}$	$+\frac{1}{2}$	1		
$-\frac{1}{2}$	$-\frac{1}{2}$	-1	$1u$	
$+\frac{3}{2}$	$-\frac{3}{2}$	0	$0g^+$	
$-\frac{3}{2}$	$+\frac{3}{2}$	0	$0u^-$	
$+\frac{1}{2}$	$-\frac{1}{2}$	0	$0u^-$	
$-\frac{1}{2}$	$+\frac{1}{2}$	0	$0g^+$	

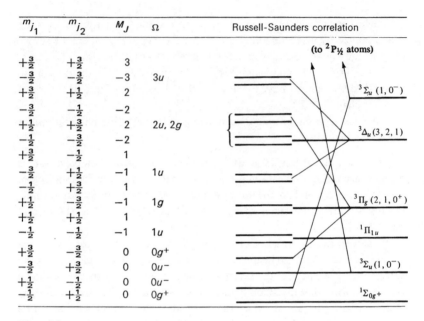

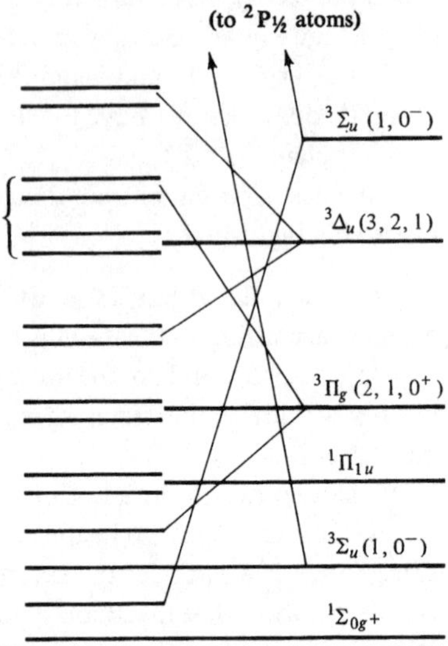

Figure 3.5
Correlation diagram for molecular electronic states of diatomic halogens.

rotation operator **J** (magnetic dipole transitions) or the quadrupole moment tensor **Q** (electric quadrupole transitions). In other words, we have to consider matrix elements of the form

$\langle\psi_1|\boldsymbol{\mu}|\psi_2\rangle$ (electric dipole),

$\langle\psi_1|\mathbf{J}|\psi_2\rangle$ (magnetic dipole),

$\langle\psi_1|\mathbf{Q}|\psi_2\rangle$ (electric quadrupole).

The matrix element must be not identically equal to zero in order for the corresponding transition to be active.

In the case of atoms, some simple selection rules were deduced on the basis of the mathematical properties of the spherical harmonic factors in the atomic wave functions; however, the wave functions in the case of molecules are more complicated, so that a simplifying approach is needed. This is provided by the use of the symmetry properties introduced in section 2. There are two basic principles that are operative in the application of these properties.

1. In order for the integral

$$\int F(\mathbf{r})\,d\mathbf{r}$$

to be not equal to zero, $F(\mathbf{r})$ cannot be an antisymmetric function, that is, a function such that $F(-\mathbf{r}) = -F(\mathbf{r})$. This argument should be self-evident from consideration of simple one-dimensional functions such as sine or cosine; an antisymmetric function possesses equal and opposite contributions from the regions $\mathbf{r}\,d\mathbf{r}$ and $-\mathbf{r}\,d\mathbf{r}$, which exactly cancel each other. Since the eigenfunctions for a molecular system must exhibit the same symmetry elements as the molecule, these functions must be either symmetric or antisymmetric with respect to a given operation; thus, we can make the more specific statement that for a given matrix element to be nonzero, its functional form must be totally symmetric.

2. The symmetry of the product of several functions possessing certain symmetry properties is the product of the individual symmetries, according to the general rule even × even = even, odd × odd = even, and even × odd = odd. For example, if we consider the reflection operator σ, whose eigenfunctions are denoted + and −, we have

$$\sigma(\psi_1^+ \psi_2^+) = \psi_1^+ \psi_2^+,$$

$$\sigma(\psi_1^- \psi_2^-) = \psi_1^- \psi_2^-,$$

$$\sigma(\psi_1^+ \psi_2^-) = -\psi_1^+ \psi_2^-.$$

Similarly, for the inversion operator i, with even eigenfunctions labeled by g and odd eigenfunctions labeled by u, we have

$$i(\psi_1^u \psi_2^u) = \psi_1^u \psi_2^u,$$

and so on.

Thus, in order to determine selection rules, we have to find combinations of initial and final wave functions and transition operators such that the product of all three is totally symmetric. In order to do this, we must know the symmetry properties of the individual operators. The electric dipole moment operator just has the form of the coordinates (x, y, z) themselves. Let us examine the behavior of these coordinates under the symmetry operations of a linear diatomic molecule. First, we shall consider the z coordinate, which is taken to be coincident with the internuclear axis. Obviously, there is no change in this coordinate with rotation around this axis (C_ϕ), so that the species of z is Σ. It is Σ^+ because there is likewise no change with reflection through a plane containing the axis (σ_v). If a center of symmetry is present, however, inversion through the center takes z into $-z$, so that the complete species would be Σ_u^+.

Looking at the coordinates x and y, we see that a C_2 operation (which is a subset of C_ϕ) takes x into $-x$, y into $-y$, so that these coordinates have Π character. The inversion operation changes the sign, as it does for z, so that the symmetry is Π_u.

The angular momentam operators, which determine magnetic dipole intensities, have the nature of the rotation operators R_x, R_y, and R_z. The R_z operator is invariant under rotation about the z axis, but changes sign under

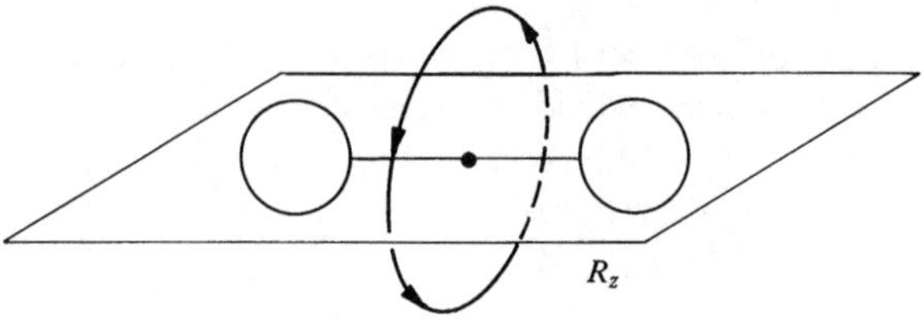

reflection through a plane containing that axis; therefore, it is Σ^-. The R_x and R_y operators are taken into negatives of themselves by a C_2 operation, which

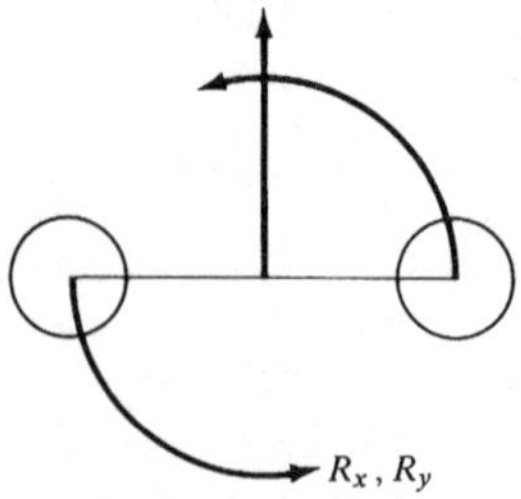

makes them Π. A little more care is required to see the effect of the inversion operation. The angular momentum vector has the form of a cross product $\mathbf{r} \times \dot{\mathbf{r}}$; the inversion operation changes the sign of both (x, y, z) and $(\dot{x}, \dot{y}, \dot{z})$, so that the net effect on the rotation is to leave the sign unchanged; therefore, R_z is Σ_g^-, and R_x and R_y are Π_g.

The elements of the quadrupole moment tensor are products of coordinates that have the forms xy, xz, yz, $x^2 + y^2$, $x^2 - y^2$, and z^2. By making use of the multiplication properties of symmetry species, we find that the species of these combinations are Σ_g^+, Π_g, and Δ_g.

Armed with the knowledge of the symmetry species of the transition operators, we can now proceed to establish selection rules for electric dipole allowed electronic transitions. What is required is that the product of the ground and excited state wave functions and the transition operator have the symmetry Σ_g^+, which is the totally symmetric representation of the $D_{\infty h}$ group. For light polarized along the internuclear axis of the molecule, we see that the ground and excited states must, first, have the same value of Λ; second, in the case of Σ–Σ transitions, only + states can be connected with + states and − states with − states, with no combinations between them, and only u states can be connected with g states. The x and y polarizations, which have Π character, can couple states differing by one unit of angular momentum, but the $u \leftrightarrow g$ selection rule still holds. Thus, we are led to the following selection rules:

$\Delta\Lambda = 0$ (parallel transition) or ± 1 (perpendicular transition);
$+ \not\leftrightarrow -$ in Σ–Σ transitions;
$u \leftrightarrow g$ but $u \not\leftrightarrow u$ and $g \not\leftrightarrow g$ for all transitions involving homonuclear molecules.

We have not yet considered the spin of the molecules. For light molecules, in which the spin is uncoupled from the orbital angular momentum, we have the restriction $\Delta\Sigma = 0$, since the dipole moment operator contains no spin coordinates at all. In heavy molecules, where Ω is the good quantum number, the selection rule is $\Delta\Omega = 0$ or ± 1, with the same $+-$ and ug restrictions.

Selection rules for magnetic dipole transitions follow from the same sort of considerations as we applied to electric-dipole transitions. For magnetic dipole transitions, we have the same $\Delta\Lambda = 0, \pm 1$ restriction, but in this case u and g levels are connected to levels of the same, rather than the opposite parity, and the selection rule for Σ–Σ transitions is $\Sigma^+ \leftrightarrow \Sigma^-$, with no transitions between Σ states of the same reflection symmetry. There are also components of the magnetic dipole interaction that break down the spin selection rules more readily than in the electric dipole case, but this must be seen from the physics of the situation; it is not obvious from symmetry considerations alone. The selection rules for electric quadrupole transitions are $\Delta\Lambda = 0, \pm 1$, and ± 2; $u \leftrightarrow u$ and $g \leftrightarrow g$.

We shall consider, as examples, the electronic spectra arising from diatomic alkalies and halogens. The lowest states of the alkalies arise from the interaction of two ^{2}S atoms; these give rise to a $^1\Sigma$ ground state and a $^3\Sigma$ state; the latter is not observed optically because of the $\Delta\Sigma = 0$ selection rule. Thus the observed transitions are to the various $^1\Pi$ and $^1\Sigma$ states arising from the ^{2}S–^{2}P atomic configurations, which are shown in figure 3.6. The letter designation of the various states is simply an ordering, with no specific connotation, except that X is always reserved for the ground state of a molecule. In the case of the halogens (figure 3.7), the only state arising from the same $2(^2P_{3/2})$ configuration as the ground $^1\Sigma_{0g}^+$ state, which can be connected with the ground state by electric dipole radiation, is the $A\,^3\Pi_{1u}$ state, and this transition is actually quite weak; the first strong transition in the halogens, which is responsible for the green, brown, and violet colors of chlorine, bromine, and iodine, respectively, is to the $B^3\Pi_{0u}^+$ state arising from the $^2P_{3/2} + {}^2P_{1/2}$ atomic states. This is an example of a rule treated more explicitly in the next section, namely, that the strongest molecular transitions are those between states arising from different separated atom states. This is quite analogous to the prohibition of intramultiplet transitions in atoms (chapter 2.5).

5 Simple Molecular Orbital Theory and Molecular Potential Curves

If we wish to have any deeper understanding of molecular energy states than simply labeling them in order by angular momentum eigenvalues, we shall need some model to work with; this will generally be the *molecular orbital model.* Let us develop the principles of this model by constructing it for the simplest possible molecular system, the one-electron system H_2^+. This system can be solved for the exact Born-Oppenheimer states (reference 5) because there are no $1/r_{12}$ terms in the Hamiltonian

$$\mathscr{H}(H_2^+) = -\frac{\hbar^2}{2m}\nabla^2 - \frac{e^2}{r_A} - \frac{e^2}{r_B} + \frac{e^2}{R_{AB}} \tag{3.5}$$

in which the internuclear distance R_{AB} is held at a fixed constant value. The exact solution for this system will not be of much direct use to us in describing more complex many-electron systems (it will, of course, be very useful for testing the accuracy of various approximations to be introduced), so we shall try to construct the wave functions for H_2^+ in terms of hydrogenlike atomic orbitals ϕ_i, which we already know,

$$\psi = \sum_i c_i\phi_i. \tag{3.6}$$

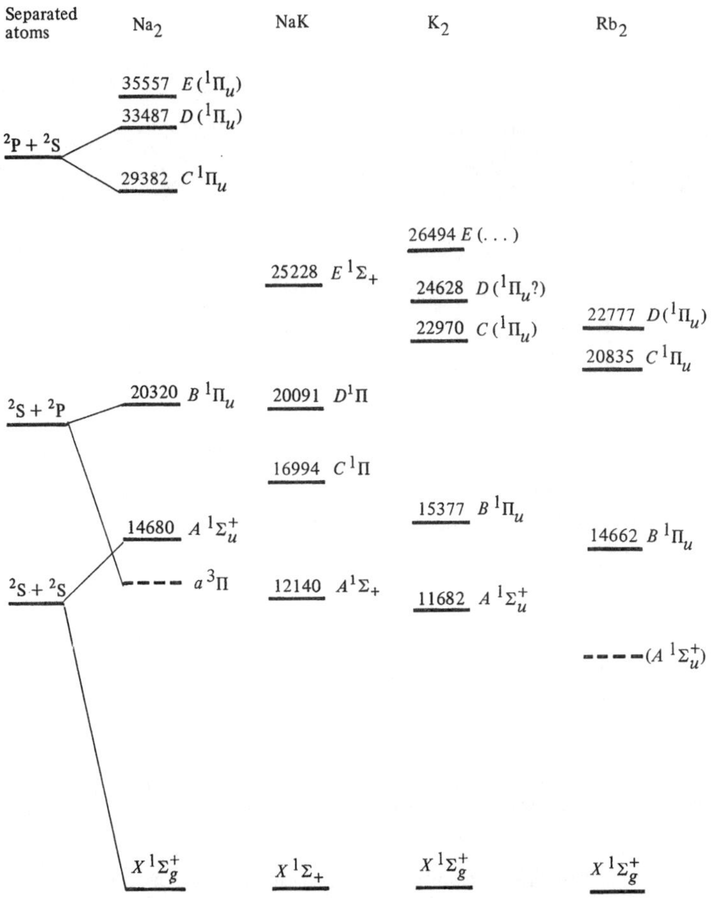

Figure 3.6
Optically allowed excitations in some diatomic alkalies. States for which the electronic quantum numbers have not been definitely established are shown in parentheses. The energies of the various observed states are taken from the compilations by Herzberg (references 2 and 3).

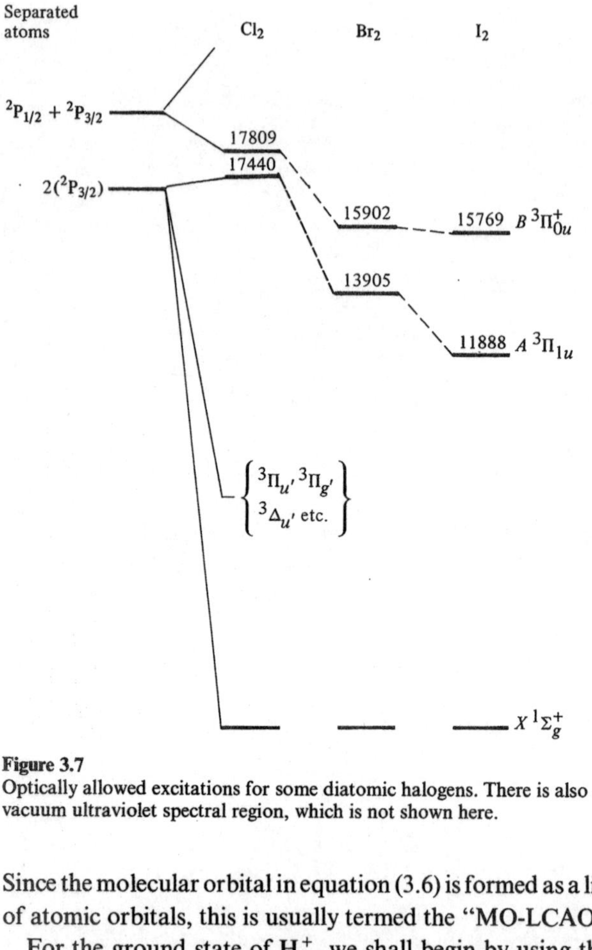

Figure 3.7
Optically allowed excitations for some diatomic halogens. There is also a rich ultraviolet and vacuum ultraviolet spectral region, which is not shown here.

Since the molecular orbital in equation (3.6) is formed as a linear combination of atomic orbitals, this is usually termed the "MO-LCAO" approximation.

For the ground state of H_2^+, we shall begin by using the hydrogen atom ground state orbitals, so that $\phi_i = \mathrm{H}(1s_i)$, where i refers to the two nuclei, A and B. These two atomic orbitals are degenerate, so that finding the molecular orbitals is just a classic example of degenerate perturbation theory, but it is not necessary even to use this formal apparatus to find the MOs. The H_2^+ molecular ion has a center of symmetry, so that the wave functions, which are the solutions to its Hamiltonian, must be either symmetric or antisymmetric with respect to inversion through that center. The only such combinations of the two atomic orbitals that satisfy this symmetry requirement, plus the normalization condition

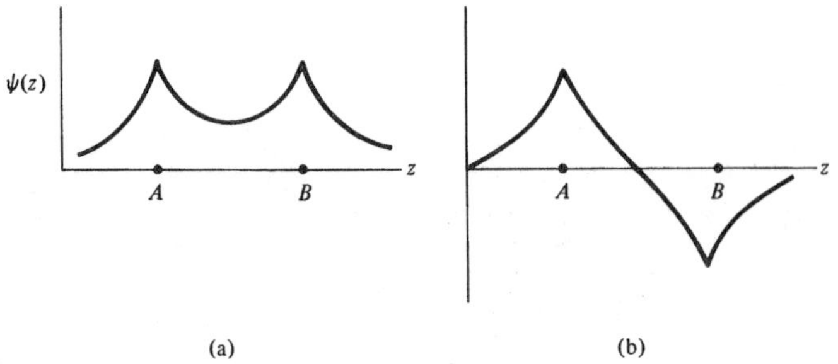

Figure 3.8
Molecular orbitals for H_2^+, shown as projections on a σ_z plane: (a) $(\sigma_g 1s)$ bonding orbital; (b) $(\sigma_u^* 1s)$ antibonding orbital.

$$\int |\psi|^2 \, d\mathbf{r} = 1$$

are

$$\psi_1 = \frac{1}{\sqrt{2}}(\phi_A + \phi_B)$$

and

$$\psi_2 = \frac{1}{\sqrt{2}}(\phi_A - \phi_B)$$

These are both designated σ orbitals, since they possess the full cylindrical symmetry around the internuclear axis. ψ_i is a σ_g orbital, because it is the combination that is even with respect to the inversion operation, and ψ_2 is σ_u. The orbitals are further designated $\sigma_g 1s$ and $\sigma_u 1s$, respectively, to indicate the atomic orbitals from which they are formed. Finally, ψ_2 is denoted $\sigma_u^* 1s$, for reasons that will be evident when we consider the energies of the orbitals. The appearance of the orbitals is shown in figure 3.8.

We can find the energies associated with each orbital by diagonalizing the secular determinant [see equation (1.15)]

$$\begin{vmatrix} H_{AA} - E & H_{AB} - ES \\ H_{AB} - ES & H_{BB} - E \end{vmatrix} = 0, \tag{3.7}$$

in which the Coulomb integrals are

$$H_{AA} = \int \psi_A^* \mathscr{H} \psi_A \, d\mathbf{r} = H_{BB},$$

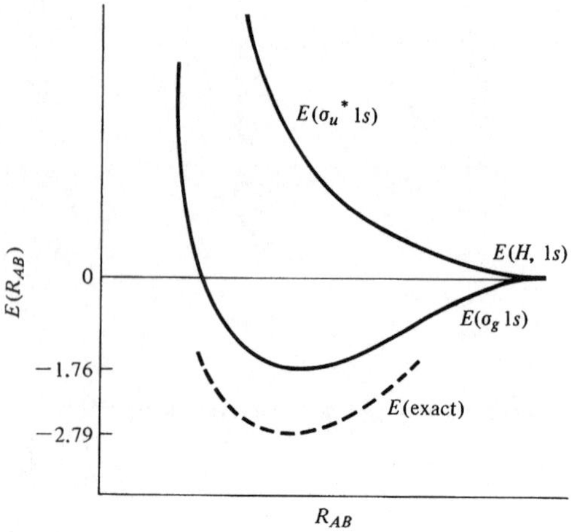

Figure 3.9
Potential energy curves for H_2^+, from equation (3.8). The lower curve is bonding, the upper curve, antibonding. The dotted curve is the exact $E(R)$ result.

the resonance integrals are

$$H_{AB} = \int \psi_A^* \mathscr{H} \psi_B \, d\mathbf{r},$$

and the overlap integral is

$$S = \int \psi_A^* \psi_B \, d\mathbf{r}.$$

The solution to equation (3.7) is $E = (H_{AA} \pm H_{AB})/(1 \pm S)$, which, when worked out explicitly (references 6, 7), is just

$$E = E(1s) - (1 \pm S)^{-1} \frac{e^2}{a_0} \left[\frac{1}{\rho} - e^{-2\rho} \left(\frac{1}{\rho} + 1 \right) \pm e^{-\rho}(1 + \rho) \right] + \frac{e^2}{a_0} \frac{1}{\rho}, \tag{3.8}$$

where $\rho = R_{AB}/a_0$. If we plot this double-valued energy function against R_{AB}, we get one curve with a potential minimum, corresponding to the eigenvector for the symmetric orbital $\sigma_g 1s$, and one curve that is a monotonically decreasing function of R_{AB} belonging to the $\sigma_u^* 1s$ orbital, as shown in figure 3.9. If the single electron is in the $\sigma_g 1s$ orbital, the H_2^+ species is stable. If, however, the electron is excited to the higher orbital, the molecule will fly apart, because the energy decreases monotonically as R_{AB} increases. This is the reason for the asterisk in the notation for this orbital—it denotes an *antibonding* orbital, which makes the molecule unstable with respect to the separated atoms.

The dissociation energy of the H_2^+ molecule, which is the depth of the minimum in the lower potential curve with respect to the separated atom limit, is 1.76 eV on the basis of equation (3.8). The dissociation energy from the exact wave functions, which is corroborated by experimental measurements, is about 2.79 eV. Thus the MO-LCAO model is not really a very good quantitative model, even for this ultrasimple system. One should not be surprised at this, however, since it was brashly asserted that the rather complex exact wave functions could be adequately approximated as a linear combination of two functions chosen essentially because we liked them. This is a case of a too "limited basis set"; if we take our linear combination as a sum of a large number of hydrogenlike atomic wave functions, we shall eventually be able to approximate the true wave functions, and thus the energy, as closely as we please. This, of course, is just a fundamental mathematical property of any complete set of functions. Experience shows that the least number of basis functions is required if we start with atomic orbitals, which reflects our molecular intuition that molecules are made up of atoms that retain at least some of their individual properties in chemical combinations. What is perhaps more surprising, when all is said and done, is that we were able to get such a good result from a single-function basis set.

The purpose of this development was actually not to calculate the dissociation energy of H_2^+, but to bring out the idea of a molecular potential curve, which is basic to all further development of molecular spectroscopy, and to set up a frame-work for constructing molecular orbitals of more complicated many-electron systems. For example, if we wished to look at the electron configuration of molecular oxygen, we would have to provide 8 molecular orbitals for the 16 electrons, and these would have to be constructed from a minimum of 8 atomic orbitals. A number of these would be atomic p orbitals, so let us see what kind of molecular orbitals can arise from linear combinations of p orbitals.

Two p_z atomic orbitals, lying along the internuclear axis, can combine in either a symmetric or an antisymmetric manner, to yield $\sigma_g 2p$ and $\sigma_u^* 2p$ orbitals, respectively (figure 3.10). These are cylindrically symmetric about the internuclear axis, and thus have angular momentum eigenvalue $\lambda = 0$, just as do the $\sigma 1s$ (and $\sigma 2s$ as well) orbitals. The p_x and p_y orbitals, on the other hand, combine to form molecular orbitals that are not cylindrically symmetric, but for which $\lambda = 1$; these are designated π orbitals. In this case,

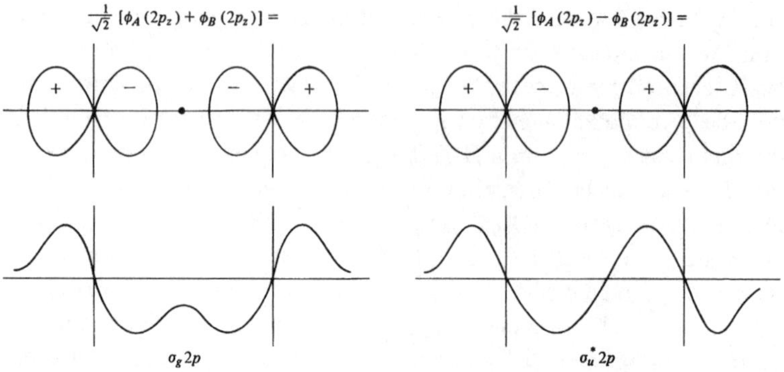

Figure 3.10
Combination of p_z atomic orbitals to form $\sigma_g 2p$ and $\sigma_u^* 2p$ molecular orbitals.

the symmetric combination, having the lower energy, is antisymmetric with respect to inversion through a center of symmetry, and vice versa; so that the orbitals are designated $\pi_u 2p$ and $\pi_g^* 2p$, respectively (figure 3.11).

The wave functions that we are plotting here are just the simple LCAO approximations. A much better idea of the true charge distributions can be gained from examining the accurate electron density contour maps shown in figure 3.12. From these, we can see that while the exact shape of each orbital varies from molecule to molecule, the symmetry properties of nodal planes, and so on, are constant and just the same as that of the simple LCAOs. We have σ, π, δ, ϕ, ... one-electron molecular orbitals for which the angular momentum eigenvalue $\lambda = 0, 1, 2, 3, \ldots$. For homonuclear diatomics, there are u and g orbitals corresponding to antisymmetric and symmetric combinations of atomic orbitals; for heteronuclear diatomics, which do not possess a center of inversion, the two combinations are simply distinguished by subscripts such as A and B.

When we come to construct many-electron molecular states on the basis of one-electron molecular orbitals, the same building-up, or *Aufbau*, principle as was used in atomic structure is applied. This involves three assumptions, namely:

1. The angular momentum along the internuclear axis adds according to a classical vector model, analogous to the Russell-Saunders coupling model for atoms. That is, the total electronic orbital angular momentum $\Lambda = \sum_i \lambda_i$, the total electronic spin angular momentum $\mathbf{S} = \sum_i \mathbf{s}_i$, and the total axial angular momentum $\mathbf{\Omega} = \mathbf{\Lambda} + \mathbf{\Sigma}$, where $\mathbf{\Sigma}$ is the z component of $\mathbf{S}$.

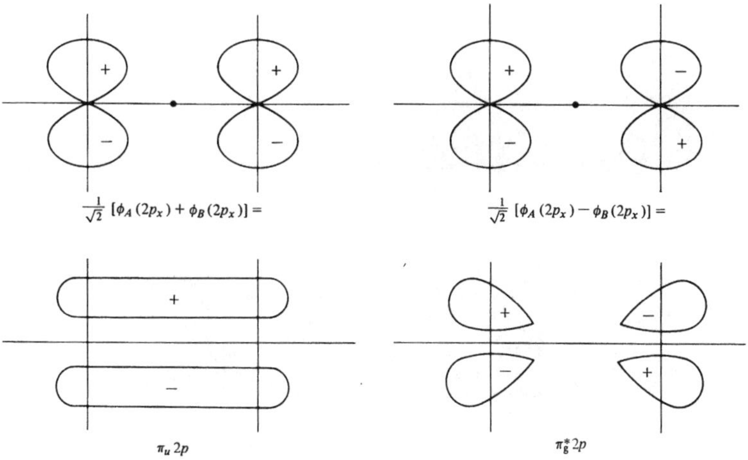

Figure 3.11
Combination of p_x atomic orbitals to form $\pi_u 2p$ and $\pi_g^* 2p$ molecular orbitals.

2. The electron repulsion terms

$$\sum_{i>j} \frac{e^2}{r_{ij}}$$

are treated as a perturbation; in contrast to the previous statement, this is a very rough approximation.
3. The Pauli exclusion principle applies.

Using these rules, we proceed to construct states from a molecular orbital configuration, in much the same way as we constructed atomic states from an atomic orbital configuration. Specifically, the ground state (X) of H_2 has the MO configuration $(\sigma_g 1s)^2$, and the electronic state $^1\Sigma_g^+$. The first excited configuration, $(\sigma_g 1s)(\sigma_u^* 1s)$, gives an unstable $b\,^3\Sigma_u^+$ state, which appears as a continuum at about 100,000 cm^{-1}. We may note that this is, of course, the same result as was obtained from counting the m states of the separated atoms in section 3; two hydrogen atoms ($^2S_{1/2}$) give $^1\Sigma_g^+$ and $^3\Sigma_u^+$ states. In the LCAO approximation, two s orbitals give $\sigma_g 1s$ and $\sigma_u^* 1s$ molecular orbitals, which can accept two electrons to give the same $^1\Sigma_g^+$ and $^3\Sigma_u^+$ states.
If we consider a totally filled configuration, such as $(\sigma_g 1s)^2(\sigma_u^* 1s)^2$, which corresponds to the unstable diatomic helium molecule, the only state that can be formed is $^1\Sigma_g^+$. This is known as a "KK closed shell"; all states formed

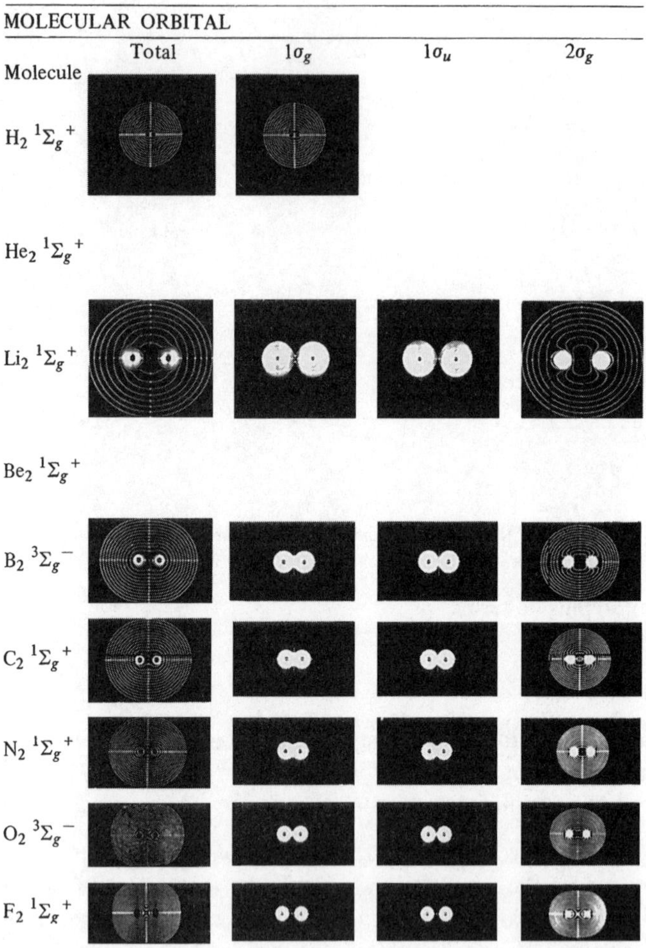

Figure 3.12
Electron density contour maps of $1\sigma_g$ through $3\sigma_u$ molecular orbitals, for H_2 through F_2. These are based on highly accurate calculations done by Dr. A. C. Wahl at the Argonne National Laboratory. [Reproduced with permission.]

$2\sigma_u$	$1\pi_u$	$3\sigma_g$	$1\pi_g$

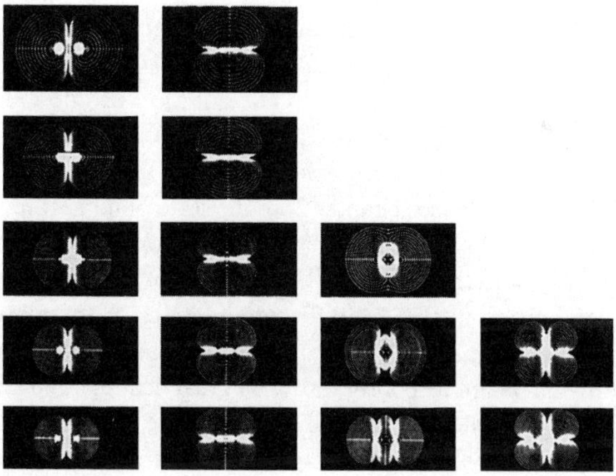

Table 3.3

λ_1	λ_2	s_1	s_2	Λ	Σ	State
1	1	+	+	This and similar states are forbidden by the Pauli principle.		
1	1	+	−	2	0	$^1\Delta$
1	1	−	+			
−1	−1	+	−	−2	0	
−1	−1	−	+			
1	−1	+	+	0	1	
−1	1	+	+			
1	−1	−	−	0	−1	$^3\Sigma^-$
−1	1	−	−			
1	−1	+	−	0	0	$^1\Sigma^+$
1	−1	−	+	0	0	

from maximally filled molecular orbitals are totally symmetric, and these orbitals need not be considered further when determining the symmetry properties of states from partially filled outer shells.

For example, the addition of a single electron in a π orbital to a system of closed inner shells produces a single $^2\Pi$ state, as in the radical CH, which has the configuration $(\sigma 1s_C)^2(\sigma 1s_H)^2(\sigma 2s_C)^2(\pi 2p_C)^1$. The addition of one more electron, to produce a $(\pi 2p)^2$ configuration, generates a number of states, which can be found by the construction in table 3.3.

In accordance with Hund's rule (see chapter 2.5), the state of highest multiplicity should be the one of lowest energy. So, in NH, for example, the ground state (X) is the $^3\Sigma^-$, with the $a\,^1\Delta$ and $b\,^1\Sigma^+$ lying higher in energy. For an inequivalent π^2 configuration, the Pauli-forbidden states in the above tabulation can exist, so that we have $^1\Sigma^+$, $^3\Sigma^+$, $^1\Sigma^-$, $^3\Sigma^-$, $^1\Delta$, and $^3\Delta$ states.

When we come to an equivalent π^3 configuration, a great deal of labor can be saved by realizing that the states are just the same as those arising from a π^1. This can be seen most easily by considering the various possibilities of adding together the λ_i. We can have either

$$\underset{\longrightarrow}{+1}\quad \underset{\longrightarrow}{+1}\quad \underset{\longleftarrow}{-1}\qquad \text{for which } \sum_i \lambda_i = +1$$

or

$$\underset{\longrightarrow}{+1}\quad \underset{\longleftarrow}{-1}\quad \underset{\longleftarrow}{-1}\qquad \text{for which } \sum_i \lambda_i = -1.$$

Thus $\Lambda = 1$, and the state arising from this configuration, is a $^2\Pi$. This is an example of the so-called hole formalism, which states that a shell, which is lacking n electrons of being completely filled, gives rise to the same states as that shell containing only n electrons. The same principle applies in atoms; the ground state of a boron atom, with a p^1 configuration, is ^2P, just the same as the halogen atoms with a p^5 configuration. The only difference is in the relative stability of the $\mathbf{J} = \mathbf{L} + \mathbf{S}$ (in atoms) or $\mathbf{\Omega} = \mathbf{\Lambda} + \mathbf{\Sigma}$ (in molecules) states; in a nearly empty shell, the state with the smaller quantum number has the lower energy, whereas in a nearly filled shell, it is the state with the larger quantum number. Thus the ground state of B or Al is $^2P_{1/2}$, while for the halogens it is $^2P_{3/2}$. By the same sort of reasoning, a π^4 configuration is equivalent to a π^0, that is, a totally symmetric $^1\Sigma^+$ state. Thus a π^4 configuration acts as a closed shell, just as a σ^2 does.

We can apply all these principles to the determination of the states of molecular oxygen. When the 16 electrons are filled in order into the available molecular orbitals, we obtain the configuration

$$KK(\sigma_g 2s)^2(\sigma_u^* 2s)^2(\sigma_g 2p)^2(\pi_u 2p)^4(\pi_g^* 2p)^2.$$

As we have seen, an equivalent π^2 configuration generates $^1\Sigma^+$, $^1\Delta$, and $^3\Sigma^-$ states. Since there is an even number of electrons of each parity, these will all be g states. These are exactly the states observed experimentally as the lowest three potential curves in figure 3.15:

$b^1\Sigma_g^+ \ldots$ 13,195.222 cm^{-1},

$a^1\Delta_g \ldots$ 7,918.1 cm^{-1},

$X^3\Sigma_g^- \ldots$ 0.0 cm^{-1}.

Notice that there are no dipole allowed transitions between the ground state of oxygen and these lowest-lying excited states. This is an example of a general rule, which we have met previously in atomic spectroscopy, that there exist no strongly allowed transitions between states arising from the same configuration. In homonuclear diatomic molecules, this rule is immediately obvious from the $u \leftrightarrow g$ selection rules, because all the states arising from a given molecular orbital configuration have the same parity. As in the case of atoms (compare chapter 2.5), the operation of this rule causes the existence of *metastable* states of molecules, which possess excess electronic energy, but which cannot dispose of it because the transition to the ground state is optically forbidden. A notable example is the $^1\Delta$ state of oxygen, which has a very long radiative half-life and is also remarkably resistant to relaxation by collisions, so that a high concentration can be produced by several methods, such as an electric discharge through oxygen gas or reaction of an alkaline solution of peroxide and hypochlorous ion. Of particular interest is that oxygen $^1\Delta$ can transfer its energy to iodine atoms, exciting them from the ground $^2P_{3/2}$ state to the excited $^2P_{1/2}$ spin-orbit state; the atoms can then be stimulated to emit this energy as laser radiation (see chapter 10).

The spin selection rule ($\Delta S = 0$) is also important, particularly for low-Z species. The net effect of the spin and parity selection rules operating together is that virtually every observed strong electronic transition in molecules can be effectively described as a one-electron orbital transition. For example, the first strong absorption in oxygen, the deep ultraviolet Schumann-Range bands assigned as $B^3\Sigma_u^- \leftarrow X^3\Sigma_g^-$, corresponds to a change in electron configuration of

$$KK(\sigma_g 2s)^2(\sigma_u^* 2s)^2(\sigma_g 2p)^2(\pi_u 2p)^4(\pi_g^* 2p)^2$$

to

$$KK(\sigma_g 2s)^2(\sigma_u^* 2s)^2(\sigma_g 2p)^2(\pi_u 2p)^3(\pi_g^* 2p)^3$$

or, briefly, a $\pi_u \rightarrow \pi_g^*$ transition. From the shapes of the orbitals, we can deduce the nature of the transition moment:

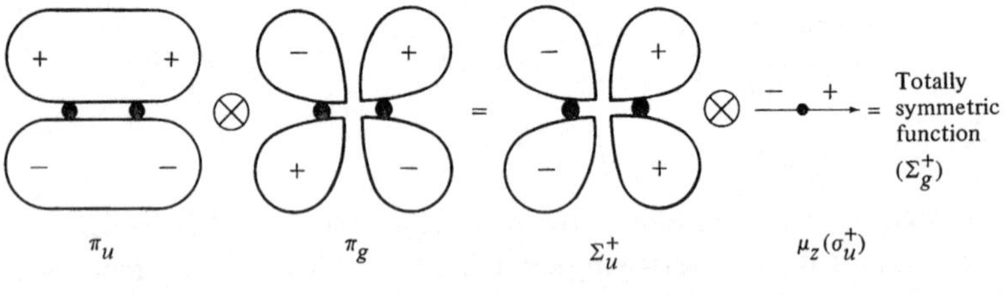

This diagram is intended to give a pictorial representation of the symmetries of the various quantities involved in the calculation of the transition probability. The product of the $(\pi_u 2p)$ and the $(\pi_g^* 2p)$ molecular orbitals has Σ_u^+ symmetry, that is, odd with respect to inversion and even with respect to reflection through an axial mirror plane. The product of this species with the dipole moment operator must be totally symmetric (Σ_g^+); in order for this to be the case, the dipole moment must have σ_u^+ symmetry, which dictates that the axial, or μ_z component, be the active one. Thus the transition will have parallel polarization, as required for $\Delta\Lambda = 0$. The reader should also follow this argument using the direct product tables for the $D_{\infty h}$ group (table A.27).

It is plain that a large number of molecular states can arise from combinations of various atomic states, especially when the latter are of high multiplicity. Usually, only a very few of these are known spectroscopically. Perhaps the most complete catalog of potential energy curves available is that compiled for the aeronomically important species N_2, NO, and O_2 by F. R. Gilmore (reference 8). These are shown in figures 3.13–3.15. Even for these extensively studied species, it is plain that many of the states, especially the repulsive ones, are characterized only very poorly or not at all. Even in the case of the simplest stable diatomic molecule, H_2, it is only recently that enough data have been compiled and correctly interpreted to allow a similar summary of the potential curves to be made; this has been done and is shown in figure 3.16.

Before we leave this topic, we should consider the possible interaction of several different states of the same molecule. A classic example of this occurrence is in diatomic lithium hydride. The ground state molecular orbital configuration arising from a hydrogen atom ($1s$) and a lithium atom ($1s^2 2s$) would be $K_{\text{Li}}(2s\sigma)^2$, which yields a ${}^1\Sigma_+$ state. The first excited state would have the configuration $K_{\text{Li}}(2s\sigma)(2p\sigma)$, with a bound ${}^1\Sigma_+$ and a repulsive ${}^3\Sigma_+$ state. Higher states would have $\sigma\pi$ configurations, and thus be ${}^1\Pi$ or ${}^3\Pi$. What we actually see, however, does not fit this simple prediction. The first two excited states above the $X\,{}^1\Sigma_+$ ground state are themselves both ${}^1\Sigma_+$ states, designated A and B, at energies of 3.2 and approximately 4.3 eV, respectively. There appears to be an extra ${}^1\Sigma$ state, and the dissociation energy of the ground state is considerably larger than that predicted by a simple MO theory. What has happened is that we have left out the possibility of states arising from the interaction, not of neutral H and Li atoms, but of ionic Li^+ (^{1}S) and H^- (^{1}S). Although these are both heliumlike atomic configurations, the Coulomb attraction energy is substantial. In electron volts, this energy has the value of e/R_e, where R_e is the experimentally determined distance between the atomic centers. If we use e = a full electronic charge and $R_e = 1.6$ Å, we have

$$E_{\text{Coulomb}} = \frac{4.8 \times 10^{-10}\ \text{esu}}{1.6 \times 10^{-8}\ \text{cm}} = 0.03\ \text{statvolts} = 9\ \text{eV}$$

when the factor of 300 between statvolts and mks volts is applied. This energy is enough to make the ${}^1\Sigma$ state arising from Li^+ and H^- more stable than any of the others arising from neutral atoms, thus leading to the situation shown in figure 3.17, in which the deep ${}^1\Sigma$ potential curve correlating with the ionic

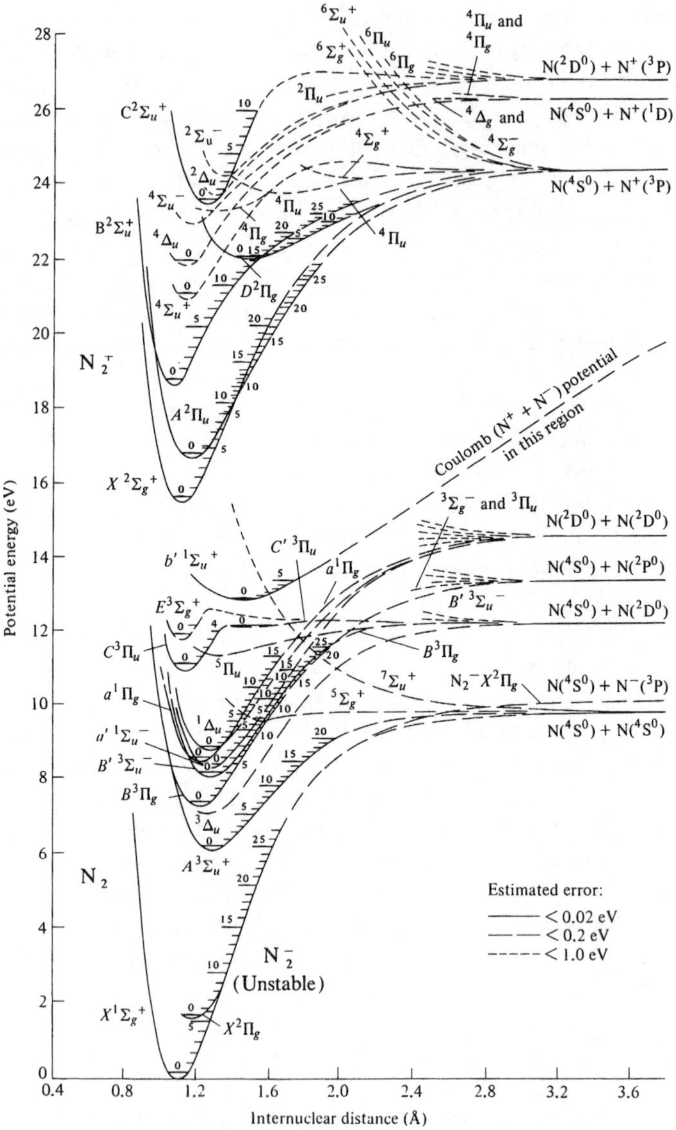

Figure 3.13
Potential energy diagram for N_2 compiled by F. R. Gilmore (reference 8). [Reproduced with the permission of F. R. Gilmore, The Rand Corporation.]

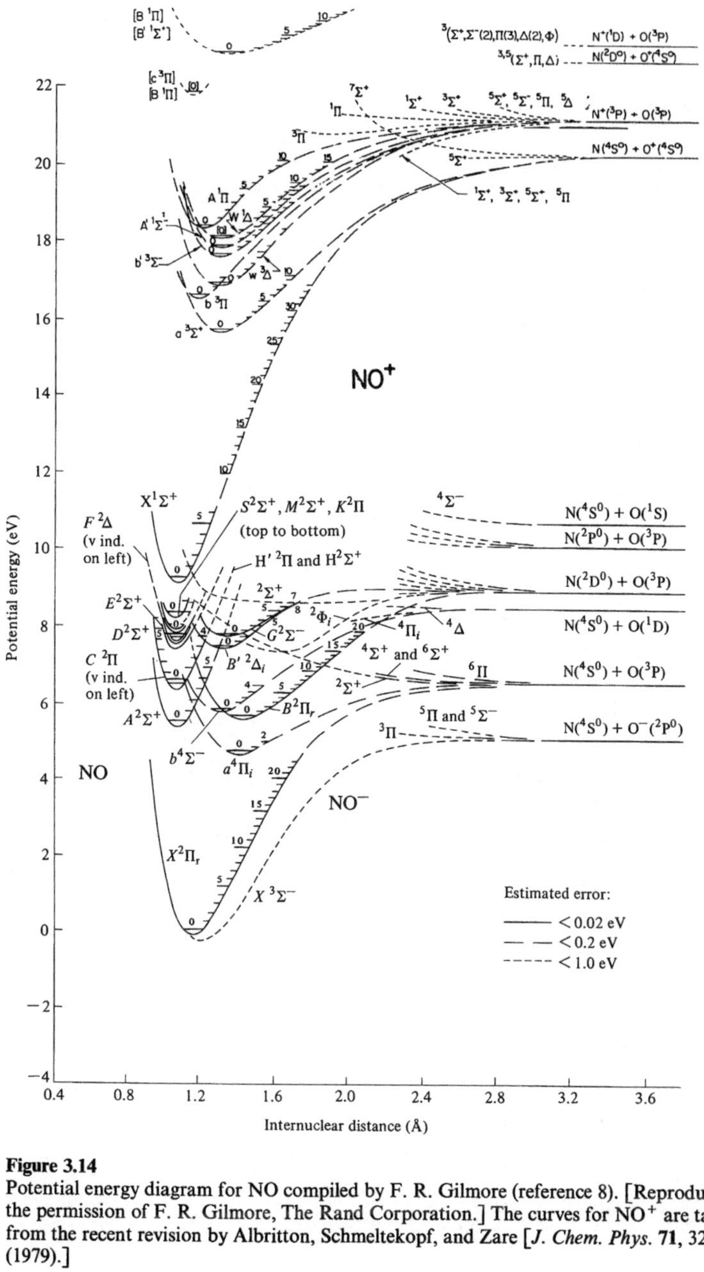

Figure 3.14
Potential energy diagram for NO compiled by F. R. Gilmore (reference 8). [Reproduced with the permission of F. R. Gilmore, The Rand Corporation.] The curves for NO^+ are taken from the recent revision by Albritton, Schmeltekopf, and Zare [*J. Chem. Phys.* **71**, 3271 (1979).]

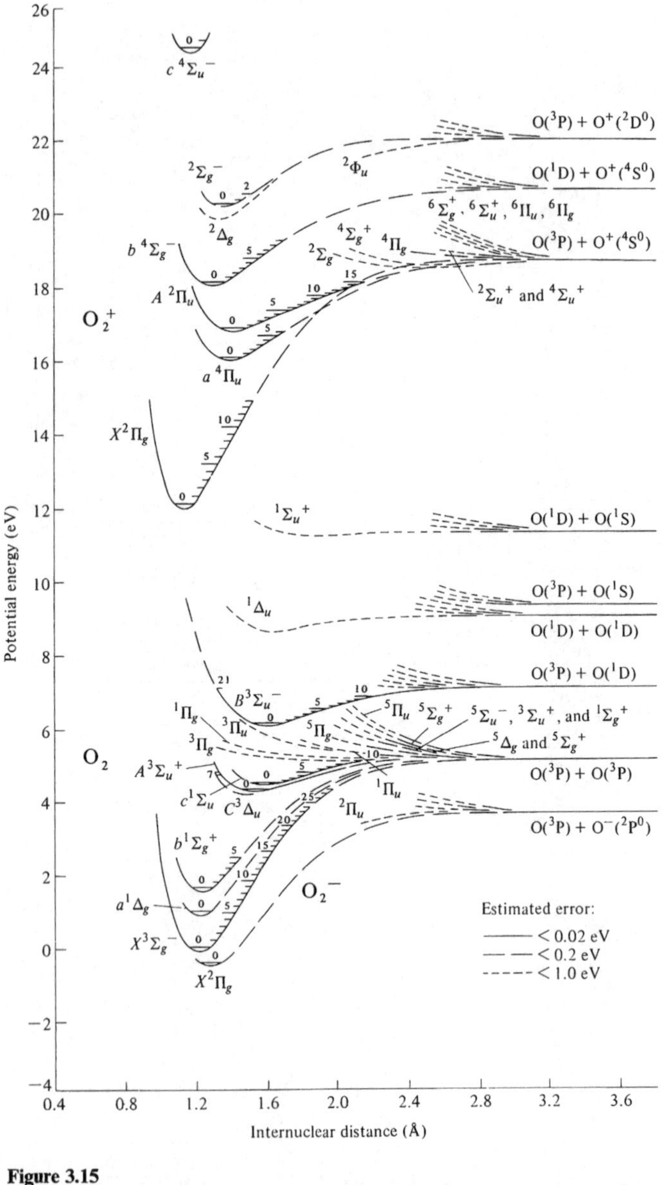

Figure 3.15
Potential energy diagram for O_2 compiled by F. R. Gilmore (reference 8). [Reproduced with the permission of F. R. Gilmore, The Rand Corporation.]

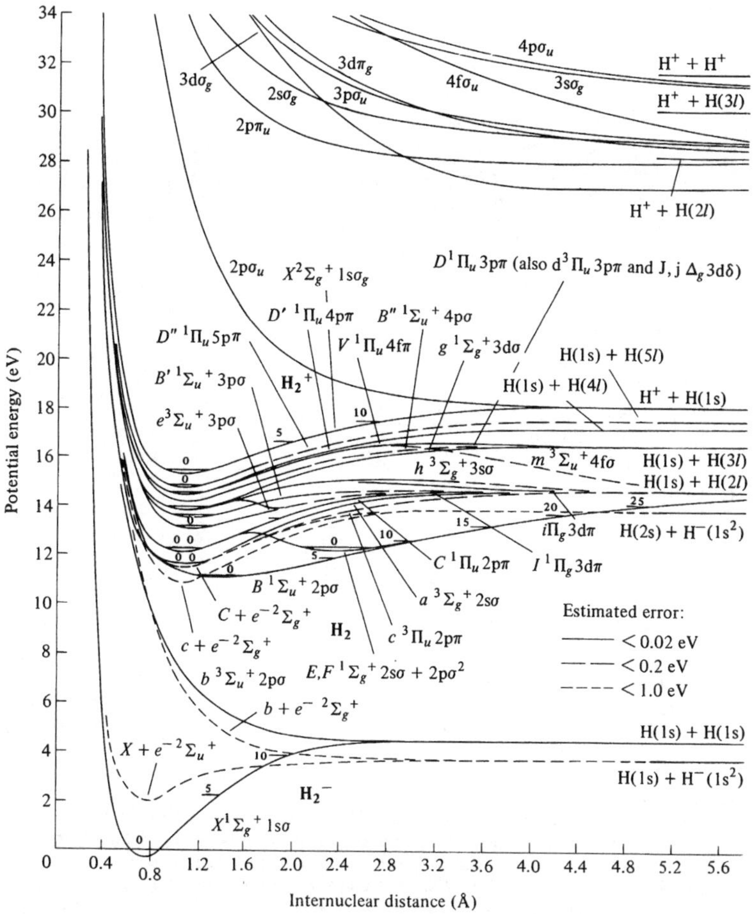

Figure 3.16
Potential energy curves for H_2^-, H_2, and H_2^+, compiled by T. Sharp. [From T. Sharp, "Potential Energy Diagram for Molecular Hydrogen and Its Ions," *Atomic Data*, Vol. 2 (Academic Press, New York, 1971), p. 119. Reproduced with permission.]

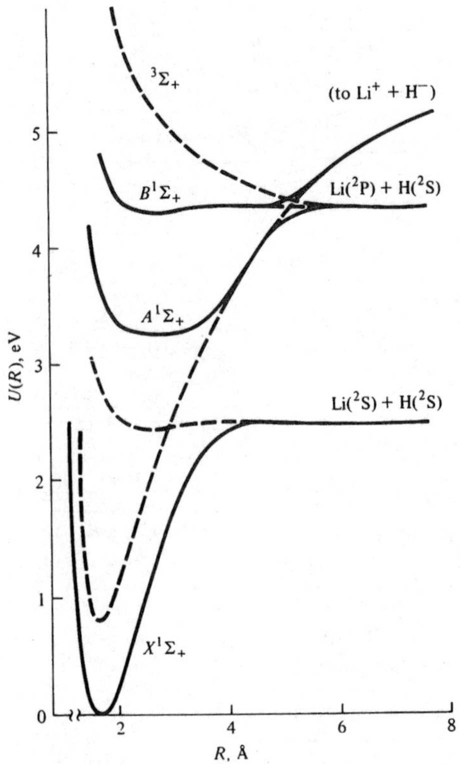

Figure 3.17
Potential energy curves for the lithium hydride molecule. The dotted curves are those estimated on a zero-order basis, neglecting interactions between states of the same symmetry. The solid curves are those obeying the noncrossing rule and are in accordance with experiment. The unusual shape of the $A\,^1\Sigma^+$ curve has been verified by detailed spectroscopic analysis. [From G. Herzberg, *Molecular Spectra and Molecular Structure*, Vol. 1: *Spectra of Diatomic Molecules* (Van Nostrand Reinhold, New York, 1950), p. 376. Reproduced with permission.]

species crosses the much shallower potential curves arising from the neutral species. Where this crossing occurs, however, there is a strong interaction between the two states, which splits them apart at the crossing point. Indeed, the crossing of two potential curves of the same symmetry never actually occurs; there are always terms in the Hamiltonian that lift the potential degeneracy and force the two curves apart. This is such a general occurrence that it has been called the "noncrossing rule" in spectroscopy. Another well-known example is that of the states of the alkali halides.

Problems

1.

(a) Write out the electron configurations for the molecules O_2^+, O_2, O_2^-, and O_2^{2-}.
(b) Determine the ground state term symbols (${}^M\Lambda_{g,u}^{\pm}$) for O_2^+, O_2, O_2^-, and O_2^{2-}. If there are two or more low-lying states, select one as that of the ground state and justify your selection.

2.

(a) Using symbols appropriate to the separated atom approximation to a molecular orbital, write the electronic configuration of lowest energy for the diatomic species C_2, NO^+, and S_2.
(b) Write the term symbols (${}^M\Lambda_{g,u}^{\pm}$) for all the electronic states derivable from the configuration of lowest energy for all three species. Which of these states will have nonzero magnetic moments?

3. Use the "Gilmore diagrams" provided for O_2, N_2, NO, and H_2:

(a) Sketch the expected low-resolution absorption spectrum to be expected for each of these four molecules between 1,000 and 10,000 Å.
(b) A photoionization experiment is carried out by shining light on a gas sample placed between two electrodes. A positive ion current is measured between the cathode and ground. For the four molecules, what is the longest wavelength of light that would be effective in producing a measurable ion current (this is called the "photoionization threshold")?
(c) The first positive bands ($B\,{}^3\Pi_g \rightarrow A\,{}^3\Sigma_u^+$) of nitrogen are observed in an "active nitrogen" discharge, in which the active species are predominantly ground state N atoms. Suggest a mechanism for the population of the excited state.

(d) When solutions of hydrogen peroxide and "Clorox" are mixed, a red chemiluminescence near 7,600 Å is observed. What might this be caused by?

References

1. M. Born and J. R. Oppenheimer, *Ann. Phys.* **84**, 457 (1927).

2. G. Herzberg, *Molecular Spectra and Molecular Structure*, Vol. 1: *Spectra of Diatomic Molecules* (Van Nostrand Reinhold, New York, 1950).

3. K. P. Huber and G. Herzberg, *Constants of Diatomic Molecules* (Van Nostrand Reinhold, New York, 1979).

4. E. Wigner and E. E. Witmer, *Z. Physik* **51**, 859 (1928).

5. O. Burrau, *Kgl. Dansk. Videnskab. Selskab.* **7**, 1 (1927).

6. G. W. King, *Spectroscopy and Molecular Structure* (Holt, Rinehart and Winston, New York, 1964), pp. 133–137.

7. H. Eyring, J. Walter, and G. W. Kimball, *Quantum Chemistry* (Wiley, New York, 1944), pp. 192–199.

8. F. R. Gilmore, "Potential Energy Curves for N_2, NO, O_2, and Corresponding Ions," RAND Corporation Memorandum R-4034-PR (June 1964).

4 Rotation and Vibration of Diatomic Molecules

The complete description of any molecular state involves a number of different degrees of freedom, each with its associated coordinate. There is translation in three-dimensional space (with coordinates X, Y, Z), rotation in space (with angular coordinates θ, ϕ, χ), vibrational motion of the nuclei (with an internuclear separation coordinate for each pair of atoms, such as R in a diatomic molecule), and motion of the electrons (with internal coordinates $\mathbf{r}$). In addition, there may be electron spin and nuclear spin variables.

The simplest starting point for considering the total energy spectrum of a molecule is to assume that all of these degrees of freedom are separable from one another. Classically, this means that the equations of motion separate into individual equations for each of the coordinates and the total energy is the sum of fixed individual contributions from each of the degrees of freedom; that is,

$$E_{\text{total}} = E_{\text{trans}} + E_{\text{rot}} + E_{\text{vib}} + E_{\text{electronic}} + E_{\text{spin}}. \tag{4.1}$$

Quantum mechanically, this requires that the terms in the Hamiltonian operator corresponding to each variable commute with all the other terms, so that the wave function for the total system can be written as a product of wave functions for each variable,

$$\Psi_{\text{total}} = \psi_{\text{trans}}(X, Y, Z)\psi_{\text{rot}}(\theta, \phi, \chi)\psi_{\text{vib}}(R)\psi_{\text{elec}}(\mathbf{r})\psi_{\text{spin}}. \tag{4.2}$$

As we shall see, some of the degrees of freedom (such as translation) are very well separated from all the others, whereas for others, the separability approximation is not a good one at all.

1 Rigid Rotor Hamiltonian, Eigenfunctions, and Spectra

We have seen, in the last chapter, that a particular electronic state of a diatomic molecule in the Born-Oppenheimer approximation is characterized by a *potential energy curve* $E(R)$, with a minimum at some R_e if the molecule is stable. Let us consider a diatomic molecule in one of its electronic states, and observe it in a set of coordinates moving with the constant velocity of the center of mass of the molecule, corresponding to its translational energy and momentum; these are called "space-fixed axes," although they are really moving at a constant velocity. The molecule, however, may be *rotating* in this coordinate system, and the rotation is described by the angular coordinates $\theta(t)$ and $\phi(t)$; the third angle χ corresponds to rotation about the internuclear

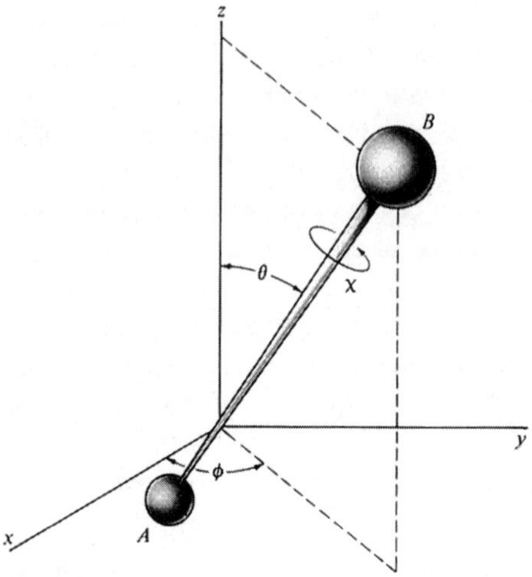

Figure 4.1
Diatomic molecule AB rotating in space-fixed axes. The origin of coordinates is at the center of mass of the molecule.

axis itself (see figure 4.1). Since nearly all the mass in a diatomic, or, for that matter, any linear, molecule is located on the axis, this rotation makes essentially no contribution to the pure rotational energy.

If we assume that the molecule is rigid, so that the internuclear separation is fixed at the equilibrium distance R_e, then the rotational motion can be separated from all the other degrees of freedom, and we can write the Schrödinger equation for rotation as

$$\mathscr{H}\psi_{\text{rot}} = -\frac{\hbar^2}{2\mu}\left[\frac{1}{R_e^2 \sin\theta}\left(\frac{\partial}{\partial\theta}\sin\theta\frac{\partial}{\partial\theta}\right) - \frac{1}{R_e^2\sin^2\theta}\frac{\partial^2}{\partial\phi^2}\right]\psi_{\text{rot}},$$

$$= E_{\text{rot}}\psi_{\text{rot}}$$

in which the reduced mass $\mu = m_A m_B/(m_A + m_B)$. But the operator for the rotational angular momentum is just

$$J^2 = -\hbar^2\left[\frac{1}{\sin\theta}\left(\frac{\partial}{\partial\theta}\sin\theta\frac{\partial}{\partial\theta}\right) + \frac{1}{\sin^2\theta}\frac{\partial^2}{\partial\phi^2}\right],$$

so we can write

$$J^2\psi_{\text{rot}} = 2\mu R_e^2 E_{\text{rot}}\psi_{\text{rot}} = \hbar^2 J(J+1)\psi_{\text{rot}},$$

since we know that the square of the total angular momentum is always quantized as $J(J+1)\hbar^2$. Therefore, the rotational energy is quantized as

$$E_{\text{rot}} = \frac{\hbar^2}{2I} = \frac{\hbar^2}{2\mu R_e^2} J(J+1) = BJ(J+1), \tag{4.3}$$

where the rotational constant $B = h/8\pi^2 c\mu R_e^2$, in units of cm^{-1}. As with other angular momentum problems, the projection of the rotational angular momentum on an external field direction, J_z, is quantized in units of M_J, with $-J \leq M_J \leq +J$, so that the degeneracy of the Jth rotational level $g_J = 2J+1$.

To find the selection rules for pure rotational transitions, we must look at the form of

$$\mathscr{H}'(t) = \boldsymbol{\mu} \cdot \mathbf{E} = \mu_x E_x + \mu_y E_y + \mu_z E_z.$$

In space-fixed axes, the components of a fixed dipole moment μ_0 rotating with the molecule are just

$$\mu_x = \mu_0 \sin\theta \cos\phi,$$

$$\mu_y = \mu_0 \sin\theta \sin\phi,$$

$$\mu_z = \mu_0 \cos\theta.$$

Thus the matrix elements involved in the dipole transition strength are just

$$\langle JM|\boldsymbol{\mu}\cdot\mathbf{E}|J'M'\rangle = |\mu_0||E| \int_0^{2\pi}\int_0^{\pi} P_J^M(\cos\theta)e^{-iM\phi} \begin{pmatrix} \sin\theta\cos\phi \\ \sin\theta\sin\phi \\ \cos\theta \end{pmatrix} \times P_{J'}^{M'}(\cos\theta)e^{+iM'\phi}\sin\theta\, d\theta\, d\phi,$$

which vanishes unless

$$J = J' \pm 1, \qquad M = M' \text{ or } M' \pm 1, \qquad \text{and } \mu_0 \neq 0.$$

The selection rules are, therefore,

$$\Delta J = \pm 1, \qquad \Delta M = 0 \text{ or } \pm 1,$$

and, furthermore, the molecule must possess a permanent dipole moment μ_0 in order to have a pure rotation spectrum.

As shown in figure 4.2a, the transitions consist of successive jumps from each rotational level to the one immediately above it (in absorption), so that

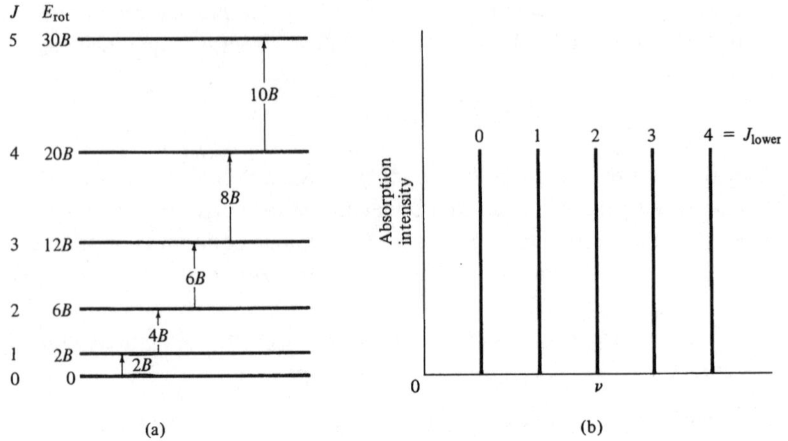

Figure 4.2
(a) Energy levels of a rigid diatomic rotor, showing dipole-allowed transitions. (b) Absorption spectrum of such a rotor.

the spectrum itself appears as a series of equally spaced lines, as in figure 4.2b. The line frequency formula is just $\nu = 2B(J+1)$, with $J = 0, 1, 2, \ldots$. Since typical values of μ are in the range of 10–100 amu (1.66×10^{-24} g), and typical values of R_e are about 1 Å (10^{-8} cm), $B = h/8\pi^2 c\mu R_e^2$ is of the order of 2 cm^{-1}, or 6×10^{10} Hz. This means that pure rotational absorption in molecules must be measured by either microwave absorption (discussed in chapter 7.2), far-infrared absorption, or Raman spectroscopy (discussed in section 6).

Actually, the absorption spectrum shown in figure 4.2b is somewhat misleading in that actual relative intensities are not properly indicated. Since the intensity of absorption of radiation [compare equation (1.35) or (1.36)] is directly proportional to the number of molecules in the absorbing state, which is, in turn, governed by simple Boltzmann statistics,

$$\begin{aligned} I_{abs} &\propto N(J_{lower}) \\ &\propto g_J e^{-E_J/kT} \\ &= (2J+1)e^{-BJ(J+1)/kT}, \end{aligned} \tag{4.4}$$

we obtain a spectrum looking like the one shown in figure 4.3. Some useful rules of thumb that may be derived from equation (4.4) are that the

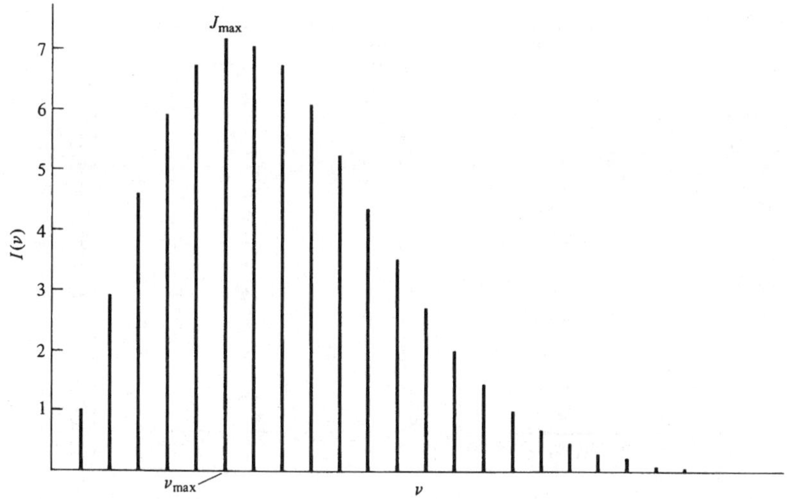

Figure 4.3
Typical pure rotational absorption band contour, for T/B approximately equal to 70 (for example, NO at 120°K or HCl at 740°K).

rotational quantum number corresponding to the most intense line J_{max} is the integer nearest to $0.59[T(°\mathrm{K})/B(\mathrm{cm}^{-1})]^{1/2}$ and the frequency of the band envelope maximum, measured relative to the band origin, is $\bar{\nu}_{max}$ (in cm^{-1}) $\approx 1.18(BT)^{1/2}$.

The rotational level populations given by equation (4.4) are still not complete, however. This is because of symmetry restrictions on nuclear spin states, which we have not considered explicitly until now. As we saw earlier (chapter 2.2), the wave function for a system of electrons must be antisymmetric with respect to exchange of any two of the electrons in the system. This is a property, not only of electrons, but of any particle with half-integral spins; included in this category are protons and neutrons. Such particles are said to obey *Fermi-Dirac statistics*. On the other hand, particles with integral (including zero) spin possess wave functions that must be symmetric with respect to particle exchange. Photons, which have one unit of intrinsic angular momentum, obey such symmetry rules, which are known as *Bose-Einstein statistics*. A consequence of these statistics is that many particles can occupy the same quantum level, while in the case of fermions, only one particle is permitted per distinguishable state. This manifests itself, for

Table 4.1

Nucleus	Number of protons	Number of neutrons	Spin I	Statistics[a]
H	1	0	$\frac{1}{2}$	FD
D(^{2}H)	1	1	1	BE
T(^{3}H)	1	2	$\frac{1}{2}$	FD
^{3}He	2	1	$\frac{1}{2}$	FD
^{4}He	2	2	0	BE
^{12}C	6	6	0	BE
^{13}C	6	7	$\frac{1}{2}$	FD
^{14}N	7	7	1	BE
^{15}N	7	8	$\frac{1}{2}$	FD
^{16}O	8	8	0	BE
^{17}O	8	9	$\frac{5}{2}$	FD
^{18}O	8	10	0	BE
^{19}F	9	10	$\frac{1}{2}$	FD
^{127}I	53	74	$\frac{5}{2}$	FD

a. Key: FD, Fermi-Dirac; BE, Bose-Einstein.

example, in the phenomenon of stimulated emission of radiation (chapter 1.7), which makes masers and lasers possible. The presence of some number of photons in a certain state of the radiation field tends to favor the appearance of as many additional photons as possible in that same state.

Atomic nuclei, which are ensembles of protons and neutrons, obey statistics according to their overall spin. Generally, odd-mass-number nuclei are fermions, and even-mass-number nuclei are bosons. The nuclear spins and statistical nature of a number of important nuclei are given in table 4.1.

It is an interesting point that the existence of superfluid helium is a consequence of the difference in statistics between ^{3}He and ^{4}He. At very low temperatures, ^{4}He can condense into a liquid in which nearly all the atoms are in the lowest possible momentum states. ^{3}He, however, obeys Fermi statistics, so that there can be only one atom per momentum state. In a macroscopic sample, levels are occupied that would not be if ordinary Boltzmann statistics applied, and the fluid performs such antics as flowing up and over the edge of its container in an attempt to find accessible lower-energy states for its atoms.

The total wave function for a diatomic molecule can be separated into electronic, rotational, and nuclear spin parts, and we have to find the effect of a nuclear interchange on the symmetry of the whole. In a heteronuclear diatomic, interchange symmetry does not apply, and the nuclear-spin weight is just $(2I_a + 1)(2I_b + 1)$—the same for each rotational level. In a homonuclear molecule, however, we require that Ψ_{total} be symmetric with respect to exchange if the nuclei are bosons, and antisymmetric with respect to exchange if the nuclei are fermions. Of the $(2I+1)^2$ possible nuclear spin states, $(2I+1)(I+1)$ are symmetric, and $(2I+1)I$ are antisymmetric; this is a more general version of the result we obtained for the symmetric (triplet) and antisymmetric (singlet) states possible for two $s = \frac{1}{2}$ electrons.

The basic symmetry property of the rotational levels is that the even J levels ($J = 0, 2, 4, \ldots$) are symmetric, and the odd J levels ($J = 1, 3, 5, \ldots$) are antisymmetric. This is easy to see, since the rotational wave functions are just the same Legendre polynomials as we had for the angular part of the hydrogen atom wave functions. Thus, $\psi_{J=0} = \text{constant}$, $\psi_{J=1} = \cos\theta$, $\psi_{J=2} = 3\cos^2\theta - 1$, and so on. The $J =$ even functions do not change sign under a C_2 operation (equivalent to exchanging the nuclei), while the $J =$ odd functions do so. When multiplied by an even electronic wave function, the parity property remains unchanged, but when multiplied by an odd electronic wave function, the parity is reversed, as we shall see.

Let us see how this works out in some particular cases. The ground electronic state of the hydrogen molecule is $^1\Sigma_g^+$, which is a symmetric function. The hydrogen nucleus, which is just a single proton, is a fermion, so that the total wave function

$$\Psi_{total} = \psi_{electron}\psi_{rotation}\psi_{nuclear\ spin}$$

must be antisymmetric overall. This means that the even J states combine with antisymmetric nuclear spin functions that have weight $(2(\frac{1}{2}) + 1)\frac{1}{2} = 1$. Similarly, the odd J states combine with symmetric nuclear-spin functions having weight $(2(\frac{1}{2}) + 1)(\frac{1}{2} + 1) = 3$. A 3:1 intensity alternation in H_2 rotational level populations is predicted and is indeed observed. Another example is molecular oxygen, which has a $^3\Sigma_g^-$ ground state. This is an antisymmetric electron wave function, so in this case the *even* J levels are antisymmetric and the *odd* J levels are symmetric. ^{16}O is a boson, so the total wave function must be symmetric overall. Thus the even J levels combine with antisymmetric nuclear spin states having a weight of zero, since $I = 0$, and odd J levels combine with symmetric nuclear spin states having a weight of 1. Thus, in this case, all the even rotational levels are simply missing—nuclear spin statistics dictate that a $^{16}O^{16}O$ molecule having $J = 0$, 2, and so on, cannot exist, at least in this universe. The same considerations apply to the $^{18}O^{18}O$ isotope of oxygen. In fact, the existence of the rare ^{17}O isotope was discovered from the presence of very weak lines located at the frequencies of the "forbidden" transitions in the ultraviolet spectrum of oxygen. On the average, at high temperatures, exactly half the levels expected for a diatomic molecule are missing by virtue of nuclear spin restrictions if the molecule is homonuclear. This is customarily accounted for by dividing the rotational statistical weights by a *symmetry number* σ, with $\sigma = 1$ for a heteronuclear molecule and $\sigma = 2$ for a homonuclear species.

The behavior of these nuclear spin statistics leads to some interesting consequences. If H_2 is cooled to its boiling point of about 20°K, in the presence of a catalytic material that can pull apart the H_2 molecules and recombine them in random spin combinations, then essentially pure $J = 0$ hydrogen will be obtained, since

$$\frac{n(J=1)}{n(J=0)} = \frac{3 \times 3e^{-2hcB/kT}}{1} = 0.15\% \text{ at } 20°\text{K}.$$

If the catalyst is removed and the hydrogen allowed to warm up, a gas with only even J states will be obtained, since the nuclear spin states are only very slowly interconverted. This gas is known as "parahydrogen." "Orthohydrogen," the gas of only odd J states, cannot be obtained in this way, but can be separated from a mixture by low-temperature gas chromatography. Another consequence is observed in the fluorescence spectrum of I_2 which will be discussed in more detail in chapter 5. If I_2 is excited to a particular even J state by monochromatic radiation, then collisions in the excited state can produce changes in rotational level of $\Delta J = \pm 2n$ exclusively, since the nuclear spin states are not affected by the collisions. This means that every other line is missing in the monochromatically excited fluorescence emission spectrum, whereas the ordinary intensity alternation observed in the absorption or broad-band emission spectrum is $\frac{7}{5}$.

2 Harmonic Oscillator Hamiltonian, Eigenfunctions, and Spectra

In treating rotation, we assumed that the molecule was rigid, with an internuclear separation fixed at R_e. We now wish to consider vibration of the molecule about the equilibrium distance R_e; to do this, we shall assume the molecule to be at rest, that is, not rotating in space-fixed axes, and investigate the vibration in the absence of rotation. In the next section, we shall permit both motions to occur simultaneously.

The potential energy curve $E(R)$, arising from the variation of total electron energy with internuclear separation, acts as the potential for the

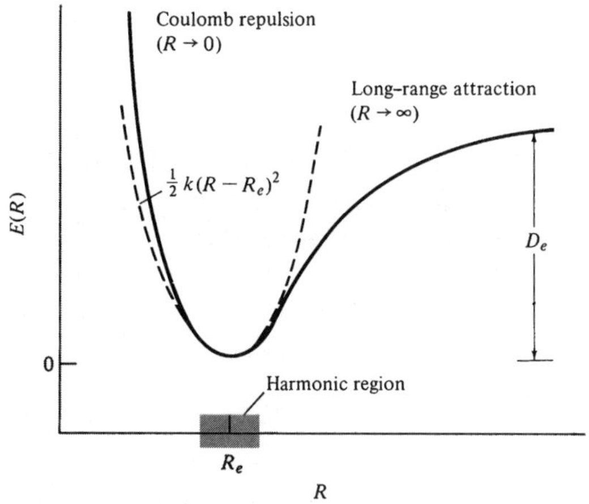

Figure 4.4
Harmonic approximation to a molecular potential curve $E(R)$, with force constant k equal to $(\partial E/\partial R)_{R=R_e}$.

Schrödinger equation governing motion of the nuclei,

$$(\mathscr{H} - E)\psi_{\text{vib}} = -\frac{\hbar^2}{2\mu}\frac{d^2\psi_{\text{vib}}}{dR^2} + [E(R) - E_{\text{vib}}]\psi_{\text{vib}} = 0. \tag{4.5}$$

Let us expand $E(R)$ in a Taylor's series about the equilibrium distance R_e, to give

$$E(R) = E_0 + (R - R_e)\left(\frac{\partial E}{\partial R}\right)_{R=R_e} + \tfrac{1}{2}(R - R_e)^2\left(\frac{\partial^2 E}{\partial R^2}\right)_{R=R_e} + \cdots.$$

The $(\partial E/\partial R)$ term vanishes at $R = R_e$ because R_e is the minimum of the potential curve, so that the first derivative is zero. The leading term in the expansion is thus the $(R - R_e)^2$ term, so that the potential is approximated by a parabolic well,

$$E(R) \approx \tfrac{1}{2}k(R - R_e)^2,$$

with force constant k equal to the curvature of the potential. Such a parabolic potential is a good approximation to a realistic potential curve near $R = R_e$, as shown in figure 4.4, but deviates badly at very small or very large internuclear separations, especially as the molecule approaches its dissociation limit.

The solution to equation (4.5), with $E(R) = \frac{1}{2}k(R - R_e)^2$, is just the harmonic oscillator, which has quantized energy levels

$$E_{vib} = \hbar\omega(v + \tfrac{1}{2}), \qquad v = 0, 1, 2, \ldots,$$

with vibration frequency $\omega = \sqrt{k/\mu}$.[1] The wave functions for the vibration are just the harmonic oscillator wave functions shown in figure 4.5.

To determine the absorption and/or emission spectrum of an oscillating molecule, we again must look at the electric dipole moment; in particular, this time we must consider how it varies with internuclear separation. If we expand the dipole moment in a Taylor's series about $R = R_e$,

$$\mu = \mu_e + \left(\frac{\partial \mu}{\partial R}\right)_{R=R_e}(R - R_e) + \frac{1}{2}\left(\frac{\partial^2 \mu}{\partial R^2}\right)_{R=R_e}(R - R_e)^2 + \cdots,$$

and look for off-diagonal matrix elements of μ, we find that the only nonvanishing ones are

$$\langle v|(R - R_e)|v'\rangle \sim \delta(v, v' \pm 1),$$

which gives the *fundamental* absorption with $\Delta v = \pm 1$,

$$\langle v|(R - R_e)^2|v'\rangle \sim \delta(v, v' \pm 2),$$

which gives much weaker *overtone* absorption with $\Delta v = \pm 2$, and so on. Thus the strength of a vibrational band in the infrared depends on the magnitude of the derivative of the dipole moment with internuclear distance. A molecule with a relatively small dipole moment may still have a large dipole derivative, and, conversely, a molecule with a very large dipole moment may have a small dipole derivative if the dipole moment is near its maximum value at $R = R_e$. For example, carbon monoxide, which has a permanent moment of only 0.11 D (1 D = 1 debye = 10^{-18} esu cm), possesses a large dipole derivative, and thus one of the strongest known infrared absorptions. A homonuclear molecule, however, for which $\mu = 0$ for all internuclear separations, has a dipole derivative that is zero everywhere, and thus no vibrational absorption at all; indeed, all homonuclear molecules, such as H_2, O_2, N_2, I_2, and so on, are transparent throughout the entire infrared region of the spectrum because they have neither vibrational nor rotational absorption activity.

1. A useful equation relating the force constant, reduced mass, and vibration constant is $k = 0.0589\mu_A\omega_e^2$, where k is in dyn/cm; μ_A is in atomic mass units; and ω_e is in cm^{-1}. Typical values for ω_e are in the range 3,000–300 cm^{-1}, in the infrared.

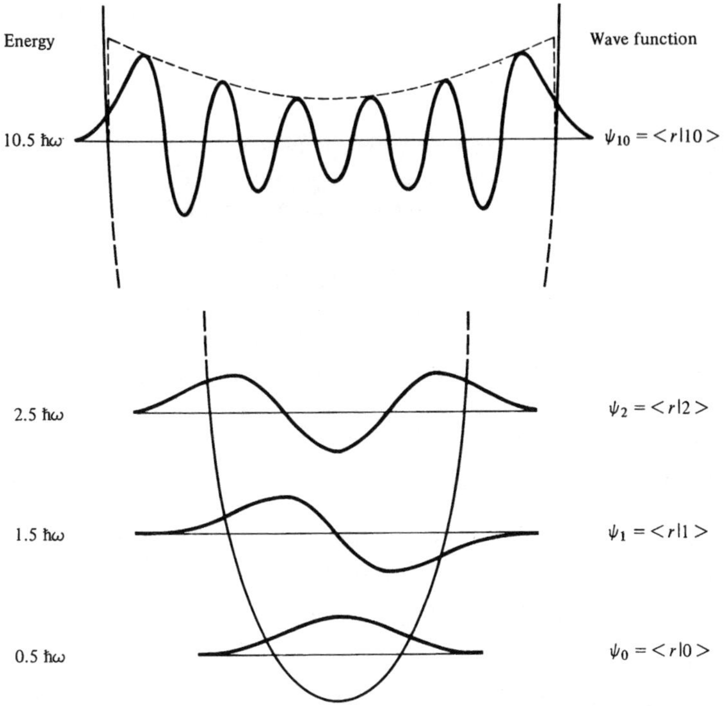

Figure 4.5
Energy levels and corresponding wave functions for a harmonic oscillator potential. Note that the quantum number of the wave function is the same as the number of nodes, that is, the number of times the function goes through zero. The time-averaged classical position of the oscillator in a high vibrational level is shown by the dashed curve; note that the envelope of the quantum mechanical wave function approaches this classical limit at large values of v.

The spectrum for vibrational absorption is exceedingly simple. In the harmonic oscillator approximation, we have just

$$h\nu_{\text{abs}} = E(v + \tfrac{1}{2}) - E(v' + \tfrac{1}{2}) = \hbar\omega,$$

so that all $\Delta v = 1$ bands occur at the same frequency.

3 Combination of Rotational and Vibrational Motions

In any real molecule, rotation and vibration are occurring simultaneously, and the spectrum will reflect this fact. The simplest way of treating this situation is to continue the assumption that the rotational and vibrational energies are additive and to develop the spectrum on the basis of this assumption and the selection rules $\Delta v = \pm 1$, $\Delta J = \pm 1$. We can have $\Delta J = 0$ with $\Delta v = \pm 1$ only in certain exceptional cases, such as when the electronic angular momentum Λ of the molecule is not equal to zero, that is, the molecule is not in a Σ state.

The spectrum will appear at a center frequency ν_0, corresponding to the pure vibrational transition, with rotational energy differences added on. Conventionally, the quantum numbers and spectroscopic constants for the lower state in a transition are labeled with a double prime, whereas those for the upper state are labeled with a single prime. When the rotational state *increases* by one unit in an absorption transition, so that $J' = J'' + 1$, the line is called an *R branch* transition; when the rotational state *decreases* by one unit, so that $J' = J'' - 1$, the line is called a *P branch* transition. When it does occur, transitions with $J' = J''$ are called the *Q branch*. The frequencies of each of the lines are just

$$\begin{aligned}\nu_P(J) &= \nu_0 + B'J'(J' + 1) - B''J''(J'' + 1)\\ &= \nu_0 + B'(J - 1)J - B''J(J + 1)\\ &= \nu_0 - (B' + B'')J + (B' - B'')J^2\end{aligned} \tag{4.6}$$

and

$$\begin{aligned}\nu_R(J) &= \nu_0 + B'J'(J' + 1) - B''J''(J'' + 1)\\ &= \nu_0 + B'(J + 1)(J + 2) - B''J(J + 1)\\ &= \nu_0 + (B' + B'')(J + 1) + (B' - B'')(J + 1)^2.\end{aligned} \tag{4.7}$$

The appearance of such a rotation–vibration band is shown in figure 4.6.

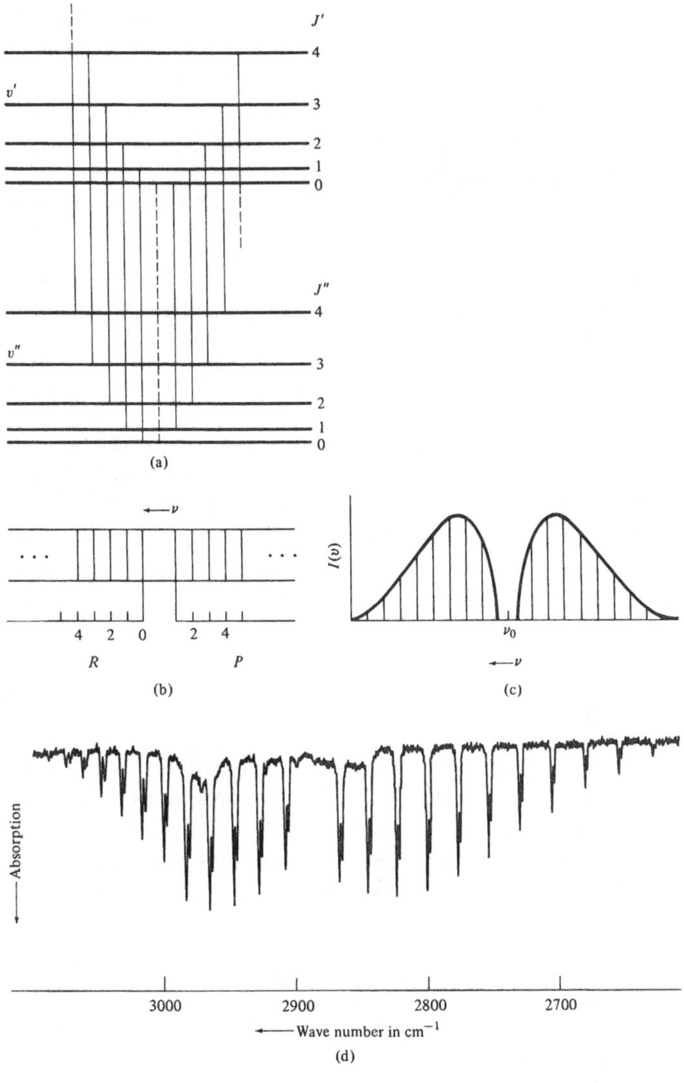

Figure 4.6
Structure of a vibration-rotation band: (a) rotational energy levels associated with upper and lower vibrational energy levels, and allowed transitions; (b) spectrum lines for this transition; (c) intensity distribution in the spectrum, using equation (4.4); (d) actual infrared absorption spectrum of HCl. Note the splitting of each line due to the $H^{35}Cl$ and $H^{37}Cl$ isotopic shift, discussed in problem 4. [Spectrum is reproduced with permission of the MIT Spectroscopy Laboratory.]

Note that, in the absence of a Q branch, there is no line at the position of the band center itself, ν_0.

If there is absolutely no interaction between vibration and rotation, then $B' = B''$, and the P and R branches consist of a series of evenly spaced lines. This would be true only in a perfectly rigid rotor, for which $B = h/8\pi^2 c\mu R_e^2$. For a vibrating molecule, however, we have

$$B_v = \frac{h}{8\pi^2 c\mu}\langle v|\frac{1}{R^2}|v\rangle. \tag{4.8}$$

The dependence of the rotational constant on the vibrational quantum number is most conveniently expressed as a power series,

$$B_v = B_e - \alpha_e(v + \tfrac{1}{2}) + \gamma_e(v + \tfrac{1}{2})^2 + \delta_e(v + \tfrac{1}{2})^3 + \cdots \tag{4.9}$$

where $B_e = B(R_e)$, which is never observed because of the zero point motion of the nuclei. As we shall see, the fact that a real molecular potential departs substantially from a harmonic oscillator as v increases makes the major contribution to the vibration–rotation interaction constants α_e, γ_e, δ_e, ... Even for a harmonic oscillator, however, α_e has a nonzero value. If we evaluate equation (4.8) with harmonic oscillator matrix elements, we find (see reference 1) that

$$\alpha_e = -\frac{6B_e^2}{\omega_e}, \tag{4.10}$$

which would cause B' to be greater than B''. This is due to the local maxima in the vibrational wave function at the classical turning points at either end of the potential, as may be noted by inspection of figure 4.5, and the fact that the $1/R^2$ in equation (4.8) weights the inner turning point more heavily than the outer one. The magnitude of this departure from a rigid rotor model may be estimated by calculating

$$\frac{\alpha_e}{B_e} = -\frac{6B_e}{\omega_e}.$$

For typical values of $B_e = 1\ \text{cm}^{-1}$ and $\omega_e = 1{,}000\ \text{cm}^{-1}$, we have a change on the order of 0.6%. The effect of this is to produce nonuniform line spacings in the P and R branches, and, in the case that a Q branch is present, to displace the $Q(J)$ lines slightly from each other (see problem 1.) However, the contribution to α_e from the anharmonicity, discussed in section 5, proves to be much larger than this particular effect, and for most molecules, B_v is a decreasing function of v.

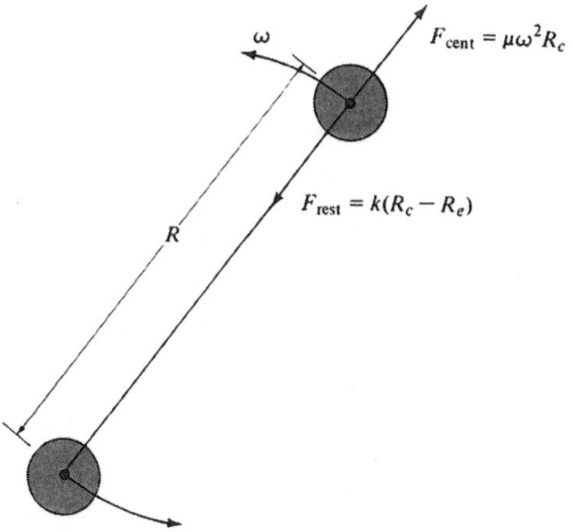

Figure 4.7
Centrifugal distortion in a rotating diatomic molecule.

4 The Nonrigid Rotor: Centrifugal Distortions

Another type of vibration–rotation interaction, which occurs even in the $v = 0$ state, is a stretching of the molecular bond length away from its equilibrium position by centrifugal force in the rotating molecule. The effect of this centrifugal stretching on the energy levels can be found by calculating the new equilibrium bond length in the presence of rotation, and this can be done by equating the centrifugal force F_c to the restoring force produced by the potential energy curve F_r, as shown in figure 4.7. Classically,

$$F_c = \mu\dot{\theta}^2 R_c = \frac{\mathbf{J}^2}{\mu R_c^3},$$

where R_c is the bond length at which the two forces are in balance, and $\mathbf{J}$ is the classical rotational angular momentum:

$$\mathbf{J} = \mu\dot{\theta}\mathbf{R}^2.$$

Since $\mathbf{J}^2$ is quantized in units of $\hbar^2 J(J+1)$,

$$F_c = \frac{\hbar^2 J(J+1)}{\mu R_c^3}.$$

The restoring force for a particle in a parabolic potential well is just that for an ideal spring,

$$F_r = k(R_c - R_e).$$

By equating F_c and F_r, we find that the displacement produced by the centrifugal force is

$$R_c - R_e = \frac{\hbar^2 J(J+1)}{\mu R_c^3 k}.$$

The rotational energy of the distorted molecule, including the displacement from equilibrium, is

$$E_{rot} = \frac{\hbar^2 J(J+1)}{2\mu R_c^2} + \tfrac{1}{2}k(R_c - R_e)^2.$$

Substituting for R_c and keeping the leading terms give

$$E_{rot} = \frac{\hbar^2 J(J+1)}{2\mu R_c^2} - \frac{\hbar^4 J^2 (J+1)^2}{2\mu^2 R_e^6 k}$$

$$= BJ(J+1) - D[J(J+1)]^2,$$

where the centrifugal distortion constant $D = \hbar^4/2\mu^2 R_e^6 k = 4B_e^3/\omega_e^2$. For typical spectroscopic constants ($B_e = 1$ cm^{-1}, $\omega_e = 1{,}000$ cm^{-1}), the centrifugal distortion contribution at $J = 30$ is of the order of 1 cm^{-1} out of a total rotational energy of 900 cm^{-1}. So the effect is small, but easily measurable in high-resolution spectroscopy.

Further vibration–rotation interaction terms are accounted for empirically by extending the series for the rotational energy,

$$E_{rot} = B_v J(J+1) - D_v[J(J+1)]^2 + H_v[J(J+1)]^3 + \cdots,$$

with

$$D_v = D_e - \beta_e(v + \tfrac{1}{2}) + \cdots,$$

and so on.

5 The Anharmonic Oscillator

As soon as the harmonic oscillator approximation for vibration was introduced, it was pointed out that the parabolic potential is a reasonably good approximation to real molecular potential curves only in the region of small displacements from equilibrium at the bottom of the potential. This is adequate for describing the $1 \leftarrow 0$ fundamental, but not for transitions involving any higher levels. As figure 4.4 shows, the potential curve rises more steeply than a parabola as $R \rightarrow 0$, as Coulomb repulsion between the electrons becomes more significant, and less steeply as $R \rightarrow \infty$. Indeed, the potential must eventually approach a limiting, finite value corresponding to separated atomic states, and generally approaches this limit with a long-range attractive behavior such as R^{-6} (arising from van der Waals or dispersion forces), rather than anything like R^2. This finite energy between the minimum of the potential and the separated atoms is the *dissociation energy* of the molecule, given the symbol D_e, which must not be confused with the centrifugal distortion coefficient D_e introduced just above.

The general term for the departure of a molecular potential curve from the ideal harmonic oscillator is *anharmonicity*. The effect of anharmonicity on the vibrational energy levels of a molecule depends on the detailed shape of the potential, of course, but it is convenient to represent the levels in terms of a series in $v + \frac{1}{2}$, much like the expansion of the rotational constant in equation (4.9):

$$G_v = \omega_e(v + \tfrac{1}{2}) - \omega_e x_e(v + \tfrac{1}{2})^2 + \omega_e y_e(v + \tfrac{1}{2})^3 + \omega_e z_e(v + \tfrac{1}{2})^4 + \cdots. \tag{4.11}$$

The center frequency of the $1 \leftarrow 0$ band is slightly shifted,

$$\begin{aligned} \nu_0 &= \omega_e(\tfrac{7}{2} - \tfrac{1}{2}) - \omega_e x_e(\tfrac{9}{4} - \tfrac{1}{4}) + \cdots \\ &= \omega_e - 2\omega_e x_e + \cdots. \end{aligned}$$

The ratio $x_e = \omega_e x_e/\omega_e$, by itself, is often called the "anharmonicity factor."

The effect of anharmonicity on vibrational energy levels and spectra is, primarily, that the energy levels are no longer exactly equally spaced, as in the harmonic oscillator, but get closer and closer together as v increases.[2] This, in turn, has several consequences.

1. The restriction that $\langle v|(R - R_e)|v'\rangle \sim \delta(v, v' \pm 1)$ holds only for a harmonic oscillator; thus the overtone bands can have appreciable intensity.

2. An exception to this statement is the A state of LiH (compare figure 3.17), in which the curvature of the potential increases rather than decreases as one goes up in energy, so that the anharmonicity x_e is negative, and the levels become more, rather than less, widely separated with increasing v. This type of behavior is also seen in quartic potentials, which have a flatter bottom and steeper sides than parabolic potentials. An extreme example of this type of behavior is the square well with infinitely steep walls, in which the energy levels go as n^2; interestingly enough, a square well potential can be represented by $E(R) = |R/a - R_e/a|^\infty$, where $2a$ is the width of the potential centered about R_e.

2. The overtones are not exactly at $2\nu_0$, $3\nu_0$, and so on, but at slightly lower frequencies.
3. The "hot bands" ($v = 1 \rightarrow v = 2$, $v = 2 \rightarrow v = 3$, and so on) are shifted slightly to the red of the fundamental ($v = 0 \rightarrow v = 1$) band.
4. There are additional terms in the vibration–rotation interaction constant α_e. An expression for α_e for an anharmonic oscillator is (see reference 1)

$$\alpha_e = \frac{6\sqrt{\omega_e x_e B_e^3}}{\omega_e} - \frac{6B_e^2}{\omega_e};$$

$\omega_e x_e$ is just about always larger than B_e, so that the first term dominates the second, and overall α_e is positive. Thus the rotational constant decreases with increasing v. This is just what one would expect from the "tilt" of the anharmonic potential, which causes the average R for each vibrational level to increase as v increases, thus leading to a larger moment of inertia and therefore a smaller B value.

From equation (4.11), we can derive these expressions for the spacings between successive vibration levels:

$$\Delta G_{v+1/2} = [(v + \tfrac{3}{2}) - (v + \tfrac{1}{2})]\omega_e - [(v + \tfrac{3}{2})^2 - (v + \tfrac{1}{2})^2]\omega_e x_e + \cdots.$$

If we cut the series off after the first anharmonicity term, we have just

$$\begin{aligned} \Delta G_{v+1/2} &= \omega_e - [v^2 + 3v + \tfrac{9}{4} - (v^2 + v + \tfrac{1}{4})]\omega_e x_e \\ &= \omega_e - 2(v + 1)\omega_e x_e. \end{aligned} \tag{4.12}$$

Thus the spacing decreases linearly with v; that is, the second differences

$$\Delta^2 G_{v+1/2} = -2\omega_e x_e$$

are a constant.

Very often, one will have reliable data on vibrational level spacings only part way up the potential, but would like to know the dissociation limit, that is, the complete depth of the potential well. If equation (4.12) holds, then a linear extrapolation of $\Delta G_{v+1/2}$ should yield the exact number of vibrational levels in the potential, and D_0^0, the dissociation energy from the $v = 0$ level, should be obtained simply by summing the $\Delta G_{v+1/2}$ values read off the linear graph:

$$D_0^0 = \sum_{v=0}^{v=v_{\max}} \Delta G_{v+1/2}.$$

The dissociation energy from the bottom of the potential, D_e, is obtained simply by adding the zero point energy to D_0^0:

$$D_e = D_0^0 + \tfrac{1}{2}\omega_e - \tfrac{1}{4}\omega_e x_e.$$

If we substitute the expression [equation (4.12)] for $\Delta G_{v+1/2}$ and carry out the sum, we find that $D_e = \omega_e^2/4\omega_e x_e$.

This procedure is called the *Birge-Sponer extrapolation* (reference 2) and has often been applied to estimates of dissociation energies. An example of such a Birge-Sponer plot is shown in figure 4.8a; note that, in this particular case, there is a departure from the linear extrapolation for very high v levels. This occurs because the molecule, which separates into excited atomic states, has a long-range attractive potential between the two atoms at large separations. Such errors are common in Birge-Sponer extrapolations, so that dissociation energies determined by this method must be regarded with caution. LeRoy and Bernstein (references 3–6) have suggested more accurate procedures for determining dissociation energies and long-range potential curves from available data. In their method, if the molecule dissociates according to a $-C/R^n$ potential curve, then the vibrational first differences are related by

$$\Delta G_v^{[2n/(n+2)]} \propto D - E_v,$$

where D is the experimental dissociation limit, and the total vibrational energy of the state is

$$E_v = \sum_{v'=0}^{v} \Delta G_{v'}.$$

This enables D to be determined accurately from the data; then the number of vibrational levels in the potential is determined from the relation

$$(D - E_v)^{[(n-2)/2n]} \propto (v_D - v).$$

where v_D is the vibrational quantum number of the topmost level in the potential. When this is applied to the B state of I_2, whose Birge-Sponer plot showed marked departure from linearity in figure 4.8a, an excellent straight line is obtained, as shown in figure 4.8b. The 10/7 exponent is derived from $n = 5$, which is the appropriate power of $(1/R)$ for a potential between a $J = \tfrac{1}{2}$ and a $J = \tfrac{3}{2}$ atom at large separations.

Empirical Molecular Potential Functions It should be evident that it would be very convenient to have some analytic function, containing only a few parameters, by means of which one could represent an anharmonic potential

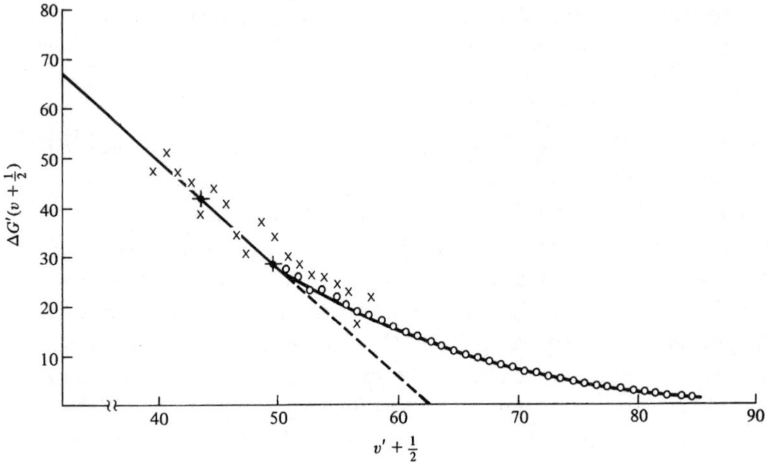

(a)

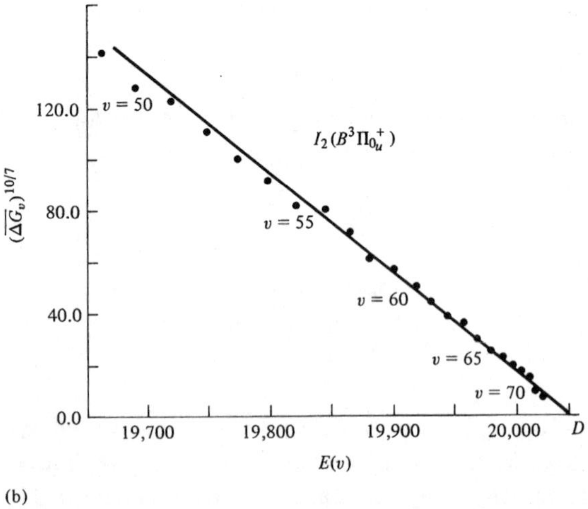

(b)

Figure 4.8
Extrapolations to the dissociation limit of vibrational-level data for the $B^3\Pi_{0u}^+$ state of I_2. (a) Birge-Sponer extrapolation showing positive curvature: ×, band head data of R. Mecke, *Ann. Phys.* **71**, 104 (1923); +, data of J. I. Steinfeld, J. D. Campbell, and N. A. Weiss, *J. Mol. Spectroscopy* **29**, 204 (1969); o, data of W. G. Brown, *Phys. Rev.* **38**, 709 (1931) and R. Mecke, *Ann. Phys.* **71**, 104 (1923). (b) Linear LeRoy-Bernstein plot of the same data from R. J. LeRoy and R. B. Bernstein, *J. Mol. Spectroscopy* **37**, 109 (1971) (reference 4). D is determined from this plot as 20,044.0 cm^{-1}; there is a total of 87 vibrational levels in the potential.

energy curve. Ideally, the potential function should be one in which the Schrödinger equation for the energy levels could be solved exactly. Such a function was introduced by P. M. Morse in 1929 (reference 7); it has the form

$$E(R) = D_e[1 - e^{-a(R-R_e)}]^2. \tag{4.13}$$

The energy eigenvalues can be solved for exactly (see reference 7) and have the form

$$G_v = \omega_e(v + \tfrac{1}{2}) - \omega_e x_e(v + \tfrac{1}{2})^2,$$

with no higher terms, just as required for the Birge-Sponer extrapolation procedure to be valid. The parameter a is related to the fundamental frequency ω_e and the dissociation energy D_e by

$$a = \omega_e \left(\frac{\pi c \mu}{\hbar D_e}\right)^{1/2},$$

where a, ω_e, and D_e are all in cm^{-1}, and the anharmonicity is given by

$$\omega_e x_e = \frac{\hbar a^2}{4\pi c \mu}.$$

When the presence of terms higher than $(v + \frac{1}{2})^2$ is indicated in G_v, then the Morse function is not a satisfactory representation of the potential. In this case the most practical procedure for obtaining a potential function directly from spectroscopic constants is a numerical procedure known as the Rydberg-Klein-Rees (RKR) method. This method is based on the semiclassical JWKB[3] method (references 5 and 8) in quantum mechanics, which, in turn, is based on the quantization of the action integral. This integral is

$$I = (2\mu)^{1/2} \oint [E - U_{\text{eff}}(r)]^{1/2}\, dR = h(v + \tfrac{1}{2}), \tag{4.14}$$

in which the path of integration is taken over one complete cycle of the motion. In the RKR procedure, E is the vibration-rotation energy of a particular level, $E = G_v + B_v J(J + 1)$; $U_{\text{eff}}(R)$ is the effective potential including the centrifugal term arising from rotation, $U_{\text{eff}}(R) = E(R) + \hbar^2 J(J + 1)/2\mu R^2 = E(R) + \kappa/R^2$; and the path of integration is over one vibration period, $R_{\min} \to R_{\max} \to R_{\min}$. In the JWKB method, one knows the effective potential

3. The initials stand for H. Jeffreys, G. Wentzel, H. A. Kramers, and L. Brillouin; occasionally, it is referred to simply as the WKB method.

for a system and uses equation (4.14) to find the energy levels; in the RKR method, one has found the energy levels as spectroscopic term values and wishes to invert the JWKB procedure to find the effective potential.

This inversion can be accomplished by the following procedure. Let us define

$$S(E,\kappa) = \frac{1}{(2\pi^2\mu)^{1/2}} \int_0^{I'} [E - U(I,\kappa)]^{1/2}\, dI,$$

where

$$U(I,\kappa) = \frac{\omega_e}{h} I - \frac{\omega_e x_e}{h^2} I^2 + \frac{\omega_e y_e}{h^3} I^3 + \cdots$$

$$+ \frac{1}{R_e^2}\kappa - \frac{\alpha_e}{h\omega_e B_e R_e^2} I\kappa - \frac{D_e}{B_e^2 R_e^4}\kappa^2 + \cdots$$

and the upper limit of the integral, I' is that value for which $E - U(I', \kappa) = 0$. If we further define $f = \partial S/\partial E$ and $g = -\partial S/\partial\kappa$, then the inner turning point of the potential we are seeking for a particular energy level, $R_{\min}(E)$, is given by

$$R_{\min} = \left[\frac{f}{g} + f^2\right]^{1/2} - f,$$

and the outer turning point $R_{\max}(E)$, by

$$R_{\max} = \left[\frac{f}{g} + f^2\right]^{1/2} + f.$$

The procedure is then, for each experimentally measured value of E, to carry through this calculation for the classical inner and outer turning points of the potential for that value and to tabulate a function $E(R)$ two points at a time. Since there are now rapid and convenient numerical procedures available (reference 9) for doing this, it would seem advisable to use the RKR potential whenever a Morse curve is found to be inadequate, rather than to attempt to construct some more complicated analytic function with a larger number of parameters.

6 Rotational and Vibrational Raman Spectra

In sections 1 and 2 of this chapter, it was pointed out that a molecule must possess a permanent dipole moment in order to show a pure rotational or rotation-vibration absorption spectrum. This requirement eliminates a large number of molecules about which we would like to have such information as internuclear distances and vibration frequencies—for example, H_2, N_2, O_2, the halogens, and so on. These molecules can, fortunately, be studied by means of inelastic light scattering, in which light of a given frequency interacts with a molecule to leave it in a different rotation and vibration state than the one in which it was originally found, the energy difference showing up as a frequency shift between the scattered light ν_{scatt} and the excitation frequency ν_{exc}, according to the relation

$$h\nu_{\text{scatt}} = h\nu_{\text{exc}} + \Delta E_{\text{rot-vib}},$$

where ΔE is positive (blue shift) if the molecule drops to a lower energy state, and negative (red shift) if the molecule is excited to a higher one. This phenomenon of inelastic light scattering is known as the *Raman effect*, after the Indian physicist Chandrasekhara Raman (1888–1970), who received a Nobel Prize for the discovery in 1930. Actually, the phenomenon itself was probably first recorded by R. W. Wood, who thought that the Raman lines were just smudges on his spectrographic plates.

Let us begin with a classical treatment of light scattering. Suppose we take a space-fixed axis system (xyz), with z-polarized light coming in from the x direction. A molecule with internal coordinates $(\xi\eta\zeta)$ is located at the origin of the (xyz) system; we let θ be the angle between ζ and z. Then

$$\mathscr{E}_\xi = -\mathscr{E}_z \sin\theta \qquad \text{and} \qquad \mathscr{E}_\zeta = \mathscr{E}_z \cos\theta,$$

where $\mathscr{E}_z$ is the amplitude of the electric field associated with the light wave. We are considering a molecule with no permanent dipole moment; however, an electric polarization is induced in the molecule by the applied electric field of magnitude

$$M_\xi = \alpha_\perp \mathscr{E}_\xi = -\alpha_\perp \mathscr{E}_z \sin\theta$$

and

$$M_\zeta = \alpha_\parallel \mathscr{E}_\zeta = \alpha_\parallel \mathscr{E}_z \cos\theta.$$

In these equations α is the polarizability of the molecule; the parallel component is along the internuclear axis, which is coincident with ζ, whereas the perpendicular component lies at right angles to this. Thus the polarizability of the molecule can be represented as an anisotropic ellipsoid, as in figure 4.9. In the (xyz) axis system, the components of the polarization are given by

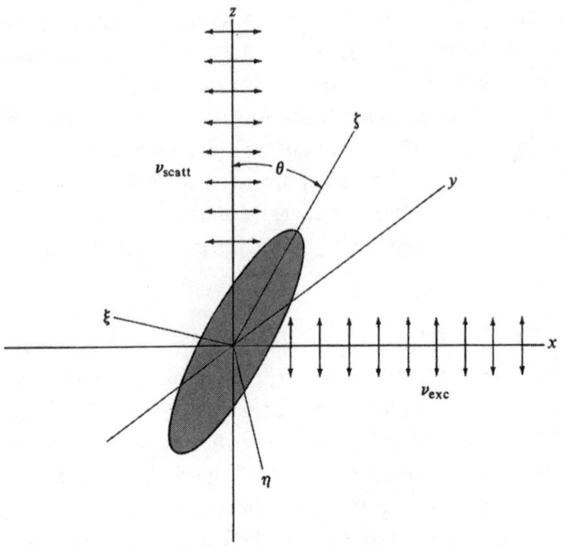

Figure 4.9
Classical model of Raman scattering. Light coming in from the x direction polarizes the molecule, here represented as an anisotropic ellipsoid. The time-varying polarization radiates in the x, y, and z directions.

$$M_z = -M_\xi \sin\theta + M_\zeta \cos\theta$$
$$= \alpha_\perp \mathscr{E}_z \sin^2\theta + \alpha_\parallel \mathscr{E}_z \cos^2\theta.$$

If the molecule is rotating, θ will be a function of time; specifically, $\theta(t) = 2\pi\nu_{\text{rot}}t$. Substituting this in, we have

$$M_z = \mathscr{E}_z\left\{\frac{\alpha_\perp[1 - \cos 2\pi(2\nu_{\text{rot}})t]}{2} + \frac{\alpha_\parallel[1 + \cos 2\pi(2\nu_{\text{rot}})t]}{2}\right\}$$

$$= \frac{(\alpha_\perp + \alpha_\parallel)}{2}\mathscr{E}_z + \frac{\alpha_\parallel - \alpha_\perp}{2}\cos 2\pi(2\nu_{\text{rot}})t\mathscr{E}_z.$$

Similarly,

$$M_x = \frac{\alpha_\parallel - \alpha_\perp}{2}\sin 2\pi(2\nu_{\text{rot}})t\mathscr{E}_z.$$

Now let us recall that the light wave is oscillating at its own particular frequency, so that

$$\mathscr{E}_z = \mathscr{E}_0 \cos 2\pi\nu_{\text{exc}} t.$$

Thus

$$M_z = A\mathscr{E}_0 \cos 2\pi\nu_{\text{exc}} t + \gamma\mathscr{E}_0 \cos 2\pi(2\nu_{\text{rot}})t \cos 2\pi\nu_{\text{exc}} t$$

and

$$M_x = \gamma\mathscr{E}_0 \cos 2\pi(2\nu_{\text{rot}})t \cos 2\pi\nu_{\text{exc}} t,$$

where $A = (\alpha_{\parallel} + \alpha_{\perp})/2$ is the isotropic part of the polarizability, and $\gamma = (\alpha_{\parallel} - \alpha_{\perp})/2$ is the anisotropic part. The radiation produced by the z component of the polarization will be propagated in the x and y directions, and, by the usual rule for multiplication of trigonometric functions, will have frequency components at ν_{exc} (elastic or *Rayleigh* scattering) and at $\nu_{\text{exc}} + 2\nu_{\text{rot}}$ and $\nu_{\text{exc}} - 2\nu_{\text{rot}}$; the latter are the Raman-scattered frequencies. Since the light intensity is proportional to the square of the associated electric field, we have

$$I_z(\nu_{\text{exc}}) \sim A^2 I_0$$

and

$$I_z(\nu_{\text{exc}} \pm 2\nu)_{\text{rot}} \sim \gamma^2 I_0.$$

Similarly,

$$I_{x,y}(\nu_{\text{exc}} \pm 2\nu_{\text{rot}}) \sim \gamma^2 I_0.$$

The fact that the scattered frequency is shifted by twice the molecular rotation frequency can be interpreted fairly straightforwardly in this classical picture: By virtue of the symmetry of the polarizability ellipsoid, the molecular polarization appears to be modulated at *twice* the actual rotation frequency.

The quantum mechanical theory of Raman scattering, which will be considered in much more detail in chapters 13.6 and 13.7, involves treating the interactions between the molecule and the radiation field to second order. The expression for the Raman-scattering probability involves terms of the form

$$\sum_v \frac{\langle f|\boldsymbol{\mu}\cdot\mathscr{E}|v\rangle\langle v|\boldsymbol{\mu}\cdot\mathscr{E}|i\rangle}{h\nu_{\text{exc}} - (E_v - E_f)},$$

where $\{v\}$ is a set of "virtual states," which appear in the perturbation expansion.

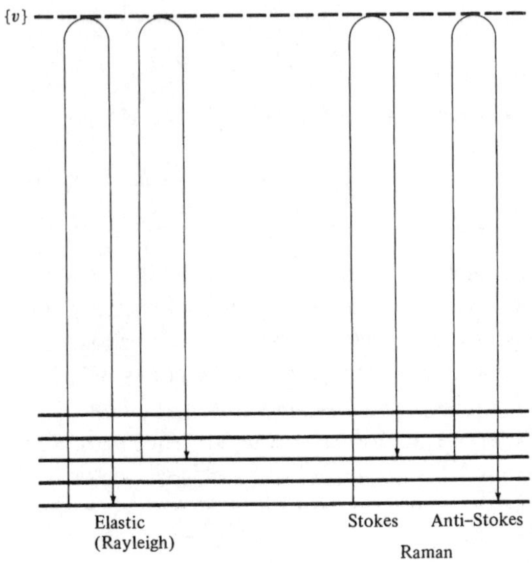

Figure 4.10
Quantum mechanical representation of Raman scattering and Rayleigh scattering.

This is the same form as appears in the perturbation treatment of the molecular polarizability, so we recover the classical result, namely, that the Raman scattering looks at the matrix elements of the molecular polarizability between initial and final molecular states, just as ordinary absorption and emission processes looked at the matrix elements of the electric dipole moment operator. The process can be visualized by figure 4.10, in which the transition is represented as taking place between two real vibration-rotation levels in the ground state of the molecule via one of a number of intermediate "virtual" levels. If the net transition is from a lower to a higher energy level, so that $\nu_{scatt} < \nu_{exc}$, the process is called *Stokes* scattering; the opposite case, in which $\nu_{scatt} > \nu_{exc}$, is termed *anti-Stokes* scattering.[4]

The selection rules for rotationally inelastic Raman scattering can be arrived at simply by considering the above expression for the scattering probability. If each matrix element of x has the usual selection rule $\Delta J = \pm 1$, and

4. These historical names derive from a terminology proposed by Sir George Stokes (1819–1903), who proposed the general rule that all luminescence takes place to longer wavelengths than the light used to excite it. As with most generalizations not based on a full understanding of the principles involved, many exceptions to this rule were soon found.

the scattering process can be conceived of as two electric dipole terms in succession, then the allowed transitions would be $\Delta J = 0$ or ± 2; $\Delta J = 0$, of course, corresponds to the purely elastic transition. Note that the $\Delta J = 2$ rule is just the quantum mechanical analogue of the classical result that the scattered wave is modulated at *twice* the classical rotation frequency. The rotational Raman spectrum, therefore, consists of a set of equally spaced lines on either side of the exciting line. The equation for the line positions can be derived simply by noting that the transition is from a level J to a level $J \pm 2$, so that the energy difference is

$$\begin{aligned} \Delta\nu_{\text{Raman}} &= B(J+2)(J+3) - BJ(J+1) \\ &= B(J^2 + 5J + 6 - J^2 - J) \\ &= B(4J+6) \\ &= 2B(2J+3), \end{aligned} \tag{4.15}$$

where $J = J_{\text{initial}}$ for the Stokes branch, and $J = J_{\text{final}}$ for the anti-Stokes branch.

Raman scattering can also involve transitions between vibrational levels. In this case, we must take into account the variation of the molecular polarizability with vibrational amplitudes, so that

$$\alpha = \alpha_0 + \left(\frac{\partial \alpha}{\partial r}\right)_{r=r_e} (r - r_e) + \cdots.$$

Classically, a vibrational Raman effect is possible because the internuclear separation $(r - r_e)$ is oscillating with a frequency $2\pi\nu_{\text{vib}}t$, which modulates the scattered light and produces new frequencies at $\nu_{\text{scatt}} = \nu_{\text{exc}} \pm \nu_{\text{vib}}$. Quantum mechanically, the selection rules for vibrational Raman scattering will be determined by the matrix elements of $(r - r_e)$, and the magnitude of the scattering by the quantity $(\partial\alpha/\partial r)$. Since all harmonic and near-harmonic oscillators possess matrix elements of $r - r_e$ off-diagonal by one vibrational quantum number, it is clear that all diatomic molecules, both homonuclear and heteronuclear, may display a vibrational Raman spectrum.

When the rotational selection rules are added on, we obtain three branches of the vibrational Raman spectrum. By analogy with the P, Q, and R branches of the infrared absorption spectrum, these are designated

O branch $(\Delta J = -2)$: $\nu_{\text{Raman}} = \nu_{\text{exc}} - \Delta G_{01} + 2B(2J - 1)$;

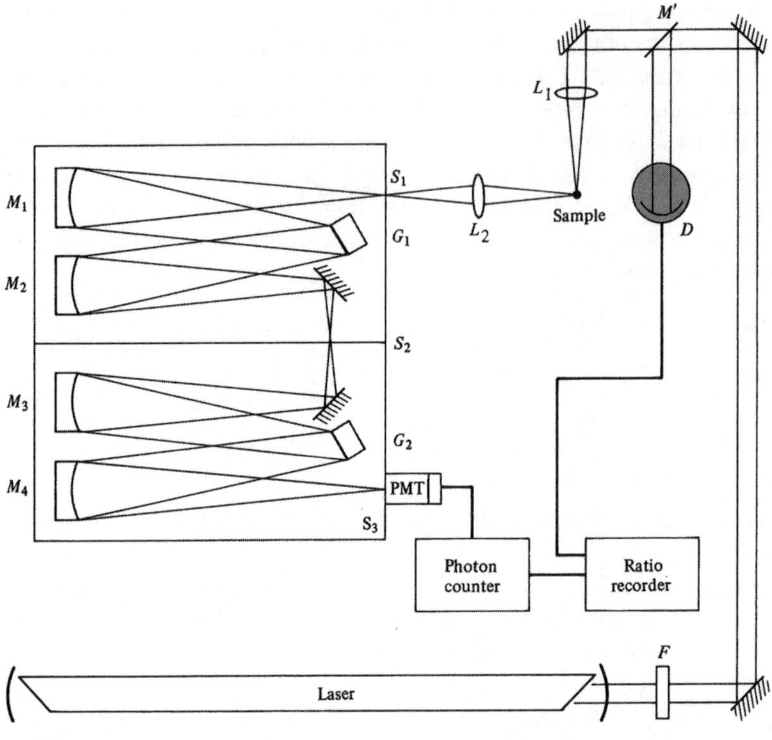

Figure 4.11
Schematic diagram of a laser Raman spectrometer: F, filter to eliminate extraneous radiation from the laser tube; M′, partially transmitting mirror that deflects a small portion of the laser radiation to a photocell D, which monitors the excitation intensity and corrects the scattered intensity for any variations; L_1 and L_2, focusing optics on the sample and on the monochromator entrance slit S_1, respectively; S_2, intermediate slit; S_3, exit slit; M_1–M_4, monochromator collimating optics; G_1 and G_2, dual gratings, which are driven in tandem to scan the spectrum.

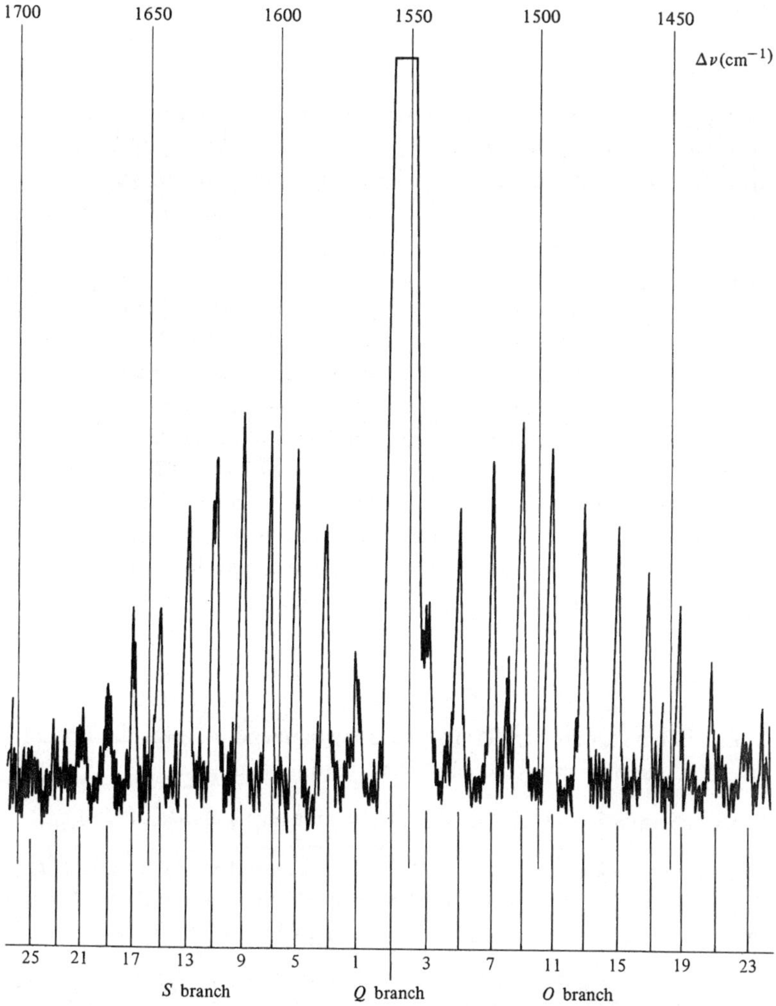

Figure 4.12
Vibration-rotation Raman spectrum of O_2 (Stokes band) in air at normal atmospheric pressure. A 0.5-W argon ion laser was used (see chapter 10.7.3), and the strongest *O* or *S* branch lines correspond to about 100 photon counts per second. Of course, the manufacturer of the instrument chose this spectrum to demonstrate its capabilities because in oxygen, in which all even-numbered rotational levels are missing because of nuclear spin statistics (see section 1), the spacing between lines is equal to $8B$, rather than the $4B$ given by equation (4.15).

Q branch ($\Delta J = 0$): $\nu_{\text{Raman}} = \nu_{\text{exc}} - \Delta G_{01}$;

S branch ($\Delta J = +2$): $\nu_{\text{Raman}} = \nu_{\text{exc}} - \Delta G_{01} - 2B(2J + 3)$.

Until recently, Raman spectroscopy was performed by a brute force technique of illuminating a large volume of material with a very large, high-current, low-pressure mercury arc. The development of the gas laser (see chapter 10) has revolutionized the technology of this kind of spectroscopy, and led to a real renaissance in the field. A typical laser-Raman apparatus is diagrammed in figure 4.11. The beam of a high-power gas laser (usually argon at 5,145 Å) is sent into the sample cell; because of the high collimation of the laser beam, the sample can be very small, so that even biological specimens can be examined. The scattered light is extracted at right angles and analyzed with a monochromator. A double monochromator, consisting of two complete focusing and dispersing elements in series, is often used to reduce the intense Rayleigh-scattered light to a level 10 orders of magnitude below its peak intensity. The Raman signals are counted, normalized by the laser intensity, and displayed. A spectrum obtainable in this way is shown in figure 4.12—the vibrational Raman spectrum of oxygen in normal air, showing complete resolution of the rotational branches.

Problems

1. Develop a formula similar to equations (4.6) and (4.7) for the Q branch lines of a diatomic molecule vibration-rotation band.

2. Extend equation (4.12) for $G_{v+1/2}$ to include terms in $\omega_e y_e$ and $\omega_e z_e$.

3. Using the relations between D_e, ω_e, and $\omega_e x_e$ for a Morse oscillator (section 5), find an expression for the number of bound vibrational levels in terms of ω_e and $\omega_e x_e$. Does this expression hold for $I_2^*(B^3\Pi_{0u}^+)$?

4. Gaseous HCl is normally a 3:1 mixture of $H^{35}Cl$ and $H^{37}Cl$. To a high approximation, the rotational energy levels of such diatomic rotators are

$$E(J)(\text{in cm}^{-1}) = J(J+1)\bar{B} - J^2(J+1)^2\bar{D},$$

where $\bar{B}$, the rotational constant, is larger by a factor 1.0015 for $H^{35}Cl$ than for $H^{37}Cl$, and the centrifugal distortion constant $\bar{D}$ is the same for both molecules within the error of its measurement.

(a) Derive an expression for the separation of the pure rotational absorption lines of $H^{35}Cl$ and $H^{37}Cl$ as a function of J', the J value for the upper state ($\Delta J = J' - J'' = +1$).
(b) What is the spacing in cm^{-1} of the two lines for which $J' = 10$?

5. The "transition moment," or the probability of transition, between two rotational levels in a linear molecule may be assumed to depend only on the permanent electric moment of the molecule and thus to be the same for all allowed pure rotational transitions. In the pure rotational *emission spectrum* of $H^{35}Cl$ gas, lines at 106.0 and 233.2 cm^{-1} are observed to have equal intensities. What is the temperature of the gas? The rotational constant B for $H^{35}Cl$ is known to be 10.6 cm^{-1}, and the ratio hc/k has the value 1.44 cm deg.

6. The harmonic oscillator wave functions for the levels $v = 0$ and $v = 1$ are

$$\psi_0 = \left(\frac{a}{\pi}\right)^{1/4} \exp\left(\frac{-a(r - r_e)^2}{2}\right),$$

$$\psi_1 = \left(\frac{4a^3}{\pi}\right)^{1/4} \Delta r \exp\left(\frac{-a(r - r_e)^2}{2}\right),$$

in which a is a constant. These functions are orthogonal and normalized to unity.

(a) Show that ψ_0 and ψ_1 are orthogonal.
(b) Calculate the average value of $(r - r_e)^{-2}$ for the state $v = 1$.

7. What would happen to the Birge-Sponer extrapolation scheme for a molecular potential correlating with ionic states of the separated atoms?

8. The energy levels of a rigid linear rotator are $E(J)$ (in cm^{-1}) $= BJ(J + 1)$. If a linear molecule is symmetric and without nuclear spin, half of the energy levels are missing. In the ground electronic and vibrational state, the odd J levels are missing in CO_2, whereas the even J levels are missing in O_2. Calculate the ratio of the separation $\Delta\nu_1$ in cm^{-1} between the *first pure rotational Raman line* (say, the first Stokes line) and the exciting line, to the separation $\Delta\nu_2$ between the second pure rotational Raman line (second Stokes line) and the exciting line for

(a) a molecule with all rotational levels present,
(b) a molecule like CO_2 (odd Js missing),
(c) a molecule like O_2 (even Js missing).

References

1. G. Herzberg, *Molecular Spectra and Molecular Structure*, Vol. 1: *Spectra of Diatomic Molecules* (Van Nostrand Reinhold, New York, 1950), p. 108.

2. R. T. Birge and H. Sponer, *Phys. Rev.* **28**, 259 (1926).

3. R. J. LeRoy, *J. Chem. Phys.* **52**, 2678, 2683 (1970); R. J. LeRoy and R. B. Bernstein, *J. Chem. Phys.* **52**, 3869 (1970).

4. R. J. LeRoy and R. B. Bernstein, *J. Mol. Spectroscopy* **37**, 109 (1971).

5. R. J. LeRoy, "Applications of Bohr Quantization in Diatomic Molecule Spectroscopy," in *Semiclassical Methods in Molecular Scattering and Spectroscopy*, M. S. Child, ed. (D. Reidel, Dordrecht, 1980), pp. 109–126.

6. J. W. Tromp and R. J. LeRoy, *Can. J. Phys.* **60**, 26 (1982).

7. P. M. Morse, *Phys. Rev.* **34**, 57 (1929).

8. J. T. Vanderslice, E. A. Mason, W. G. Maisch, and E. R. Lippincott, *J. Mol. Spectroscopy* **3**, 17 (1959).

9. R. N. Zare, *J. Chem. Phys.* **40**, 1934 (1964); R. N. Zare, "Programs for Calculating Relative Intensities in the Vibrational Structure of Electronic Band Systems," University of California Radiation Laboratory Report UCRL-10925 (1963).

5 Electronic Spectra of Diatomic Molecules

1 "Ro-vibronic" Energy Levels

A transition between two different electronic states of a diatomic molecule generally involves simultaneous changes in vibrational energy and rotational energy. From the analysis of such transitions, the equilibrium internuclear distances and dissociation energies of the excited molecular states can be determined. In this chapter, we shall see how such an analysis may be carry out.

For each molecular state, the spectroscopic term value contains electronic, vibrational, and rotational contributions:

$$T_e + G_v + F_v(J).$$

The electronic term T_e measures the energy of the minimum of the potential curve for a particular state above the minimum of the ground-state curve; for the ground state itself, of course, $T_e = 0$. The vibrational terms G_v are given by expressions of the type of equation (4.11); the exact form of the rotational terms $F_v(J)$ depends on the electronic quantum numbers of the particular state, and will be discussed in more detail in section 5. The actual spectrum consists of a large number of lines, grouped into vibrational bands, with frequency given by

$$h\nu = (T_e' - T_e'') + G_{v'} + F_{v'}(J') - G_{v''} - F_{v''}(J''), \tag{5.1}$$

where the single prime refers to the upper state and the double prime, to the lower state; thus $T_e'' = 0$ if the lower state is the ground state. We shall first consider the overall vibrational structure of the transition, as may be seen with a low-resolution spectrometer, and then go on to consider the rotational fine structure of the bands.

2 The Franck-Condon Principle

The first question we wish to consider is which vibrational levels are to be coupled in an electronic transition. In an infrared vibrational transition within a single electronic state, we had a fairly strict selection rule, $\Delta v = \pm 1$, even for anharmonic potentials. This arose because the matrix elements of $R - R_e$ were nonvanishing only along the diagonals just above and below the main diagonal, with very small terms elsewhere in the $\langle v|R - R_e|v'\rangle$ matrix. There is no reason to expect such a rule to be operative when the two vibrational levels involved belong to completely different molecular potential curves.

To derive the selection rules for vibronic transitions, we must go back to the Born-Oppenheimer separation of the molecular wave functions (chapter 1.3). This separation had us write the wave function as

$$\Psi = \psi_e(\mathbf{r}, R)\psi_{vib}(R),$$

where the internuclear separation R was a fixed parameter in the electronic wave function. The intensity of an electric dipole transition will, as usual, be proportional to the square of the matrix element between initial state $\langle i|$ and final state $|f\rangle$,

$$\mu_{if} = \int \Psi^{*\prime}\boldsymbol{\mu}\Psi'' \, d\mathbf{r}\, dR$$

where

$$\boldsymbol{\mu} = \boldsymbol{\mu}_e + \boldsymbol{\mu}_{nuc} = -\sum_i e\mathbf{r}_i + \sum_\alpha eZ_\alpha \mathbf{R}_\alpha.$$

Let us proceed to evaluate μ_{if}:

$$\mu_{if} = \int \psi_{e'}^* \psi_{vib'}^* (\boldsymbol{\mu}_e + \boldsymbol{\mu}_{nuc}) \psi_{e''} \psi_{vib''} \, d\mathbf{r}\, dR$$

$$= \int \left(\int \psi_{e'}^* \boldsymbol{\mu}_e \psi_{e''} \, d\mathbf{r} \right) \psi_{vib'}^* \psi_{vib''} \, dR + \int \psi_{e'}^* \psi_{e''} \, d\mathbf{r} \int \psi_{vib'}^* \boldsymbol{\mu}_{nuc} \psi_{vib''} \, dR.$$

The overlap integral of the two electronic wave functions in the second term of this expression will vanish because these two functions are members of the same complete, orthonormal set. If we define the electronic transition moment

$$M_e(R) = \int \psi_{e'}^* \boldsymbol{\mu}_e \psi_{e''} \, d\mathbf{r},$$

then the overall transition moment is given by the average of $M_e(R)$ over the vibrational wave functions in the upper and lower states,

$$\mu_{if} = \int M_e(R) \psi_{vib'}^* \psi_{vib''} \, dR.$$

If we assume that the R dependence of M_e is small, then it can be taken out of the integral, and we obtain

$$\mu_{if} = \overline{M_e(R)} \int \psi^*_{\text{vib}'} \psi_{\text{vib}''} \, dR. \tag{5.2}$$

Thus the relative intensity of a transition between any two vibrational states is given by the square of the vibrational overlap integral,

$$\left| \int \psi^*_{\text{vib}'} \psi_{\text{vib}''} \, dR \right|^2 = |\langle v' | v'' \rangle|^2,$$

which is known as the *Franck-Condon factor*, $q_{v'v''}$, after J. Franck and E. U. Condon, who first developed this principle (reference 1). Mathematically, the Franck-Condon factor is just the overlap between the vibrational wave functions appropriate to the upper and lower states of the electronic transition. A sum rule for the Franck-Condon factors is easily derived:

$$\sum_{v''} q_{v'v''} = \sum_{v''} |\langle v' | v'' \rangle|^2$$
$$= \sum_{v''} \langle v' | v'' \rangle \langle v'' | v' \rangle.$$

Using the completeness relation [equation (1.4)], we can eliminate the sum over $|v''\rangle \langle v''|$ and obtain

$$\sum_{v''} q_{v'v''} = \langle v' | v' \rangle = 1.$$

The physical interpretation of the Franck-Condon factor goes back to the original basis of the Born-Oppenheimer separation, which is that the nuclei are moving much more slowly than the electrons. In the "time" required for an electronic transition to occur [an ill-defined concept, but for the purposes of this argument, something on the order of $h/(E_{\text{upper}} - E_{\text{lower}}) \sim 10^{-14}$ sec or so], the nuclei are very nearly stationary. Thus, on the potential curve in figure 5.1, the molecule must execute a "vertical" transition, so that it finds itself in an excited electronic state, but with the same internuclear separation as it had in the ground electronic state. Thus the only regions of the excited state potential that are accessible in the transition are those for which the vibrational wave function of the ground state has a finite density. Furthermore, if the vibrational wave functions have several nodes, there will be interference between the two, leading to irregular variations in the Franck-Condon factors. An analogous argument holds for emission spectra: The

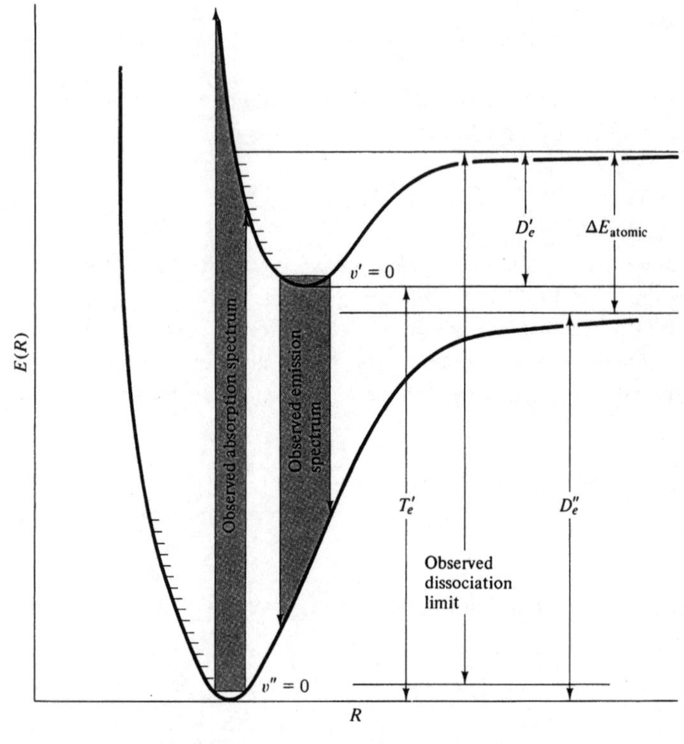

Figure 5.1
"Vertical transitions" permitted by the Franck-Condon principle between two electronic states. Also shown schematically is the relation among the dissociation energy of the ground state D_e'', that of the excited state, D_e', the electronic term T_e', the optical dissociation limit, and the difference between separated atom term values, ΔE_{atomic}.

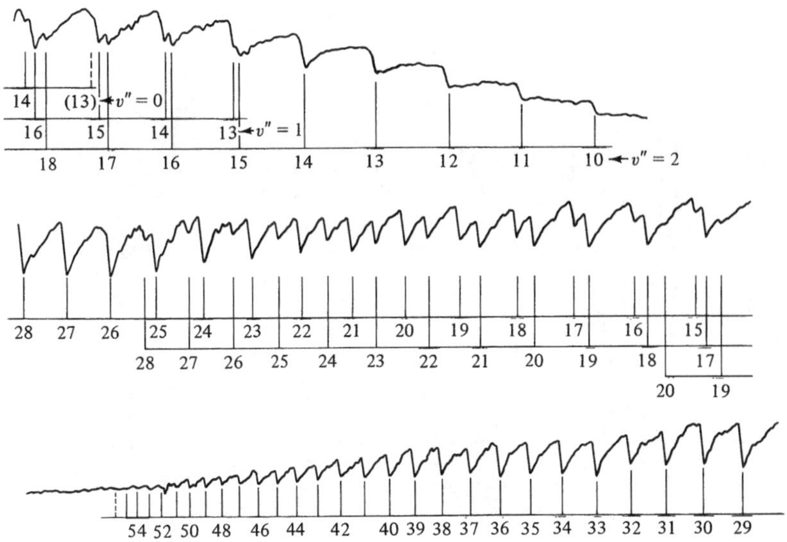

Figure 5.2
A very-low-resolution visible absorption spectrum of molecular iodine vapor, recorded on a single-beam Heath monochromator. Long progressions arising from $v'' = 0$, 1, and 2 in the ground state can be seen nearly to the dissociation limit at 4,995 Å, but the 0–0 band is not observable due to its very small Franck-Condon factor.

only regions of the lower state that will be accessible are those for which the vibrational wave function of the upper state, from which the molecule is radiating, has a finite density.

Let us look at a few examples of the Franck-Condon principle in operation. If there is a very small change in the equilibrium internuclear distance in going from the ground to the excited state, then only a small number of bands will be seen in absorption—typically, only the 0–0 band will have appreciable intensity. If, on the other band, there is a large difference in R_e between the two states, then a long progression in v' will be observed. An instance of this is the $B^3\Pi_{0u}^+ \leftarrow X^1\Sigma_{0g}^+$ transition in I_2, which is shown in figure 5.2. In this transition, R_e goes from 2.665 Å in the ground state to 3.0276 Å in the excited state, and about 30 members of the $v'' = 0$ progression are observable. This series of bands is also overlapped by additional progressions arising from the $v'' = 1$ and 2 levels, which are thermally populated at room temperature. The same principle operates in emission; figure 5.3 shows part

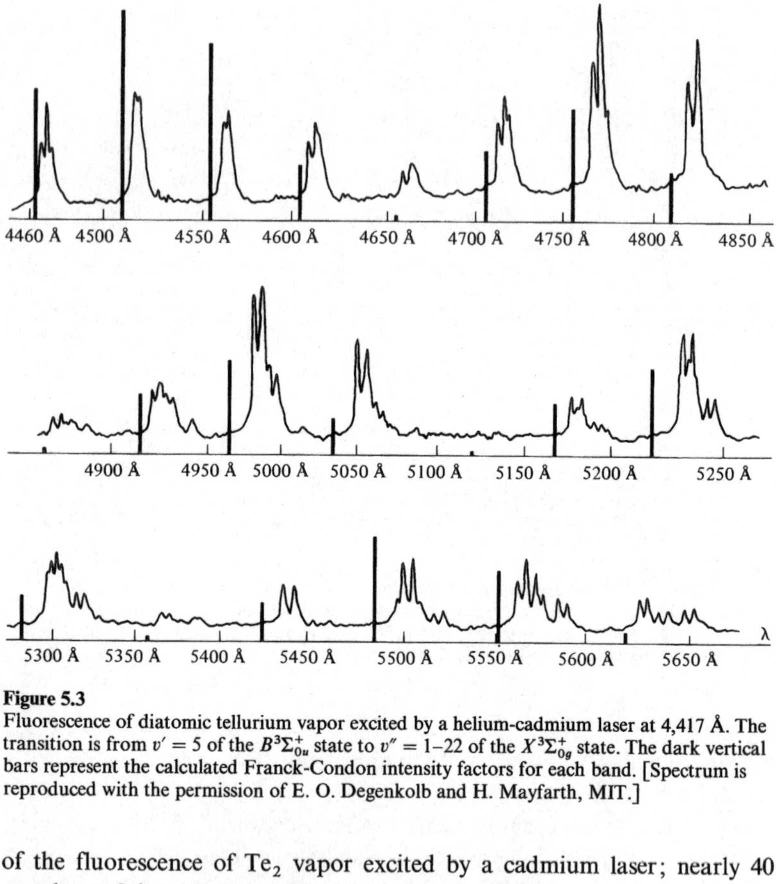

Figure 5.3
Fluorescence of diatomic tellurium vapor excited by a helium-cadmium laser at 4,417 Å. The transition is from $v' = 5$ of the $B^3\Sigma^+_{0u}$ state to $v'' = 1$–22 of the $X^3\Sigma^+_{0g}$ state. The dark vertical bars represent the calculated Franck-Condon intensity factors for each band. [Spectrum is reproduced with the permission of E. O. Degenkolb and H. Mayfarth, MIT.]

of the fluorescence of Te_2 vapor excited by a cadmium laser; nearly 40 members of the $v' = 5$ progression can be observed.

3 Vibrational Band Analysis

The first task in interpreting an electronic spectrum is the assignment of the upper and lower vibrational quantum numbers for each band. If we neglect the rotational energy differences in equation (5.1), then the frequencies of each band are given by

$$h\nu(v', v'') = (T'_e - T''_e) + (G'_v - G''_v). \tag{5.3}$$

Very often, a spectrum will consist of a large number of bands in no clear

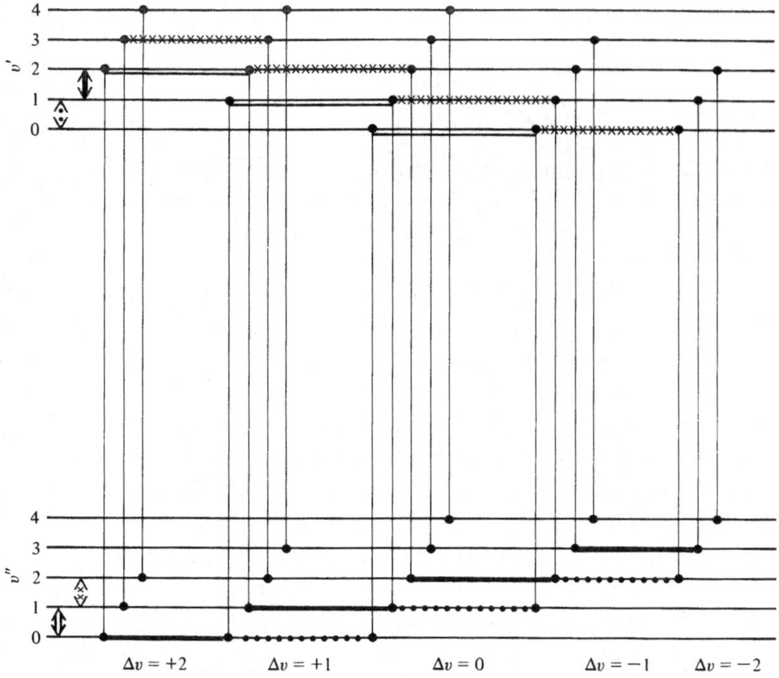

Figure 5.4
Basis for the Deslandres table analysis scheme. The bands are arranged in *sequences* having the same v. The energy differences between pairs of bands connected by similarly marked lines are all equal, and equal to the correspondingly marked vibrational spacings. That is, pairs connected by solid lines measure Δ'_{12}; pairs connected by dotted lines measure Δ'_{01}; pairs connected by double lines measure Δ''_{01}; and pairs connected by crosses measure Δ''_{12}.

order. In such a case, a scheme of analysis known as the *Deslandres table* proves very useful. The basis of this scheme is that the frequency of each band represents a difference between specific energy terms in the upper and lower states. The energy difference between bands originating from some given level and terminating on two different levels must be characteristic of the particular pair of levels in common, regardless of the third level involved. Thus, when all the bands are correctly assigned, the differences between appropriate pairs must be equal to within the accuracy of measurement. This is illustrated in figure 5.4. If a pair of numbers (i,j) represents the upper and lower vibrational quantum numbers, then for the ground-state combination differences

$$\Delta''_{jl} = (i,j) - (i,l) = (m,j) - (m,l) = \cdots$$

and for the upper-state combination differences

$$\Delta'_{ik} = (i,j) - (k,j) = (i,n) - (k,n) = \cdots.$$

Thus if the band positions are correctly placed in a table arranged by v' and v'' values, the columns formed by energy differences between successive pairs in a row will all be equal, and the rows formed by successive pairs in a column will all be equal.

As an example, we give the Deslandres table (table 5.1) for the $A\,^2\Pi$–$X\,^2\Sigma^+$ "red" system of CN. This is an extensive band system observed between 450 and 1,500 nm in discharges and also in the solar atmosphere. The entries for each band are the observed band head, in vacuum cm^{-1}, and immediately below, in italics, the Franck-Condon factor for that band. Between each pair of columns and rows are the calculated first differences, Δ''_{jl} and Δ'_{ik}, respectively. Notice that when the bands are correctly assigned, as they are here, the corresponding differences are equal, within experimental error. (The precision of these data is about $\pm 1\ \text{cm}^{-1}$, with the weaker and longer-wavelength bands having the largest errors. Much more precise observational data on this system are now available: see reference 3.) Since these differences correspond directly to the $\Delta G_{v+1/2}$ values ($\Delta'_{12} = \Delta G'_{1.5}$, $\Delta''_{23} = \Delta G''_{2.5}$, and so on), they can be used directly for a least-squares fitting of the spectroscopic constants in each electronic state. For this band system, the results are (all in cm^{-1}):

$$A\,^2\Pi: \quad T'_e = 9118.0, \qquad \omega'_e = 1812.55, \qquad \omega_e x'_e = 12.61;$$

$$X\,^2\Sigma: \qquad \omega''_e = 2068.74, \qquad \omega_e x''_e = 13.13.$$

At this point, readers should make sure that they understand how these values have been obtained from the data presented in the table, and then go on to do problem 3. We shall return to this band system as an example in subsequent sections.

Isotopic Shifts of Vibrational Bands When the array of bands forms a well-defined block, as in the example given previously, the upper left-hand corner entry in the table can be unambiguously assigned as the 0–0 band. When the Franck-Condon factors are unfavorable, however, this part of the array will be missing (as in the case of I_2: refer to figure 5.2), and absolute vibrational numbering can be quite difficult. In such a case, the use of iso-

Table 5.1
Deslandres table for CN $A^2\Pi - X^2\Sigma^+$ band system

v' \ $v'' = 0$	0	Δ''_{01}	1	Δ''_{12}	2	Δ''_{23}	3	Δ''_{34}	4	Δ''_{45}	5
0	9,150.63	2,047.27	7,103.36								
	4.99(−1)[a]		*3.71(−1)*								
Δ'_{01}	1,786.62		1,785.44								
1	10,937.25	2,048.45	8,888.8	2,015.5	6,873.3						
	3.20(−1)		*4.56(−2)*		*3.50(−1)*						
Δ'_{12}	1,761.48		1,768.01								
2	12,698.73	2,041.92	10,656.81		—		6,642.7				
	1.26(−1)		*2.40(−1)*		*1.22(−2)*		*2.10(−1)*				
Δ'_{23}	1,736.01		1,735.84								
3	14,434.74	2,045.09	12,392.65		—						
	3.99(−2)		*1.95(−1)*		*9.89(−2)*						
Δ'_{34}	1,710.43		1,710.44								
4	16,145.17	2,042.08	14,103.09	2,015.74	12,087.35						
	1.12(−2)		*9.47(−2)*		*1.82(−1)*						
Δ'_{45}	1,684.95		1,684.89		1,684.84						
5	17,830.12	2,042.14	15,787.98	2,015.79	13,772.19	1,989.44	11,782.75				
	2.97(−3)		*3.62(−2)*		*1.34(−1)*		*1.23(−1)*				
Δ'_{56}	1,659.07		1,659.46		1,659.27		1,659.25				
6	19,489.19	2,041.75	17,447.44	2,015.98	15,431.46	1,989.46	13,442.00	1,963.00	11,479.00		
	7.58(−4)		*1.21(−2)*		*6.79(−2)*		*1.42(−1)*		*5.79(−2)*		
Δ'_{67}	1,631.81		1,633.91		1,633.99		1,633.86		1,633.73		
7	21,121.0	2,039.65	19,081.35	2,015.90	17,065.45	1,989.59	15,075.86	1,963.13	13,112.73		
	1.90(−4)		*3.75(−3)*		*2.82(−2)*		*9.55(−2)*		*1.22(−1)*		
Δ'_{78}			1,606.61		1,606.67		1,606.71		1,606.67		
8			20,687.96	2,015.84	18,672.12	1,989.55	16,682.57	1,963.17	14,719.40	1,936.68	12,782.72
			1.11(−3)		*1.04(−2)*		*4.88(−2)*		*1.10(−1)*		*8.40(−2)*
Δ'_{89}			1,580.44		1,582.98		1,582.93		1,582.92		1,582.79
9			22,268.4	2,013.3	20,255.10	1,989.60	18,265.50	1,963.18	16,302.32	1,936.81	14,365.51
			3.17(−4)		*3.57(−3)*		*2.13(−2)*		*6.92(−2)*		*1.08(−1)*
$\Delta'_{9,10}$							1,556.63		1,556.56		1,556.47
10							19,822.13	1,963.25	17,858.88	1,936.90	15,921.98
							8.41(−3)		*3.55(−2)*		*8.41(−2)*

a. Franck-Condon factors are given in modified exponential notation; for example, $4.99(-1) = 4.99 \times 10^{-1}$. Data are from S. N. Suchard, *Spectroscopic Data*, Vol. 1, Part A (IFI/Plenum, New York, 1975), pp. 285–290.

topic substitution can be very helpful because this shifts the band positions in a way that is dependent on the absolute value of the vibrational quantum number. Since the vibration frequency of an oscillator is (see chapter 4.2)

$$\nu_{\text{osc}} = \left(\frac{1}{2\pi}\right)\sqrt{\frac{k}{\mu}},$$

the frequency of an oscillator with the same force constant but different mass is just

$$\frac{\tilde{\nu}_{\text{osc}}}{\nu_{\text{osc}}} = \sqrt{\frac{\mu}{\tilde{\mu}}} = \rho, \qquad (5.4)$$

where the quantities with a tilde refer to the isotopically varied molecule, and ρ is the "isotopic ratio."

The result of a rather lengthy calculation shows that the term values for the isotopically substituted molecule are given by

$$\tilde{G}_v = \rho\omega_e(v + \tfrac{1}{2}) - \rho^2\omega_e x_e(v + \tfrac{1}{2})^2 + \rho^3\omega_e y_e(v + \tfrac{1}{2})^3 + \cdots.$$

Thus, since $h\nu = T_e' + G_v' - G_v''$,

$$h\nu = T_e' + \rho\omega_e'(v' + \tfrac{1}{2}) - \rho^2\omega_e x_e'(v' + \tfrac{1}{2})^2 + \rho^3\omega_e y_e'(v' + \tfrac{1}{2})^3 + \cdots$$
$$- [\rho\omega_e''(v'' + \tfrac{1}{2}) - \rho^2\omega_e x_e''(v'' + \tfrac{1}{2})^2 + \rho^3\omega_e y_e''(v'' + \tfrac{1}{2})^3 + \cdots],$$

so that

$$\begin{aligned} h\Delta\nu_{\text{isotopic}}(v', v'') = {} & (1 - \rho)[\omega_e'(v' + \tfrac{1}{2}) - \omega_e''(v'' + \tfrac{1}{2})] \\ & - (1 - \rho^2)[\omega_e x_e'(v' + \tfrac{1}{2})^2 - \omega_e x_e''(v'' + \tfrac{1}{2})^2] \\ & + (1 - \rho^3)[\omega_e y_e'(v' + \tfrac{1}{2})^3 - \omega_e y_e''(v'' + \tfrac{1}{2})^3] + \cdots. \end{aligned} \qquad (5.5)$$

Usually, we know the quantum numbers of the lower state; in an absorption spectrum, the intensities of the bands will reflect the thermal equilibrium populations. The assignment is then just a matter of fitting values of v' to the isotopic shifts in equation (5.5).

An example of the application of this method is the iodine absorption spectrum shown in figure 5.2. For many years, the vibrational numbering of the upper B state had been ambiguous. It was finally concluded that the traditional numbering was too high by one unit, on the basis of matching calculated Franck-Condon factors to observed resonance fluorescence intensities (reference 4); almost simultaneously, this reassignment was proved correct by measurements of the absorption spectrum of a different isotope

Table 5.2
Observed and calculated shifts of vibrational band origins as between $^{127}I_2$ and $^{129}I_2$[a]

Band (v'–v'')	$\Delta\nu$ calculated from equation (5.5): With v' increased by 1 unit	$\Delta\nu$ calculated from equation (5.5): With assignment as given	$\Delta\nu$ observed (cm^{-1})
13–1	9.03	8.41	8.42 ± 0.05
14–1	9.60	9.02	9.00
16–2	9.02	8.50	8.43
9–3	3.17	2.45	2.47

a. Data from R. L. Brown and T. C. James, *J. Chem. Phys.* **42**, 33 (1965).

of iodine, with the results shown in table 5.2 (reference 5). Isotope shifts also appear in the rotational line structure of a band. An example of this has been given in problem 4 of chapter 4; complete formulas may be found in Herzberg (reference 6).

Determination of Dissociation Energies: The Case of H_2 Let us consider briefly the question of determining bond dissociation energies from spectra. As discussed previously, the Birge-Sponer extrapolation procedure (chapter 4.5) is only an approximate method, based as it is on the assumption of a single linear anharmonicity. A much more accurate procedure is based on the energy relation indicated in figure 5.1. If the *dissociation limit*, that is, the position in the spectrum at which sharp lines are no longer observed and continuous absorption begins to appear, can be measured accurately, and ΔE_{atomic} determined from atomic spectra, then D''_e is found from the relation

$$h\nu(\text{dissociation limit}) + G''_{0.5} = D''_e + \Delta E_{\text{atomic}}.$$

One system for which there has been considerable interest in finding D''_e by this method is molecular hydrogen, for which there is some hope of carrying out a quantum mechanical calculation of the dissociation energy with high accuracy. The determination was carried out by Herzberg and Monfils in 1960 (reference 7), who found a value of 36,113.6 cm^{-1} for D''_e of H_2. A calculation by Kołos and Wolniewicz (reference 8), however, gave a value of 36,118 cm^{-1}. This is actually a very serious discrepancy, small though it may appear. The calculations were carried out using a variational method (reference 9), for which there exists a theorem that the energy eigenvalue obtained by this method can never be lower than the true eigenvalue, but must always lie above it. The calculation, however, gave a binding energy for H_2 that was apparently deeper than that measured. There was no ambiguity possible in ΔE_{atomic} because the energy of any state of the hydrogen atom can be calculated exactly. Thus either the experimental determination

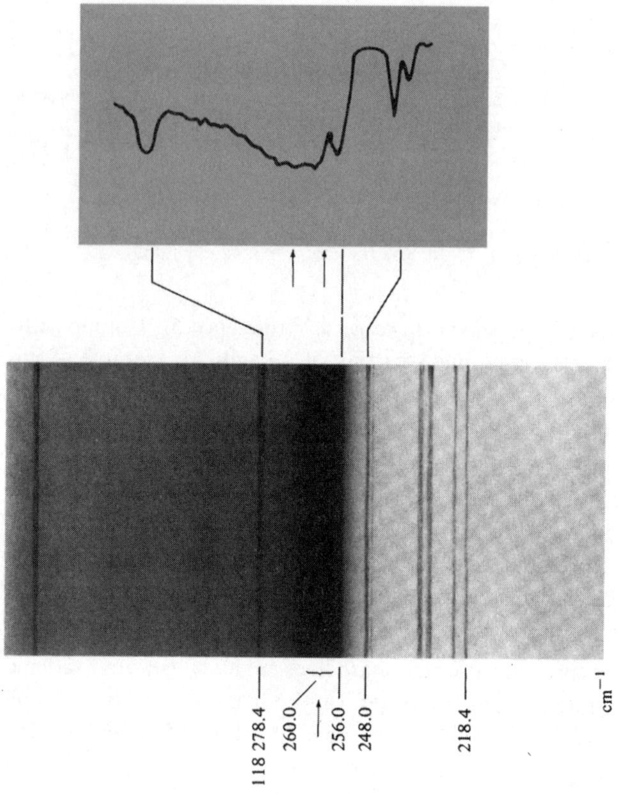

Figure 5.5
The $J'' = 1$ dissociation limit of H_2. Bottom panel: The photographic spectrum. Top panel: A photometer enlargement of a portion of the plate. The actual dissociation edge is between the two arrows of the enlargement; it is obscured by a diffuse line at 118,260 cm^{-1}. [Spectrum by G. Herzberg, *J. Mol. Spectroscopy* **33**, 151 (1970). Reproduced with permission.]

was wrong or a fundamental theorem of quantum mechanics was in question. Upon redoing the experiment, Herzberg indeed found (reference 10) that the dissociation limit was as predicted, that is, between 36,116.3 and 36,118.3 cm^{-1}. The difficulty in measuring the limit accurately is evident from figure 5.5. The actual limit, which is the boundary between the light and dark regions on the spectrum plate, is overlapped by other absorption lines, making its exact position hard to determine. This is perhaps the only case in which a calculation has proved capable of giving more accurate answers than high-resolution spectroscopy.

4 Rotational Fine Structure of Bands

When examined under high resolution, each of the bands making up the vibrational structure of an electronic spectrum is seen to be composed of a large number of discrete lines. These lines are due to the rotational energy differences $F_{v'}(J') - F_{v''}(J'')$ in equation (5.1), which we have neglected so far. When we have discussed the rotational structure, we shall have a nearly complete description of diatomic molecular spectra; spin and hyperfine splittings still remain to be treated.

We shall begin by considering the simplest possible case, that of a molecule in a $^1\Sigma$ (or $\Omega = 0$) state, so that there is no angular momentum other than rotation. For such a case, the rotational energy term is given by an expression of the form of equation (4.9),

$$F_v(J) = [B_e - \alpha_e(v + \tfrac{1}{2}) + \gamma_e(v + \tfrac{1}{2})^2 + \delta_e(v + \tfrac{1}{2})^3]J(J + 1)$$
$$- [D_e - \beta_e(v + \tfrac{1}{2})][J(J + 1)]^2 + \cdots,$$

which includes the vibration-rotation interaction effects. The important feature of this case is that there is only a single rotational quantum number J to worry about.

The customary electric dipole selection rules on J apply, namely, $\Delta J = 0$ or ± 1. This gives rise to three *branches* in each band:

Branch	J'	Change of J in absorption	Change of J in emission
P	$J'' - 1$	Decrease by 1	Increase by 1
Q	J''	No change	No change
R	$J'' + 1$	Increase by 1	Decrease by 1

Note that the P, Q, R notation applies to the line positions themselves; whether the rotational angular momentum of the molecule actually increases or decreases depends on whether the line is observed in emission or absorption.

From equation (5.1) for the line frequencies, the position of each line in each branch of each band is given by

$$\nu_P(J) = \nu_0 - (B' + B'')J + (B' - B'')J^2,$$

$$\nu_Q(J) = \nu_0 + (B' - B'')J + (B' - B'')J^2,$$

$$\nu_R(J) = \nu_0 + 2B' + (3B' - B'')J + (B' - B'')J^2.$$

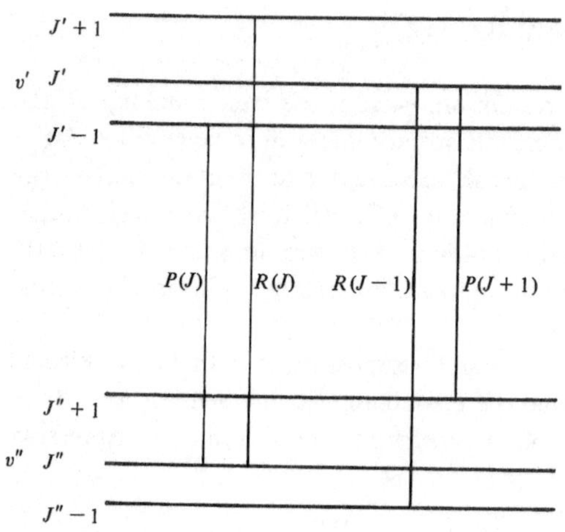

Figure 5.6
Rotational lines for obtaining combination differences.

ν_0 is termed the *band origin*. Actually, for transitions in which $\Omega = 0$ in both the upper and the lower states, the Q branch is absent, and only two branches appear in each band.

The rotational analysis of a band proceeds along lines similar to the Deslandres table scheme for vibrational analysis. That is, *combination differences* are sought between those pairs of lines that will measure the separation of particular rotational sublevels. With reference to figure 5.6, the excited state combination differences are

$$\begin{aligned}\Delta_2 F'(J) &= R(J) - P(J)\\ &= B'(J+1)(J+2) - B'(J-1)J\\ &= B'(J^2 + 3J + 2 - J^2 + J)\\ &= B'(4J+2) = 4B'(J+\tfrac{1}{2}),\end{aligned} \tag{5.6a}$$

and the ground-state combination differences are

$$\begin{aligned}\Delta_2 F''(J) &= R(J-1) - P(J+1)\\ &= B''(J+1)(J+2) - B''(J-1)J\\ &= 4B''(J+\tfrac{1}{2}).\end{aligned} \tag{5.6b}$$

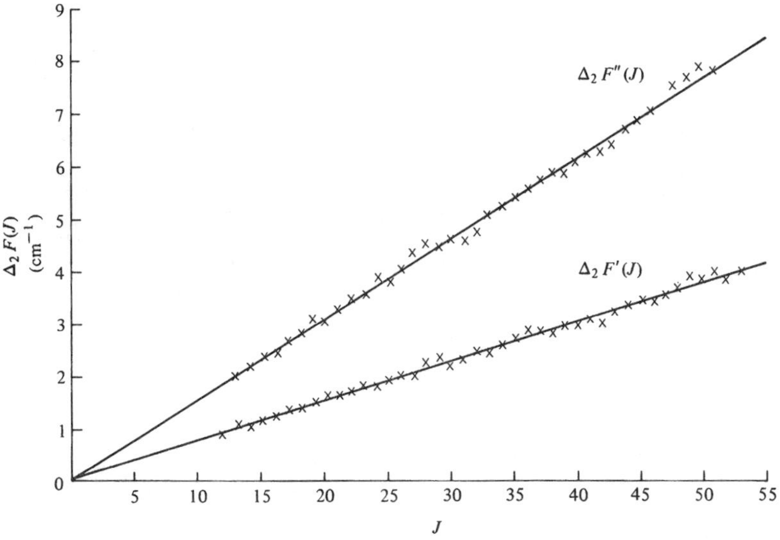

Figure 5.7
$\Delta_2 F(J)$ curves for upper and lower states of the 43–0 band of the iodine $B^3\Pi^+_{0u} \leftarrow X^1\Sigma^+_g$ absorption spectrum. [From J. D. Campbell, S. B. Thesis, MIT (1967). Reproduced with permission.]

Thus, if one plots $\Delta_2 F(J)$ against a set of integers, a straight line of slope B will be obtained, with a zero intercept at $J = -\frac{1}{2}$. This last point is useful in establishing the absolute rotational numbering of lines in a band; very often, as in the example shown in figure 5.7, it will be relatively easy to count successive lines in a band, but the first few lines may be obscured or difficult to resolve. In such a case, moving the $\Delta_2 F(J)$ plot up and down until the intercept occurs at $-\frac{1}{2}$ serves to establish the correct values of J.

One reason that low J lines may be difficult to resolve is that they often tend to run together in a *band head.* This comes about in the following manner. Let us consider the line frequency to be a continuous variable, and differentiate it with respect to J. For example, for the R branch

$$\frac{d\nu_R(J)}{dJ} = (3B' - B'') + 2(B' - B'')J = 0$$

whenever the frequency attains a maximum or minimum value. Solving for this extremum, we get

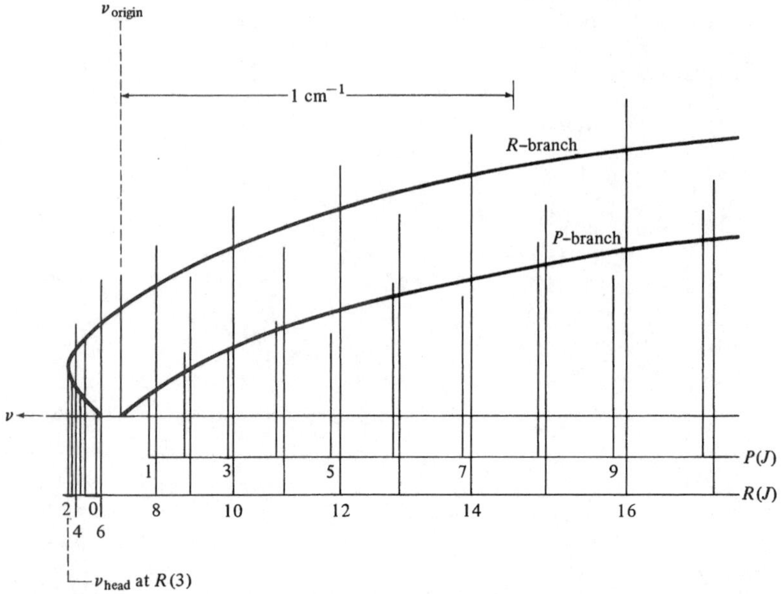

Figure 5.8
Band head formation in the hypothetical 0–0 band of the iodine $B \leftarrow X$ transition. Note that ν_{head} is slightly displaced from ν_{origin}. The intensity distribution shown corresponds to room temperature. Frequency increases toward the left, so wavelength increases toward the right. The band is thus "shaded toward the red"; that is, the lines move out to wavelengths longer than that of the band head. The 7:5 nuclear spin intensity alternation is also indicated.

$$J_R^* = -\frac{3B' - B''}{2(B' - B'')},$$

where J_R^* is the nearest integer value to the ratio of B values. Similarly,

$$J_P^* = +\frac{B' + B''}{2(B' - B'')}$$

is the value at which the P branch "turns around."

For example, in the iodine visible absorption spectrum, we would have $B_0' = 0.0289 \text{ cm}^{-1}$ and $B_0'' = 0.0373 \text{ cm}^{-1}$ for the 0–0 band. Solving for the band heads, we obtain $J_R^* = 3$, and $J_P^* = -4$; that is, the P branch never turns around. The detailed appearance of this band is shown in figure 5.8. The intensity distribution in the rotational band is governed by the same combination of rotational Boltzmann factor and nuclear spin statistical weights as has been previously discussed (chapter 3.1).

5 Angular Momentum Coupling Cases

When we come to analyze the rotational level structure of molecules that posses electronic orbital and/or spin angular momentum, the situation can be a good deal more complicated than for the $^1\Sigma$ case just treated. The difficulty arises from the fact that, while the molecule is rotating in space, the electrons are rotating within the molecule. What we see of the electron angular momentum depends on how this internal angular momentum couples to the molecular rotation. While under these circumstances, none of the individual angular momenta is really a constant of the motion, experience has shown that there are several limiting coupling cases that serve to describe fairly well the large majority of molecules. These are known as Hund's coupling cases (a)–(e)—see reference 6; cases (a)–(d) are of the most practical interest, and they will be described here. The notation we use is that recommended by Hougen in his 1970 report (reference 11).

Hund's Case (a) In this coupling scheme, depicted in figure 5.9, the strongest interaction is the electrostatic correlation that couples the electron spin and orbital angular momenta individually to the molecular axis. As previously described, the projection of L on the axis is denoted Λ; the projection of S is Σ; and the total $\Omega = \Lambda + \Sigma$. The molecular rotational angular momentum $\mathbf{R}$, which is, of course, perpendicular to the plane containing the molecular axis, couples to Ω to form the total angular momentum $\mathbf{J}$. The conserved quantities in this coupling scheme are J^2, J_z (one particular component), and Ω. The quantum number J can take on the values $\Omega + R$, where $R = 0$, 1, 2, . . . , but can never be less than Ω. The rotational energy levels are just $(h^2/8\pi^2\mu R_e^2)\mathbf{R}^2 = B[J(J+1) - \Omega^2]$. The total energy in excess of the vibronic contribution is $B[J(J+1) - \Omega^2] + A\Omega^2 = BJ(J+1) + (A - B)\Omega^2$, where A is essentially the molecular spin-orbit coupling constant. In this last form, the energy expression is very similar to that for a symmetric top (see chapter 7.1 and also Herzberg, reference 6, pp. 115ff).

An example of the energy levels in case (a) coupling is shown for a $^3\Pi$ state in figure 5.10.

Hund's Case (c) Closely related to case (a) coupling is case (c). In this case, the good quantum numbers are still J, J_z, and Ω, but the spin-orbit coupling

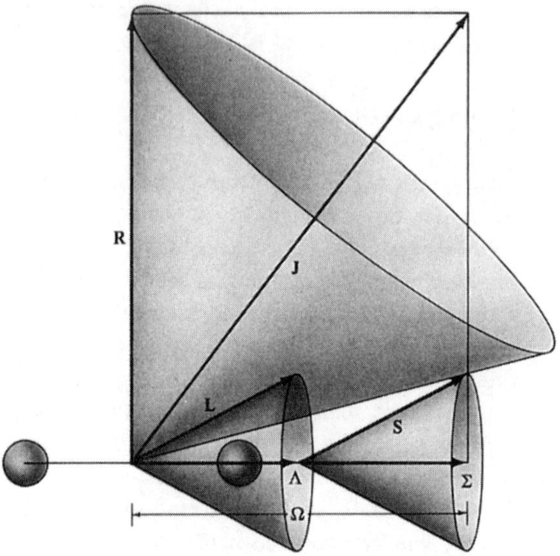

Figure 5.9
Hund's coupling case (a).

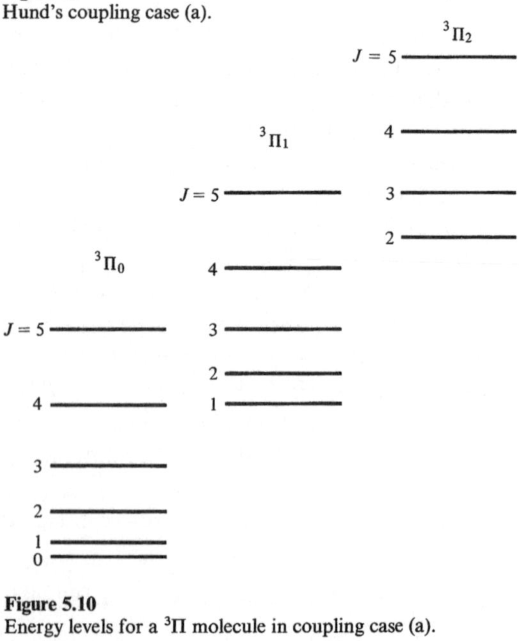

Figure 5.10
Energy levels for a $^3\Pi$ molecule in coupling case (a).

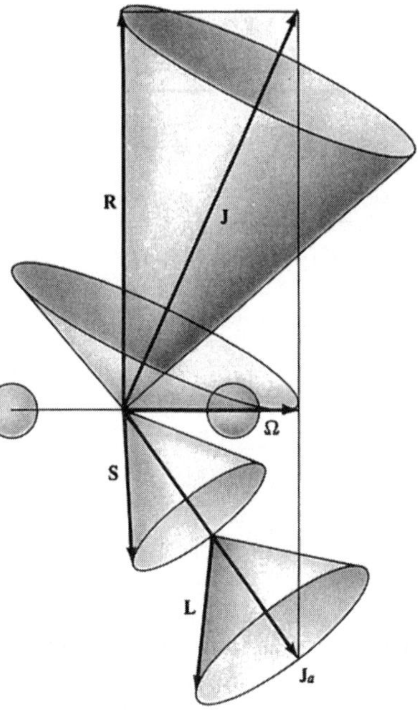

Figure 5.11
Hund's coupling case (c).

is so large that **L** and **S** are coupled to form J_a, the total electronic angular momentum, which then couples to the axis with projection Ω. This scheme is shown in figure 5.11; for this sort of coupling, Λ and Σ are not well defined. The energy expressions for this coupling case are the same as for case (a), except that the constant *A* is so large that the manifold of levels appears as several distinct electronic states, rather than as a splitting of rotational levels in a single state. An example of this sort of coupling is the diatomic halogens, especially the heavier species such as Br_2 and I_2.

Hund's Case (b) When $\Lambda = 0$ but $S \neq 0$ in a diatomic molecule, then the coupling case depicted in figure 5.12 generally applies. In this scheme, *S* does not couple to the internuclear axis; actually, the electron spin remains in a fixed orientation in space, while the molecule rotates under it. The good quantum numbers are *S* and the total angular momentum *J*, which can take on all integer values from $R + S$ to $R - S$; that is, each rotational level is

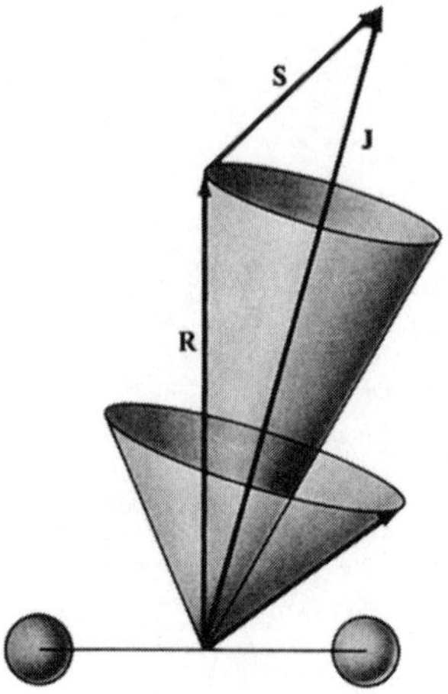

Figure 5.12
Hund's coupling case (b) for $\Lambda = 0$.

split into $2S + 1$ components. This coupling case is found most commonly in multiplet Σ states (CN, H_2^+, HgH, NH, O_2, and so on). In the less common case that $\Lambda \neq 0$, but S is still uncoupled from the axis, then R and Λ are coupled to form an angular momentum N, and N and S couple to give J.

The energy levels in case (b) coupling are, for the example of a $^2\Sigma$ state,

$$
\begin{aligned}
F_1(R) &= B_v R(R+1) + \tfrac{1}{2}\gamma R \qquad (J = R + \tfrac{1}{2}), \\
F_2(R) &= B_v R(R+1) - \tfrac{1}{2}\gamma (R+1) \qquad (J = R - \tfrac{1}{2}).
\end{aligned}
\tag{5.7}
$$

γ is the spin-rotation coupling constant; it includes a contribution from the direct magnetic $\mathbf{S}\cdot\mathbf{R}$ interaction, which is small, and an electron-coupled part of the form

$$\sum_i \mathbf{S}\cdot\mathbf{l}_i,$$

which, although it enters as a second-order perturbation, usually makes the larger contribution.

For a $^3\Sigma$ molecule, we have to include a magnetic $\mathbf{s}_1\cdot\mathbf{s}_2$ coupling, so that

$$
\begin{aligned}
F_1(R) &= B_v R(R+1) + \gamma(R+1) - \frac{2\lambda(R+1)}{2R+3} \qquad (J = R+1), \\
F_2(R) &= B_v R(R+1) \qquad (J = R),
\end{aligned}
\tag{5.8}
$$

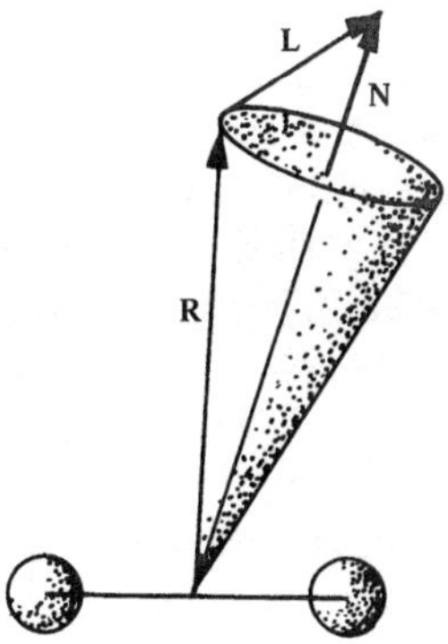

Figure 5.13
Hund's coupling case (d).

$$F_3(R) = B_v R(R+1) - \gamma R - \frac{2\lambda R}{2R-1} \qquad (J = R - 1),$$

where λ is the spin-spin coupling constant.

Hund's Case (d) This case applies when the orbital angular momentum **L** is nearly uncoupled from the internuclear axis. **L** and rotational angular momentum **R** couple to form **N**, as shown in figure 5.13; if spin angular momentum **S** is additionally present, this would couple with **N** to form total angular momentum **J**. This coupling case is most commonly encountered in Rydberg states of molecules, in which the orbital electron is excited to high principal quantum number and spends most of its time at large distances from the nuclei. Case (e), in which **L** and **S** are strongly coupled to each other but not to the internuclear axis, does not correspond to a physically significant situation.

While these coupling cases provide an adequate description for most molecules, it must be realized that they are just idealized limiting cases, and that, upon close enough inspection, some deviations from the simple energy expressions can usually be found. What each limiting coupling case actually provides is the representation in which the molecular Hamiltonian, which includes no rotation, is most nearly diagonal. There will usually be small off-diagonal terms that must be treated as perturbations, if the most accurate formulas for the energy levels are to be found. Let us consider a few of these interactions.

1. The interaction of nuclear rotation and the electronic orbital angular momentum via a **R** · **L** interaction leads to a splitting of each rotational level known as Λ *doubling*. When this is very large, so that **L** is completely uncoupled from the internuclear axis, we obtain the limiting case (d). An example of Λ doubling is found in Π states (Λ = 1), perturbed by a nearby Σ state. The rotational energies, including centrifugal distortion, are given by

$$F_{\pm}(J) = B_v J(J+1) - D_v[J(J+1)]^2 \pm qJ(J+1).$$

When the angular momentum Λ can be attributed to a single orbital electron with angular momentum l, as in CH or the excited states of the alkaline earth hydrides (CaH, HgH, and so on), then the Λ-doubling constant q is given by the "Van Vleck pure precession formula,"

$$q = \frac{2B_v^2 l(l+1)}{(E_\Pi - E_\Sigma)}. \tag{5.9}$$

Since B_v is of the order of 5 cm^{-1} for the hydrides, and the energy denominator may be a few hundred or a few thousand cm^{-1}, the Λ doubling is quite small—of the order of a fraction of a cm^{-1}.

2. When the **R** · **L** and **S** · **N** interactions become comparable with the spin-orbit **L** · **S** interaction, these two angular momenta uncouple, and Ω is no longer a good quantum number. This can happen, for example, in case (c) coupling at high rotational velocities. Typically, the 0^+ component of a $^3\Pi$ manifold will take on some triplet character, as evidenced by a J-dependent Zeeman effect or Faraday effect activity (see sections 8 and 10).

3. The spin-rotation interaction **S** · **R** has the effect of uncoupling the spin angular momentum from the internuclear axis. As the magnitude of the interaction increases, a case (a) molecule goes over into a case (b) molecule. For a nonrotating $^2\Pi$ molecule, the part of the Hamiltonian that includes rotation and spin-orbit coupling is

$$\mathscr{H}' = B[\mathbf{J} - \mathbf{L} - \mathbf{S}]^2 + A\mathbf{L}\cdot\mathbf{S}, \tag{5.10}$$

where A is, as before, the spin-orbit coupling parameter. The eigenstates in the $|\Lambda S \Sigma\rangle$ representation are $|1\,\frac{1}{2}\,\pm\frac{1}{2}\rangle$ and $|-1\,\frac{1}{2}\,\pm\frac{1}{2}\rangle$. In a $|\Omega JM\rangle$ representation, Ω can take on the values $+\frac{3}{2}$, $+\frac{1}{2}$, $-\frac{1}{2}$, and $-\frac{3}{2}$. The energy levels are

$$E = E_0 + B[(J+\tfrac{1}{2})^2 - 1] \pm \tfrac{1}{2}[A(A-4B) + 4B^2(J+\tfrac{1}{2})^2]^{1/2}, \tag{5.11}$$

where E_0 is the pure electronic + vibrational energy. Case (a) coupling is obtained when $A \gg BJ$, so that

$$E_{\text{rot}} = B[J(J+1) + \tfrac{1}{4}] \pm \tfrac{1}{2}A,$$

the $\pm\tfrac{1}{2}A$ splitting corresponds to the energy difference between the $^2\Pi_{1/2}$ and $^2\Pi_{3/2}$ states. Case (b) coupling is obtained when $A \ll BJ$, so that

$$E_{\text{rot}} = BR(R+1) = \begin{cases} B(J+\tfrac{1}{2})(J+\tfrac{3}{2}) & \text{for } J = R - \tfrac{1}{2} \\ B(J-\tfrac{1}{2})(J+\tfrac{1}{2}) & \text{for } J = R + \tfrac{1}{2}. \end{cases}$$

The $\pm\tfrac{1}{2}\gamma R$ term, arising from the $\mathbf{R}\cdot\mathbf{S}$ interaction, is added on as a perturbation correction.

Approximates energy level expressions for molecules near one or the other of these limiting cases, and for intermediate coupling cases, can be found in chapter 5 of Herzberg (reference 6).

6 Selection Rules in Electronic Transitions

Knowing the energy levels for a system is, of course, only half the job in interpreting its spectrum. We must also know the selection rules governing transitions between the various levels. We have already obtained, in chapter 3.4, the selection rules between overall electronic states, and in chapter 4.3, the rules between rotational states in a vibrational band. For electronic states, the allowed electric dipole transitions are

$$\Delta\Lambda = 0, \pm 1 \quad \text{and} \quad \Delta\Sigma = 0: \quad \text{cases (a) and (b)}$$

or

$$\Delta\Omega = 0, \pm 1: \quad \text{case (c)},$$

$$\Delta L = 0, \pm 1: \quad \text{case (d)}.$$

Also, + states cannot combine with – (in Σ–Σ bands), and u states can combine only with g, and vice versa (in homonuclear molecules). For rotation, we had $\Delta J = \pm 1$ in cases (a) and (c), and in addition $\Delta J = 0$ was allowed if Λ was not equal to zero in at least one of the states involved. The rotational selection rules are $\Delta R = 0, \pm 1$ in case (b), and $\Delta R = 0$ in case (d); that is, a transition to a Rydberg state is nearly a pure electronic transition.

We must now combine these selection rules to obtain the complete structure of a molecular band spectrum. The problem in doing this is that the electronic selection rules have been defined in a molecule-fixed coordinate system, whereas the rotational selection rules have been defined in a space-fixed coordinate system. We must refer both of these to the same set of axes; it will be convenient to use space-fixed axes, which means that we must derive the symmetry properties of the electronic wave functions in this axis system.

The two operations that we have to consider are, for Σ states, reflection of the wave function through a plane containing the internuclear axis, and, for homonuclear molecules, inversion through the center of symmetry. We shall state the following without proof; for further details, see reference 11.

1. The operation of reflecting through a plane containing the symmetry axis, in the molecule-fixed system, is equivalent to inversion of coordinates in the space-fixed system.
2. The operation of coordinate inversion in the molecule-fixed system is equivalent to interchange of nuclei in the space-fixed system.

Inversion in space-fixed axes (i') has eigenvalues $+1$ and -1. Its effect on a particular molecular state depends upon both the intrinsic electronic symmetry and the rotational state. For Σ^+ states, we have

$$i'|J\rangle = (-1)^J|J\rangle;$$

for Σ^- states, we have

$$i'|J\rangle = (-1)^{J+1}|J\rangle.$$

Thus, a Σ^+ has the oddness or evenness of its rotational quantum number, whereas a Σ^- state has oddness or evenness opposite to its rotational quantum number. In $\Lambda \neq 0$ states, where there are two degenerate (or nearly so) $+$ and $-$ components, each J state will possess both an odd and an even member.

The effects of the nuclear interchange operator $\dot{X}_N$ also depend on both the rotational state and the symmetry properties of the electronic wave function. For homonuclear molecules, we have to take into account the u–g inversion symmetry, as well as the $+$–$-$ reflection symmetry for Σ states. The results of applying the X_N operator to these various possibilities are summarized in the following table (s = symmetric, a = antisymmetric):

	Symmetry
$X_N|J\Lambda_g^+\rangle = (-1)^J|J\Lambda_g^+\rangle$	s
$X_N|J\Lambda_g^-\rangle = (-1)^{J+1}|J\Lambda_g^-\rangle$	a
$X_N|J\Lambda_u^+\rangle = (-1)^{J+1}|J\Lambda_u^+\rangle$	a
$X_N|J\Lambda_u^-\rangle = (-1)^{J+2}|J\Lambda_u^-\rangle = (-1)^J|J\Lambda_u^-\rangle$	s

These results are easy to remember once it is noticed that each "odd" factor in the wave function—odd J number, reflection symmetry, or odd parity—contributes an odd factor, expressed by multiplication by -1, to the overall symmetry.

The selection rules are easily determined, once the behavior of the dipole moment under the i' and X_N operations is established. Clearly, inverting all coordinates changes the sign of the dipole moment, so that μ has $(-)$ character, and in order to have totally symmetric transition moment matrix elements, we must have

$$+ \leftrightarrow - \quad \text{but} \quad - \not\leftrightarrow - \quad \text{and} \quad + \not\leftrightarrow +.$$

Evaluating the effect of the nuclear interchange operator is a little more subtle. Since this operation will be of interest *only* when the two nuclei are identical, the operation under these circumstances obviously leaves the dipole moment unchanged; thus μ has (s) character, and we have the selection rules

$$s \leftrightarrow s \quad \text{and} \quad a \leftrightarrow a \quad \text{but} \quad a \not\leftrightarrow s.$$

The most convenient way of carrying out the bookkeeping of the symmetry properties of all the levels, and picking out the allowed branches in the spectrum, is to use the diagrams introduced by Herzberg (reference 6). These diagrams simply count the first few rotational levels, with one marker for each degenerate sublevel, and each one indicated as to the appropriate symmetry. Some examples are given in figure 5.14.

Case (a) Coupling	Case (c) Coupling	As Components of a $^3\Pi$						
$^1\Sigma^+$	0^+	$^3\Pi_{0^+}$	(+)	(−)	(+)	(−)	(+)	(−)
		$J = R$:	0	1	2	3	4	5
$^1\Sigma^-$	0^-	$^3\Pi_{0^-}$	(−)	(+)	(−)	(+)	(−)	(+)
$^1\Pi$	$\Omega = 1$	$^3\Pi_1$		(+)(−)	(−)(+)	(+)(−)	(−)(+)	(+)(−)
		$J = R + \Omega$:		1	2	3	4	5
$^1\Delta$	$\Omega = 2$	$^3\Pi_2$			(+)(−)	(−)(+)	(+)(−)	(−)(+)

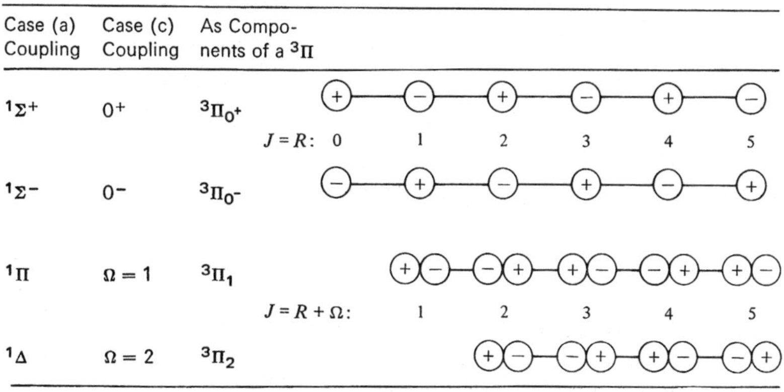

Figure 5.14
Parity of rotational levels in $^1\Sigma$, $^1\Pi$, and $^1\Delta$ electronic states.

In case (b) coupling, it is necessary only to indicate the multiplicity of each R level, since the spin adds nothing to the symmetry properties:

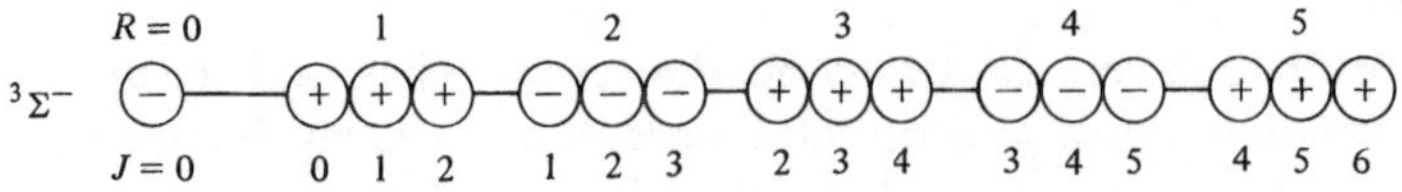

To determine the allowed branches of a particular transition, all one needs to do is to align the diagrams corresponding to the two electronic states involved, and connect levels consistent with the selection rules on symmetry and J. For example, in a $^1\Sigma^+ - {}^1\Sigma^+$ transition,

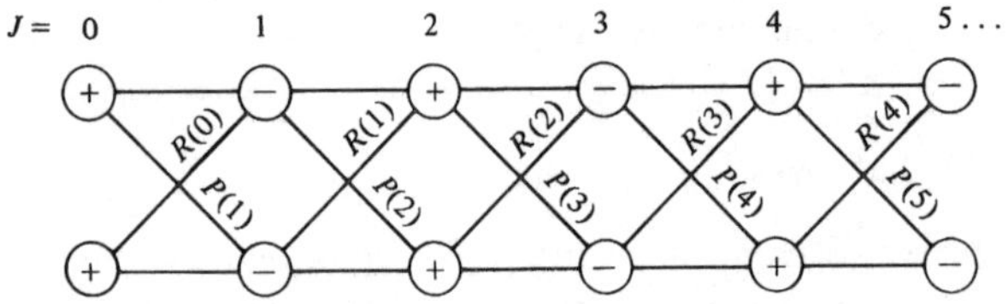

Note that the symmetry selection rules automatically rule out the Q branch ($\Delta J = 0$) in this transition. If one tried to do the same thing for a hypothetical $^1\Sigma^+ - {}^1\Sigma^-$ transition, one would find that it would be impossible to satisfy both the symmetry selection rule and the $\Delta J = \pm 1$ selection rule at the same time.

For a $^1\Pi - {}^1\Sigma^-$ transition, we would have

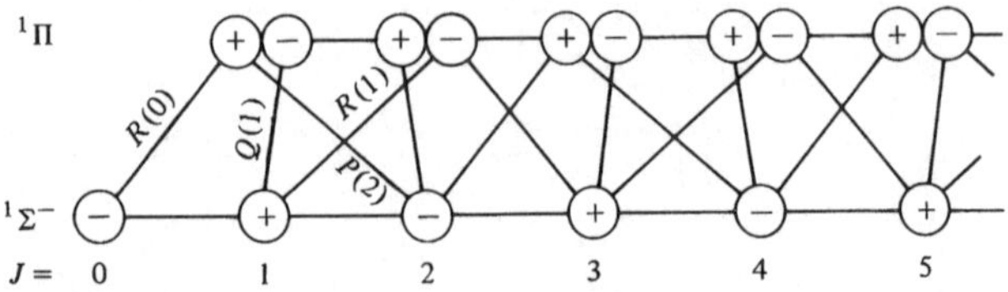

Note that it is now possible to have a Q branch. Also, the $P(1)$ line is missing, because there is no $J = 0$ state in the $^1\Pi$. The absence of this line is a characteristic indicator of this type of electronic transition.

We can illustrate the type of spectrum obtained in a homonuclear system with the $^3\Pi_u \rightarrow {}^3\Pi_g$ transition in N_2, which corresponds to the so-called second positive bands. This is a case in which all of our selection rules must come into play. These are s$\leftrightarrow$s and a$\leftrightarrow$a, $+ \leftrightarrow -$, $\Delta J = 0, \pm 1$ [except that $\Delta J \neq 0$ for the $(\Omega = 0) \leftrightarrow (\Omega = 0)$ part of the band], and $\Delta\Sigma = 0$, so that no intercombination bands are seen. The various branches are illustrated in figure 5.15. Altogether, the eight predicted branches can be observed in this spectrum, although the Q_2 and Q_3 branches turn out to be relatively weak.

Diagrams for a number of other cases are given by Herzberg (reference 6). The corresponding diagrams for magnetic dipole and electric quadrupole transitions are left as a problem at the end of this chapter.

We conclude this section with some examples, in figure 5.16, of high-resolution diatomic molecular spectra, taken from the literature, which illustrate the various types of bands that can occur.

7 Perturbations and Predissociation

Thus far, we have been considering each electronic state as an isolated entity. We conceded that a particular state may not conform exactly to an ideal angular momentum coupling case, but, by and large, each state is describable by a unique set of good quantum numbers. In many instances, this is a good description. In many others, however, the presence of a large number of nearby states affects the energy levels of the one we are looking at. A glance at figures 3.13–3.16 will show that most electronic states of typical molecules, with the exception of the ground state, are not isolated, but are embedded in and intertwined with a large number of other states. It would be surprising if the presence of all these neighboring states did not manifest itself in some way. Indeed, some of the most interesting phenomena in molecular spectroscopy are a result of such manifestations.

The small terms in the molecular Hamiltonian, which hindered our search for simple descriptions of molecular states and wave functions, and which we therefore discarded, are the ones that produce the perturbations we are about to consider. What we are really saying is that the molecules, which have not discarded any terms in their own Hamiltonians (being somewhat more circumspect in this regard than most spectroscopists), are not really described in terms of the Λs, Ωs, and so on, with which we choose to label them. If we insist on using such descriptions, then it will be necessary from time to time to use a linear combination of several approximate states to

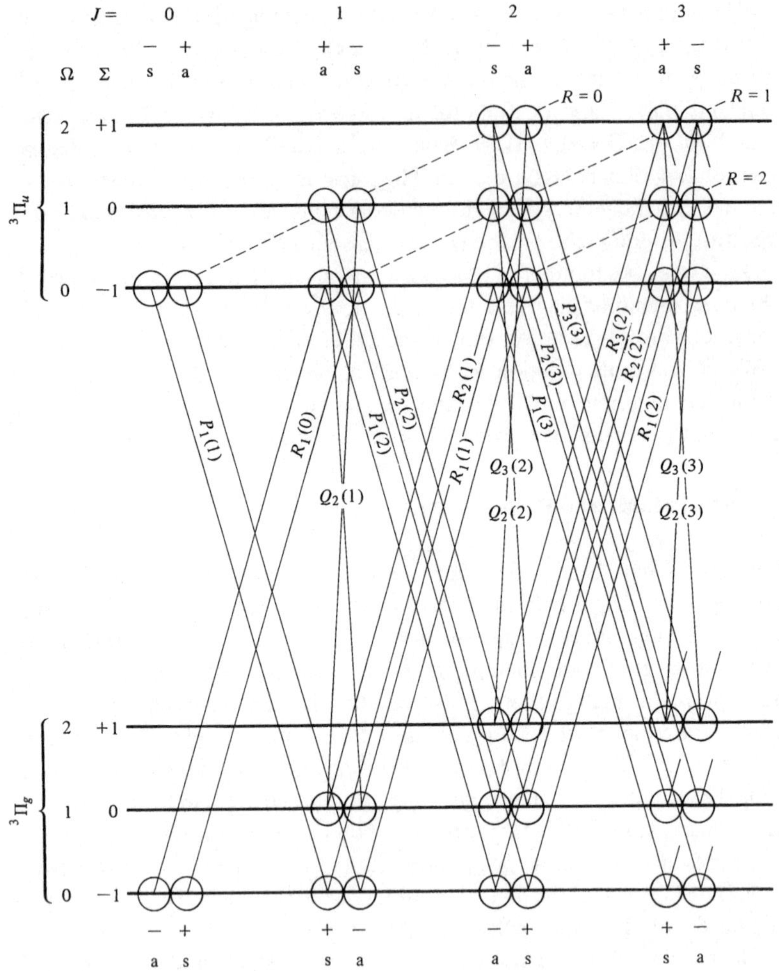

Figure 5.15
Allowed branches in a $^3\Pi_u$–$^3\Pi_g$ transition.

arrive at a satisfactory description of a real molecule. Since one particular state usually makes the dominant contribution to a real system, it is convenient to think of the situation in terms of one idealized state perturbed by another idealized state. One should never forget, however, that in every case the molecule has solved its own Schrödinger equation exactly, and is probably laughing at our attempts at attaining to some approximate solution.

The terms in the Hamiltonian that we have omitted and that are effective in producing perturbations may be distinguished as vibrational or rotational:

$$\mathscr{H}' = \mathscr{H}'_{\text{vib}} + \mathscr{H}'_{\text{rot}}.$$

The vibrational terms are the extra terms neglected in the Born-Oppenheimer separation; from equation (3.4), they have the form

$$\mathscr{H}'_{\text{vib}} = -\sum_A \frac{\hbar^2}{2M_A}[2\nabla_A\nabla_A + \nabla_A^2]. \tag{5.12}$$

The rotational terms are the ones that we have already considered as causing departures from the ideal angular momentum coupling cases, that is, $\mathbf{R}\cdot\mathbf{L}$, $\mathbf{R}\cdot\mathbf{S}$, and so on. Physically, these represent torques introduced on the molecule in carrying out the transformation from molecule-fixed to space-fixed axes.

The work of Van Vleck, Kramers, Mulliken, and others has shown that the vibrational perturbations are diagonal in all the important molecular quantum numbers; that is,

$$\langle \Omega S \Lambda J v p \mathscr{S} | \mathscr{H}'_{\text{vib}} | \Omega' S \Lambda' J' v' p' \mathscr{S}' \rangle$$

$$\sim q_{vv'} W'_v \,\delta(\Omega,\Omega')\,\delta(S,S')\,\delta(\Lambda,\Lambda')\,\delta(J,J')\,\delta(p,p')\,\delta(\mathscr{S},\mathscr{S}'),$$

where p is the parity eigenvalue (± 1); $\mathscr{S}$ is the symmetry eigenvalue (s or a); $q_{vv'}$ is the Franck-Condon factor; and W'_v is the matrix element of (5.12). An example of the operation of this sort of perturbation is the avoided crossing between the two ${}^1\Sigma^+$ states in lithium hydride, discussed in chapter 3.5. This sort of splitting was also hinted at in chapter 3.1, when the comment was made that, although W'_v may be very small in magnitude, these terms may yet be very effective in splitting degenerate states.

In contrast with the vibrational perturbations, rotational perturbations couple states of different Ω and Λ:

$$\langle \Omega S \Lambda J v p \mathscr{S} | \mathscr{H}'_{\text{rot}} | \Omega' S' \Lambda' J' v' p' \mathscr{S}' \rangle$$

$$\propto q'_{vv} W'_r \,\delta(J,J')\,\delta(S,S')\,\delta(p,p')\,\delta(\mathscr{S},\mathscr{S}')\,\delta(\Omega,\Omega' \pm 1)\,\delta(\Lambda,\Lambda' \pm 1).$$

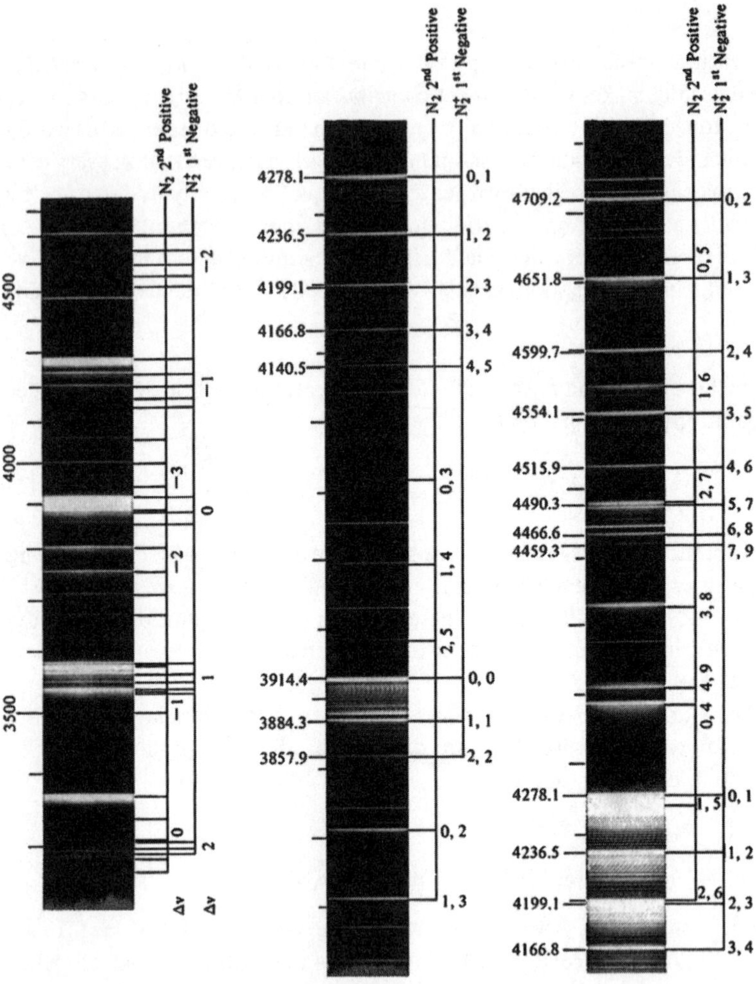

(a) (b)

Figure 5.16
Some typical diatomic molecular spectra. (a) Emission spectrum of nitrogen between 3,300 and 4,700 Å, showing the N_2 second positive ($C^3\Pi_u$–$B^3\Pi_g$) and N_2^+ first negative ($B^2\Sigma_u^+$–$X^2\Sigma_g^+$) bands. (b) Higher resolution scan of the $\Delta v = 0$, -1, and -2 sequences of the first negative bands. Note in particular the 3:1 intensity alternation in the rotational structure of the (0, 1) band near 4,250 Å and the (1, 3) band near 4,600 Å. The resolution is not quite sufficient, however, to bring out the triplet structure indicated in figure 5.15. [Reproduced with the permission of D. C. Tyte and R. W. Nicholls from their "Identification Atlas of Molecular Spectra. 3. The N_2^+ $B^2\Sigma_u^+$–$X^2\Sigma_g^+$ First Negative System of Nitrogen" (Center for Research in Experimental Space Science, Department of Physics, York University, Toronto

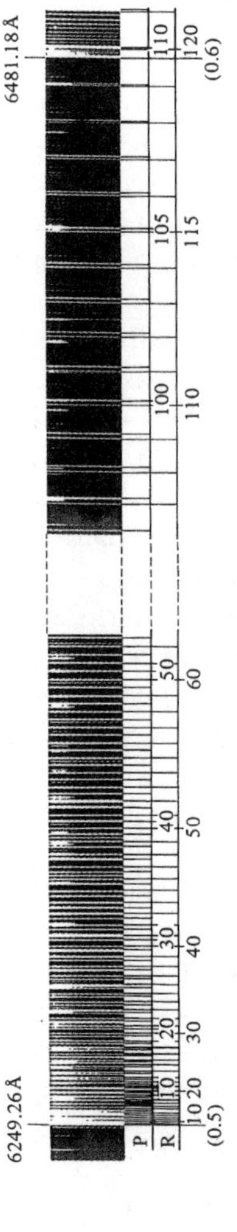

(c)

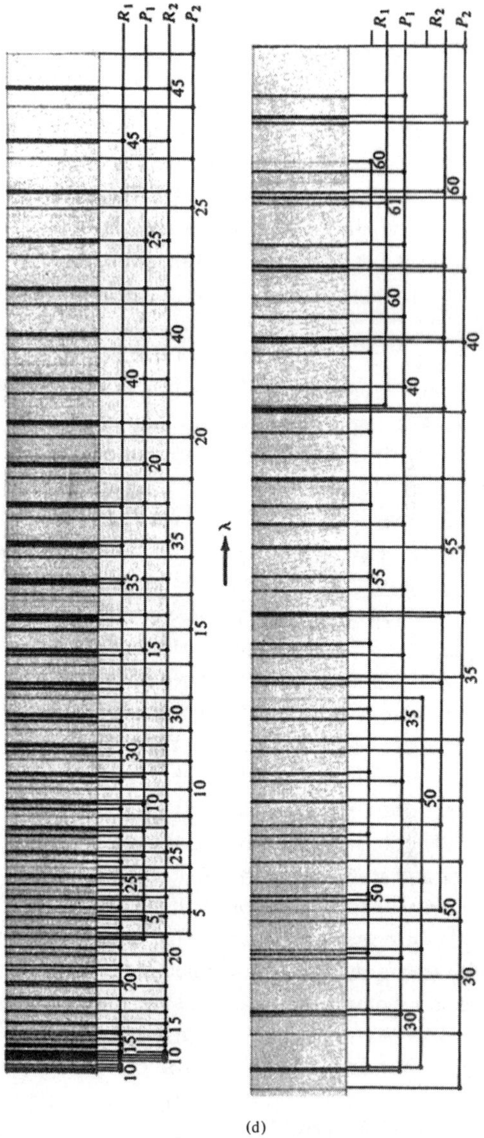

(d)

Ontario, 1965).] (c) Spectrum of iodine monofluoride. This is a typical 0^+–0^+ transition, containing only a simple *P* branch and *R* branch structure. [R. A. Durie, *Can. J. Phys.* **44**, 339 (1966). Reproduced with the permission of the National Research Council of Canada.] (d) Spectrum of rhodium carbide. This is a $^2\Sigma$–$^2\Sigma$ transition, so the *Q* branch is absolutely missing. Each *P* branch and *R* branch has two subbands, corresponding to the spin splitting. This is in all respects a classically simple spectrum, except that a perturbation appears near $P_1(60)$ and $R_2(50)$, manifesting itself as a break in the linear $\Delta_2F(J)$ plot and a doubling of some of the lines; see section 7. [A. Lagerqvist and R. Scullman, *Ark. Fys.* **32**, 481 (1966). Reproduced with permission.]

The effects of such a perturbation are seen as the line shift and doubling in spectra such as that shown in figure 5.16d. These selective effects result from overlapping of the stacks of rotational levels of the two electronic states whose Ω and Λ values satisfy the relation given. As levels of the same J value approach one another in energy, the perturbation becomes active in this local region.

As an example of this type of perturbation, consider the well-analyzed $A^2\Pi$–$X^2\Sigma$ "red" bands of the CN radical (see reference 3, also section 3). The rotational and spin-orbit Hamiltonian is given by equation (5.10). In case (a) coupling, the constants of the motion are J^2, S^2, J_z, S_z, and L_z, with eigenvalues $J(J+1)$, $S(S+1)$, Ω, Σ, and Λ, respectively. The cross terms in (5.10), particularly the products $J_\pm L_\mp$, $L_\pm S_\mp$, have nonvanishing matrix elements between the $^2\Pi$ and $^2\Sigma$ states, and thus mix these states. The mixing occurs where levels of the same J in different vibronic stacks have approximately equal energy; these crossing are depicted in figure 5.17. For example, $v' = 7$ $(A^2\Pi_{1/2})$ and $v'' = 11$ $(X^2\Sigma)$ cross at $J(J+1) \approx 1{,}800$, or between $J = 43$ and 45. If we tabulate the actual rotational line positions, and compare them with those calculated as $\nu_{\text{unperturbed}} = B_v J(J+1)$, we see (figure 5.18) that, as we go through that crossing region, the actual frequencies are shifted by up to 3 or 4 cm^{-1}.

The interactions we have considered up to now involve two bound molecular states. There is another whole class of effects that involves interaction between a bound state and a repulsive state crossing through it, known as *predissociation*. When absorption of light occurs to a repulsive state of a molecule, the process of *photodissociation* occurs, and the molecule dissociates into free atoms in a time of the order of 10^{-13} sec. When absorption takes place to a bound state that, however, has some of the character of a nearby repulsive state, a slower process of dissociation may occur with a "nonradiative" rate given by the "Fermi golden rule" (reference 12),

$$k_{\text{NR}} = \frac{2\pi}{\hbar^2} |\langle f|\mathscr{H}'|i\rangle|^2 \rho(E), \tag{5.13}$$

where $\rho(E)$ is the density of continuum levels in the repulsive state, and

$$\langle f|\mathscr{H}'|i\rangle = q_{vE} \langle \Omega \pm 1\, S\, \Lambda \pm 1\, lEp\mathscr{S} | \mathscr{H}'_{\text{rot}} | \Omega S \Lambda J v p \mathscr{S} \rangle,$$

in which the selection rules for rotational perturbation have been assumed; E is the translational energy of the separating atoms (replacing the quantum number v for vibration); and l is the orbital angular momentum associated with the trajectory of the atoms (equal to the quantum number J of rotation

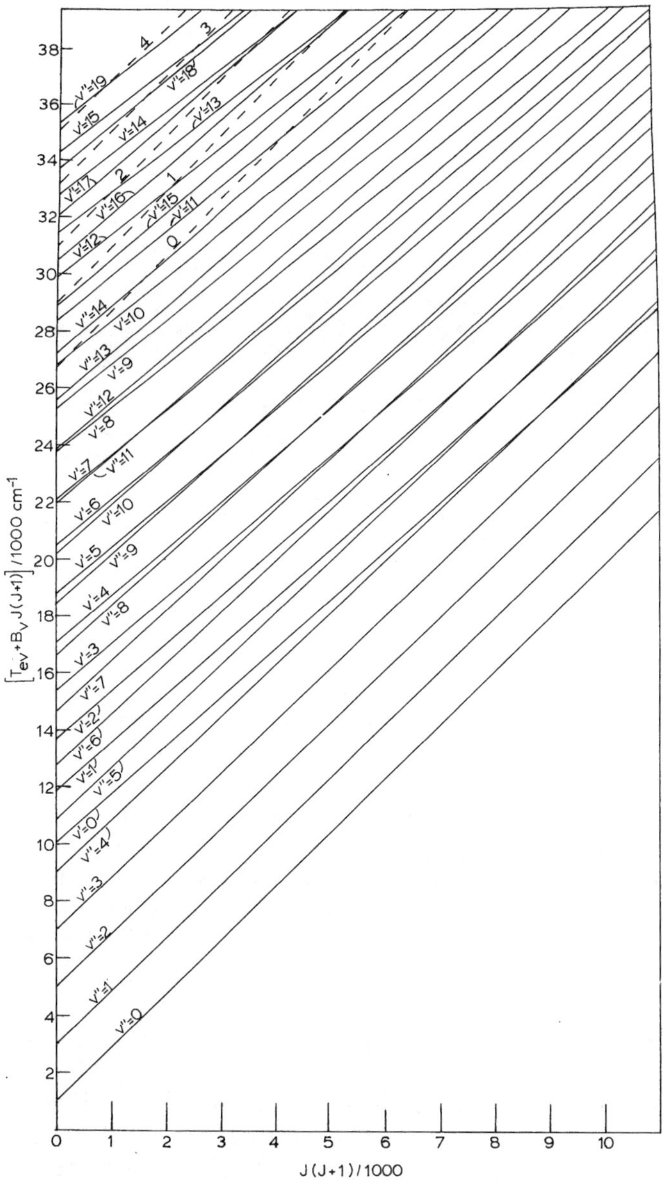

Figure 5.17
Overall view of the energy levels for CN $X^2\Sigma^+$ (v''), $A^2\Pi$ (v'), and $B^2\Sigma^+$. The B state is indicated by the broken lines, with the single number denoting the vibrational level. [From A. J. Kotlar, R. W. Field, J. I. Steinfeld, and J. A. Coxon, J. Mol. Spectroscopy **80**, 86 (1980). Reproduced with permission.]

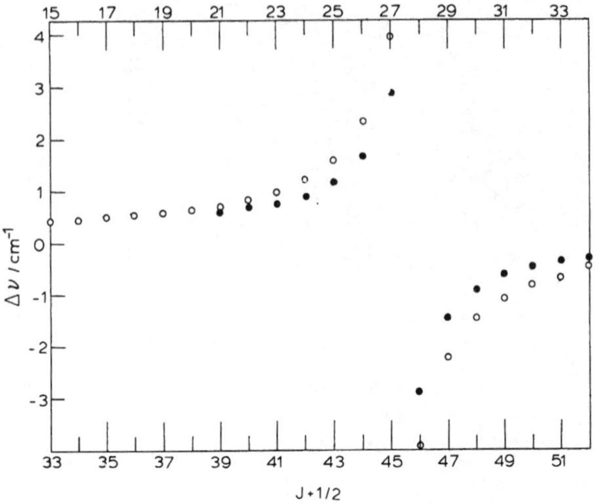

Figure 5.18
$\Delta\nu = (\nu_{\text{perturbed}} - \nu_{\text{unperturbed}})$ versus $J + \frac{1}{2}$ for rotational levels of CN $A\,^2\Pi_{1/2}$ $(v' = 7)$; data are from A. J. Kotlar, R. W. Field, J. I. Steinfeld, and J. A. Coxon, *J. Mol. Spectroscopy* **80**, 86 (1980).

in the original molecule). Predissociation may be observed in two ways. If k_{NR} is of the order of 10^9 sec^{-1} or greater, so that the lifetime with respect to dissociation of the excited state is 10^{-9} sec or less, then the uncertainty broadening of the absorption line begins to be larger than the Doppler width, and the line appears *diffuse*. Also, the *fluorescence yield* of the excited state is given by

$$\phi_F = \frac{k_{\text{fl}}}{(k_{\text{fl}} + k_{\text{NR}})}, \tag{5.14}$$

so as k_{NR} becomes much greater than k_{fl}, no fluorescence is seen. When predissociation becomes appreciable above a certain energy in the excited state that corresponds to the crossing point of the repulsive curve, a breaking off of fluorescence is seen, and at the same time the absorption lines begin to appear diffuse. When fluorescence is observed from an excited state, it is generally assumed that predissociation is not occurring, but examples have been found, such as the *B* state of iodine, in which fluorescence and spontaneous predissociation occur at roughly the same rate (reference 13). Formaldehyde has also been found to behave in this way (see chapter 9).

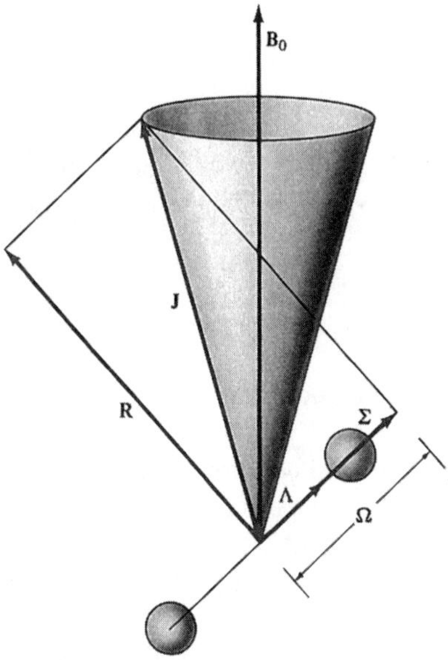

Figure 5.19
Vector precession diagram for the interaction of a Hund's case (a) molecule with a weak external magnetic field.

The subject of spectroscopic perturbations is a large and fascimating one, and has been treated in detail in a recent monograph by Field and Lefebvre-Brion (reference 14), to which the reader is referred for additional information.

8 Zeeman Effect in Molecules[1]

The interaction of the magnetic moment of a molecule with an external magnetic field has the same form as equation (2.11) for atoms, but differs in detail:

$$W'_{\text{Zeeman}} = -\boldsymbol{\mu}\cdot\mathbf{B}_0 = -\langle\boldsymbol{\mu}_{\text{rot}}\cdot\mathbf{B}_0 + \boldsymbol{\mu}_{\text{orbital}}\cdot\mathbf{B}_0 + \boldsymbol{\mu}_{\text{spin}}\cdot\mathbf{B}_0\rangle.$$

The average is over rotation in space-fixed axes. A suitable vector model for magnetic field interaction is shown in figure 5.19. The case depicted is that of

1. An excellent discussion of the molecular Zeeman effect is given in Chapter 11 of Townes and Schawlow (Reference 15).

a very weak magnetic field, in which the total molecular angular momentum **J** precesses around the field vector $\mathbf{B}_0$. The interactions of the individual angular momentum components with the field are all weaker than their coupling to one another, so that we remain in the case (a) coupling scheme. For a $^1\Sigma$ molecule with no nuclear spin, $J = R$, and the only interaction is with the rotational magnetic moment of the molecule. The result, after averaging over rotation, is

$$W'_{\text{Zeeman}} = -\frac{e\hbar}{2Mc} B_0 g_{\text{rot}} \frac{M_J}{\sqrt{J(J+1)}}. \tag{5.15}$$

The presence of the total molecular mass M in the denominator of this expression ensures that the Zeeman splittings will be too small to be observable in ordinary optical spectroscopy. The rotational g factor g_{rot} in equation (5.15) must be calculated from the electronic properties of the molecule; it cannot be obtained by a simple vector model analysis such as sufficed for calculation of the Landé g factor for atoms. Physically, g_{rot} arises from unequal rates of rotation of the negative electronic and positive nuclear charge distributions; a large value of g_{rot} implies that the electrons are "slipping" with respect to the motion of the nuclei. For example, the rotational g factor of H_2 has been found (by molecular beam techniques, to be discussed in chapter 10) to be $+0.88$, which corresponds to nearly free rotation of the two protons inside a nearly stationary electron cloud.

If Λ and/or Σ are nonzero, we obtain the result

$$W'_{\text{Zeeman}} = -\frac{e\hbar}{2mc} B_0 \left\{ \frac{(\Lambda + 2\Sigma)\Omega M_J}{J(J+1)} + \frac{m}{M} g_{\text{rot}} \frac{M_J}{\sqrt{J(J+1)}} \right\}. \tag{5.16}$$

Note the $J(J+1)$ factor in the denominator of equation (5.16). This reflects the fact that the Zeeman interaction tends to be averaged out at higher and higher rotational velocities. The effect on the splitting is shown in Figure 5.20; more and more lines are squashed into a narrower total frequency spread as J increases.

Experimental methods utilizing the Zeeman effect in molecules include

1. direct observation of resonances between Zeeman sublevels by magnetic resonance absorption, for example in NO and O_2;
2. molecular beam magnetic resonance, to be discussed in chapter 10.4; and
3. direct optical observation of the splittings, although this is severely limited by the "squashing" effect shown in figure 5.20.

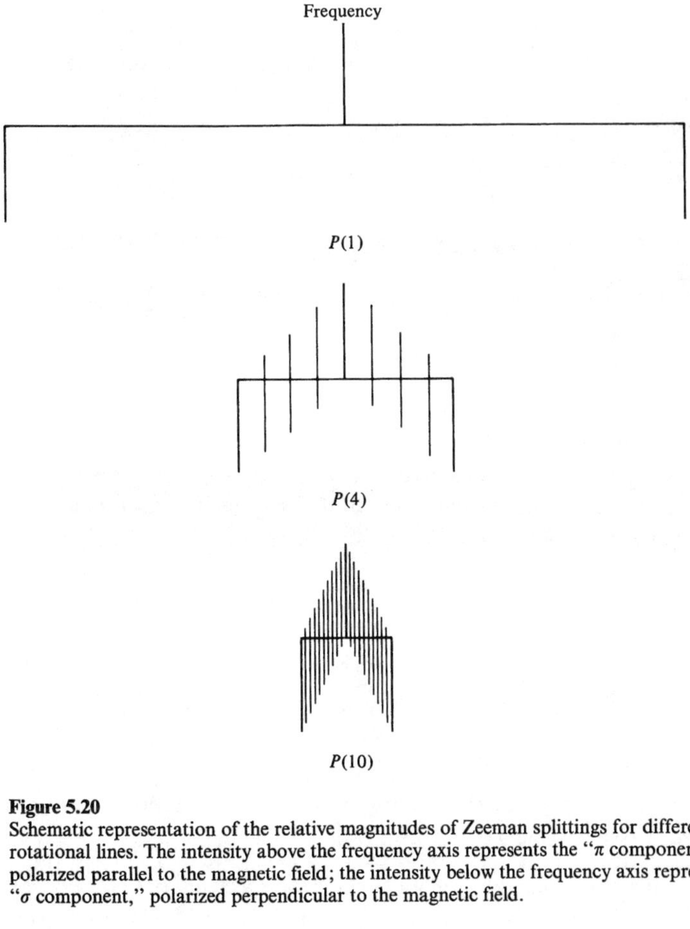

Figure 5.20
Schematic representation of the relative magnitudes of Zeeman splittings for different rotational lines. The intensity above the frequency axis represents the "π component", polarized parallel to the magnetic field; the intensity below the frequency axis represents the "σ component," polarized perpendicular to the magnetic field.

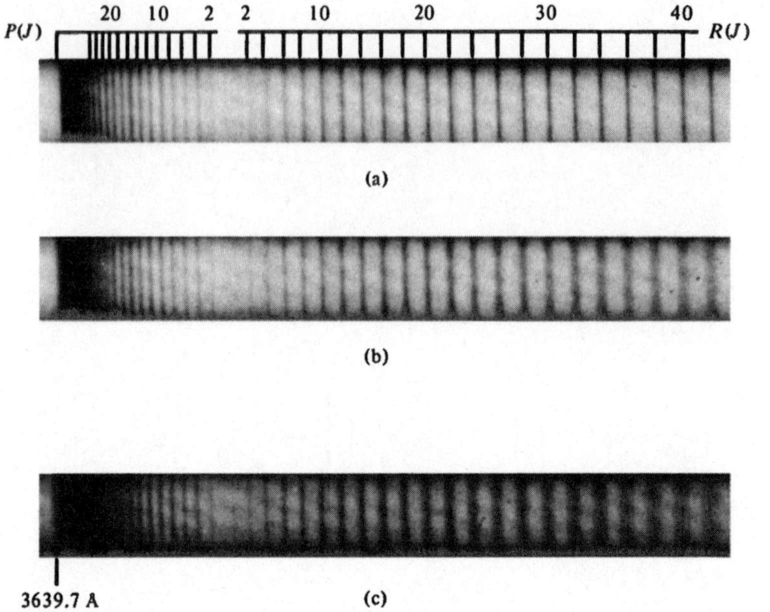

Figure 5.21
J-dependent Zeeman splitting in the ultraviolet absorption spectrum of CS_2. The 07–02 band is shown, with (a) $B_0 = 0$, (b) $B_0 = 2{,}300$ G, and (c) $B_0 = 5{,}300$ G. [Spectrum is reproduced with the permission of A. E. Douglas and E. R. V. Milton; see *J. Chem. Phys.* **41**, 357 (1964).]

4. Zeeman splittings can also be observed in high-resolution microwave spectroscopy (see chapter 7).

An example of the optical Zeeman effect is the spectrum of CS_2 (a linear, although not diatomic, molecule), shown in figure 5.21. This spectrum demonstrates very clearly the triplet nature of the excited state. Note that the width of each line pattern, instead of decreasing with J as predicted by equation (5.16), increases with J. This is an example of the Ω uncoupling mentioned previously (section 5). The upper state has a total electronic angular momentum of zero, as inferred from the simple structure of the rotational lines in the band; but it is actually the $\Omega = 0$ component of a triplet manifold, and, as the rotation of the molecule increases, the spin angular momentum is partially uncoupled and becomes available to interact with the external field.

9 Stark Effect in Molecules[2]

We had previously observed that, in atoms, the Zeeman effect had proved much more useful than the Stark effect. Pretty much the opposite is true in molecules. The energy of interaction of a molecule with an external electric field is given by the appropriate analogue of equation (2.10),

$$W'_{\text{Stark}} = -\langle \boldsymbol{\mu}_e \cdot \mathbf{E} \rangle,$$

where once again the interaction is averaged over molecular rotation. The explicit form for this averaging is

$$W^{(1)}_{\text{Stark}} = \iint \psi^*_{JM} \mu_e E \cos\theta \, \psi_{JM} \sin\theta \, d\theta \, d\phi.$$

If the permanent electric dipole moment μ_e is zero, then there is no first-order Stark effect. The only energy changes that are obtained are small second-order contributions proportional to the polarizability, the square of the electric field strength, and a complicated function of J and M. If both μ_e and $\Lambda \neq 0$, we obtain

$$W^{(1)}_{\text{Stark}} = \frac{\mu_e E M \Lambda}{J(J+1)}. \tag{5.17}$$

If $\Lambda = 0$, however (that is, the molecule is in a Σ state), there is again no first-order effect, but there is a second-order interaction of the form

$$W^{(2)}_{\text{Stark}} = E^2 \sum_{J'M'} \frac{|\langle JM|\mu_e|J'M'\rangle|^2}{E_{JM} - E_{J'M'}} = \frac{\mu_e^2 E^2 [J(J+1) - 3M^2]}{2hBJ(J+1)(2J-1)(2J+3)}. \tag{5.18}$$

In this case, only levels with different absolute values of M are split (in contrast with Zeeman splitting, where the complete M degeneracy is broken). An energy level pattern of the type shown in figure 5.22 is obtained.

By far the most widely used application of the Stark effect is in Stark-modulated microwave spectroscopy, which will be discussed in chapter 7.2. Molecular beam electric resonance and quadrupole deflection experiments, discussed in chapter 10.4, have also yielded a variety of interesting results. Direct optical detection of Stark splittings has been limited by the same sort of resolution problems that affected Zeeman studies. However, dipole mo-

2. An excellent discussion of the molecular Stark effect is given in chapter 10 of Townes and Schawlow (reference 15).

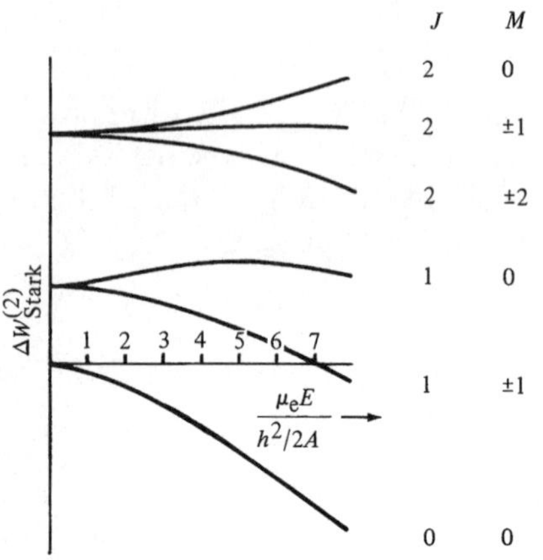

Figure 5.22
Stark splitting for $J = 0, 1, 2$ of a $^1\Sigma$ molecule.

ments of excited states of NH, OH, and CH have been determined by Dalby and his coworkers (reference 16) by examining the emission spectra of gas discharges in the high-field regions near electrodes. Freeman, Klemperer, and Lombardi (references 17, 18) have used the Stark effect for the determination of the dipole moments of the excited states of a number of larger molecules.

10 Magnetooptic Rotation (Faraday Effect) in Molecules

Closely related to the Zeeman effect is the phenomenon of magnetooptic rotation. It may be recalled at this point that molecules having no symmetry elements whatsoever (that is, belonging to the C_1 group: see chapter 6) possess the property of being able to rotate the plane of polarization of a light beam passing through an ensemble of such molecules. Since molecules of this sort generally possess very complex spectra, we have not considered the property of optical rotation in this book; for further details, the reader is referred to the text by Kauzmann (reference 19) or Djerassi (reference 20). The same sort of property can be induced in a number of small molecules, however, by the action of a magnetic field.

The experimental arrangement for observing the effect is shown in figure 5.23. Plane-polarized light is passed through the sample, contained in the

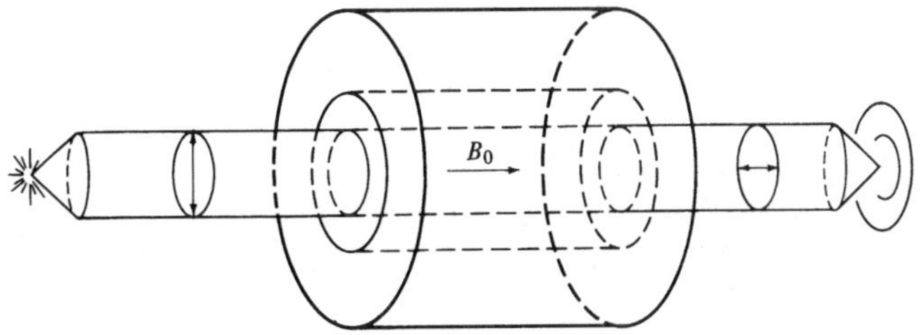

Figure 5.23
Experimental arrangement for observing the Faraday effect (magnetooptic rotation).

axial magnetic field produced by a surrounding solenoid. The light is viewed through an analyzing polaroid oriented at 90° to the first element. A "transmission" spectrum of bright lines will be seen against a dark background, indicating that the plane of polarization has been partially rotated in the neighborhood of the molecular absorption lines.

Let us see how the magnetic rotation can come about. The electric field vector of a light wave polarized in the x direction can be decomposed into right and left circularly polarized components,

$$E_x = \tfrac{1}{2}(E_+ + E_-),$$

with

$$E_+ = E_0 e^{i(\omega t - kz)}$$

and

$$E_- = E_0 e^{-i(\omega t - kz)}.$$

The wave vector $k = 2\pi/\lambda = 2\pi\nu n/c$, where n is the refractive index of the medium through which the light passes. In the case that $n_+ \neq n_-$, then, after the light passes a distance l through the medium, the plane of polarization will be rotated by an angle $\theta = (\pi l/\lambda_0)(n_- - n_+)$, with $\lambda_0 = c/\nu$.

A magnetic field has the property that it can render a medium containing molecules that have a Zeeman effect *dichroic*, that is, having different indices of refraction for right and left circularly polarized radiation. The following qualitative argument will indicate the nature of this effect. Consider a molecule that has a Zeeman effect, as shown in figure 5.24. When a field is applied

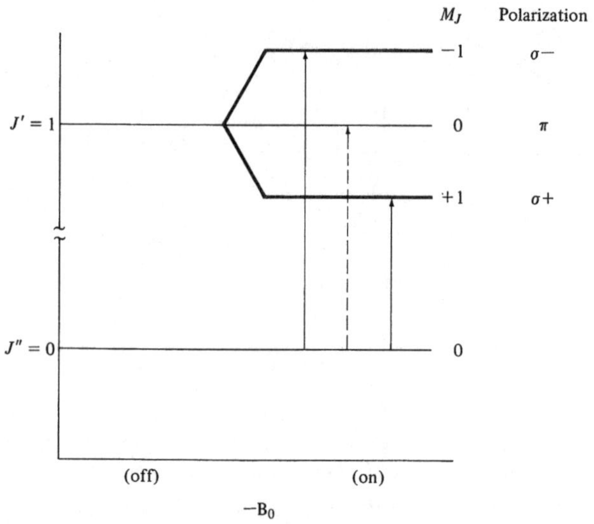

Figure 5.24
Zeeman effect for an $R(1)$ transition.

to such a molecule, the $\Delta M = +1, 0$, and -1 transitions are separated from one another. But the $\Delta M = +1$ transition is right circularly polarized, and the $\Delta M = -1$ transition is left circularly polarized (see chapter 2.8 for a discussion of this point). Now recall that the refractive index shows a *dispersion* in the vicinity of an absorption line. Since the absorption lines are slightly displaced from one another by the magnetic splitting, the corresponding refractive index variations will be too, so that n_+ and n_- will no longer cancel each other out. We shall obtain the situation depicted in figure 5.25, which will lead to a rotation in the immediate vicinity of the absorption line.

There are several areas of spectroscopy in which magnetooptic rotation (MOR) may be useful. Very often, a weak singlet–triplet transition may be buried in a strong singlet–singlet transition, especially in large molecules. The latter will be completely inactive in MOR, and the singlet–triplet structure will be observed without interference. Also, since the strength of the MOR is proportional to the Zeeman splitting, which goes as $[J(J+1)]^{-1}$, only low J lines will be active in the effect. This has the effect of picking out band origins in complex, overlapped spectra; use has been made of this

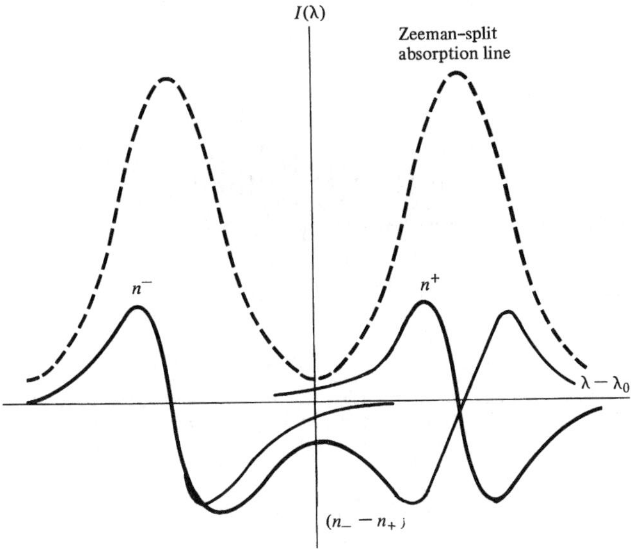

Figure 5.25
Magnetically induced dichroism in a Zeeman-split line.

property in the analysis of the ICl spectrum, whose MOR spectrum is shown in figure 5.26. An exception to this occurs when the Zeeman activity is induced by rotational Ω uncoupling; then the strength of the MOR intensity increases, rather than decreases, with *J*. Examples showing this behavior are CS_2 and I_2. This same technique, under the name of magnetic circular dichroism (MCD), is also a very useful method in the spectroscopy of large, complex molecules, such as proteins.

11 Photofragment Spectroscopy of Continuum States

Thus far, we have been exclusively concerned with the spectroscopy of bound molecular states with well-defined vibrational and rotational levels. A glance at figures 3.13–3.16 will show, however, that for typical molecules, there are at least as many repulsive states as bound states. Most of these are completely uncharacterized, however, because the absorption to these states, when it is seen, is a completely structureless continuum that defies conventional analysis. A method that permits accurate mapping of many of these

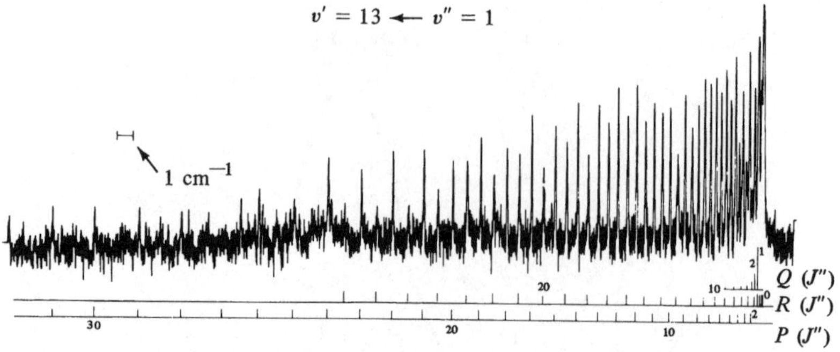

Figure 5.26
Magnetooptic rotation spectrum in the 13–1 band of IC1. Note that most of the intensity in the vibrational band appears right at the band origin. [Spectrum is reproduced with the permission of W. H. Eberhardt, W. C. Cheng, and H. Renner; see *J. Mol. Spectroscopy* **3**, 664 (1959).]

dissociating states is the photofragment spectrometer developed by K. R. Wilson and his associates (references 21, 22), now used in many laboratories.

The principle of this method is illustrated for the case of chlorine in figure 5.27. If a diatomic molecule is irradiated with light in a continuous absorption region, at energies lying above an atomic dissociation limit, the molecule will photodissociate into atomic fragments. The energies involved in the process are related by the equation

$$h\nu_0 + E_{\text{par}} = D_e'' + E(\text{atomic excitation}) + \tfrac{1}{2}\mu v_f^2,$$

where E_{par} is the internal energy in the absorbing molecule;[3] the atomic excitation energy takes account of any energy differences between ground state atoms and actual dissociation products; and v_f is the velocity with which the atoms separate. If the light energy is applied in a short, intense burst (such as a pulsed laser beam), and the molecules are dissociated in a collision-free region with a well-defined geometry (such as a molecular beam apparatus), then the flight time of the atomic fragments from the dissociation region to a suitable detector is a direct function of the amount of excess energy released in the dissociation. An apparatus capable of carrying out such measurements is shown in figure 5.28. Typical results of a photo-

3. E_{par} includes the zero-point vibrational energy of the molecule and, in addition, at any finite temperature, a distribution of rotational and vibrational energies. E_{par}^0, at absolute zero, would be the zero-point energy $\tfrac{1}{2}\hbar\omega_0$ alone.

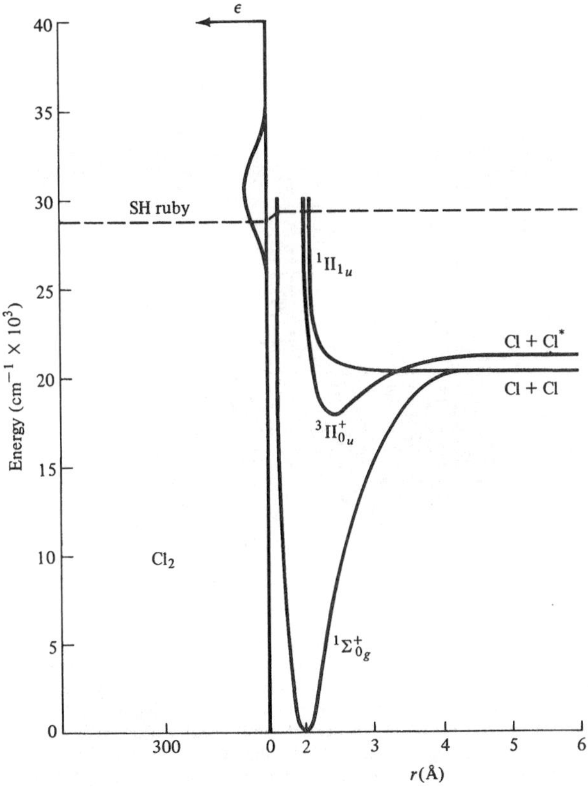

Figure 5.27
Cl_2 absorption spectrum and potential curves. Second-harmonic ruby laser light, at 28,810 cm^{-1}, falls in the Cl_2 uv absorption continuum, which is generally assigned to the $^1\Pi_{1u}$ state dissociating to ground state $Cl(^2P_{3/2})$ atoms. The $^3\Pi^-_{0u}$ contributes to absorption in the visible and dissociates to one ground state $Cl(^2P_{3/2})$ and one excited $Cl^*(^2P_{1/2})$ atom. The dotted energy line jogs between the two panels because it indicates the photon energy $h\nu$ on the left and the total molecular energy $h\nu + E_{par}$ on the right. E_{par} is the average room temperature thermal energy of the molecules before photon absorption measured from the bottom of the ground state potential curve. [From K. R. Wilson, "Photofragment Spectroscopy of Dissociative Excited States," in *Symposium on Excited State Chemistry*, J. N. Pitts, Jr. ed. (Gordon and Breach, New York, 1970). Reproduced with permission.]

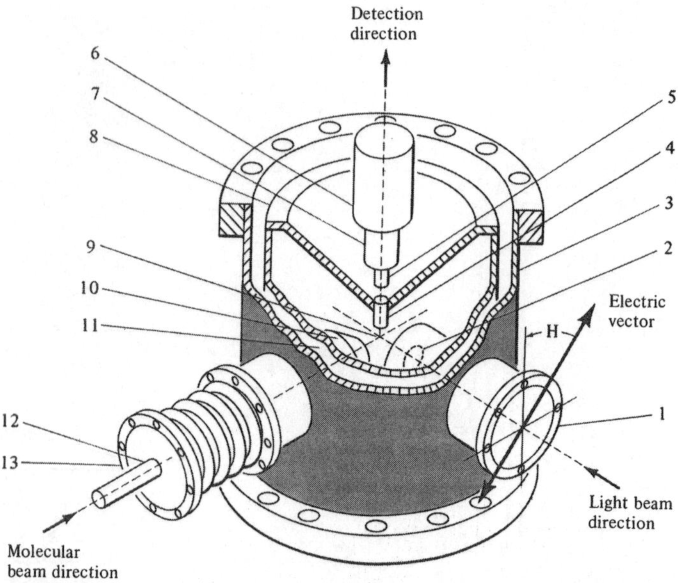

Figure 5.28
Cutaway drawing of photofragment spectrometer. The beam of molecules to be photodissociated enters from the left and is crossed perpendicularly by pulses of polarized light from a laser. The photodissociation fragments, which recoil upward, are detected by a mass spectrometer as a function of mass of photon energy, time after the laser pulse, and direction of recoil, Θ, measured from the electric vector of the light. The interaction region and the mass spectrometer are in separately pumped chambers connected by a small liquid nitrogen- (LN_2-) cooled tube, which collimates the fragments. The numbered components are (1) port for laser beam, (2) lens to match diameter of laser beam to that of molecular beam, (3) outer wall of bakable ultrahigh vacuum chamber, (4) LN_2-cooled fragment collimating tube, (5) mass spectrometer electron bombardment ionizer, (6) mass spectrometer electron multiplier, (7) quadrupole mass spectrometer, (8) LN_2-cooled partition between interaction and detection chamber, (9) interaction volume, (10) LN_2-cooled molecular beam collimator, (11) LN_2-cooled inner wall of interaction chamber, (12) molecular beam oven with capillary slits, (13) molecular beam port. [From K. R. Wilson, "Photofragment Spectroscopy of Dissociative Excited States," in *Symposium on Excited State Chemistry*, J. N. Pitts, Jr., ed. (Gordon and Breach, New York, 1970). Reproduced with permission.]

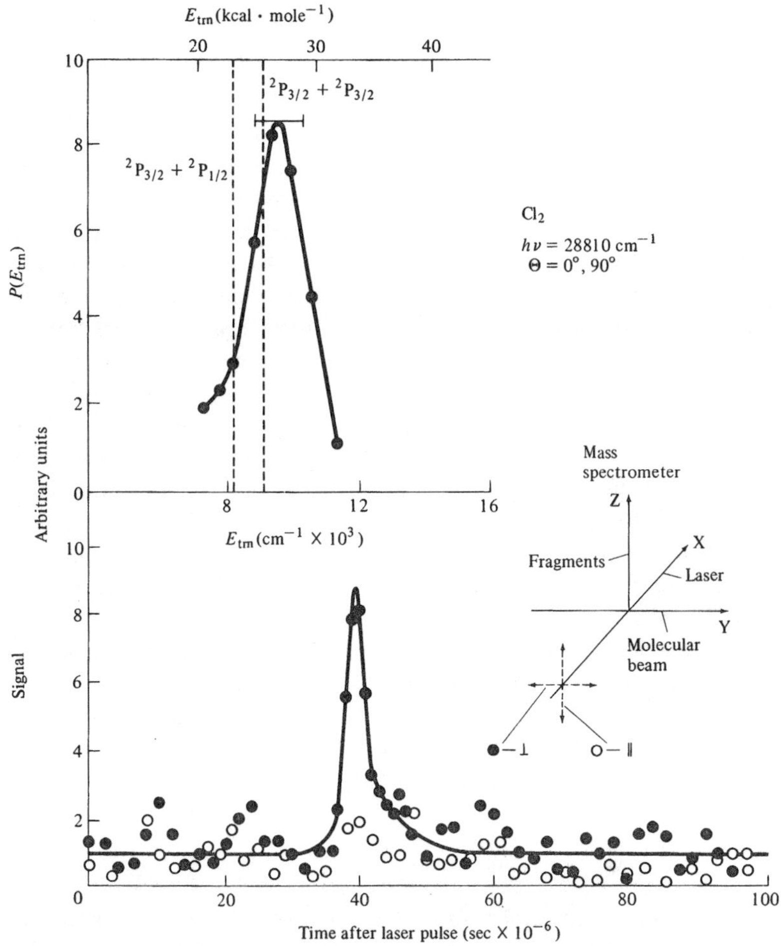

Figure 5.29
Cl_2 photofragment spectrum at 28,810 cm^{-1}. As shown in the distribution of mass spectrometer signal versus time after laser pulse, the majority of the Cl atoms recoil near laboratory angle $\Theta = 90°$ (average of 36 laser pulses, indicated by •), ⊥ to the electric vector of the light rather than near $\Theta = 0°$ (average of 20 laser pulses indicated by ○), ∥ to the electric vector. This implies that in the upper state $\Omega = 1$. The temperature of the molecular beam oven is 420°K, the fragment flight path is 5.0 cm, the mass detected is $^{35}Cl^+$, and the mass spectrometer delay is 10.8 μsec. When the $\Theta = 90°$ signal distribution is transformed into a distribution (as shown in the insert) in total translational energy E_{trn}, the predicted energy for dissociation into ground state $Cl(^2P_{3/2})$ atoms falls within the estimated error bar of the peak, while $Cl(^2P_{3/2}) + Cl^*(^2P_{1/2})$ falls outside. Thus measurements of the direction of recoil and of translational energy are both consistent with the assignment of $^1\Pi_{1u}$ for the upper state. [From K. R. Wilson, "Photofragment Spectroscopy of Dissociative Excited States," in *Symposium on Excited State Chemistry*, J. N. Pitts, Jr., ed. (Gordon and Breach, New York, 1970). Reproduced with permission.]

fragment spectrometry experiment, for chlorine, are shown in figure 5.29. In this case, the results obtained are in accord in every detail with theoretical expectations. For diatomic molecules, the final translational energy of the fragments can indicate to which of several possible different repulsive potential curves, having different final atomic states, the absorption is occurring. This has been applied in the case of diatomic iodine (reference 23), for which this experiment also provided the first unambiguous proof that spontaneous predissociation was actually occurring after excitation to energies lying below the nominal dissociation limit (reference 24). For polyatomic molecules, this method is very helpful in determining how much of the excess energy remains in internal degrees of freedom of the fragments produced in the dissociation and how much is released as translational energy of separation of the fragments. Y. T. Lee has employed this technique to study molecular fragmentation resulting from multiple infrared photon excitation (reference 25; also see chapter 13.4).

With the development of tunable uv and visible wavelength lasers (chapter 10), a host of new techniques is available for the study of diatomic molecule spectroscopy. A few of these, which will be touched upon later in this text, are optical-optical double resonance (chapter 12.7; also see reference 26) and multiphoton ionization (chapter 13.5).

Problems

1. In the vacuum ultraviolet absorption spectrum of the O_2 molecule at 300°K, there appears a v' progression of bands that converges to a well-defined limit at 1759 Å (56,850 cm^{-1}). The dissociation products are one O atom in its ground state ($^3P_2$) and one O atom in an excited state, which can be either $^1D_2$ or $^1S_0$. These latter two states are, respectively, 1.967 and 4.190 eV (15,870 and 33,801 cm^{-1}) above the $^3P_2$ state. From thermochemical studies, the heat of dissociation of O_2 into two O atoms in the $^3P_2$ state is known to lie in the range 100–150 kcals/mole.

(a) Sketch the potential curves in cm^{-1} for O_2 corresponding to the foregoing data, roughly to scale. $r_e' = 1.42$ Å, $r_e'' = 1.21$ Å.
(b) What must be the electronic state of one O atom when the excited molecular state dissociates if the other O atom is in the $^3P_2$ state?
(c) From the spectroscopic data, compute the dissociation energy D_0'' in cm^{-1} of the O_2 molecule in its ground electronic state into two O atoms ($^3P_2$).

(d) The following constants in cm^{-1} hold for the upper and lower states of O_2: $D_0' = 23{,}133$; harmonic frequency: $\omega_e' = 819$; $\omega_e'' = 1{,}580$; $\omega_e x_e' = 22$; $\omega_e x_e'' = 12$. Compute T_e, the separation in cm^{-1} of the two pure electronic states, that is, the separation of the potential minima of the two states.

2. The ground state and a low-lying excited electronic state of the BeO molecule have the following properties:

Term symbol	$^1\Sigma^+$	$^1\Pi$
Electronic energy, T_e/cm^{-1}	0	9,405.6
ω_e/cm^{-1}	1,487.3	1,144.2
$\omega_e x_e/cm^{-1}$	11.8	8.4
$r_e/10^{-8}$ cm	1.33	1.46

Note that the electronic energy T_e is the energy from the minimum of one curve to the minimum of the other; this is not equal to the vibrational origin of the 0–0 band.

(a) Construct a Deslandres table of the vibrational band origins of the $^1\Pi$–$^1\Sigma^+$ system, for $v'' = 0$–3 and $v' = 0$–5. Which of these vibrational bands would you expect to be the most intense when the system is observed in absorption? Comment on the relative intensities that you would expect for the other bands in your table.

(b) In the rotational structure of the individual vibronic bands in this system, what branches would you expect to observe? In which branch would you expect to observe a band head? Calculate the transition in J that will give rise to a line at the band head and the distance in cm^{-1} from the band head to the vibrational band origin.

(c) What would you guess about the MO configurations corresponding to these two states? (*Hint:* Note that BeO is isoelectronic with C_2, so that the MOs may be expected to be somewhat similar, except that the *g*–*u* property will be lost, and the orbitals will be distorted toward the higher nuclear charge of the O atom.) Would you suspect the presence of any other excited electronic states *below* the $^1\Pi$ state? If so, what would its term symbol be?

3. The following bands are observed in the second positive system of nitrogen (units are reciprocal centimeters corrected to vacuum):

35,522	29,940	25,913
35,453	29,654	25,669
33,852	29,010	25,354
33,751	28,819	25,003
33,583	28,559	24,627
32,207	28,267	24,414
32,076	27,949	24,137
31,878	27,451	23,800
31,643	27,226	23,414
30,590	26,942	23,016
30,438	26,621	
30,212	26,274	

Arrange these in a Deslandres table, and find values for ω_e'', $\omega_e x_e''$, ω_e', and $\omega_e x_e'$. (*Important Suggestion:* Look at the pattern of bands first, before doing anything else. Do any natural groupings seem to suggest themselves? It may help to draw a "stick spectrum" of the band origins, to scale, in order to pick out these patterns. Remember that bands having the same Δv fall along diagonals on the Deslandres table.)

Is there any suggestion of a cubic term in either state? If so, derive an expression for the third difference, $\Delta^3 G_{v+1/2}$, including terms in $\omega_e y_e(v+\frac{1}{2})^3$ in $G_{v+1/2}$, and estimate $\omega_e y_e$.

4. The selection rules on electronic-rotational states in space-fixed axes are the following:

For magnetic dipole transitions

$+ \leftrightarrow +$, $- \leftrightarrow -$, $+ \not\leftrightarrow -$

$s \leftrightarrow s$, $a \leftrightarrow a$, $s \not\leftrightarrow a$

$\Delta\Lambda = \pm 1$ if $\Delta\Sigma = 0$

$\Delta\Lambda = 0$ if $\Delta\Sigma = \pm 1$ (why?)

$\Delta J = 0, \pm 1$, but $J = 0 \not\leftrightarrow J = 0$

For electric quadrupole transitions

$+ \leftrightarrow +$, $- \leftrightarrow -$, $+ \not\leftrightarrow -$

$s \leftrightarrow s$, $a \leftrightarrow a$, $s \not\leftrightarrow a$

$\Delta J = 0, \pm 1, \pm 2$

except $J = 0 \not\leftrightarrow J = 0$, $J = \frac{1}{2} \not\leftrightarrow J = \frac{1}{2}$, $J = 1 \not\leftrightarrow J = 0$

(Convince yourself that these are consistent with the MD and EQ selection rules on electronic states previously given above for molecule-fixed axes.)

Show the allowed branches for the following transitions:

$^1\Pi_g$–$^1\Sigma_g^+$ (magnetic dipole)
$^1\Sigma_g^+$–$^3\Sigma_g^-$ (magnetic dipole)
$^1\Delta_g$–$^3\Sigma_g^-$ (magnetic dipole)
$^1\Sigma_g^+$–$^1\Sigma_g^+$ (electric quadrupole)

5. The first strong electronic band system of carbon monoxide (the ground state vibrational frequency of which is observed at 2,140 cm^{-1} in the infrared) appears in absorption at room temperature at about 1,550 Å in the vacuum ultraviolet. The system shows a progression with a spacing of 1,480 cm^{-1}. The vibronic bands show a single set of unperturbed P, Q, and R branches degraded to the red. Analysis by combination differences of these branches gives $B'_e = 1.61$ cm^{-1}, $B''_e = 1.93$ cm^{-1}. In each band, the lines nearest the origin are $P(2)$, $Q(1)$, and $R(0)$.

(a) Deduce all you can about the two electronic states involved in the transition from these data and your general knowledge of the properties of carbon monoxide.
(b) Sketch the lower portions of the potential curves in cm^{-1} for CO, roughly, to scale, from these data [find the harmonic force constant, and use the potential $U(r) = \frac{1}{2}k(r - r_e)^2$]. Use the Franck-Condon principle to find the strongest vibronic bands in the spectrum.
(c) Interpret the electronic terms of the two states in terms of the most likely MO configurations of each.
(d) The dipole moment of the ground state of CO is about 0.1 D (debye; 1 debye $= 10^{-18}$ esu-cm). Show how an optical Stark effect experiment can be used to find the dipole moment of the excited states; estimate the magnitudes of the splittings for the $P(2)$, $Q(1)$, and $R(0)$ lines, for an applied field of 10,000 V/cm (1 statvolt = 300 ordinary volts), and an assumed excited state moment of 1.0 debye.

6. Draw two Morse curves defined by the following constants:

$T''_e = 0$ cm^{-1}	$T'_e = 10{,}000$ cm^{-1}
$D''_e = 10{,}000$ cm^{-1}	$D'_e = 5{,}000$ cm^{-1}
$R''_e = 1.44 \times 10^{-8}$ cm	$R'_e = 1.54 \times 10^{-8}$ cm
$\omega''_e = 582$ cm^{-1}	$\omega'_e = 308$ cm^{-1}

and determine the two strongest transitions originating from $v' = 19$. Use a reduced mass of $\mu = 6.857$ amu.

Plot $V'(x) - V''(x)$ versus x. Use this curve to determine

(a) the long and short wavelength limits of all bound-bound transitions in this system that possess significant intensity;
(b) the long and short wavelength limits of strong bands from $v' = 19$;
(c) the plot of the x value(s) sampled versus transition energy for the progression of bands ($v' = 19, v''$) for $v'' = 0$ through the maximum v'' level that can be reached via a nonnegligible Franck-Condon factor from $v' = 19$.

References

1. J. Franck, *Trans. Faraday Soc.* **21**, 536 (1925); E. U. Condon, *Phys. Rev.* **32**, 858 (1928).

2. S. N. Suchard, *Spectroscopic Data* (IFI/Plenum, New York, 1975).

3. A. J. Kotlar, R. W. Field, J. I. Steinfeld, and J. A. Coxon, *J. Mol. Spectroscopy* **80**, 86 (1980).

4. J. I. Steinfeld, R. N. Zare, L. Jones, M. Lesk, and W. Klemperer, *J. Chem. Phys.* **42**, 25 (1965).

5. R. L. Brown and T. C. James, *J. Chem. Phys.* **42**, 33 (1965).

6. G. Herzberg, *Molecular Spectra and Molecular Structure*, Vol. 1: *Spectra of Diatomic Molecules* (Van Nostrand Reinhold, New York, 1950).

7. G. Herzberg and A. Monfils, *J. Mol. Spectroscopy* **5**, 482 (1960).

8. W. Kołos and L. Wolniewicz, *J. Chem. Phys.* **41**, 3663, 3674 (1964), **43**, 2429 (1965), **45**, 509 (1966), **48**, 3672 (1968), **49**, 404 (1968), **51**, 1417 (1969).

9. L. Pauling and E. B. Wilson, Jr., *Introduction to Quantum Mechanics* (McGraw-Hill, New York, 1935), pp. 180ff.

10. G. Herzberg, *J. Mol. Spectroscopy* **33**, 147 (1970).

11. J. Hougen, *The Calculation of Rotational Energy Levels and Rotational Line Intensities in Diatomic Molecules*, N.B.S. Monograph 115 (US Government Printing Office, Washington, DC, June 1970).

12. L. I. Schiff, *Quantum Mechanics*, 3rd ed. (McGraw-Hill, New York, 1968), p. 285.

13. A. Chutjian and T. C. James, *J. Chem. Phys.* **51**, 1242 (1969).

14. H. Lefebvre-Brion and R. W. Field, *Perturbations in the Spectra of Diatomic Molecules* (Academic Press, New York, 1984).

15. C. H. Townes and A. Schawlow, *Microwave Spectroscopy* (McGraw-Hill, New York, 1955).

16. T. A. R. Irwin and F. W. Dalby, *Can. J. Phys.* **43**, 1766 (1965); D. H. Phelps and F. W. Dalby, *Phys. Rev. Letters* **16**, 1 (1966).

17. D. E. Freeman, W. Klemperer, and J. R. Lombardi, *J. Chem. Phys.* **40**, 604 (1964), **45**, 52, 58 (1966), **46**, 2746, 3482 (1967).

18. J. R. Lombardi, *J. Chem. Phys.* **50**, 3780 (1969), **51**, 1228 (1969).

19. W. Kauzmann, *Quantum Chemistry* (Academic Press, New York, 1957), pp. 616ff, 703ff.

20. C. Djerassi, *Optical Rotatory Dispersion* (McGraw-Hill, New York, 1960).

21. K. R. Wilson, "Photofragment Spectroscopy of Dissociative Excited States," in *Symposium on Excited State Chemistry*, J. N. Pitts, Jr., ed. (Gordon and Breach, New York, 1970).

22. G. E. Busch, K. E. Cornelius, R. T. Mahoney, R. I. Morse, D. W. Schlosser, and K. R. Wilson, *Rev. Sci. Instr.* **41**, 1066 (1970).

23. R. J. Oldman, R. K. Sander, and K. R. Wilson, *J. Chem. Phys.* **54**, 4127 (1971).

24. G. E. Busch, R. T. Mahoney, R. I. Morse, and K. R. Wilson, *J. Chem. Phys.* **51**, 837 (1969).

25. Y. T. Lee and Y. R. Shen, *Physics Today* **33**(11), 52 (1980).

26. R. A. Gottscho, *Coherent Focus on Science*, Vol. 2 (Coherent, Inc., Palo Alto, CA, 1978).

6 Rudimentary Group Theory

Tiger, Tiger burning bright
In the forests of the Night
What immortal Hand or Eye
Can frame thy fearful Symmetry?
William Blake

In the discussion to this point, we have been dealing exclusively with atoms and diatomic molecules. These species possess a very high degree of symmetry (spherical or cylindrical, respectively), and thus the effects of symmetry operations on an atom or molecule are particularly easy to see. As we come to treat polyatomic molecules, however, a formal language is required to deal with their symmetry properties and operations; this language is group theory. In attempting to present this material, one is faced with a dilemma: whether to develop a full exposition, including detailed derivations and examples, or to limit oneself to a brief overview. The decision taken here, which is consistent with the approach taken throughout this book, and is also evident from the title of this chapter, is the latter. We shall introduce the concepts and language of group theoretical methods, give a few simple examples of its use (in section 6), and refer extensively to the numerous excellent textbooks on this subject (references 1–10).

In effect, we have already made use of group theory, in the discussion of diatomic molecule wave functions in chapters 3.2 and 3.4, and of spin statistics in chapter 4. In deriving selection rules, for example, we are interested in evaluating integrals of the form

$$\int \psi_1^*(\mathcal{O}p)\psi_2\, d\mathbf{r},$$

where the operator may be the rotational or vibrational part of the Hamiltonian, if we are interested in energy term values, or the electric dipole moment, if we are interested in selection rules, and so on. An integral of this type will vanish if its argument is antisymmetric in space; therefore, the only integrals that we need concern ourselves with are those in which the product of the two wave functions, $\psi_1^*\psi_2$, has the same symmetry as the operator in the integral. This enables us to reduce otherwise very large and complex matrix expressions to much simpler ones, by simply setting all the matrix elements of the wrong symmetry combinations equal to zero. A corollary of this principle is that the wave functions for any system must belong to the same symmetry group as its Hamiltonian. This can be seen by letting $\psi_1 = \psi_2$, and the operator be $\mathscr{H}$, in the integral we have been considering. We know that this integral is the energy of the system, which is generally not equal to zero. Thus $(\psi^*\psi)$ and $\mathscr{H}$ must have the same symmetry, and we can always choose a linear combination of the individual ψs that will have the correct symmetry.

1 Symmetry Elements E, C_n, σ, and i

The traditional treatments of molecular symmetry make use of the molecular point group, which is the group of rotations and reflections originally devised to describe systems at rest in space, such as crystalline solids. For molecules undergoing dynamic motions, such as rotations and vibrations, a more convenient representation is the molecular symmetry group, which is the group of inversions and nuclear permutations. (The relations between these groups are thoroughly developed in reference 10.) Here we shall make use of the more familiar point group notation.

The point symmetry properties of a molecule are characterized by the effects of four basic types of operations on the molecule. These are the identity, or "leaving-it-alone," rotation, reflection, and inversion. They can be defined in terms of their effect on a set of coordinates.

Identity, E: $E(x, y, z) = (x, y, z)$.
Rotation, C_p^z: $C_p^z(r, \theta, z) = (r, \theta + (2\pi/p), z)$.

For example, $C_2^z(x, y, z) = (-x, -y, z)$; the rotation axis is always designated the z axis, by convention.

Reflection, σ_{ij}: $\sigma_{xz}(x, y, z) = (x, -y, z)$, and so on.
Inversion, i: $i(x, y, z) = (-x, -y, -z)$.

The effect of multiplying symmetry operators, that is, performing two operations in succession, is best evaluated by geometrical inspection, as in the examples in figure 6.1. A reflection plane that is perpendicular to a principal rotation axis is designated a σ_h (horizontal) plane; a reflection plane that contains such a rotation axis is designated a σ_v (vertical) plane. Thus, if the rotation axis is taken to be the z axis, σ_{xz} and σ_{yz} are σ_v planes and σ_{xy} is a σ_h plane.

Table 6.1

Point group	Symmetry elements	Examples
Uniaxial groups		
C_1	E only	All molecules that possess no discernible symmetry element fall into this group. This is actually the most well-populated group, encompassing molecules from propane up through lysozyme and DNA. Unfortunately, the absence of symmetry in these molecules means that few simplifications can be applied to the analysis of their spectra.
C_2	C_2	H_2O_2, hemoglobin.
$C_3, C_4, \ldots$	$C_3, C_4, \ldots$	Any triatomic molecule of general shape with three unequal bond lengths and angles. Also, the tiger cited at the head of this chapter would have C_s symmetry. 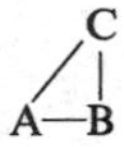
C_{2h}	$C_2 + \sigma_h$	Glyoxal, 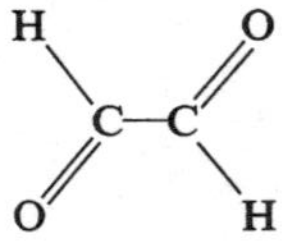
$C_{3h}, \ldots$	$C_3 + \sigma_h, \ldots$	
$C_{1v} = C_{1h} = C_s$	σ_v only	See above, C_s.
C_{2v}	$C_2 + \sigma_v$	Formaldehyde, 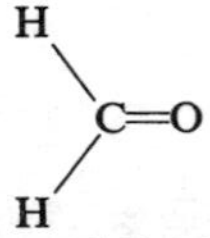
C_{3v}	$C_3 + \sigma_v$	Ammonia, NH_3; methyl chloride, CH_3Cl.
⋮	⋮	
$C_{\infty v}$	$C_\infty + \sigma_v$	Any heteronuclear diatomic molecule.

Dihedral groups		
$D_1 = C_2$	C_2 only	See above, C_2.
D_2	3 mutually perpendicular C_2s	
D_3	$C_3 + 3\ C_2$s	Ethane, partially staggered conformation. 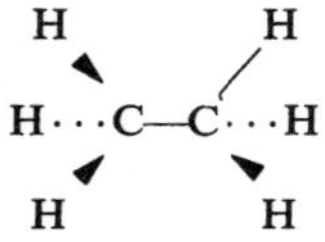
$D_{1h} = C_{2v}$	$C_2 + \sigma_h$	See above, C_{2v}
D_{2h}	$3\ C_2\text{s} + \sigma_h$	Ethylene. 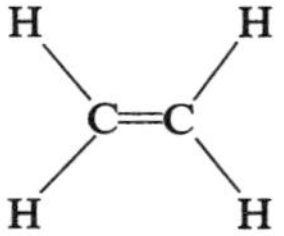
D_{3h}	C_3, $3\ C_2$s, σ_h	BCl_3.
D_{6h}	C_6, $6\ C_2$s, σ_h	Benzene.
$D_{\infty h}$	$C_\infty + \sigma_h$	Any homonuclear diatomic molecule.
D_{2d}	$3\ C_2\text{s} + 2\ \sigma_v$s	Allene. 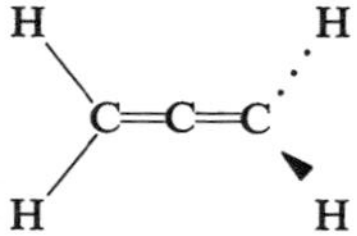
D_{3d}	C_3, $3\ C_2$s, $3\ \sigma_v$s	Staggered ethane. 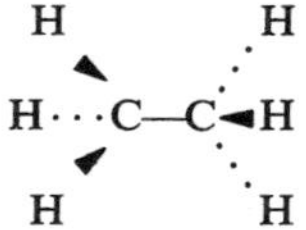
Cubic groups		
T_d	$3\ C_2$s, $4\ C_3$s, $6\ \sigma$s	Methane.
O_h	$3\ C_4$s, $4\ C_3$s, $6\ C_2$s, $9\ \sigma$s, i	Cubane, SF_6.
I_h	$6\ C_5$s, $10\ C_3$s, $15\ C_2$s, . . .	A few rare polyboron lattices.

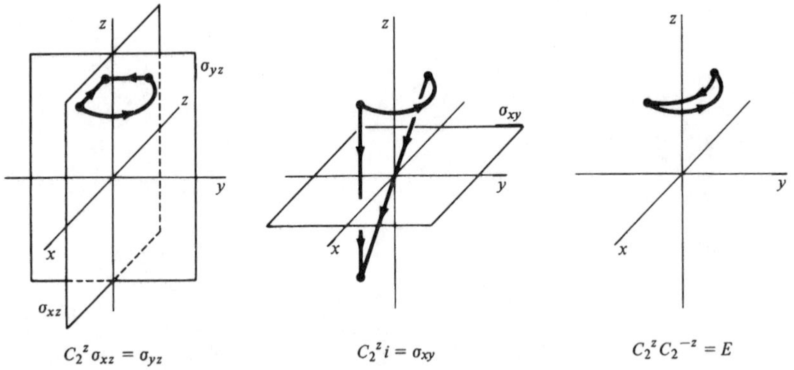

Figure 6.1
Products of symmetry operations in the D_{2h} point group.

2 Point Groups for Molecules

The symmetry of a molecule is described by all those symmetry elements it possesses. The various combinations have been designated by the point group notation, which is summarized in table 6.1. Note that as the symmetry increases, the presence of certain elements dictates the presence of certain others. For example, in the D_{6h} group to which benzene belongs, it is sufficient to specify the sixfold rotation axis and the horizontal reflection plane. As shown in figure 6.2, this causes the simultaneous presence of six additional twofold axes (perpendicular to the C_6 axis), six vertical reflection planes, a C_2 axis and a C_3 axis coincident with the C_6 axis, and a center of inversion i.

3 Group Properties and Multiplication Tables

The molecular symmetry elements, which we have been discussing, from a mathematical group. In order to form such a group, each of the h operations contained in any of the point groups must possess the following properties:

1. *Closure*, which requires that the product of any two elements in the group be another element of the group, or

$$R_i R_j = R_k,$$

where i, j, and k are all contained in h.

2. Multiplication of any two elements must be *associative*; that is,

$R_1(R_2R_3) = (R_1R_2)R_3.$

3. There must be a unique *identity* operation E, which has the property $ER = RE = R$.
4. Each element R must have an *inverse* R^{-1}, with the property $RR^{-1} = R^{-1}R = E$.

For the symmetry elements of the point groups, multiplication of two elements is understood to mean performing the two symmetry operations in succession. Note that we have not required that the multiplication be *commutative*, that is, that $R_1R_2 = R_2R_1$. This property is not a group criterion in general; those groups whose elements do commute are a special category, called *Abelian* groups. In those instances when commutation was implicit in the above definitions, they have been stated, namely, that the identity commutes with every other element, and that every element commutes with its own inverse.

We can illustrate these properties with the group C_{2v}, which contains the the elements E, C_2, σ_{xz}, and σ_{yz}. The *order* of the group is equal to the total number of elements, namely, 4. A table of the 16 possible products of pairs of elements can be constructed by inspecting the results of each transformation, as in the examples in section 1; this table has the form

	E	C_2	σ_{xz}	σ_{yz}
E	E	C_2	σ_{xz}	σ_{yz}
C_2	C_2	E	σ_{yz}	σ_{xz}
σ_{xz}	σ_{xz}	σ_{yz}	E	C_2
σ_{yz}	σ_{yz}	σ_{xz}	C_2	E

First, note that the product of any two elements of the group, in any order, is another member of the group, as required by the closure criterion. In fact, the two elements E and C_2 are seen to form a little *subgroup* of their own. Since the identity operator E appears along the principal diagonal, it follows that each operation is its own inverse.

It will also be useful very shortly to distinguish different *classes* of elements, A set of group elements A, B, C, ... are all in the same class if operating on one of these elements with any of the other elements in the entire group in the sequence

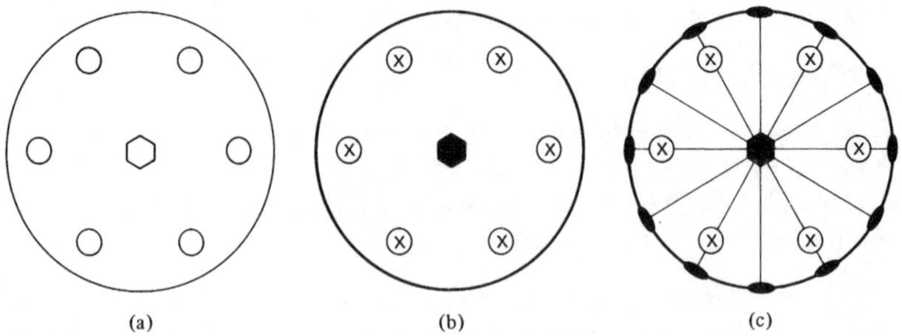

Figure 6.2
Generation of symmetry elements in the D_{6h} group. (a) Symmetry of a C_6 axis perpendicular to the page. (b) Addition of a horizontal mirror plane perpendicular to the C_6; × indicates above the plane, o below the plane. (c) These same elements are then seen to possess six C_2s (perpendicular to the C_6), six σ_vs, (containing the C_2s and the C_6), a C_2 and a C_3 coincident with the C_6, and an i in the center of the figure.

$$RAR^{-1}, RBR^{-1}, \ldots$$

yields one of the other members of the class, that is,

$$RAR^{-1} = B, RBR^{-1} = A, \ldots.$$

If we try this out on the members of the C_{2v} group, remembering that each element is its own inverse, we get, for example,

$$\sigma_{xz}C_2\sigma_{xz} = \sigma_{xz}\sigma_{yz} = C_2,$$

$$C_2\sigma_{yz}C_2 = C_2\sigma_{xz} = \sigma_{yz},$$

and we very rapidly discover that each element is in a class by itself. This turns out to be a specific case of a more general property. If we look back at the multiplication table, we see that it is symmetric to reflection across the main diagonal. This means that the product of any two elements in either order gives the same result; that is, multiplication is *commutative* in this group, which is, therefore, Abelian. It is easy to show that, in an Abelian group, each element is a distinct class, since any product of the form $RAR^{-1} = RR^{-1}A = EA = A$. So in the C_{2v} group, as in all Abelian groups, the number of classes, s, is equal to the order of the group h.

A definite physical significance can be attached to the distinct classes of a symmetry group. Each of the symmetry operations in a given class is essentially equivalent and distinct from the symmetry operations belonging to other classes. In groups of low symmetry, such as C_{2v}, each operation is in a class by itself, but groups containing threefold or higher-symmetry axes possess classes containing more than one symmetry element. For example, in the D_{6h} group illustrated in figure 6.2, the three C_2 axes passing through the symmetry elements belong to one distinct class, and the three C_2' axes passing between the symmetry elements belong to a different class.

4 Representations and Character Theory

The most convenient way of manipulating symmetry operations in a group is by means of *representations*. These are sets of numbers that obey the same multiplication rules as the symmetry operations themselves. For example, in the C_{2v} group, the following sets of numbers all reproduce the multiplication table for the operations that they represent, as given in the previous section:

Representation	E	C_2	σ_{xz}	σ_{yz}
Γ_1	+1	+1	+1	+1
Γ_2	+1	+1	−1	−1
Γ_3	+1	−1	+1	−1
Γ_4	+1	−1	−1	+1

Representations need not be scalar numbers, either. Sets of matrices can constitute a representation, as in the following examples:

$$\Gamma_5: \quad \begin{pmatrix} +1 & 0 \\ 0 & +1 \end{pmatrix} \begin{pmatrix} +1 & 0 \\ 0 & +1 \end{pmatrix} \begin{pmatrix} +1 & 0 \\ 0 & -1 \end{pmatrix} \begin{pmatrix} +1 & 0 \\ 0 & -1 \end{pmatrix},$$

$$\Gamma_6: \quad \begin{pmatrix} +1 & 0 \\ 0 & +1 \end{pmatrix} \begin{pmatrix} +1 & 0 \\ 0 & -1 \end{pmatrix} \begin{pmatrix} +1 & 0 \\ 0 & -1 \end{pmatrix} \begin{pmatrix} +1 & 0 \\ 0 & +1 \end{pmatrix}.$$

One could go on like this forever, making up new sets of numbers and matrices that duplicate the multiplication properties of the group. However, most of these will be *reducible* to a finite, basic set of representations, known as the *irreducible representations*. The number of irreducible representations of a group is equal to the number of classes, s, of elements in the group; so for the C_{2v} group, there will be only four irreducible representations.

A representation is reducible if it can be decomposed, element by element, into a sum of two or more other representations. In the case of matrix representations, the decomposition may be in terms of a direct sum, which is constructed as follows:

$$\text{the direct sum of } \mathbf{A} \text{ and } \mathbf{B} = \underset{(n\times n)}{\mathbf{A}} \oplus \underset{(m\times m)}{\mathbf{B}} = \underset{(n+m)\times(n+m)}{\begin{pmatrix} \mathbf{A} & 0 \\ 0 & \mathbf{B} \end{pmatrix}}.$$

Thus, in the preceding example, Γ_5 is a direct sum of the 1×1 matrices $\Gamma_1 \oplus \Gamma_2$, and $\Gamma_6 = \Gamma_1 \oplus \Gamma_4$. Similarly, multiplying Γ_1 through Γ_4 by some constant c, to get other than 1's in the representations, would mean that the new representation could simply be reduced into c of the original ones. Γ_1 through Γ_4 are the four irreducible representations of the C_{2v} group.

Representations are tabulated in terms of their characters. The character of a representation is the set of numbers, one for each operation in the group, that are the algebraic sums of the diagonal elements of the matrices making up the representation. For one-dimensional representations, the character is just the set of numbers themselves. For the higher-dimensional representations in the previous example, we would have

$$\chi_{\Gamma_5} = (2, 2, 0, 0)$$

and

$$\chi_{\Gamma_6} = (2, 0, 0, 2).$$

Notice that the characters of the reducible representations are just the sums of the characters of the irreducible representations into which each one is decomposed. The reduction of Γ_5 and Γ_6 could have been done from the characters alone, but it is much easier to pick out the components of the direct sum from the explicit form of the representation than from the characters alone.

There is a standard notation, known as the "Placzek notation," for the various irreducible representations of the point groups. This notation is applied to molecular states of various symmetries in the same way as Σ, Π, Δ, ... is used for diatomic molecules. The notation is defined in terms of the characters of the corresponding representations:

$$\chi_A(E) = \chi_B(E) = 1,$$

$$\chi_E(E) = 2,^1$$

$$\chi_F(E) = 3,$$

$$\chi_A(C_p) = +1, \quad \chi_B(C_p) = -1,$$

$$\chi_{A_1}(\sigma_v) = +1, \quad \chi_{A_2}(\sigma_v) = -1,$$

$$\chi_{A'}(\sigma_h) = +1, \quad \chi_{A''}(\sigma_h) = -1,$$

$$\chi_u(i) = +1, \quad \chi_g(i) = -1.$$

The two- and three-dimensional representations appear only in those point groups containing threefold or higher rotation axes. None of the irreducible representations of the point groups for molecules is higher than three dimensional. The F representation is occasionally called T, especially in solid state spectroscopy.

The labeling of the representations for the C_{2v} group, using this notation, can easily be done by referring to the character table. We see that

$$\Gamma_1 = A_1, \qquad \Gamma_2 = A_2, \qquad \Gamma_3 = B_1, \qquad \text{and} \qquad \Gamma_4 = B_2.$$

Character tables for the molecular point groups may be found in any of the references listed at the chapter's end.

5 Direct Products of Representations

In the use of group theory to determine spectroscopic selection rules, which is the most important application we shall consider, it will be necessary to determine the symmetry classification of the product of two quantities (wave functions, dipole operator, and so on). When we did this for diatomic molecules, we used the mnemonic that each element of oddness or evenness multiplied through as -1 or $+1$, respectively. When we come to the more complex symmetry elements of polyatomic molecules, however, this simple scheme is not sufficient.

The representation of a product of two functions, belonging to irreducible representations Γ_a and Γ_b, is the *direct product* Γ_{ab}, defined as

1. Please be sure to distinguish between E, the identity operation, and E, the symbol for a doubly degenerate representation. This is just one of the many notational ambiguities that has unfortunately crept into spectroscopic usage.

$$
\begin{matrix}
\Gamma_a & \otimes & \Gamma_b & = & \Gamma_{ab} \\
\begin{pmatrix} 1 & 2 & 3 \\ 4 & 5 & 6 \\ 7 & 8 & 9 \end{pmatrix} & & \begin{pmatrix} a & b & c & d \\ e & f & g & h \\ i & j & k & l \\ m & n & o & p \end{pmatrix} & & \left(\begin{array}{ccc|ccc|c} 1a & 2a & 3a & 1b & 2b & 3b & \\ 4a & 5a & 6a & 4b & 5b & 6b & \cdots \\ 7a & 8a & 9a & 7b & 8b & 9b & \\ \hline 1e & 2e & 3e & & & & \\ 4e & 5e & 6e & & \ddots & & \\ 7e & 8e & 9e & & & & \\ \hline & \vdots & & & & & \end{array}\right). \\
(n \times n) & & (m \times m) & & (mn \times mn)
\end{matrix}
$$

For one-dimensional representations, of course, the direct product is just the simple arithmetical product. In the reference textbooks on group theory cited at the beginning of this chapter, it is shown that the character of a direct product representation is just the product of the character of the individual representations; that is,

$$\chi_{\Gamma_{ab}}(R) = \chi_{\Gamma_a}(R) \cdot \chi_{\Gamma_b}(R)$$

for each operation R in the group. For the simple one-dimensional representations, since the character and the representation are one and the same number, this is self-evidently true.

For an example of how this works, let us again look at a portion of the character table of the C_{2v} group.

	E	C_2	σ_{xz}	σ_{yz}
A_2	1	1	-1	-1
B_1	1	-1	1	-1
$A_2 \otimes B_1$	1	-1	-1	1

But by inspection, this last row is just identical to the character for B_2; thus, we can say that $A_2 \otimes B_1 = B_2$. In this was we construct the *direct product tables* so dear to the hearts of spectroscopists. For the C_{2v} group, the complete product table is

C_{2v}	A_1	A_2	B_1	B_2
A_1	A_1	A_2	B_1	B_2
A_2	A_2	A_1	B_2	B_1
B_1	B_1	B_2	A_1	A_2
B_2	B_2	B_1	A_2	A_1

Direct products involving the two- and three-dimensional representations are often reducible to a direct sum of several irreducible representations. As an example, consider the direct product $E \otimes E$, in the C_{3v} group. The product character would then be

C_{3v}	E	$2C_3$	$3\sigma_v$
E	2	-1	0
E	2	-1	0
$E \otimes E$	4	$+1$	0

This character does not correspond to any of the irreducible representations of C_{3v}, but it can be reduced to $A_1 \oplus A_2 \oplus E$, as shown by the characters:

C_{3v}	E	$2C_3$	$3\sigma_v$
A_1	1	1	1
A_2	1	1	-1
E	2	-1	0
$A_1 \oplus A_2 \oplus E$	4	1	0

Direct product tables for all the important point groups are given in appendix A.

6 Selected Applications

A very useful and straightforward application of this group theoretical apparatus is the determination of allowed transitions in polyatomic molecules. To do this, we must first establish the symmetry species of the various transition operators. This must be done for each operator (electric dipole, magnetic dipole, electric quadrupole, and so on) in each symmetry group, and it is done most readily by geometric inspection. For example, consider the electric dipole moment in the C_{2v} group, as shown in figure 6.3. We can construct the character table for each of the three components of the dipole moment by simply entering the effect of each operation on that component; for example, $C_2(x) = -x$, so that a -1 is entered for the character of x under C_2 in C_{2v}. In this way, we find

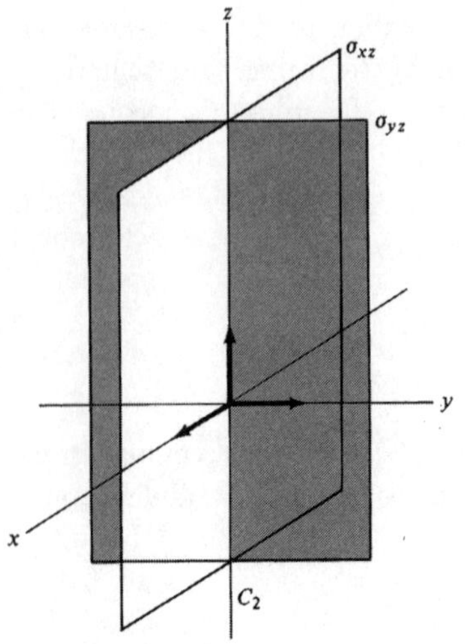

Figure 6.3
The three components of the electric dipole moment (x, y, and z) and the symmetry elements of the C_{2v} point group.

	E	C_2	σ_{xz}	σ_{yz}	Equivalent representation
x	1	-1	1	-1	B_1
y	1	-1	-1	1	B_2
z	1	1	1	1	A_1

The equivalent representations for the various components of the electric dipole, magnetic dipole, and electric quadrupole operators are included along with the direct product tables in appendix A. For a complete set of tables of characters and basis functions, the reader is referred to the references at the chapter's end.

We can now illustrate the method of finding selection rules. For example, suppose we wished to know which states could be connected by electric dipole radiation polarized perpendicular to the plane of the molecule, that is, in the x direction. To be able to have $\Gamma(\psi_1 \mu_x \psi_2) = A_1$, with $\Gamma(x) = B_1$, we need $\Gamma(\psi_1 \psi_2) = B_1$, because $B_1 \otimes B_1 = A_1$. We can get B_1 by either of the direct products $A_1 \otimes B_1$ or $A_2 \otimes B_2$. Thus the allowed transitions for x polarized radiation are $A_1 \leftrightarrow B_1$ and $A_2 \leftrightarrow B_2$.

Group theoretical arguments are also very helpful in establishing nuclear spin statistical weights for rotational levels of polyatomic molecules. We have already considered this problem for homonuclear diatomic molecules (see chapter 4.1), using the nuclear exchange operation of the $D_{\infty h}$ group. For the more general case of polyatomics, the nuclear spin statistical weight of a given ro-vibronic level can be obtained by evaluating the direct product of the ro-vibronic representation Γ_{evr} and the nuclear spin representation Γ_{ns} to give

$$\Gamma_{\text{total}} = \Gamma_{\text{evr}} \otimes \Gamma_{\text{ns}}.$$

The nuclear spin statistical weight g_{ns} is then found by counting up the number of terms that belong to the irreducible representations of Γ_{total} that are permitted by the Pauli exclusion principle. For ammonia (NH_3), which belongs to the point group C_{3v} and possesses three indistinguishable hydrogen nuclei, each with nuclear spin $I = \frac{1}{2}$ and obeying Fermi-Dirac statistics, this treatment gives $g_{\text{ns}} = 4$ for all levels with K an integer multiple of 3 ($K = 0, 3, 6, \ldots$) and $g_{\text{ns}} = 2$ for all the other K levels (see chapter 7, figure 7.6). Results for a variety of symmetric top molecules may be found in the article by Weber (reference 11).

Group theoretical methods are also extremely helpful in determining the proper symmetry coordinates for small-amplitude vibrations in polyatomic molecules (see chapter 8.3), and in numerous other applications discussed in references 1–10.

Problems

1. Given a molecule of formula XY_4, in which certain Y atoms may or may not be equivalent by symmetry:

(a) Write down the point groups to which the molecule XY_4 could possibly belong. Omit groups of order less than 4. Your answer should contain at least six groups of order 4 or higher.
(b) Draw structures of XY_4 belonging to the two point groups of highest order and list the symmetry operations for each group.

2. Given a molecule XY_6 of point group O_h, which for purposes of visualization may be considered as placed within a cube, X being at the cube center and each Y at the center of one of the six cube faces. Suppose that certain Y atoms are replaced by different Z atoms, to be described. State in each

case to what point group (for example, O_h, D_{5h}, and so on) the resulting XY_nZ_{6-n} molecule beongs:

(a) XY_5Z.
(b) XY_4Z_2 (Zs opposite, that is, Z-Y-Z).
(c) XY_4Z_2 (Zs not opposite).
(d) XY_3Z_3 (two Zs opposite).
(e) XY_3Z_3 (no Zs opposite).

3. The symbol R is used in the following to represent several different operations, which are appropriately defined. Make a general statement about the results obtained when each operation is performed on the function shown opposite it.

Operator	**Function**
(a) $R\zeta \rightarrow -\zeta$	$\psi_v(\zeta)$ for the harmonic oscillator
(b) $R(x, y, z) \rightarrow (-x, -y, -z)$	$\psi_{J,M}(\theta, \phi)$ for the linear rotator
(c) $R = z$ component of angular momentum	$\psi_{J,M}(\theta, \phi)$ for the linear rotator
(d) $R =$ interchange of like nuclei	ψ_{mol} for $^{12}C_2{}^2H_2$
(e) $R\psi_j(x, y, z) \rightarrow -\psi_j(x, y, z)$	$\bar{F}_j$, the average value of the function $F(x, y, z)$ for the state j
(f) $R\psi_j(xyz) \rightarrow -\psi_j$ $R\psi_k(xyz) \rightarrow +\psi_k$	Transition probability P_{jk} from state j to state k induced by means of the dipole moment $M(x, y, z)$
(g) $R(x, y, z) \rightarrow -x, -y, -z$	M_0, the dipole moment of a linear molecule
(h) $R =$ interchange of two electrons	ψ_{mol} for $^3\Sigma_g^-$ state of $^{16}O_2$
(i) $R =$ interchange of like nuclei	ψ_{mol} for $^{12}C^{17}O_2$ (nuclear spin for ^{17}O is $\frac{5}{2}$)

References

1. L. M. Falicov (notes compiled by A. Luehrmann), *Group Theory and Its Physical Applications* (University of Chicago Press, Chicago, 1966).

2. F. A. Cotton, *Chemical Applications of Group Theory* (Interscience, New York, 1963).

3. G. G. Koster, J. O. Dimmock, R. G. Wheeler, and H. Statz, *Properties of the Thirty-Two Point Groups* (MIT Press, Cambridge, MA, 1963).

4. M. Hamermesh, *Group Theory and Its Application to Physical Problems* (Addison-Wesley, Reading, MA, 1962).

5. M. Tinkham, *Group Theory and Quantum Mechanics* (McGraw-Hill, New York, 1964).

6. J. W. Leech and D. J. Newman, *How to Use Groups* (Methuen, London, 1969).

7. J. R. Ferraro and J. S. Ziomek, *Introductory Group Theory and Its Application to Molecular Structure* (Plenum, New York, 1969).

8. C. L. Perrin, *Mathematics for Chemists* (Interscience-Wiley, New York, 1970), chapter 8.

9. R. L. Flurry, Jr., *Symmetry Groups: Theory and Chemical Applications* (Prentice-Hall, Englewood Cliffs, NJ, 1980).

10. P. R. Bunker, *Molecular Symmetry and Spectroscopy* (Academic Press, New York, 1979).

11. A. Weber, *J. Chem. Phys.* **73**, 3952 (1980).

7 Rotational Spectra of Polyatomic Molecules

1 Rigid Body Hamiltonian, Eigenfunctions, and Spectra

In our treatment of polyatomic molecules, we shall follow the same route as was taken for diatomic molecules; namely, we shall consider first rigid body rotation in space-fixed axes, then small internal vibrations of the nuclei in the ground electronic states, and finally all the degrees of freedom coupled together in electronic spectra. We shall begin with a discussion of rotational spectra for the simpler classes of polyatomic molecules; for a full discussion of this topic, the reader is referred to the text by Townes and Schawlow (reference 1).

The rotational energy of a rigid body in free space is purely kinetic, so

$$E_{\text{rot}} = \tfrac{1}{2}\sum_i m_i(\dot{x}_i^2 + \dot{y}_i^2 + \dot{z}_i^2). \tag{7.1}$$

With a great deal of algebraic manipulation, it can be shown that this energy is equivalent to that for motion about three *principal axes*, each with its associated moment of inertia, in space. The three axes intersect at the center of mass of the molecule. If ω_j is the angular velocity about each of these axes, then the rotational energy is given by

$$\begin{aligned} E_{\text{rot}} &= \left(\sum_i m_i r_{ia}^2\right)\omega_a^2 + \left(\sum_i m_i r_{ib}^2\right)\omega_b^2 + \left(\sum_i m_i r_{ic}^2\right)\omega_c^2 \\ &= I_a\omega_a^2 + I_b\omega_b^2 + I_c\omega_c^2, \end{aligned} \tag{7.2}$$

where I_a, I_b, and I_c are the moments of inertia referred to the three axes. It is conventional to choose $I_a < I_b < I_c$.

As a simple example, we shall calculate the moments of inertia for the molecule BF_3. Let us assume that we already know that the molecule is planar, with three fluorine atoms arranged around a central boron atom at angles of 120°, and a B—F bond distance of 1.30 Å (figure 7.1). The center of mass of the molecule is clearly at the B atom. One inertial axis is perpendicular to the plane of the molecule; since this axis will possess the largest moment of inertia, we shall designate it the (c) axis. A second axis will pass through one of the B—F bonds, and the third will be mutually perpendicular to the first two. The moment of inertia about the (c) axis, I_c, is

$$I_c = 3m_F r_{BF}^2$$

numerically, with $m_F = 19$ atomic mass units (amu) and $r_{BF} = 1.30$ Å,

$$I_c = 96.3 \text{ amu Å}^2 = 1.6 \times 10^{-38} \text{ g cm}^2.$$

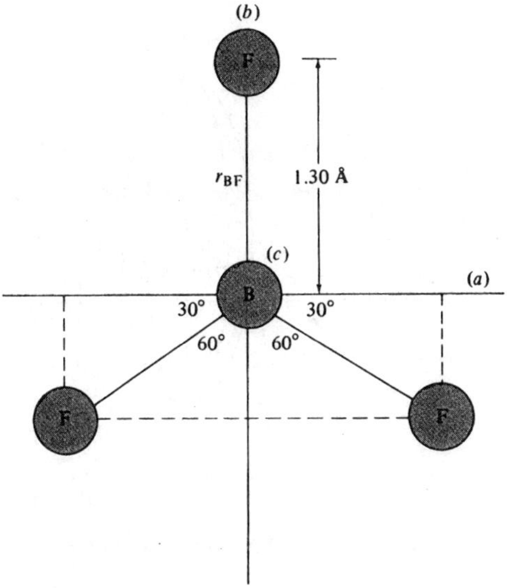

Figure 7.1
Determination of the moments of inertia for boron trifluoride (BF_3), given a known molecular structure.

For the other moments of inertia, we have

$$I_a = m_F r_{BF}^2 + 2m_F(r_{BF}\sin 30°)^2$$
$$= m_F r_{BF}^2(1 + 2(\tfrac{1}{2})^2) = \tfrac{3}{2}m_F r_{BF}^2$$

and

$$I_b = 2m_F(r_{BF}\sin 60°)^2$$
$$= m_F r_{BF}^2\left(2\left(\frac{\sqrt{3}}{2}\right)^2\right) = \tfrac{3}{2}m_F r_{BF}^2.$$

Note that two of the moments of inertia, I_a and I_b, are equal to each other and to exactly one-half I_c. Relations such as these between inertial constants are a consequence of the symmetry properties of the molecule, which we shall now proceed to consider.

In general, a C_p rotation axis always coincides with one of the principal axes. A reflection plane contains two of the principal axes. If $p \geq 3$, then

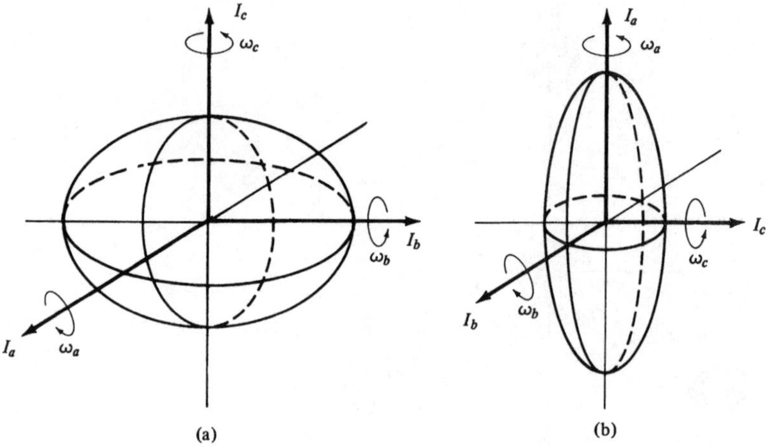

Figure 7.2
Principal rotation axes for (a) an oblate symmetric top and (b) a prolate symmetric top.

two of the three moments of inertia are equal. There exist two ways in which this can happen. We can have $I_a = I_b < I_c$, in which case we have an *oblate symmetric top*, as shown in figure 7.2a. An oblate top is of the general shape of a hockey puck, a flying saucer, or a very short, fat man; molecular examples would be cyclobutane, benzene, coronene, and phthalocyanine. If $I_a < I_b = I_c$, we have a *prolate symmetric top*, as shown in figure 7.2b. A prolate top is of the general shape of a football, a cigar, or a whirling dervish; molecular examples would be methyl bromide and ammonia. If the molecule has more than one rotation axis with $p \geq 3$, as occurs in the cubic point groups, then we have a *spherical top*; examples are tetrahedral molecules such as methane and octahedral molecules such as SF_6. If the molecule possesses no rotation axes with $p \geq 3$, then all three moments of inertia are unequal, and the molecule is an *asymmetric top*; most molecules fall into this category. Finally, if $I_a = 0$, then $I_b = I_c$, and we have a simple linear rotor, which has already been treated in chapter 4.

Although only a lamentably small fraction of all known molecules are symmetric or spherical tops, there is a somewhat more sizable group that falls into the category of *near-prolate* tops, with $I_a < I_b \sim I_c$, or *near-oblate* tops, with $I_a \sim I_b < I_c$. Examples of near-prolate tops would be C_{2v} molecules such as formaldehyde, shown in figure 7.3a, and D_{2h} molecules such as ethylene, shown in figure 7.3b. The convenience of having a molecule be

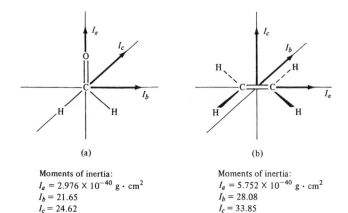

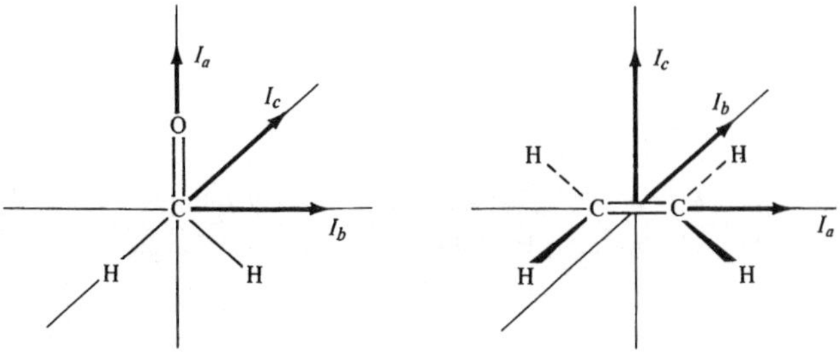

(a) (b)

Moments of inertia:
$I_a = 2.976 \times 10^{-40}$ g · cm^2
$I_b = 21.65$
$I_c = 24.62$

Moments of inertia:
$I_a = 5.752 \times 10^{-40}$ g · cm^2
$I_b = 28.08$
$I_c = 33.85$

Figure 7.3
Examples of near-prolate symmetric tops. (a) Formaldehyde: the two σ_v planes contain the A and B axes and the A and C axes. (b) Ethylene: the σ_h plane contains the B and C axes, and the two σ_v planes contain the A and B axes and the A and C axes.

a symmetric or near-symmetric top will become evident immediately upon consideration of the quantized energy levels corresponding to equation (7.2).

We can find these energy levels by just rewriting equation (7.2) in terms of the angular momenta $J_i = I_i\omega_i$, to yield

$$E_{\text{rot}} = \frac{1}{2}\left(\frac{J_a^2}{I_a} + \frac{J_b^2}{I_b} + \frac{J_c^2}{I_c}\right).$$

Since $J^2 = J_a^2 + J_b^2 + J_c^2$, we can add and subtract whichever of these components may be convenient. For a prolate top, we do this with J_a and J_c:

$$E_{\text{rot}} = \frac{1}{2}\left(\frac{J^2}{I_b} - \frac{J_a^2}{I_b} - \frac{J_c^2}{I_b} + \frac{J_a^2}{I_a} + \frac{J_c^2}{I_c}\right);$$

since $I_b = I_c$, this becomes

$$E_{\text{rot}} = \frac{1}{2}\left(\frac{J^2}{I_b} - \frac{J_a^2}{I_b} + \frac{J_a^2}{I_a}\right). \tag{7.3}$$

We know that the total J^2 is quantized in units of $J(J+1)\hbar^2$ because we are dealing with rotation in three-dimensional space. Any one of the individual components corresponding to angular momentum along a top axis is quantized in units of $K^2\hbar^2$, so that equation (7.3) becomes

$$E_{\text{rot}} = \frac{\hbar^2 J(J+1)}{2I_b} + \frac{\hbar^2 K^2}{2}\left(\frac{1}{I_a} - \frac{1}{I_b}\right)$$
$$= BJ(J+1) + (A-B)K^2, \tag{7.4}$$

with $A = h/8\pi^2 cI_a$ and $B = h/8\pi^2 cI_b$ (in units of cm^{-1}). For an oblate top, we do the same thing with J_b and J_c to obtain

$$E_{\text{rot}} = \frac{1}{2}\left(\frac{J^2}{I_b} + \frac{J_c^2}{I_c} - \frac{J_c^2}{I_b}\right).$$

By the same quantization arguments, this becomes

$$E_{\text{rot}} = \frac{\hbar^2 J(J+1)}{2I_b} + \frac{\hbar^2 K^2}{2}\left(\frac{1}{I_c} - \frac{1}{I_b}\right)$$
$$= BJ(J+1) + (C-B)K^2, \tag{7.5}$$

with $C = h/8\pi^2 cI_c$. For a spherical top, we simply substitute $I_a = I_b = I_c$ to obtain

$$E_{\text{rot}} = \frac{J^2\hbar}{2I_b} = \frac{\hbar^2 J(J+1)}{2I_b} = BJ(J+1). \tag{7.6}$$

The quantum number K, being a projection of J on the principal axis, can take on values ranging from J to $-J$ in unit steps. Thus, to find the degeneracies, we first note that each J level has $(2J+1)$ K states, corresponding to the different possible projections of J on a molecule-fixed axis. Each (J, K) level then has the usual $(2J+1)M_J$ levels, corresponding to the different possible projections of J on a space-fixed axis. Thus the statistical weight for each J level is $(2J+1)^2$, rather than the $(2J+1)$ appropriate to a linear rotor. In a spherical top molecule, all of the $(2J+1)^2$ levels are truly degenerate, in the absence of vibration-rotation interactions.

Nuclear spin statistics also influence the weights of the various rotational levels of a polyatomic molecule, but the formulas for calculating these are quite a bit more complicated than for simple linear or diatomic systems. For details, the reader is referred to the text of Herzberg [reference 2a] or the original papers by Wilson (reference 3).

No such simple closed form expressions as equations (7.4), (7.5), or (7.6) can be written for asymmetric tops. Some approximate energy level expressions, for low J, have been tabulated by Herzberg [reference 2b], and computer calculations of asymmetric rotor energy levels are also available (references 4, 5). An asymmetric rotor can be characterized by a parameter

Table 7.1

J_{K_p,K_0}	Rotational energy
0_{00}	0
1_{10}	$A + B$
1_{11}	$A + C$
1_{01}	$B + C$
2_{20}	$2A + 2B + 2C + 2\sqrt{(B - C)^2 + (A - C)(A - B)}$
2_{21}	$4A + B + C$
2_{11}	$A + 4B + C$
2_{12}	$A + B + 4C$
2_{02}	$2A + 2B + 2C - 2\sqrt{(B - C)^2 + (A - C)(A - B)}$
3_{30}	$5A + 5B + 2C + 2\sqrt{4(A - B)^2 + (A - C)(B - C)}$
3_{31}	$5A + 2B + 5C + 2\sqrt{4(A - C)^2 - (A - B)(B - C)}$
3_{21}	$2A + 5B + 5C + 2\sqrt{4(B - C)^2 - (A - B)(A - C)}$
3_{22}	$4A + 4B + 4C$
3_{12}	$5A + 5B + 2C - 2\sqrt{4(A - B)^2 + (A - C)(B - C)}$
3_{13}	$5A + 2B + 5C - 2\sqrt{4(A - C)^2 - (A - B)(B - C)}$
3_{03}	$2A + 5B + 5C - 2\sqrt{4(B - C)^2 + (A - B)(A - C)}$

$$\kappa = \frac{2B - A - C}{A - C}, \tag{7.7}$$

which can take on values between $+1$ and -1. If κ is near -1, then the rotor is near-prolate, and equation (7.4), with a nearly good quantum number K_p, is applicable; the small asymmetry produces a splitting of energy levels for which $K_p > 0$. Similarly, if κ is near $+1$, then the rotor is near-oblate, and we can use equation (7.5) with a quantum number K_0; again, there will be an asymmetry doubling for $K_0 > 0$. The rotation states are labeled as J_{K_p,K_0}, where K_0 and K_p are the K values that the molecule would have in the limiting oblate and prolate cases, respectively. Some closed form solutions for levels with $J < 4$ of asymmetric rotors have been calculated (references 6, 7) and are given in table 7.1. A correlation diagram showing how these energy expressions vary between the prolate and oblate limits is shown in figure 7.4.

2 Microwave Spectroscopy of Rotational Levels

Rotational spectra can be investigated by means of far-infrared absorption, microwave absorption and emission, and Raman scattering. The first two methods involve electric dipole transitions with a time-dependent interaction

$$\mathscr{H}' = -\boldsymbol{\mu} \cdot \mathscr{E}(t).$$

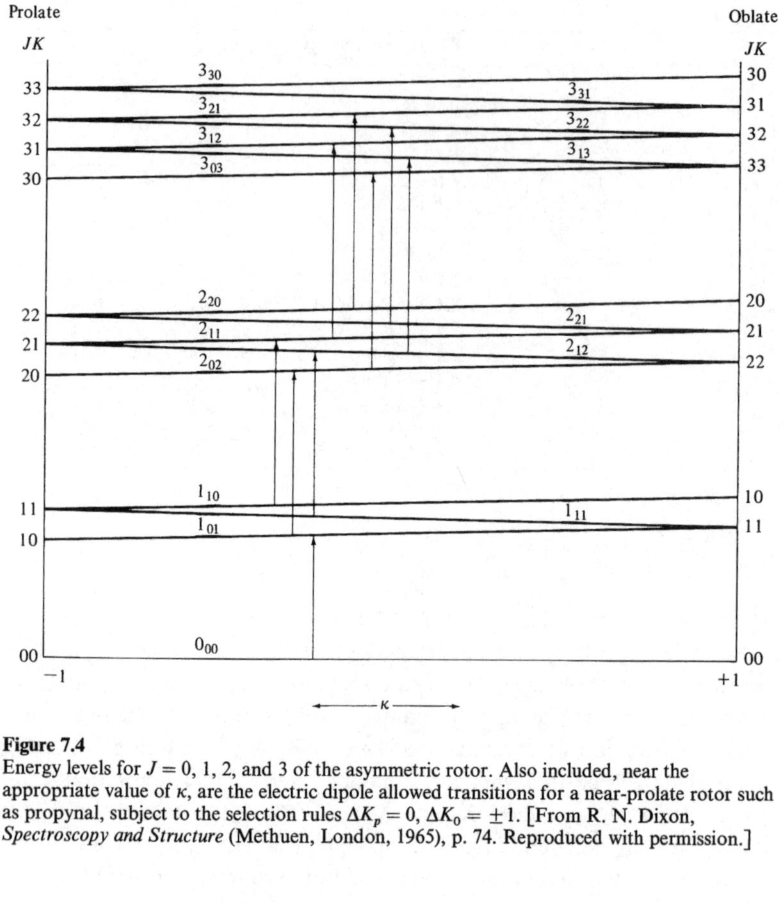

Figure 7.4
Energy levels for $J = 0$, 1, 2, and 3 of the asymmetric rotor. Also included, near the appropriate value of κ, are the electric dipole allowed transitions for a near-prolate rotor such as propynal, subject to the selection rules $\Delta K_p = 0$, $\Delta K_0 = \pm 1$. [From R. N. Dixon, *Spectroscopy and Structure* (Methuen, London, 1965), p. 74. Reproduced with permission.]

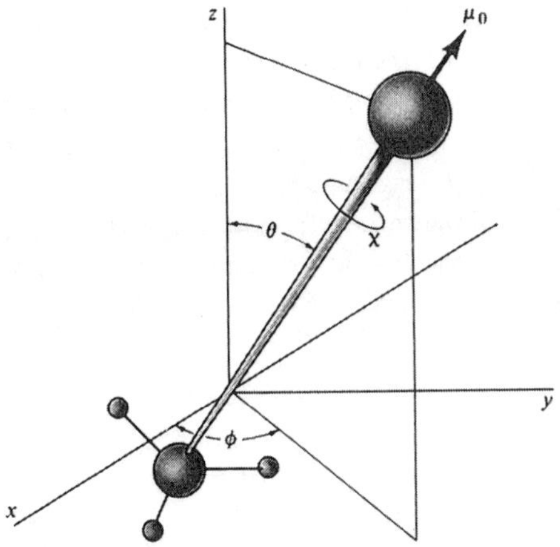

Figure 7.5
The rotation angles θ, ϕ, and χ, and the permanent dipole moment μ_0 for a prolate symmetric top, such as a methyl halide molecule.

Thus, in order to possess a far-infrared or a microwave spectrum, the molecule must first possess a permanent electric dipole moment μ_0. The selection rules then depend on where μ_0 lies relative to the principal axes of the molecule.

The wave function for a rotating symmetric top molecule is

$$\psi_{JKM} \propto P_M^J(\cos\theta)e^{iK\chi}e^{iM\phi},$$

where the angles are defined in figure 7.5. In a symmetric top, μ_0 is coincident with the rotation axis of highest symmetry, also known as the *figure axis*. This gives

$$\mathscr{H}' \sim |\mu_0||E_0|\cos\theta,$$

and since

$$\langle P_M^J(\cos\theta)|\cos\theta|P_{M'}^{J'}(\cos\theta)\rangle \sim \delta(J, J' \pm 1)$$

and

$$\int e^{iK\chi}e^{-iK'\chi}\,d\chi \sim \delta(K, K'),$$

we have the selection rules $\Delta J = \pm 1$ and $\Delta K = 0$. If the dipole moment is perpendicular to a principal axis, so that it rotates around in space, there will be an additional factor of $\cos \chi$ in the above, which gives the selection rule $\Delta K = \pm 1$. In a near-prolate top, such as the molecule propynal (see below) in which the dipole moment is very nearly along the a axis, it is possible to satisfy both sets of selection rules, namely, $\Delta K_p = 0$ and $\Delta K_0 = \pm 1$, simultaneously, as an inspection of figure 7.4 will show. Experimentally, the spectrum of propynal consists of one line at $2B$ (about 9,500 MHz), three lines at $4B$ (about 19,000 MHz), five lines at $6B$ (about 28,500 MHz), and so on, in agreement with the selection rules.

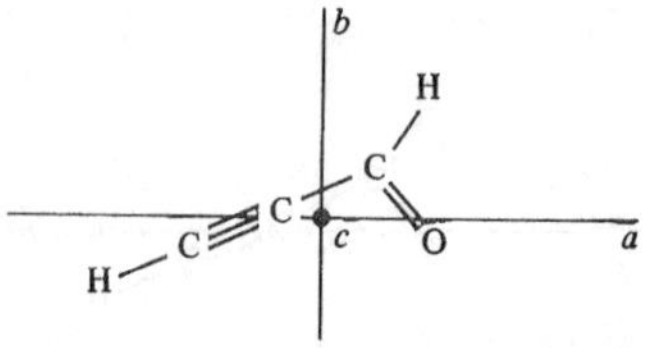

In the rotational Raman spectrum, analysis similar to that carried out in chapter 4.6 gives the selection rules $\Delta J = 0, \pm 1$, and ± 2, and $\Delta K = 0$. Spherical top molecules have such a high symmetry, by virtue of the multiple threefold axes, that they possess neither an electric dipole nor an electric quadrupole moment, and thus show neither pure rotational absorption nor Raman spectra; the rotational constants must be inferred from analysis of vibrational bands under high resolution (see chapter 8.5, especially figure 8.4).

By far the most widely used method of measuring rotational spectra is Stark-modulated microwave spectroscopy. In such a spectrometer, a microwave source such as a klystron tube, sends microwave radiation down a long waveguide containing the sample at very low pressure. A square-wave-modulated electric field is applied to the sample by means of a conducting septum, insulated from the walls, running down the middle of the wave guide. Since the molecule has to have a permanent electric dipole moment in order to have any absorption at all, it will also have a Stark effect, that is, a shift of the energy levels with applied electric field, as described in chapter 5.9. (This will generally be a second-order effect, since most molecules do not possess electronic orbital angular momentum in their ground states.) Since this shift is J, K, and M dependent, the resonant frequency corresponding to transitions between a pair of energy levels can be moved in and out of coincidence with the applied microwave frequency when the field is switched on and off, and a small amount of energy will be absorbed. The absorption is monitored on a lock-in detector, which is sensitive only to the difference between two signal levels modulated at the same frequency as some reference voltage, which in this case is the Stark modulation itself. This technique provides a very sensitive way of measuring small changes in absorption. The resulting absorption line positions, plotted against the square of the modulation amplitude, give the rotational constants A, B, and C; the slope of the lines can be used to find the magnitude of the dipole moment. Since the source of the microwave radiation, namely, the klystron tube, is a coherent oscillator, its frequency can easily be determined to an accuracy of better than one part per million, thus giving very accurate values for the molecular constants obtained in this way.

3 Additional Topics in Rotational Spectroscopy

The extremely high-frequency resolution available from microwave and radio frequency oscillators, referred to previously, makes it possible to observe a number of higher-order interactions that split rotational lines. Several of these effects will be considered in this section; other methods for observing these interactions include molecular beam spectroscopy (chapter 10.4), microwave Fourier transform spectroscopy (chapter 12.3), and microwave optical double-resonance spectroscopy (chapter 12.7; also see reference 8).

3.1 Nuclear Hyperfine Structure

The spin angular momentum of nuclei such as ^{1}H, ^{13}C, ^{19}F, and so on can couple to the rotational angular momentum, producing splittings in the rotational energy levels (see references 9 and 10; also see chapters 8 and 9 of reference 1). Since such nuclei possess electric quadrupole as well as spin magnetic moments, the hyperfine energy expression must include both of these interactions. This expression can be written as

$$E_{\text{hfs}} = -eqQf(I, J, F) + \tfrac{1}{2}M_{\perp}[J(J+1) - F(F+1) + I(I+1)], \qquad (7.8)$$

where eqQ is the nuclear quadrupole moment multiplied by the field gradient along the internuclear axis, $f(I, J, F)$ is an angular momentum coupling coefficient known as Casimir's function, and $M_{\perp}$ is the magnetic hyperfine spin-rotation coupling constant. Note the similarity of the last term in equation (7.8) to the expression for hyperfine splitting in atoms, equation (2.9). An example of hyperfine structure in a rotational line is shown in figure 12.9b.

3.2 Molecular Zeeman Effect

While most microwave spectrometers employ Stark modulation, additional information can be obtained by performing the measurements in the presence of a magnetic field. An expression for the Zeeman energies of a linear molecule with nonzero electronic angular momentum was already given as equation (5.16). For a symmetric top molecule with no orbital or spin angular momentum, the energy expression to first order in J and second order in magnetic field strength B is given by (see references 11 and 12 and chapter 11 of reference 1)

$$E(J,K,M) = E_0(J,K) - \mu_N MB\left[g_\perp + (g_\parallel - g_\perp)\frac{K^2}{J(J+1)}\right] \tag{7.9}$$

$$- \frac{B^2}{3}\left[\frac{3M^2 - J(J+1)}{(2J-1)(2J+3)}\right]\left[(\chi_\perp - \chi_\parallel) - \frac{3M^2}{J(J+1)}(\chi_\perp - \chi_\parallel)\right],$$

where $g_\perp$ and $g_\parallel$, $\chi_\perp$ and $\chi_\parallel$ are the molecular g values and the magnetic susceptibility tensor elements perpendicular and parallel to the rotational symmetry axis, respectively; μ_N is the nuclear magneton. The measurements must be performed at high magnetic fields ($B \gtrsim 2 \times 10^4$ G) for the splittings to be observable. The molecular quadrupole moment can be calculated from the measured g values and the susceptibility tensor.

3.3 *l* Doubling

In addition to coupling with nuclear spin, the rotational angular momentum can interact with several types of vibrational motion in the molecule. Molecular vibrations will be treated in detail in the following chapter, but there are two particular interactions that are important in microwave spectroscopy and will be mentioned at this point. One is *l* doubling, which occurs in linear polyatomic molecules, and which results from the interaction of the molecular rotation with a doubly degenerate bending mode ν_2 (see chapter 2 of reference 1). This splits each rotational level into two, with the splitting given by

$$\nu_l = \tfrac{1}{2}q_l(v_2 + 1)J(J+1), \tag{7.10}$$

where the l doubling constant is approximately

$$q_l \simeq 2.6B_e^2/\omega_2. \quad (7.11)$$

For example, in HCN, with rotational constant $B_e = 44{,}316$ MHz $= 1.477$ cm^{-1} and bending vibration frequency $\omega_2 = 711.7$ cm^{-1}, q_l has the value 224.4 MHz. Since transitions between the l doublet members are dipole allowed, this gives rise to microwave absorption lines in this molecule, even though the rotational transitions themselves lie in the far-infrared. For example, the l doublet transition for $J = 10$ of HCN is at 24,660 MHz.

3.4 Inversion Doubling

A somewhat different type of vibration–rotation interaction occurs in molecules possessing a double-minimum potential with a barrier greater than the vibration frequency in that potential. The classic example is ammonia (see chapter 12 of reference 1), where the barrier to inversion between the "right"- and "left"-handed forms of the pyramidal NH_3 molecule is about 2,075 cm^{-1}, while the frequency of vibration along the normal coordinate containing this barrier is 950 cm^{-1} (ν_2). The actual wave functions for this system are constructed as symmetric and antisymmetric linear combinations of the "right" and "left" basis states,

$$\psi_0 = \frac{1}{\sqrt{2}}(u_L + u_R), \qquad \psi_1 = \frac{1}{\sqrt{2}}(u_L - u_R).$$

The energies E_0 and E_1 are split by a frequency difference

$$\nu_{inv} = \nu_2/\pi A^2,$$

where

$$A = \exp\left\{\frac{2\pi}{h}\int_0^{s^*}[2\mu(V(s) - E_{vib})]^{1/2}\,ds\right\}. \quad (7.12)$$

In equation (7.12), s^* is the inner turning point of the $v_2 = 0$ level in one of the potential minima, and μ is the reduced mass for motion along that normal coordinate (see chapter 8.2). The frequency ν_{inv} can be thought of (somewhat imprecisely) as the frequency at which the molecule "tunnels" through the potential barrier. In reality, the ground state of the ammonia molecule is the appropriate symmetric or antisymmetric superposition state (ψ_0 or ψ_1); in related systems, such as amino acids, in which the inversion

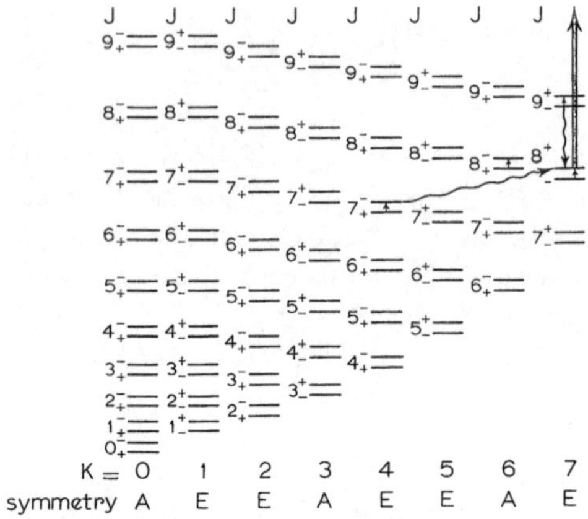

Figure 7.6
Rotational levels of ammonia for $J \leq 9$, $K \leq 7$. The symmetries of each of the inversion levels is indicated, but the (J, K) dependence of the splittings (which are exaggerated with respect to the rotational spacings) is not explicitly shown. Also indicated is the set of levels employed in microwave infrared double-resonance experiments in ammonia [see reference 13].

barrier is much higher in energy, the molecule is effectively "frozen" into one of the asymmetric configurations on the time scale of our observations.

For the specific case of ammonia, each (J, K) level is split into a symmetric $(+)$ component and antisymmetric $(-)$ component, as indicated in figure 7.6. The energy differences can be represented reasonably well by a simple equation,

$$\nu = \nu_0 - aJ(J+1) - (a-b)K^2,$$

with coefficients $\nu_0 = 23{,}787.0$, $a = 151.3$, and $(a-b) = -211.0$ MHz. Since transitions between the inversion levels ($\Delta J = \Delta K = 0$, $+ \leftrightarrow -$) are dipole allowed, there is associated with each rotational level of ammonia an inversion transition in the range 17–40 GHz, which is a convenient region for microwave measurements. Additional aspects of this spectrum are considered in problem 5.

Problems

1. The corners of a cube are numbered 1, 2, 3, 4 clockwise around the top face of the cube, and 5, 6, 7, 8 clockwise around the bottom face, corner 5

lying under corner 1, corner 6 under corner 2, and so on. A face center is denoted by the two numbers of the corners between which a face diagonal can be drawn which passes through that face center (for example, either 13 or 24 would denote the center of the top face).

(a) The structures of several kinds of AB_4 molecules are described as follows with the above numbering system. The A atom is placed at the center of the cube, and the A—B bond distances are given by the cube dimensions. However, the B atoms are not necessarily all equivalent, the actual equivalence being determined by the symmetry elements remaining in the AB_4 structures:

(i) $AB_4(1, 2, 3, 4)$,
(ii) $AB_4(1, 3, 5, 7)$,
(iii) $AB_4(1, 3, 6, 8)$,
(iv) $AB_4(1, 5, 16, 18)$,
(v) $AB_4(13, 36, 68, 18)$.

Give the point group symbol for each AB_4 structure (for example, C_{3v}, D_{6h}, and so on).
(b) Classify the five molecules of part (a) as to type of molecular rotator (linear, symmetrical top, and so on).
(c) Which of the five molecules of part (a) will give a pure rotational spectrum in the far-infrared or microwave region? Which will give a pure rotational Raman effect?

2. For the following molecules, give the point group symbol, dipole moment (zero or nonzero), and type of rotator (linear, symmetric top, and so on):

(a) dichloromethane (CH_2Cl_2);
(b) planar nitrogen tetroxide (N_2O_4), N's and O's equivalent;
(c) nitrogen tetroxide with the plane of one NO_2 at right angles to the other, N's and O's equivalent;
(d) ethane (C_2H_6) "opposed form," that is, with σ_h;
(e) ethane, "staggered form";
(f) SF_4, nonplanar, all F's equivalent;
(g) SF_4, nonplanar, two pairs of equivalent F's.

3.

(a) The selection rules for the pure rotational Raman effect in a symmetrical top molecule are $\Delta J = 0, \pm 1, \pm 2; \Delta K = 0$. Derive the algebraic expressions for the frequencies in cm^{-1} of the pure rotational Raman spectrum of a

symmetrical top whose rotational constants in cm^{-1} are A and B, the latter being the double-moment constant. Assume the spectrum to be excited by frequency ν_0 in cm^{-1}. State explicitly what values J and K may take in your expressions.
(b) Assume that A, B and T (in °K) have such values that the molecules of a given J are uniformly distributed over the various K values (0, 1, 2, ...) consistent with J and the selection rules. The most populated level has a value of J given roughly by $0.59(T/B)^{1/2}$. What is the frequency shift in cm^{-1} $(\nu_0 - \nu_{J,K})$ for the most intense line in the S branch ($\Delta J = +2$) and for the most intense line in the R branch ($\Delta J = +1$) for a molecule in which $A = 1.5$ cm^{-1}, $B = 0.75$ cm^{-1}, and $T = 300$°K?

4. Using the asymmetric rotor energy level formulas in section 1, calculate the first part ($0 \to 1$, $1 \to 2$, $2 \to 3$) of the pure rotation spectrum of formaldehyde, for which the inertial constants are given in figure 7.3a. Does this spectrum lie in the microwave or the far-infrared region?

5. An explicit form for the double-minimum potential of the ammonia inversion motion has been suggested by Newton and Thomas:

$$\frac{V(s)}{hc} = \left\{\left[\frac{(0.377)^2 - s^2}{0.536 + s^2}\right]^2 - 1\right\} \times 3.17 \times 10^4 \text{ cm}^{-1},$$

s in Ångstroms. Using equation (7.12), calculate A and ν_{inv} for the ground ($E_{\text{vib}} = \frac{1}{2}\nu_2$) and first excited ($E_{\text{vib}} = \frac{3}{2}\nu_2$) vibrational states of ammonia. How does your computed value compare with the experimental value of ν_0? By how much is the $v_2 = 1$ state of ammonia split?

References

1. C. H. Townes and A. Schawlow, *Microwave Spectroscopy* (McGraw-Hill, New York, 1955).

2. (a) G. Herzberg, *Molecular Spectra and Molecular Structure*, Vol. 2: *Infrared and Raman Spectra of Polyatomic Molecules* (Van Nostrand Reinhold, New York, 1945), pp. 26, 38, 50. (b) G. Herzberg, op. cit., p. 46.

3. E. B. Wilson, Jr., *J. Chem. Phys.* **3**, 276 (1935).

4. R. H. Schwendeman, "Tables for the Rigid Asymmetric Rotor," National Standard Reference Data Series, National Bureau of Standards Monograph No. 12 (U.S. Government Printing Office, Washington, DC, June 1968).

5. H. C. Allen, Jr., and P. C. Cross, *Molecular Vib-Rotors* (Wiley, New York, 1963).

6. G. M. Barrow, *Introduction to Molecular Spectroscopy* (McGraw-Hill, New York, 1962).

7. R. N. Dixon, *Spectroscopy and Structure* (Methuen, London, 1965).

8. J. I. Steinfeld, ed., *Laser and Coherence Spectroscopy* (Plenum, New York, 1978), chapter 1.

9. W. H. Flygare and V. W. Weiss, *J. Chem. Phys.* **45**, 8 (1966).

10. E. F. Pearson, C. L. Norris, and W. H. Flygare, *J. Chem. Phys.* **60**, 1761 (1974).

11. R. L. Shoemaker and W. H. Flygare, *J. Am. Chem. Soc.* **91**, 5417 (1969).

12. C. L. Norris, E. F. Pearson, and W. H. Flygare, *J. Chem. Phys.* **60**, 1758 (1974).

13. T. Shimizu and T. Oka, *J. Chem. Phys.* **53**, 2536 (1970); T. Shimizu and T. Oka, *Phys. Rev.* **A2**, 1177 (1970).

8 Vibrational Spectra of Polyatomic Molecules

1 Lagrangian Mechanics

Each individual nucleus in an N-atomic molecule had three degrees of freedom when it was part of a free atom; therefore the resulting molecule must have a total of $3N$ degrees of freedom. Three of these are overall translation in space-fixed axes, and three more (or two, in the case of linear molecules) are free rotation. This leaves $3N - 6$ (or $3N - 5$, for a linear molecule) possible modes of vibration. The problem we have to solve in order to discover these modes of vibration is a formidable one: that of N coupled mass points, with Cartesian coordinates $\{\zeta\}_N$, connected by forces

$$f_i = -\frac{\partial E}{\partial \zeta_i}.$$

If the kinetic energy is independent of ζ_i, the forces can be written as the gradient of a potential V, so that

$$f_i = -\frac{\partial V}{\partial \zeta_i}.$$

Molecular vibration is one problem that can be analyzed almost completely by means of classical mechanics; essentially, this is because of some unique properties of harmonic and near-harmonic oscillators, in that the expectation values of most dynamical variables of interest exactly follow the classical equations of motion without dispersion (reference 1). The most convenient form of classical mechanics with which to analyze vibrations is the Lagrangian method, which is fully described in appendix B.

We begin with the Lagrangian $L = T - V$, where

$$T = \frac{1}{2}\sum_i m_i(\dot{\zeta}_i)^2$$

and

$$V = V_0 + \sum_i \left(\frac{\partial V}{\partial \zeta_i}\right)_0 (\zeta_i - \zeta_i^0) + \frac{1}{2}\sum_{ij}\left(\frac{\partial^2 V}{\partial \zeta_i \partial \zeta_j}\right)_0 (\zeta_i - \zeta_i^0)(\zeta_j - \zeta_j^0) + \cdots.$$

The sums are understood to be taken over $(i,j) = 3N - 6$ or $3N - 5$, as the case may be. Since the vibrations all take place around the minimum of the potential, the first derivatives of V with respect to each ζ_i are all zero. We can choose $V_0 = 0$, and, if we neglect all higher terms in the Taylor's series expansion of V, we obtain the simple harmonic form of the potential,

$$V = \frac{1}{2}\sum_{ij} b_{ij}\bar{\zeta}_i\bar{\zeta}_j,$$

where $\bar{\zeta}_i$ is the displacement of the ith internal coordinate from its equilibrium position, and so on. As is shown in appendix B, Newton's laws of motion for this coupled mass system are equivalent to the equation

$$\frac{d}{dt}\left(\frac{\partial L}{\partial \dot{\zeta}_i}\right) - \left(\frac{\partial L}{\partial \zeta_i}\right) = 0. \tag{8.1a}$$

Now

$$\frac{\partial L}{\partial \dot{\zeta}_i} = \frac{\partial T}{\partial \dot{\zeta}_i} = m_i\dot{\zeta}_i$$

and

$$\frac{\partial L}{\partial \zeta_i} = -\frac{\partial V}{\partial \zeta_i} = -\frac{1}{2}\sum_j b_{ij}\bar{\zeta}_j;$$

in the case of each derivative, the sum over i is eliminated. Thus we obtain a set of $3N - 6$ (or 5) equations of the form

$$\frac{d}{dt}(m_i\dot{\zeta}_i) + \frac{1}{2}\sum_j b_{ij}\bar{\zeta}_j = 0,$$

or

$$\ddot{\zeta}_i + \sum_j \left(\frac{b_{ij}}{2m_i}\right)\bar{\zeta}_j = 0. \tag{8.1b}$$

Formally, this set of equations can be solved by diagonalizing the determinant of

$$\left|\frac{b_{ij}}{m_i} - \lambda\mathbf{1}\right|$$

and writing each coordinate as

$$\bar{\zeta}_i(t) = \zeta_i^0 \sin(t\lambda^{1/2}/2\pi + \delta_i),$$

where the amplitudes ζ_i^0 and phases δ_i are determined from the initial conditions on the set of equations. However, since much more physical insight is obtained from a normal mode analysis using the appropriate internal coordinates, we shall proceed in this direction instead.

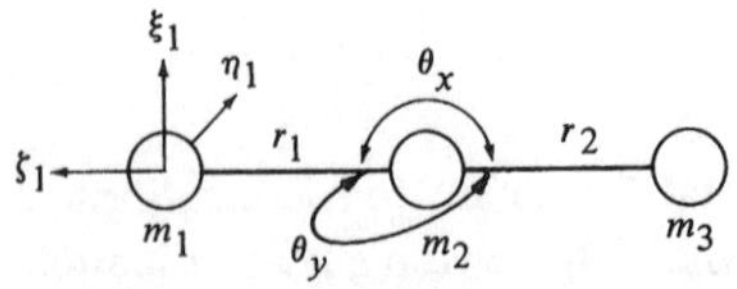

Figure 8.1
Internal coordinates for a linear triatomic molecule.

The quantum mechanical analogue of the development up to this point would be the $(3N-6)$-dimensional Schrödinger equation

$$\mathscr{H}\Psi = E\Psi,$$

with

$$\mathscr{H} = T + V = -\frac{\hbar^2}{2}\sum_i \frac{1}{m_i}\frac{\partial^2}{\partial \zeta_i^2} + \frac{1}{2}\sum_{ij} b_{ij}\bar{\zeta}_i\bar{\zeta}_j. \tag{8.2}$$

The wave function is separable only if normal coordinates are employed, and so we shall proceed to develop these and leave the introduction of quantization of the vibrational energy levels for somewhat later.

2 Normal Coordinate Analysis of a Linear Triatomic Molecule

Internal Coordinates In order to illustrate the details of normal coordinate analysis, we shall consider the simplest possible example of a vibrating polyatomic molecule, namely, a linear triatomic molecule, such as CO_2, N_2O, or HCN. Each of the three atoms making up the molecule, with mass m_1, m_2, m_3, respectively, can be located by means of Cartesian coordinates ξ_i, η_i, ζ_i, and so on, centered at each atom; however, it will be more advantageous to use, instead of these, the obvious *internal coordinates* describing relative motion of the three atoms. In order to find the number of internal coordinates, we subtract from $3N = 9$ the three translational and the two rotational degrees of freedom, giving $9 - 3 - 2 = 4$. A general prescription for finding internal coordinates has been given by Decius (reference 2). For our simple triatomic, we have (see figure 8.1)

$$s_1 = \Delta r_1, \quad s_2 = \Delta r_2 \qquad \text{(stretching vibrations)}$$

and

$$\left.\begin{aligned} s_{3x} &= (r_1 r_2)^{1/2}\,\Delta\theta_x \\ s_{3y} &= (r_1 r_2)^{1/2}\,\Delta\theta_y \end{aligned}\right\} \quad \text{(bending vibrations).}$$

The potential energy of the vibrations will be expressed in terms of these internal coordinates. We wish to do the same with the kinetic energy, but in such a way that the energy of free translation and rotation in space is eliminated. We can do this by transforming to a new set of mass-weighted displacement coordinates in which, for any small displacement from equilibrium positions, the center of mass of the molecules remains motionless and the inertial axes of the molecule remain fixed in space. In the case we are considering, such a set of displacement coordinates for stretching motion along the molecular axis would be

$$z_1 = -\frac{(m_2+m_3)}{M}s_1 - \frac{m_3}{M}s_2,$$

$$z_2 = \frac{m_1}{M}s_1 - \frac{m_3}{M}s_2,$$

$$z_3 = \frac{m_1}{M}s_1 + \frac{m_1+m_2}{M}s_2,$$

with $M = m_1 + m_2 + m_3$. We can see immediately that these coordinates satisfy the first criterion: The displacement of the center of mass of the molecule is given by

$$\begin{aligned} \bar{z} &= \frac{1}{M}(m_1 z_1 + m_2 z_2 + m_3 z_3) \\ &= \frac{1}{M^2}(-\,m_1(m_2+m_3)\Delta r_1 - m_1 m_3 \qquad \Delta r_2 \\ &\qquad + m_3 m_1 \qquad \Delta r_1 - m_2 m_3 \qquad \Delta r_2 \\ &\qquad \underline{+ m_2 m_1 \qquad \Delta r_1 + m_3(m_1+m_2)\Delta r_2)} \\ &= 0. \end{aligned}$$

The condition that the inertial axes of the molecule remain fixed is obviously also satisfied, since no change in θ in involved in the z coordinate.

The appropriate displacements for transverse motion meet similar requirements, but have more complicated forms; for example, for the x motion of the first atoms, we would have (see reference 3)

$$x_1 = -\frac{r_1 r_2 m_2 m_3}{m_1 m_2 r_1^2 + m_1 m_3 (r_1 + r_2)^2 + m_2 m_3 r_2^2} \left(\frac{r_2}{r_1}\right)^{1/2} s_{3x},$$

$$x_2 = \frac{r_1 r_2 m_2 m_3}{m_1 m_2 r_1^2 + m_1 m_3 (r_1 + r_2)^2 + m_2 m_3 r_2^2} \frac{r_1 + r_2}{(r_1 r_2)^{1/2}} s_{3x},$$

and

$$x_3 = -\frac{r_1 r_2 m_1 m_2}{m_1 m_2 r_1^2 + m_1 m_3 (r_1 + r_2)^2 + m_2 m_3 r_2^2} \left(\frac{r_1}{r_2}\right)^{1/2} s_{3x}.$$

The kinetic energy of the vibration is given by the mass-weighted velocities in each of these displacements,

$$T = \frac{1}{2}\sum_i m_i(\dot{x}_i^2 + \dot{y}_i^2 + \dot{z}_i^2). \tag{8.3a}$$

If we substitute in the internal coordinates, this becomes

$$T = \frac{1}{2}\sum_{r,t} \mu_{rt} \dot{s}_r \dot{s}_t, \tag{8.3b}$$

which would appear to be a much more complex expression than equation (8.3a). However, when the substitutions are made and the expression simplified, this reduces to

$$T = \frac{m_1(m_2 + m_3)}{2M}\dot{s}_1^2 + \frac{m_3(m_2 + m_1)}{2M}\dot{s}_2^2 + \frac{m_1 m_3}{M}\dot{s}_1 \dot{s}_2 + \frac{m_1 m_2 m_3}{2N}(\dot{s}_{3x}^2 + \dot{s}_{3y}^2), \tag{8.4}$$

where

$$N = (m_1 m_2 r_1^2 + m_1 m_3 (r_1 + r_2)^2 + m_2 m_3 r_2^2)/r_1 r_2.$$

Suppose, for a moment, that the molecule we are analyzing is HCN. The first term in equation (8.4) would correspond to a "C-H stretch," that is, the hydrogen atom moving against the CN radical; the second term would correspond to a "C-N stretch," that is, the nitrogen atom moving against the CH radical. There is a single cross term in the kinetic energy, that between these two stretching motions. The bending vibration, corresponding to internal coordinates s_{3x} and s_{3y}, is seen to be isolated from the stretching motion.

If we write an expression for the potential energy analogous to equation (8.3b), it would be

$$V = \frac{1}{2}\sum_{r,t} f_{rt} s_r s_t.$$

Symmetry requirements dictate that there be no "bend-stretch interaction" force constants, so that the potential energy will have the same form as the kinetic energy in equation (8.4). Thus, we set f_{13}, f_{31}, f_{23}, and f_{32} equal to zero.

Normal Coordinates We now seek a solution in terms of normal coordinates, which will provide us with the vibration frequencies of the molecule. These are coordinates in which there are no cross terms in either the kinetic or potential energy expressions, so that

$$T = \frac{1}{2}\sum_k \mu_k \dot{q}_k^2$$

and

$$V = \frac{1}{2}\sum_k f_k q_k^2.$$

The normal coordinates are obtained from a transformation of the non-normal internal coordinates, that is,

$$q_k = a_{1k}s_1 + a_{2k}s_2 + a_{3k}s_3. \tag{8.5}$$

In terms of these normal coordinates, the Lagrangian [equation (8.1a)] becomes

$$\ddot{q}_k + \left(\frac{f_k}{\mu_k}\right) q_k = 0, \qquad \text{or} \qquad \ddot{q}_k = -\lambda_k q_k.$$

Since we now have just three uncoupled linear differential equations, we can find a solution for each one in turn, independent of the others, of the form

$$q_k(t) = q_k^0 \sin(t\lambda_k^{1/2}/2\pi + \delta_k),$$

with

$$\lambda_k = f_k/\mu_k = 4\pi^2 v_k^2,$$

so that the observed frequencies of vibration are just

$$\nu_k = \frac{1}{2\pi}\sqrt{\frac{f_k}{\mu_k}}.$$

Since the phases δ_k are the same for all of the normal coordinates, the atoms all go through their equilibrium positions simultaneously (reference 4).

In order to find these frequencies in terms of the masses and force constants of the molecule, we substitute the definition of the q_k, equation (8.5), into the equation of motion $\ddot{q}_k = -\lambda_k q_k$, to obtain a set of simultaneous linear equations:

$$\begin{aligned}
(f_{11} - \lambda_k\mu_{11})a_{1k} + (f_{12} - \lambda_k\mu_{12})a_{2k} + (f_{13} - \lambda_k\mu_{13})a_{3k} &= 0,\\
(f_{21} - \lambda_k\mu_{21})a_{1k} + (f_{22} - \lambda_k\mu_{22})a_{2k} + (f_{23} - \lambda_k\mu_{23})a_{3k} &= 0,\\
(f_{31} - \lambda_k\mu_{31})a_{1k} + (f_{32} - \lambda_k\mu_{32})a_{2k} + (f_{33} - \lambda_k\mu_{33})a_{3k} &= 0.
\end{aligned} \tag{8.6}$$

The dynamics of the problem, as expressed in equation (8.4) for the kinetic energy, tells us that μ_{13}, μ_{23}, μ_{31}, and μ_{32} are all equal to zero; we have already agreed to take the corresponding force constants equal to zero as well. This means that the last term in the first two equations and the first two terms in the last equation drop out. As we already know, this system of equations has a nontrivial solution if the determinant of the coefficients is equal to zero. Incorporating the simplifications we are able to make, this has the form

$$\begin{vmatrix} f_{11} - \lambda\mu_{11} & f_{12} - \lambda\mu_{12} & 0 \\ f_{12} - \lambda\mu_{12} & f_{22} - \lambda\mu_{22} & 0 \\ 0 & 0 & f_{33} - \lambda\mu_{33} \end{vmatrix} = 0. \tag{8.7}$$

This is the same sort of secular determinant as we already have seen in connection with degenerate perturbation theory discussed in chapter 1.

The eigenvalue of this determinant that corresponds to the bending vibration ν_2 can be solved directly; we obtain

$$\lambda_2 = f_{33}/\mu_{33},$$

or

$$\begin{aligned}
4\pi^2\nu_2^2 &= f_\delta \frac{1}{m_1 m_2 m_3 r_1 r_2}[m_1 m_2 r_1^2 + m_1 m_3 (r_1 + r_2)^2 + m_2 m_3 r_2^2]\\
&= \frac{f_\delta}{r_1 r_2}\left[\frac{r_1^2}{m_3} + \frac{r_2^2}{m_1} + \frac{(r_1 + r_2)^2}{m_2}\right].
\end{aligned} \tag{8.8}$$

The bending vibration is uncoupled from any of the stretching vibrations, in conformity with our assertion that the "bend-stretch" interaction force constants in the potential energy were zero. If we had not made this assertion to begin with, then the two types of vibration would not have been uncoupled.

For the stretching vibrations, we obtain the quadratic equation

$$(f_{11}-\lambda\mu_{11})(f_{22}-\lambda\mu_{22})-(f_{12}-\lambda\mu_{12})^2=0,$$

or

$$\lambda^2(\mu_{11}\mu_{22}-\mu_{12}^2)-\lambda(\mu_{11}f_{22}+\mu_{22}f_{11}-2\mu_{12}f_{12})+(f_{11}f_{22}-f_{12}^2)=0,$$

so that

$$\lambda=\{\mu_{11}f_{22}+\mu_{22}f_{11}-2\mu_{12}f_{12}\pm[(\mu_{11}f_{22}+\mu_{22}f_{11}-2\mu_{12}f_{12})^2 - 4(\mu_{11}\mu_{22}-\mu_{12}^2)(f_{11}f_{22}-f_{12}^2)]^{1/2}\}[2(\mu_{11}\mu_{22}-\mu_{12}^2)]^{-1}. \tag{8.9}$$

Observe that there are two roots to this equation, corresponding to two observable frequencies, but that three force constants are to be determined. Obviously, from determination of the two vibration frequencies in a single molecule, it will be impossible to obtain a unique set of values for these three constants. There are either of two options available to us at this point.

1. In a linear *XYZ* molecule, we can omit the force constant f_{12}, thus obtaining the *simple valence force field* approximation; that is, we assume that there is a restoring force only between neighboring atoms, between which there exists a chemical bond in the conventional sense. Then equation (8.9) reduces to

$$\lambda=\frac{\mu_{11}f_{22}+\mu_{22}f_{11}\pm[(\mu_{11}f_{22}+\mu_{22}f_{11})^2-4(\mu_{11}\mu_{22}-\mu_{12})^2f_{11}f_{22})^{1/2}]}{2(\mu_{11}\mu_{22}-\mu_{12})^2}.$$

When the appropriate reduced masses are substituted from equation (8.4), this reduces to the form of the normal mode frequencies as usually presented (references 3, 5):

$$\lambda_1+\lambda_3=f_{11}\left(\frac{1}{m_1}+\frac{1}{m_2}\right)+f_{22}\left(\frac{1}{m_2}+\frac{1}{m_3}\right),$$

$$\lambda_1\lambda_3=\frac{m_1+m_2+m_3}{m_1m_2m_3}f_{11}f_{22}.$$

2. In a linear XY_2 molecule, we have the simplifications that $m_1 = m_3$, and, more important, that $f_{11} = f_{22} = f$. Thus we can express the two roots of equation (8.9), corresponding to the two normal-mode stretching frequencies, in terms of only two force constants, the bond-stretching force constant f and the interaction force constant f_{12}. After making these substitutions, and also

$$\mu_{11} = \mu_{22} = \frac{m_1(m_2 + m_1)}{2(2m_1 + m_2)} = \mu,$$

equation (8.7) becomes

$$\lambda = \frac{(\mu \pm \mu_{12})(f \pm f_{12})}{\mu^2 - \mu_{12}^2},$$

which gives the two roots

$$\begin{aligned} \lambda_1 &= 4\pi^2\nu_1^2 = \frac{1}{m_1}(f_{11} + f_{12}), \\ \lambda_3 &= 4\pi^2\nu_3^2 = \frac{1}{m_1}\left(\frac{m_2 + 2m_1}{m_2}\right)(f_{11} - f_{12}). \end{aligned} \tag{8.10}$$

In this way, we obtain an estimate of the interaction force constant, which can be used as a correction to the normal-mode frequencies calculated on the basis of the simple valence force field assumption. The study of isotopically substituted molecules can also furnish the additional data required to make independent estimates of interaction force constants.

We would expect the interaction force constant to be quite a bit smaller than the force constant between directly bonded atoms. If this is the case, then equation (8.10) tells us that one of the normal mode stretching vibrations (ν_3) has a considerably higher frequency than the other (ν_1), especially if the mass of the end atoms (m_1) is appreciably greater than that of the central atom (m_2). We can see why this is so if we investigate the form of each of the normal modes of vibration. These can be found by going back to the original set of linear equations, equation (8.6), which can be written compactly as

$$\sum_j (f_{ij} - \lambda_k \mu_{ij}) a_{jk} = 0 \qquad \begin{cases} i = 1, 2, 3 \\ k = 1, 2, 3 \end{cases}. \tag{8.11}$$

If we substitute the expressions for the reduced masses μ, the force constants f, and the eigenvalue normal mode frequencies λ_k, as calculated in the foregoing development, for $k = 1$, 2, and 3 in turn, we can solve for the a_{jk}. The a_{jk} are the coefficients of the internal coordinates introduced earlier in this section; that is, a_{1k} is the coefficient of Δr_1, a_{2k} that of Δr_2, and a_{3k} that of $(r_1 r_2)^{1/2}\,\Delta\theta$, in the kth normal mode of vibration. These will give us the eigenvectors corresponding to each eigenvalue, from which we can find the form of the normal mode vibrations. For example, in our case of linear symmetric XY_2, we obtain

$$\begin{array}{cccc} k & a_{1k} & a_{2k} & a_{3k} \\ \begin{matrix} 1 \\ 3 \\ 2 \end{matrix} & \multicolumn{3}{c}{\begin{pmatrix} (2\mu_{11})^{-1/2} & (2\mu_{11})^{-1/2} & 0 \\ [2(\mu_{11}+2\mu_{12})]^{-1/2} & -[2(\mu_{11}+2\mu_{12})]^{-1/2} & 0 \\ 0 & 0 & [2(\mu_{11}+2\mu_{12})]^{-1/2} \end{pmatrix}} \end{array}.$$

If we draw the atoms in the molecule, and attach to each atom a small vector in the direction of each normal coordinate displacement, with the magnitude of the vector proportional to the eigenvector component for the corresponding normal frequency, we obtain a picture of each vibration, as shown in figure 8.2.

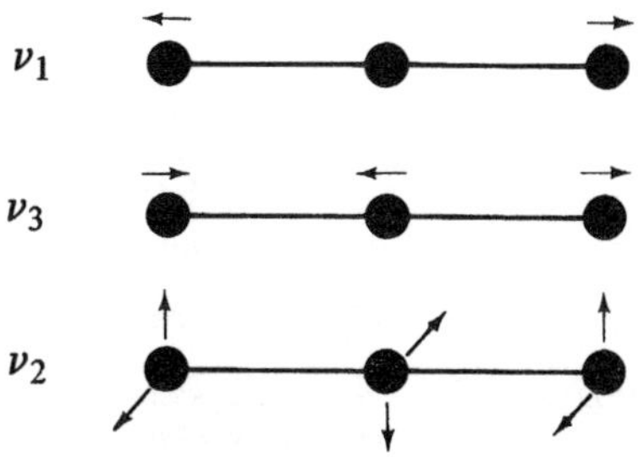

Figure 8.2
Normal mode vibrations for a linear triatomic molecule.

Thus we see that ν_2 is a bending vibration, as expected. ν_1 is a *symmetric stretch* and ν_3 is an *antisymmetric stretch*. It is now clear why the frequency of the antisymmetric stretching vibration is higher than that of the symmetric stretching vibration; in the latter, the central atom remains at rest at all times, whereas in the former, all three atoms move, and with greater amplitude. Thus the antisymmetric stretching vibration has the greater average value of the kinetic energy. By the *virial theorem*,[1] the average values of the kinetic and potential energies of an oscillator are equal; thus the total energy, which is the sum of the kinetic and potential terms, is larger for the antisymmetric stretching vibration (ν_3) than for the symmetric stretching vibration (ν_1). In the case of CO_2, the actual values are 2,349 and 1,388 cm^{-1}, respectively (reference 5). Caution must be exercised in applying this principle to other systems, however. The frequency of the bending vibration ν_2 is lower than that of either of the stretching vibrations (667 cm^{-1} in CO_2) simply because the force constants for bending are generally much smaller than those for the bond-stretching motions.

The quantum mechanical formulation of the normal vibrations follows quite analogously. Replacing equation (8.2), which is expressed in terms of the internal coordinates ζ_i, we have the equation

$$\left[-\frac{\hbar^2}{2}\sum_i \frac{\partial^2}{\partial q_i^2} + \frac{1}{2}\sum_i \lambda_i q_i^2\right]\Psi = E\Psi,$$

which is expressed in terms of the normal coordinates q_i, as found in the preceding development, and in which no cross terms appear. This permits us to separate the wave function and the energy as

$$\Psi = \prod_i \phi_i(q_i)$$

and

$$E = \sum_i E_i,$$

1. A very useful result of classical mechanics, which has applications to many spectroscopic problems, is the virial theorem. This theorem states that in a system of N particles interacting through central forces, the average value of the kinetic energy is

$$\bar{T} = -\tfrac{1}{2}\overline{\sum_{i=1}^{N} \nabla V \cdot \mathbf{r}_i}.$$

(For a proof, see reference 6.) When the potential V is a simple power law function of r,

$$V = Ar^{n+1}$$

(that is, the force law goes as r^n), the virial theorem becomes

$$\bar{T} = \frac{n+1}{2}\bar{V}.$$

For the vibrational problem just discussed, $V = \frac{1}{2}k(r - r_e)^2$, so that $n = 1$ and $\bar{T} = \bar{V}$. Another important case is that of a single charged particle in a Coulomb force field, which depends on the inverse square of the coordinate r, for which $\bar{T} = -\frac{1}{2}\bar{V}$. Since the total energy $E = \bar{T} + \bar{V}$, we also have $E = \frac{1}{2}\bar{V}$.

where

$$E_i = h\nu_i(v_i + \tfrac{1}{2}) \qquad \text{and} \qquad \nu_i = \lambda_i^{1/2}/2\pi.$$

3 "FG" Matrix Method and Symmetry of Normal Vibrations

As we have seen in the previous section, the algebra of normal coordinate calculations becomes formidable even for a simple three-atomic system. A considerable simplification can be effected by reducing the required manipulations to matrix form. This is done in two complementary ways: by the use of the so-called Wilson FG matrix method (reference 7), which we shall deal with first, and also by judicious use of group theory, in which we make liberal use of the symmetry of the molecule itself, which in turn is embodied in the form of the matrix describing its vibrations.

In order to get some feeling for the nature of these calculations, we shall apply this method to the simple linear triatomic molecule, for which we have obtained the explicit solutions in the previous sections. The detailed forms of the normal coordinate, force constant, and reduced mass matrices can all be found in the texts by Herzberg (reference 5) and Wilson, Decius, and Cross (reference 7). For practical calculations, this method has been programmed for electronic computers (reference 8). Despite all this, some physical insight can be obtained from even this brief consideration.

As the terminology might imply, the FG matrix method proceeds by setting up two matrices which embody the parameters of the molecule, namely, the **F** (for force constant) matrix and the **G** (for geometry) matrix. The **F** matrix is just the 3 × 3 matrix of force constants for the oscillator, which, for the linear triatomic case, has the form

$$\begin{pmatrix} f_{11} & f_{12} & 0 \\ f_{12} & f_{22} & 0 \\ 0 & 0 & f_{33} \end{pmatrix}.$$

This is in accordance with our assumption of only a single interaction (nonvalence) force constant, f_{12}. The **G** matrix is constructed in the following manner. For each normal mode t, we define an "effectiveness vector" $\mathbf{s}_{t\alpha}$ such that

1. the direction of $\mathbf{s}_{t\alpha}$ points in the direction in which moving atom α causes the largest increase in the internal coordinate s_t; and

2. the magnitude of $\mathbf{s}_{t\alpha}$ equals the maximum change in s_t produced by moving atom α a unit distance.

For example, in our linear triatomic molecule

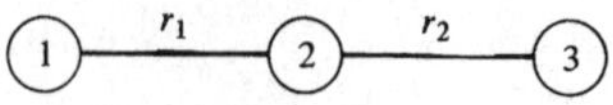

with coordinates

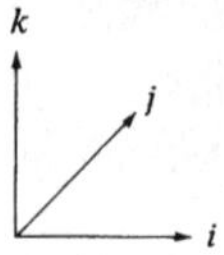

changing r_1 changes the relative positions of atoms 1 and 2, so that

$$s_{11} = -i, \qquad s_{12} = +i, \qquad s_{13} = 0,$$

while changing r_2 changes the positions of atoms 2 and 3, so that

$$s_{21} = 0, \qquad s_{22} = -i, \qquad s_{23} = +i.$$

The elements of the **G** matrix are then defined by

$$G_{tt'} = \sum_{\alpha} \frac{1}{m_\alpha} \mathbf{s}_{t\alpha} \cdot \mathbf{s}_{t'\alpha}. \tag{8.12}$$

Thus

$$G_{11} = \frac{1}{m_1}(\mathbf{s}_{11} \cdot \mathbf{s}_{11}) + \frac{1}{m_2}(\mathbf{s}_{12} \cdot \mathbf{s}_{12}) + \frac{1}{m_3}(\mathbf{s}_{13} \cdot \mathbf{s}_{13})$$

$$= \frac{1}{m_1} + \frac{1}{m_2} = \frac{m_1 + m_2}{m_1 m_2},$$

$$G_{12} = \frac{1}{m_1}(\mathbf{s}_{11} \cdot \mathbf{s}_{21}) + \frac{1}{m_2}(\mathbf{s}_{12} \cdot \mathbf{s}_{22}) + \frac{1}{m_3}(\mathbf{s}_{13} \cdot \mathbf{s}_{23})$$

$$= -\frac{1}{m_2} = G_{21},$$

and

$$G_{22} = \frac{1}{m_1}(\mathbf{s}_{21} \cdot \mathbf{s}_{21}) + \frac{1}{m_2}(\mathbf{s}_{22} \cdot \mathbf{s}_{22}) + \frac{1}{m_3}(\mathbf{s}_{23} \cdot \mathbf{s}_{23})$$

$$= \frac{1}{m_2} + \frac{1}{m_3} = \frac{m_2 + m_3}{m_2 m_3}.$$

Once more, the **G** element corresponding to bending G_{33} stands by itself, so that the **G** matrix has the form

$$\begin{pmatrix} G_{11} & G_{12} & 0 \\ G_{12} & G_{22} & 0 \\ 0 & 0 & G_{33} \end{pmatrix}.$$

Let us concentrate on just the 2×2 submatrix corresponding to the stretching vibrations, construct a matrix of the reduced masses defined by equation (8.4), and take the product

$$(\mathbf{G}\boldsymbol{\mu}) = \begin{pmatrix} \dfrac{m_1 + m_2}{m_1 m_2} & -\dfrac{1}{m_2} \\ -\dfrac{1}{m_2} & \dfrac{m_1 + m_2}{m_1 m_2} \end{pmatrix} \begin{pmatrix} \dfrac{m_1(m_2 + m_3)}{M} & \dfrac{m_1 m_3}{M} \\ \dfrac{m_1 m_3}{M} & \dfrac{m_1(m_2 + m_3)}{M} \end{pmatrix}$$

$$= \begin{pmatrix} 1 & 0 \\ 0 & 1 \end{pmatrix}.$$

That is, all we have done is to work our way through a rather complicated calculation with the simple result that $\mathbf{G}\boldsymbol{\mu} = \mathbf{1}$, or $\mathbf{G} = \boldsymbol{\mu}^{-1}$; that is, the **G** matrix is just the inverse of the reduced mass matrix. If the reduced mass matrix has already been found, **G** can be found immediately by taking its inverse; many times, however, it has proved easier to construct **G** directly, in terms of the "effectiveness vectors" introduced above.

In actual practice, the **G** matrix elements may simply be taken from the compilation in appendix VI of reference 7. In order to introduce the reader to the use of these tables, we reexpress the results found above in the notation of Wilson, Decius, and Cross. For the linear triatomic with masses and internal coordinates just shown, we have

$$G_{11} = G^2_{r_1 r_2} = \frac{1}{m_1} + \frac{1}{m_2},$$

$$G_{22} = G^2_{r_2 r_3} = \frac{1}{m_2} + \frac{1}{m_3},$$

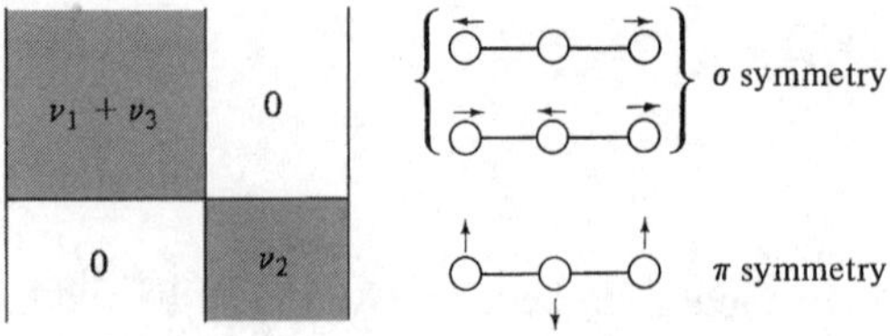

Figure 8.3
Secular determinant for a linear triatomic molecule.

$$G_{12} = G^1_{rr} = \frac{1}{m_2}\cos\theta = -\frac{1}{m_2},$$

since the equilibrium value of θ is 180° and $\cos 180° = -1$. The **G** matrix element for the bending modes is found to be

$$G_{33} = G^3_{\phi\phi} = \left(\frac{1}{r_1}\right)^2 \frac{1}{m_1} + \left(\frac{1}{r_2}\right)^2 \frac{1}{m_3} + \left(\frac{1}{r_1^2} + \frac{1}{r_2^2} - \frac{2}{r_1 r_2}\cos\theta\right)\frac{1}{m_2}$$

$$= \frac{1}{r_1^2 r_2^2}\left[\frac{r_1^2}{m_3} + \frac{r_2^2}{m_1} + \frac{(r_1 + r_2)^2}{m_2}\right].$$

Note the appearance of this expression in equation (8.8).

To find the normal mode frequencies, we reexpress (8.11) in matrix form,

$$\mathbf{F}\mathbf{a}_k - \lambda_k \boldsymbol{\mu}\mathbf{a}_k = 0.$$

Premultiplying by **G** gives

$$\mathbf{G}\mathbf{F}\mathbf{a}_k - \lambda_k \mathbf{G}\boldsymbol{\mu}\mathbf{a}_k = 0,$$

or, since $\mathbf{G}\boldsymbol{\mu} = 1$,

$$\mathbf{G}\mathbf{F}\mathbf{a}_k - \lambda_k \mathbf{a}_k = 0;$$

that is, the normal-mode eigenvalues are found from the secular determinant,

$$|\mathbf{G}\mathbf{F} - \lambda\mathbf{1}| = 0. \tag{8.13}$$

We may note that in linear XYZ molecules with no bend-stretch interaction, such as we have been considering, this secular determinant breaks up into two subdeterminants, as shown in figure 8.3.

As a further example, let us consider the case of a linear four-atomic molecule of $D_{\infty h}$ symmetry, such as acetylene:

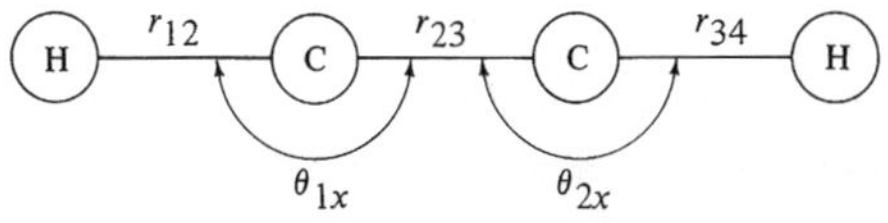

Of the $3n - 5 = 7$ vibrational motions, there will be three stretching vibrations, and four bending vibrations—two pairs of doubly degenerate bends about each dihedral angle. The symmetry coordinates for the stretching vibrations are constructed from the internal coordinates,

C—H vibrations: $\Delta r_{12}, \Delta r_{34}$,

C—C vibration: Δr_{23},

and similarly for the bending vibrations, $\Delta\theta_{1x}$, $\Delta\theta_{1y}$, $\Delta\theta_{2x}$, and $\Delta\theta_{2y}$.

While the normal coordinates can be found in terms of these internal coordinates, the problem is considerably simplified if symmetry coordinates are used, as we shall see. The symmetry coordinates must transform as the point group of the molecule ($D_{\infty h}$; see table A.27); thus, since the inversion operation i transforms Δr_{12} into Δr_{34}, and vice versa, we must take linear combinations of these two variables to give

$$\mathscr{S}_1 = \left(\frac{1}{\sqrt{2}}\right)(\Delta r_{12} + \Delta r_{34}), \quad i\mathscr{S}_1 = +\mathscr{S}_1 \qquad (\sigma_g \text{ symmetry}), \tag{8.14a}$$

and

$$\mathscr{S}_2 = \left(\frac{1}{\sqrt{2}}\right)(\Delta r_{12} - \Delta r_{34}), \quad i\mathscr{S}_2 = -\mathscr{S}_2 \qquad (\sigma_u \text{ symmetry}), \tag{8.14b}$$

$$\mathscr{S}_3 = \Delta r_{23} \text{ by itself, and is of } \sigma_g \text{ symmetry.} \tag{8.14c}$$

The symmetry coordinates for the bending vibrations are fairly easily seen to be

$$\mathscr{S}_4 = \tfrac{1}{2}(\Delta\theta_{1x} + \Delta\theta_{1y} - \Delta\theta_{2x} - \Delta\theta_{2y}) \qquad (\pi_g), \tag{8.14d}$$

and

$$\mathscr{S}_5 = \tfrac{1}{2}(\Delta\theta_{1x} + \Delta\theta_{1y} + \Delta\theta_{2x} + \Delta\theta_{2y}) \qquad (\pi_u). \tag{8.14e}$$

The set of transformations (8.14a)–(8.14e) defines a matrix $\mathbf{U}$ such that

$$\mathscr{S} = \mathbf{Ur},$$

where $\mathbf{r}$ is the vector array of internal coordinates.

The **G** matrix elements tabulated in appendix VI of reference 7 are expressed in terms of the internal coordinates of the molecule, since they are defined in terms of local groups of atoms. For the linear symmetric four-atomic molecule we are considering, these are

$$G_{11} = G^2_{r_{12}r_{23}} = \frac{1}{m_H} + \frac{1}{m_C},$$

$$G_{22} = G^2_{r_{23}r_{23}} = \frac{1}{m_C} + \frac{1}{m_C} = \frac{2}{m_C},$$

$$G_{12} = G^1_{r_{12}r_{23}} = G^1_{r_{23}r_{34}} = \frac{1}{m_C}\cos 180° = -\frac{1}{m_C},$$

$$G_{33} = G^2_{r_{23}r_{34}} = \frac{1}{m_H} + \frac{1}{m_C},$$

$$G_{44} = G_{55} = G^3_{\phi\phi} = \frac{1}{r_{12}^2}\frac{1}{m_H} + \frac{1}{r_{23}^2}\frac{1}{m_C} + \left(\frac{1}{r_{12}^2} + \frac{1}{r_{23}^2} - \frac{2}{r_{12}r_{23}}\cos 180°\right)\frac{1}{m_C}.$$

In order to refer these quantities to a specified molecular geometry, we must reexpress the **G** matrix in terms of the symmetry coordinates by the transformation

$$\mathcal{G} = \mathbf{UGU}^{-1}, \tag{8.15}$$

where $\mathbf{U}^{-1}$ is the transposed **U** matrix. As we shall see, this will reduce the **G** matrix to its simplest block diagonal form. Application of equation (8.15) to the stretching vibrations gives

$$\mathcal{G}(\sigma_g) = \begin{pmatrix} \dfrac{1}{m_H} + \dfrac{1}{m_C} & -\dfrac{\sqrt{2}}{m_C} \\ -\dfrac{\sqrt{2}}{m_C} & \dfrac{2}{m_C} \end{pmatrix}$$

and

$$\mathcal{G}(\sigma_u) = \left(\frac{1}{m_H} + \frac{1}{m_C}\right).$$

If we assume a simple valence force field, that is,

$$f_{11} = f(\text{C—H}), \quad f_{22} = f(\text{C}\equiv\text{C}), \quad f_{12} = 0, \quad \text{and} \quad f_{33} = f(\text{C—H}),$$

then λ_1 and λ_2 are found from

$$\begin{vmatrix} f(\mathrm{C{-}H})\left(\dfrac{1}{m_{\mathrm{C}}}+\dfrac{1}{m_{\mathrm{H}}}\right)-\lambda & -\sqrt{2}f(\mathrm{C{\equiv}C})\dfrac{1}{m_{\mathrm{C}}} \\ -\sqrt{2}f(\mathrm{C{-}H})\dfrac{1}{m_{\mathrm{C}}} & 2f(\mathrm{C{\equiv}C})\dfrac{1}{m_{\mathrm{C}}}-\lambda \end{vmatrix}=0$$

and

$$\lambda_3 = 4\pi^2c^2\nu_3^2 = f(\mathrm{C{-}H})\left(\frac{1}{m_{\mathrm{H}}}+\frac{1}{m_{\mathrm{C}}}\right).$$

For the bending vibrations,

$$\mathscr{G}(\pi_g, \pi_u) = \begin{pmatrix} \dfrac{r_{23}}{r_{12}}\dfrac{1}{m_{\mathrm{H}}}+\dfrac{r_{23}}{r_{12}}\left(1+2\dfrac{r_{12}}{r_{23}}\right)^2\dfrac{1}{m_{\mathrm{C}}} & 0 \\ 0 & \dfrac{r_{23}}{r_{12}}\dfrac{1}{m_{\mathrm{H}}}+\dfrac{r_{23}}{r_{12}}\dfrac{1}{m_{\mathrm{C}}} \end{pmatrix}.$$

Because we have made the transformation to symmetry coordinates, the secular determinant and the normal mode vibrations of the molecule take on the block-diagonal appearance given in figure 8.4.

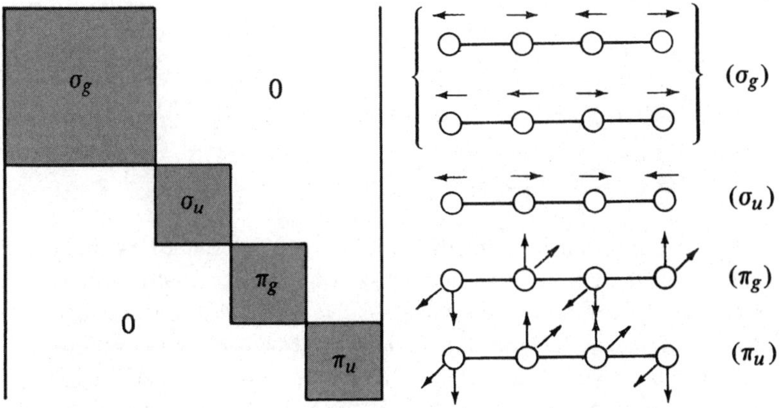

Figure 8.4
Secular determinant and form of normal mode vibrations for a linear four-atomic molecule.

In problem 2, the reader is asked to carry out explicitly the transformation between the **G** and $\mathscr{G}$ matrices, and then to rederive the $\mathscr{G}$ matrix independently using the "effectiveness vector" approach described earlier in this section.

For larger molecules, the generalized FG matrix method can be used in a machine calculation (reference 8). The simplification of the matrices introduced by the use of symmetry coordinates becomes increasingly important as the size of the molecule increases. For $N > 12$ atoms or so, even this approach becomes impractical, and then a "group frequency" model must be used (see, for example, reference 4). In this model, each identifiable functional group (C—H, O—H, C═O, and so on) is assigned a frequency on the basis of semiempirical rules; the effect of near-neighbor groups is taken into account by a set of standard corrections to these frequencies. This works reasonably well for higher-frequency normal modes, but lower-frequency vibrations involving coordinated motions of the entire molecular framework cannot be predicted very well by group frequency arguments.

4 Selection Rules for Vibrational Transitions

As we have previously seen in the case of diatomic molecules (chapter 4.2), the intensity of an infrared absorption is proportional to the square of the strength of the dipole derivative function,

$$|\langle\psi_i|\frac{\partial\boldsymbol{\mu}}{\partial R}(R - R_e)|\psi_f\rangle|^2.$$

For a particular normal mode, this becomes

$$\left(\frac{\partial\mu}{\partial q}\right)^2 |\langle\psi_i|(q - q^e)|\psi_f\rangle|^2,$$

so that the selection rules for the transition are determined by the matrix elements of the normal coordinate displacement. In a diatomic molecule, of course, the one normal coordinate is the internuclear distance displacement itself, $r - r_e$. In order for the matrix element to be nonvanishing, it must be totally symmetric—that is, the direct product $\Gamma(\psi_i) \otimes \Gamma(q) \otimes \Gamma(\psi_f)$ must contain the σ_g representation if the molecule is linear, or the A_{1g} representation for nonlinear system. Let us see how this operates in the simple case of linear, centrosymmetric CO_2.

The ground state of CO_2 has all vibrational quantum numbers equal to zero. It is, therefore, designated the 000 state, and has symmetry Σ_g^+. The coordinate along the internuclear axis, which we shall designate z, has Σ_u^+

symmetry; the coordinates perpendicular to the axis, which we shall designate x and y, have π_u symmetry. (See chapter 6.6 for a discussion of how the symmetry species are established.) Now, consider the symmetric stretching vibration ν_1, which is a σ_g vibration. The possible matrix elements would be

$$(z \text{ polarized}): \quad \Sigma_g^+ \otimes \Sigma_u^+ \otimes \Sigma_g^+ = \Sigma_u^+,$$

$$(x, y \text{ polarized}): \quad \Sigma_g^+ \otimes \Pi_u \otimes \Sigma_g^+ = \Pi_u.$$

Since neither of these contains the Σ_g^+ representation, we conclude that this vibration is *infrared inactive*. This is particularly easy to see on an intuitive basis, of course; CO_2 possesses zero dipole moment, and stretching both C—O bonds equally will leave it with a zero dipole moment throughout the vibration, so that absorption of radiation could not occur. This is just a case of the group theory reflecting the obvious symmetry properties of the molecule. For the asymmetric stretching vibration, however, we have

$$\Sigma_g^+ \otimes \Sigma_u^+ \otimes \Sigma_u^+ = \Sigma_g^+,$$

so that the vibration is infrared active. It is termed a parallel band because the change in dipole moment is along the principal (rotation) axis of the molecule; this point will become significant in the next section, in which rotational structure is discussed. The bending vibration involves motion in the x and y coordinates, for which the matrix elements are

$$\Sigma_g^+ \otimes \Pi_u \otimes \Pi_u = \Sigma_g^+ \oplus \Delta_g.$$

This is also an infrared active vibration and is designated a perpendicular band.

We can also consider selection rules for Raman transitions in a similar way. If we say that the dipole moment induced in the molecule by the electric field of the light wave is

$$\boldsymbol{\mu}_{\text{induced}} = \boldsymbol{\alpha} \cdot \mathscr{E}$$

and that the interaction responsible for transitions is

$$\mathscr{H}' = \mathscr{E} \cdot \boldsymbol{\mu}_{\text{induced}} = \mathscr{E} \cdot \boldsymbol{\alpha} \cdot \mathscr{E},$$

then it is evident that we wish to examine the matrix elements of the polarizability tensor $\boldsymbol{\alpha}$.[2] The transformation properties of this tensor are the same

2. For a discussion of operations with tensors, see reference 6, p. 146ff.

as those of the six coordinate products, x^2, y^2, z^2, xy, xz, and yz. These components have symmetries $\Sigma_g(x^2, y^2, z^2)$, $\Pi_g(xz$ and $yz)$, and $\Delta_g(x^2, y^2$, and $xy)$. Carrying through an analysis similar to that previously given, we see that the symmetric stretching vibration ν_1 is Raman active, but that the other vibrations are Raman inactive. Note that the only Raman active vibration in this molecule is the one that is infrared inactive, and vice versa. This is a general property of centrosymmetric molecules—there is no vibration that is simultaneously infrared and Raman active. The reason for this is easy to see. All molecules with a center of symmetry have the inversion operation as part of the point groups, for which the eigenvalues are either u or g. Since the dipole operators are all u, and the polarizability (quadrupole) operators are all g, no pair of states can simultaneously possess nonvanishing matrix elements of both operators. In molecules that do not have a center of symmetry, this restriction does not apply.

Let us consider one further example, that of a nonlinear molecule—in particular, BF_3, with belongs to the D_{3h} point group. The normal coordinates for such a molecule are shown in figure 8.5. The ground state, designated 0000, is A_1'; that is, it possesses full D_{3h} symmetry. Thus we seek direct products of the coordinate and/or quadrupole operators with the excited vibrational states, which are also a_1' symmetry, in order to have an allowed transition. The coordinate operators are x and y (species e') and z (species a_2''). The quadrupole operators are, therefore (see appendix A), of $a_1' \oplus e'$ symmetry (xy, x^2, y^2), e'' symmetry (xz, yz), or just a_1' (z^2). The selection rules will be as given in table 8.1. Note that, in this case, the degenerate (e) vibrations are active in both infrared absorption and Raman scattering.

As in the case of diatomic molecules, the anharmonicity of the vibrations can slightly relax the $\Delta v = 1$ selection rule to produce *overtone* bands at approximately twice the fundamental frequency. In addition, absorption from thermally populated low-lying vibrations can lead to what are called *hot bands*. For example, in BF_3, the ν_4 level lies at 482 cm^{-1}. This level can absorb energy corresponding to the ν_2 vibration to produce a state $\nu_2 + \nu_4$. This transition is designated $\nu_2 + \nu_4 - \nu_4$; the absorption falls at 711 cm^{-1}, while the ν_4 fundamental is at 719.5 cm^{-1}. This shift is due to the anharmonicity between the two different vibrations. Because of the presence of several normal modes in a polyatomic molecule, we can also have *difference* bands, such as the 001–100 transition in CO_2, whose frequency (about 960 cm^{-1}) corresponds to the difference between the ν_3 level at 2,349 cm^{-1} and the ν_1 level at 1,388 cm^{-1}; in emission, this band is responsible for the CO_2 laser activity (see chapter 10). There can also be a form of overtone

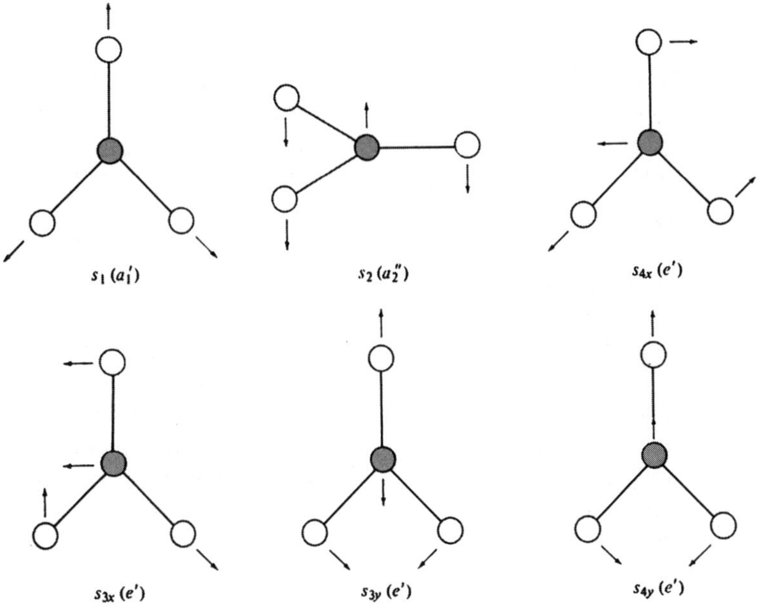

Figure 8.5
Normal mode vibrations for a D_{3h} molecule such as BF_3. There are four fundamentals; the (e') vibrations are doubly degenerate. It may be instructive for the reader to verify that the species of each vibration conforms to the character of the symmetry, as specified in chapter 6.4.

Table 8.1

Vibration	Infrared	Raman
ν_1	$\left\{\begin{array}{l} a_1' \otimes a_2'' = a_2'' \\ a_1' \otimes e' = e' \end{array}\right.$ $\therefore$ infrared *inactive*	$a_1' \otimes a_1' = a_1'$ $\therefore$ Raman *allowed*
ν_2	$a_2'' \otimes a_2'' = a_1'$ infrared *allowed*, parallel band	$\left\{\begin{array}{l} a_2'' \otimes a_1' = a_2'' \\ a_2'' \otimes e' = e'' \\ a_2'' \otimes e'' = e' \end{array}\right.$ Raman *inactive*
ν_3, ν_4	$e' \otimes e' = e' \oplus a_1' \oplus a_2'$ infrared *allowed* and Raman *allowed*	

known as *combination* bands, in which two different vibrations change their quantum numbers simultaneously. An example would be the 110–000 transition in CO_2, at 2,076 cm^{-1} (approximately equal to 667 cm^{-1} + 1,388 cm^{-1}). The hot band in BF_3 discussed could be considered a "combination-difference" band.

5 Rotational Structure of Vibrational Bands

In a linear molecule, such as HCN or CO_2, we can distinguish the two cases of parallel and perpendicular absorption bands. For a parallel band, the transition moment lies along the internuclear axis. The molecule then has the same form of matrix elements as a simple diatomic molecule, so that only P and R branches, corresponding to $\Delta J = -1$ and $+1$, appear in the absorption spectrum with frequencies given by equations (4.6) and (4.7). For a perpendicular band, however, the transition moment is at right angles to the axis, which lets a Q branch ($\Delta J = 0$) be allowed. The two fundamentals of HCN, shown in figure 8.6, are typical of such spectra.

In a nonlinear molecule, the situation can get a good deal more complex. In a symmetric top, a parallel band would correspond to a transition moment directed along the principal axis of K quantization. The selection rules are $\Delta K = 0$ and $\Delta J = 0$ or ± 1. We obtain three branches, designated ${}^QP_K(J)$, ${}^QQ_K(J)$, and ${}^QR_K(J)$. For a perpendicular band, we have $\Delta K = \pm 1$ instead; this leads to six branches, namely, ${}^RP_K(J)$, ${}^RQ_K(J)$, ${}^RR_K(J)$, and ${}^PP_K(J)$, ${}^PQ_K(J)$, ${}^PR_K(J)$. The frequencies of the lines can be found from the customary relations between term values,

$$\nu = F'_{v'}(J', K') - F''_{v''}(J'', K'') + (G_{v'} - G_{v''}),$$

where

$$F_v(J, K) = B_v J(J+1) + (A_v - B_v)K^2$$

Thus, we would have for the subband origins,

$$\nu_0^{P\text{-type}} = \nu_{\text{origin}} + (A' - B') - 2(A' - B')K + [(A' - B') - (A'' - B'')]K^2,$$

$$\nu_0^{R\text{-type}} = \nu_{\text{origin}} + (A' - B') + 2(A' - B')K + [(A' - B') - (A'' - B'')]K^2.$$

For each branch, we would have

$$\nu[P(J)] = \nu_0 - (B' + B'')J + (B' - B'')J^2,$$

$$\nu[Q(J)] = \nu_0 + (B' - B'')J + (B' - B'')J^2,$$

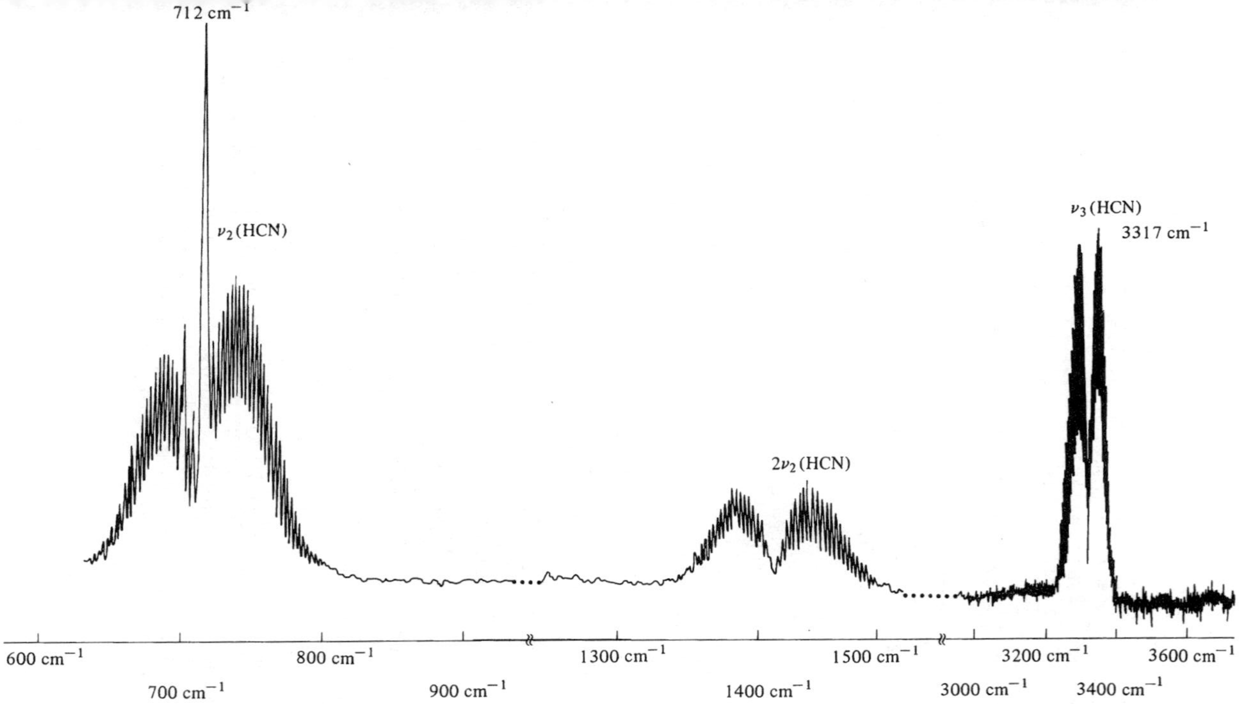

Figure 8.6
Portions of the infrared absorption spectrum of HCN. The ν_2 vibration, which is a bending vibration of π symmetry, is a perpendicular band and therefore has allowed *P*, *Q*, and *R* branches. The ν_3 [C—H stretching] vibration has Σ symmetry and is therefore a parallel band with *P* and *R* branches only. Note that the $2\nu_2$ overtone band is of $\Pi \otimes \Pi = \Sigma^+ \oplus \Delta$ symmetry; only the Σ^+ component is allowed, so that a parallel band with no *Q* branch is observed.

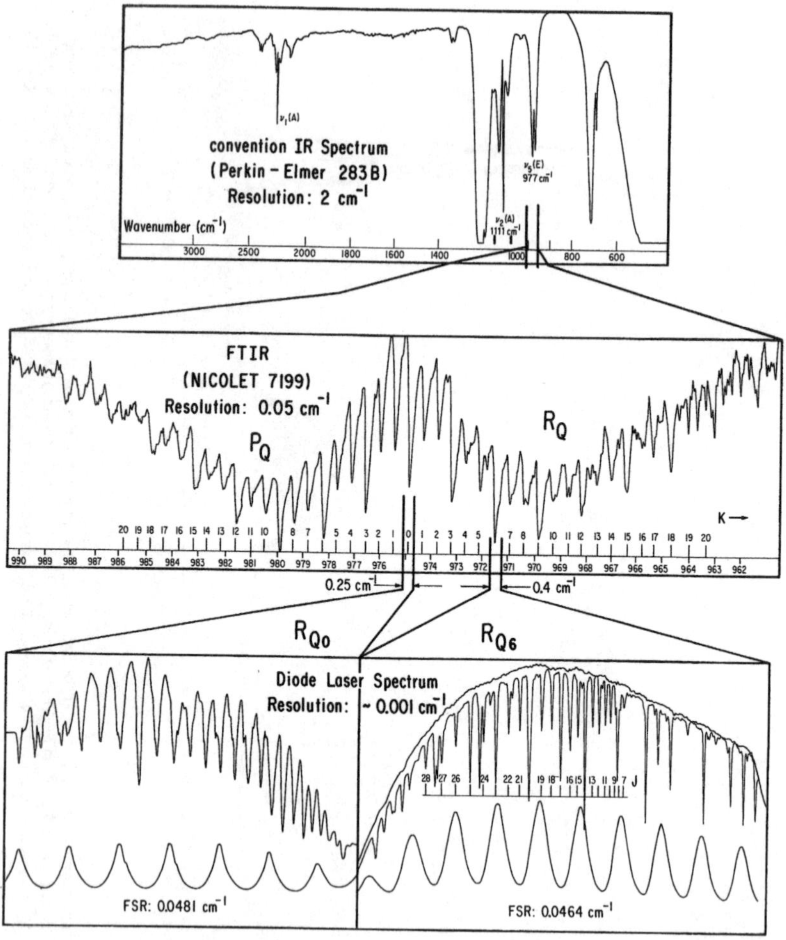

Figure 8.7
Infrared spectrum of CDF_3 at increasing levels of resolution. Top panel: conventional grating infrared spectrometer scan. Middle panel: structure of ν_5 at a resolution of 0.05 cm^{-1}, recorded on a Fourier transform infrared spectrometer. Lower panel: diode laser scans of R_{Q_0} and R_{Q_6} subbands at 0.001 cm^{-1} resolution. Also shown on the lower panels are etalon traces used for calibrating the diode, with free spectral range (FSR) given. [Spectrum courtesy of Zhu Qingshi, MIT Spectroscopy Laboratory.]

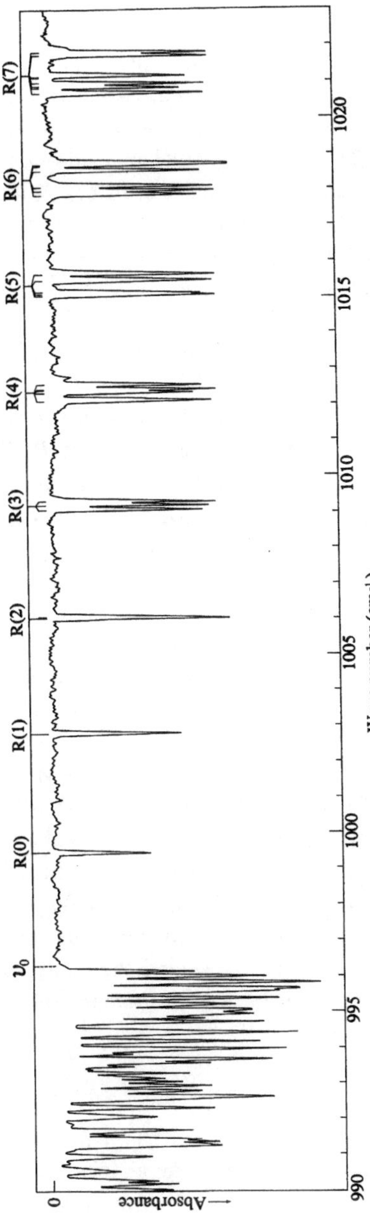

Figure 8.8
Portion of the Fourier transform infrared spectrum of the ν_4 band of CD_4. The congested region at the left-hand end is a portion of the Q branch. The splittings in the $R(J)$ lines occur in the excited vibrational state: rotational level $J' + 1$ is split into J' components, not all of which may be resolved. [Spectrum courtesy of H. Flicker and M. Buchwald, Los Alamos National Laboratory.]

$$\nu[R(J)] = \nu_0 + (B' + B'')(J+1) + (B' - B'')(J+1)^2.$$

Depending on the values of the four constants A', A'', B', and B'', a wide variety of different band shapes is possible. The spectrum of CDF_3, shown in figure 8.7, illustrates some of these. Under low resolution, the *a* vibrations, such as ν_1 and ν_2, show an apparent "*PQR*" structure, while an *e* vibration, such as ν_5 at 977 cm^{-1}, shows what appear to be *P* and *R* branches. With a factor of 50 increase in resolution, using a Fourier transform infrared spectrometer, these branches are revealed as a series of ^PQ and ^RQ subband heads, evenly spaced in quantum number *K*. A further improvement in resolution by a factor of 50 is required to resolve these subbands into Doppler-limited rotation-vibration transitions; to obtain this resolution, a tunable diode laser spectrometer (see chapter 10) must be employed.

In a spherical top molecule, we have just the B' and B'' constants, with selection rules $\Delta J = 0, \pm 1$. This would give to first order a simple *PQR* structure, with evenly spaced lines in the *P* and *R* branches: a linear molecule's spectrum. In molecules of this type, interaction between rotation and vibration becomes extremely important, however. This tends to split the lines into "clusters," which can give rise to a very complex substructure in heavier molecules. This splitting (due primarily to Coriolis interactions) is illustrated in figure 8.8, which shows a portion of the ν_4 band of CD_4. These interactions, which are important in both tetrahedral molecules, such as methane, and octahedral ones, such as SF_6 and UF_6, have been very thoroughly analyzed in recent years (reference 9).

When dealing with asymmetric rotors, for which closed form expressions for rotational energy levels are not available, the customary approach is to attempt to analyze the spectrum in terms of a near-prolate or near-oblate symmetric rotor Hamiltonian. After such a first approximation is made, and tentative assignments obtained, the system is analyzed using a Watson Hamiltonian which incorporates all asymmetry and distortion constants. A detailed treatment is beyond the scope of this book, and the reader is instead referred to several standard texts on this subject (references 10–12).

Problems

1. The harmonic oscillator (mass m) in two dimensions has a potential energy V expressed in polar coordinates r, θ, of the general form $2V = k_1 r^2 \cos^2\theta + k_2 r^2 \sin^2\theta$, where k_1 and k_2 are force constants. For the special case $k_1 = k_2 = k$, the oscillator has a single frequency $\nu = (k/m)^{1/2}/2\pi$, and

its Schrödinger equation has solutions of the form

$$\psi_{v,l} = N_{v,l} \exp\left(\frac{-\alpha r^2}{2}\right) \exp(il\theta)\, P(r),$$

where v, l are quantum numbers ($v = 0, 1, 2, \ldots, \infty$; $\pm l = 0, 2, 4, \ldots, v$ for v even, $\pm l = 1, 3, 5, \ldots, v$ for v odd); $N_{v,l}$ is a normalization constant; $\alpha = 4\pi^2 m\nu/h$; $P(r)$, a polynomial in r only, depends for its form on the values of v and l and is an even function for even v, odd for odd v.

(a) The energy levels of this two-dimensional oscillator are $E_v = (v + 1)h\nu$. What is the degeneracy of the vth level?
(b) Find $N_{v,l}$ for $v = 1$, $l = +1$, for which $P(r) = \sqrt{\alpha}r$.
(c) Show that any two ψs of the same v but different l are orthogonal.
(d) Find the average value of the angular momentum p_θ for any state v, l.
(e) Find the average value of r^{-2} for the state $v = 1$, $l = +1$.

Substitute the results of the above in the equation for the relation between E and p_θ in the plane rotor and find E for the state $v = 1$, $l = +1$. Explain the difference between this value of E and that given by the equation of part (a).

2. The linear centrosymmetric four-atomic molecule (example: acetylene) was treated in section 3.

(a) Construct the **U** matrix of transformations between internal and symmetry coordinates for this system, from the definitions given by equations (8.14a)–(8.14e).
(b) Using the transformation given by (8.15), derive the $\mathcal{G}$ matrix elements from the Wilson, Decius, and Cross **G** matrix elements given in section 3.
(c) Rederive the $\mathcal{G}$ matrix elements independently using the "effectiveness vector" approach defined by equation (8.12).

3. The carbon suboxide molecule, C_3O_2, is believed to be linear and symmetric.

(a) Classify the normal vibrations of C_3O_2 according to the symmetry species of point group $D_{\infty h}$.
(b) Draw the modes of vibration of C_3O_2.
(c) How many polarized lines should appear in the vibrational Raman spectrum? How many fundamental infrared bands should have P, Q, and R branches?
(d) The ground electronic state of C_3O_2 is nondegenerate. What is the degeneracy of ψ_{mol} when C_3O_2 is in its ground electronic state with $v_1 = v_2 =$

$v_3 = v_4 = v_5 = v_6 = 0$, $v_7 = 1$, and the rotational quantum number $J = 5$? The normal coordinate Q_7 is antisymmetric to simultaneous interchange of all pairs of equivalent nuclei. What is the degeneracy of ψ_{mol} for the above state with $J = 6$ instead of 5?

4. The cyclobutane molecule, $(CH_2)_4$, has the carbon atoms in a four-membered ring. Point groups to which the cyclobutane molecule might belong include $D_{4h}(C_4^z, C_2^x, \sigma_h^{xy}, \sigma_v^{xz}, 2\sigma_d, i)$, $D_{2d}(S_4^z, C_2^x, 2\sigma_d)$, and $C_{4v}(C_4^z, \sigma_v^{xz}, 2\sigma_d)$. More than enough symmetry elements are given in the parentheses to specify each point group.

(a) List the sets of equivalent nuclei for cyclobutane structures of D_{4h}, D_{2d}, and C_{4v} symmetry and give the characteristic symmetry elements for each set.
(b) How many totally symmetric vibrations are present in each structure?
(c) Give the type of rotator for each structure (linear, symmetric top, asymmetric top, spherical top).
(d) Which structures (if any) have infrared active totally symmetric vibrations?

5. Suppose that there are two possible structures for ethylene, planar D_{2h} and nonplanar D_{2d}.

(a) Work out the infrared active and the Raman active vibrational species for each point group.
(b) Work out the distribution of fundamental vibrational frequencies among the different species of the two structures and compare the results. What would you look for spectroscopically to decide between the two structures?
(c) What kind of rotational fine structure would you expect in the vibrational infrared bands of the D_{2h} model? What kind for the D_{2d} model?

6. Make a complete analysis of the spectrum of BF_3 given in the following table and prove the symmetry of the molecule. (Chemical evidence gives a start.)

$B^{11}F_3$ (cm^{-1})	$B^{10}F_3$ (cm^{-1})	Raman[a]	Infrared[a]
480.4	482.0	m	s
691.3	719.5	—	s
888	888	s	—
1,445.9	1,497	—	vs
1,831	1,928	—	w
2,903.2	3,008.2	—	w

a. vs = very strong; s = strong; m = medium; w = weak.

Note that the 888 cm^{-1} band has the same value for both isotopic species. This is a critical point in clinching the symmetry.

7. The molecule 1, 1-dichloroethylene ($C_2H_2Cl_2$) is planar with C_{2v} symmetry, the C=C bond coinciding with the C_2 axis. Take this axis as the z axis and the plane containing the molecule as the (x, z) plane.

(a) How many vibrational frequencies are in each symmetry species?
(b) How many polarized and how many depolarized Raman lines (fundamentals) are there?
(c) How many infrared bands (fundamentals) are there of type A, type B, and type C, respectively? The C_2 axis is the axis of the intermediate moment of inertia I_b.
(d) The normal coordinate for one vibration in this molecule is determined completely by symmetry and the requirement of zero angular momentum. Plot qualitatively the molecular dipole moment as a function of this normal coordinate through one complete vibrational cycle. (That is, Q goes from zero to Q_{max}, zero, $-Q_{\text{max}}$ and zero is succession.)
(e) Are there any vibrations in this molecule whose fundamental frequency ($\Delta v = 1$) is forbidden in the infrared, but whose first overtone ($\Delta v = 2$) is allowed (given sufficient electrical anharmonicity to permit $\Delta v = 2$)?

8. The positions of the atoms in the molecule N_4S_4 have been determined by x-ray diffraction. In terms of a set of Cartesian coordinates x, y, z placed within the molecule, these are

N_1: $x = b, y = 0, z = 0$;	N_2: $0, b, 0$;
N_3: $-b, 0, 0$;	N_4: $0, -b, 0$;
S_1: $a, -a, a$;	S_2: $-a, a, a$;
S_3: $a, a, -a$;	S_4: $-a, -a, -a$.

Here the numbers a and b are unrelated parameters of the order of a few angstroms in size.

(a) To what point group does the molecule belong?
(b) What is the distribution of vibrational degrees of freedom among the vibrational symmetry species (that is, irreducible representations)?
(c) How many different vibrational frequencies does the molecule have?
(d) How many frequencies should appear in the Raman spectrum as polarized Raman lines and how many as depolarized Raman lines?

(e) How many frequencies should appear in the infrared absorption spectrum as fundamentals ($\Delta v = +1$) and how many of these should coincide with Raman lines?

(f) Assume the parameter $a = 3$ Å $= 1.5b$. Compute the moments of inertia in atomic mass units times angstroms squared. What kind of rotator is the molecule (linear rotator, spherical top, oblate or prolate symmetric top, oblate or prolate asymmetric top)?

(g) The symmetry number of a molecule is defined as the number of indistinguishable configurations that can be obtained by rotating the rigid molecule. Two configurations are rendered "indistinguishable" when the numbering of otherwise identical nuclei is removed. What are the symmetry numbers of molecules belonging to the following groups:

C_{2v}, D_{2d}, D_{6h}, T_d, O_h?

9.

(a) Let $\psi_a \equiv \psi_{v_1=1}(Q_1)\psi_{v_2=0}(Q_2)$ and $\psi_b \equiv \psi_{v_1=0}\psi_{v_2=2}$ be the normalized harmonic oscillator wave functions of a polyatomic molecule corresponding to excited vibrational states of unperturbed energies E_a and E_b. If these two states are in Fermi resonance, the perturbation treatment in chapter 1.4 can be applied. Assume that the interaction energies H'_{aa} and H'_{bb} are zero and that H'_{ab} arises from one or more anharmonic terms in the potential function. In a certain molecule, the levels ψ_a and ψ_b are observed to be in Fermi resonance, the transitions to the *perturbed* levels being observed at 1,400 and 1,500 cm^{-1}, whereas the level $\psi_{v_1=0}\psi_{v_2=1}$ has an energy of 740 cm^{-1} above the zero level (see the accompanying diagram). Deduce the unperturbed E_a value from the above data. (H'_{ab} is to be evaluated from the data, not by integration.)

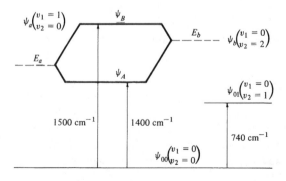

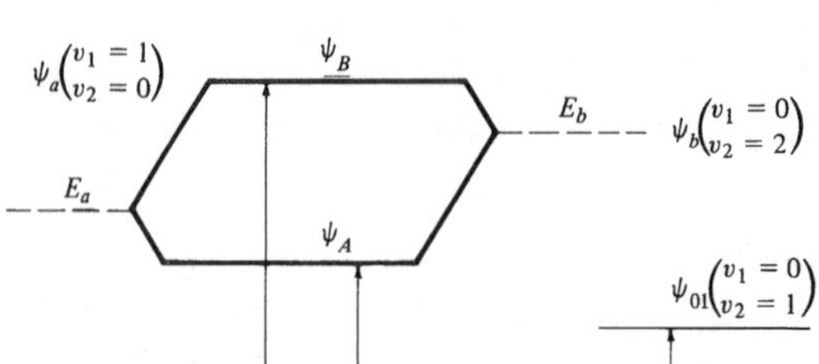

(b) The *intensity* of the Raman line for the transition from the ground state $\psi_{0,0}$ to the unperturbed state ψ_a in the absence of Fermi resonance is proportional to the square of the matrix element

$$\langle\psi_{0,0}|\frac{\partial\alpha}{\partial Q_1}Q_1|\psi_a\rangle,$$

where $\partial\alpha/\partial Q_1$ is a nonzero constant. The corresponding matrix element

$$\langle\psi_{0,0}|\frac{\partial\alpha}{\partial Q_2}Q_2|\psi_b\rangle$$

is zero because $\Delta v_2 = 2$. Find the ratio of the intensities of the two Raman lines for the transitions from state $|\psi_{00}\rangle$ to states $|\psi_a\rangle$ and $|\psi_b\rangle$.

References

1. K. Gottfried, *Quantum Mechanics*, (Benjamin, Menlo Park, CA, 1966), pp. 260ff.
2. J. C. Decius, *J. Chem. Phys.* **17**, 1315 (1949).
3. R. N. Dixon, *Spectroscopy and Structure* (Methuen, London, 1965), chapter 5.
4. N. B. Colthup, L. H. Daly, and S. E. Wiberley, *Introduction to Infrared and Raman Spectroscopy*, 2nd ed. (Academic Press, New York, 1975).
5. G. Herzberg, *Molecular Spectra and Molecular Structure*, Vol. 2: *Infrared and Raman Spectra of Polyatomic Molecules* (Van Nostrand Reinhold, New York, 1945).
6. H. Goldstein, *Classical Mechanics* (Addison-Wesley, Reading, MA, 1950).
7. E. B. Wilson, Jr., J. C. Decius, and P. C. Cross, *Molecular Vibrations* (McGraw-Hill, New York, 1954).
8. J. H. Schachtschneider and R. G. Snyder, *Spectrochim. Acta* **19**, 117 (1963).
9. W. G. Harter, *Phys. Rev.* **A24**, 192 (1981).
10. H. C. Allen, Jr., and P. C. Cross, *Molecular Vib-Rotors* (Wiley, New York, 1963).
11. D. Papoušek and M. R. Aliev, *Molecular Vibrational-Rotational Spectra* (Elsevier, Amsterdam, 1982).
12. J. R. Ferraro and J. S. Ziomek, *Introductory Group Theory and Its Application to Molecular Structure*, 2nd ed. (Plenum, New York, 1969).

9 Electronic Spectra of Polyatomic Molecules

1 Energy Levels and Spectra

The principles that govern the electronic spectra of polyatomic molecules are exactly the same as those that govern the electronic spectra of diatomic molecules, which we have previously taken up in chapter 5. The frequency of any spectrum line is given by $h\nu = E' - E''$, where E' and E'' are term values containing separable electronic, vibrational, and rotational contributions. The differences arise from the distinction that, in diatomics, there is only a single totally symmetric vibration, whereas in an N-atomic molecule, we can have $3N - 6$ different vibrations of various symmetries, as described in chapter 8. Furthermore, the rotational energy levels of most diatomic molecules have a simple $BJ(J+1)$ form, whereas in polyatomics, we can have much more complex symmetric or asymmetric rotor expressions, as shown in chapter 7. Thus, the complete form corresponding to equation (5.1) for diatomics is, in this case,

$$E = T_e + \left[\sum_i \omega_i \left(v_i + \frac{d_i}{2}\right) + \sum_i \sum_{k \geq i} \omega x_{ik} \left(v_i + \frac{d_i}{2}\right)\left(v_k + \frac{d_k}{2}\right) + \cdots\right]$$
$$+ \tfrac{1}{2}(B_v + C_v)J(J+1) + [A_v - \tfrac{1}{2}(B_v + C_v)]\omega(J, K). \tag{9.1}$$

This expression includes contributions from the anharmonicities of the vibration and from vibration–rotation interaction in that the rotational constants are taken to be functions of the v_i. The d_i are the degeneracies of each vibration, which must be taken into account when calculating the correct zero-point energy; the $\omega(J, K)$ term is whichever symmetric or asymmetric rotor energy expression is appropriate to the molecule being considered.

In some of the simpler polyatomic species, the spectrum is very similar in appearance to that of a diatomic molecule, with easily identifiable P, Q, and R branches. As the molecule increases in size and complexity, however, the spectrum becomes progressively more complex. What may appear to be a "line" in such a spectrum is often a confluence of a large number of overlapping transitions. Finally, when we get to molecules of the size of benzene (which will be discussed more thoroughly in a following section) or larger, there is no longer even the appearance of a series of lines—what we see is just an irregular absorption contour with superimposed intensity fluctuations. A line-by-line assignment and analysis of such a spectrum is out of the question; the best that can be done is a computer generation of the spectrum shape and comparison with the experimental curve, in a manner to be described shortly.

What accounts for the change from a clean, regular band spectrum found in diatomics and triatomics to the complex contours found in larger species? Basically, it is because the number of energy levels rises sharply as the number of atoms in the molecule increases. This is due to two factors: First, a larger number of vibrational levels is possible, and second, the increase in the moment of inertia with molecular size means that a much larger number of rotational levels can be populated in the ground state. As we shall see in the following chapter, a number of factors combine to produce a line width of the order of 0.03 cm^{-1}. A typical "large" molecule may have as many as 1,000 distinct spectroscopic transitions per cm^{-1}, so that there may be 30 or more lines per unit line width. Thus, the lines run together to form a quasi-continuous contour. With some of the newer techniques to be discussed in the following chapters, resolution of individual transitions may be possible, at least over limited portions of the spectrum.

In the following sections, we shall consider in more detail the nature of the electronic states of small polyatomic molecules, and then go on to consider two such molecules—formaldehyde and benzene—in detail.

2 Electronic States of Polyatomic Molecules: Walsh's Rules

In chapter 3, we classified the electronic states of diatomic molecules according to their behavior under the symmetry operations of the $C_{\infty v}$ or $D_{\infty h}$ groups, which were those appropriate for linear molecules. We now wish to do the same thing for polyatomic species; in this case, however, we must construct orbitals and find electronic states that are appropriate to the particular point group of the nuclear framework of the molecule, as discussed in chapter 6. The methods of finding such *symmetry orbitals* are described in a number of texts, including that by Herzberg (reference 1). Let us consider a specific example—that of an XY_2 molecule, which can have either a linear ($D_{\infty h}$) or a bent (C_{2v}) configuration. The correlation of these orbitals with each other and with the parent atomic orbitals is shown in figure 9.1, for molecular orbitals arising primarily from the $1s$, $2s$, and $2p$ atomic orbitals of the separated atoms.

In most cases, the symmetry properties of the molecular orbitals can be determined by inspection, following the rules given in chapter 6.4. The algebraic form of the wave functions follow the same operational require-

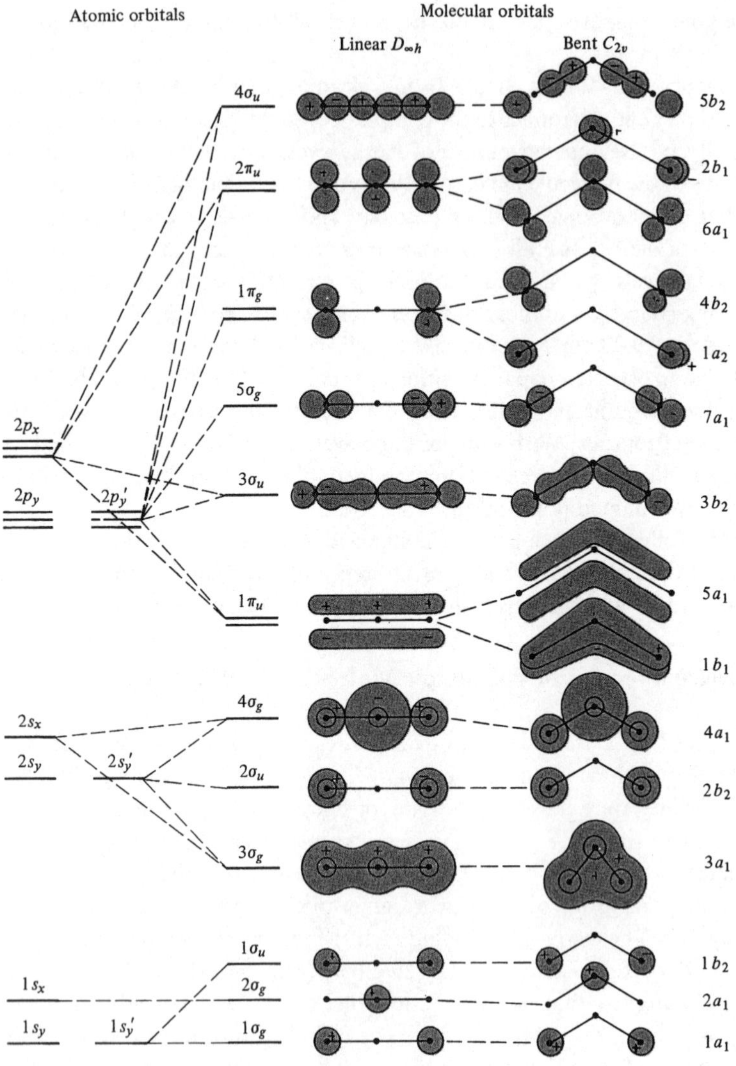

Figure 9.1
Orbital correlations for XY_2 molecules. The contours and node patterns are indicated, as well as the inner nodal surfaces for orbitals built out of 2*s* atomic orbitals.

Table 9.1

Molecular orbital	LCAO	Eigenvalue under inversion at X
$1\sigma_g$	$1/\sqrt{2}(1s_y + 1s'_y)$	$+1$
$2\sigma_g$	$1s_x$	$+1$
$1\sigma_u$	$1/\sqrt{2}(1s_y - 1s'_y)$	-1
$3\sigma_g$	$1/2(2s_y + 2s'_y) + 1/\sqrt{2}(2s_x)$	$+1$
$2\sigma_u$	$1/\sqrt{2}(2s_y - 2s'_y)$	-1
$4\sigma_g$	$1/2(2s_y + 2s'_y) - 1/\sqrt{2}(2s_x)$	$+1$
$1\pi_u$	$1/2(2p_y^z + 2p_y^{z'}) + 1/\sqrt{2}(2p_x^z)$	-1

ments, as for some of the lower-lying orbitals for linear symmetric ($D_{\infty h}$) XY_2 (see table 9.1). One particular point to note is the smooth correlation between the orbitals in the linear and bent configurations. All the σ_g orbitals go over into a_1 orbitals, all the σ_u orbitals into b_2 orbitals, and so on. The reflection and inversion operations of $D_{\infty h}$ are replaced by the C_2 and σ_v operations of C_{2v}.

For diatomic molecules, we saw that the electronic energy was a smoothly varying function of the one structural parameter, namely, the internuclear distance R. This dependence was expressed as a potential energy curve, $E(R)$. For polyatomics, the important consideration is the dependence of the orbital energies on such structural parameters as bond angles, which determine whether the molecule is linear or bent, planar or pyramidal, and so on. Such dependences have been worked out by A. D. Walsh, who, in an extensive series of papers (reference 2), presented orbital correlations for a number of polyatomic structural types. These correlations are based on semiempirical chemical arguments, but have been quite successful in accounting for the shapes of many small polyatomic molecules. An example of a Walsh diagram, for XY_2 molecules, is given in figure 9.2. The use of such diagrams can be made apparent by several examples.

The carbon dioxide molecule is constructed of a carbon atom, with a $(1s)^2(2s)^2(2p)^2$ atomic configuration, and two oxygen atoms with $(1s)^2(2s)^2(2p)^4$ atomic configurations. Leaving aside the $1s$ core electrons, the molecule has a total of 16 valence electrons to be placed in the molecular orbitals of figure 9.2. Let us put them into orbitals $3\sigma_g$ through $1\pi_g$, as is shown. When this is done, we see that any attempt to change the O—C—O bond angle to any value other than 180° would involve an increase in the energies of all the occupied orbitals; thus we conclude that the stable geometry for CO_2 should be linear, and the ground state, having a

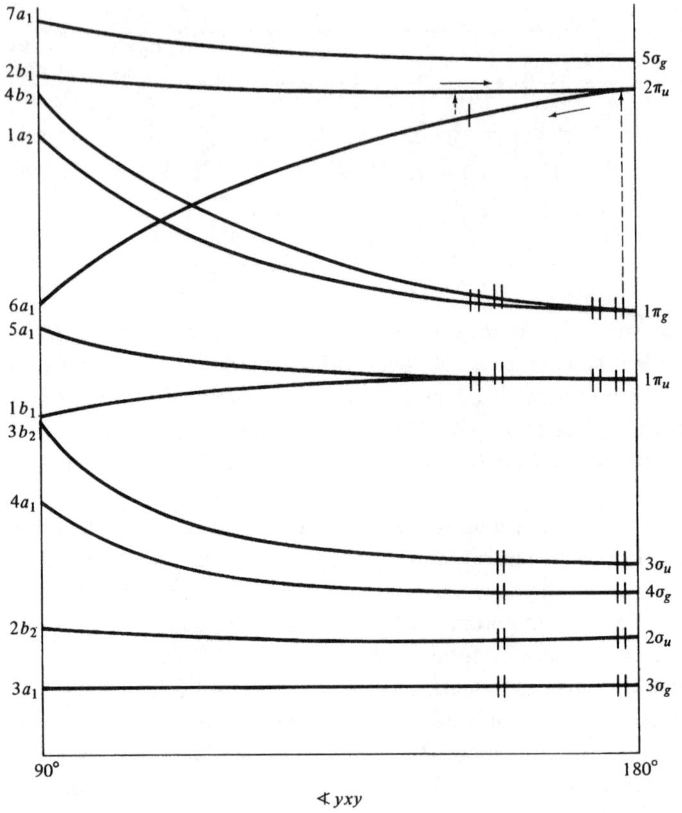

Figure 9.2
Walsh diagram for XY_2 molecules. The electrons at the right-hand end of the figure are representative of a molecule like CO_2, whereas those toward the center correspond to NO_2. Also indicated are the first orbital excitations for each species.

$$(3\sigma_g)^2(2\sigma_u)^2(4\sigma_g)^2(3\sigma_u)^2(1\pi_u)^4(1\pi_g)^4$$

configuration, is a ${}^1\Sigma_g^+$ state. NO_2, on the other hand, is a 17-valence-electron system. The extra electron would go into the ${}^3\pi_u$ orbital, but the Walsh diagram shows the energy of this orbital to be a rapidly decreasing function of the O—N—O bond angle, much more rapidly than the energies of the other occupied orbitals increase. Thus the lowest-energy structure will be a bent one, as shown. The molecular electronic configuration is

$$(3a_1)^2(2b_2)^2(4a_1)^2(3b_2)^2(1b_1)^2(5a_1)^2(1a_2)^2(4b_2)^2(6a_1)^1$$

for a ${}^2A_1$ ground state.

The Walsh diagrams also serve to systematize, if not predict, the orbital excitations in these molecules. In CO_2, the first excitation would be from the $1\pi_g$ to the $2\pi_u$ orbital. Since the energy of the latter decreases sharply with bond angle, we would expect a bent excited state. Also, the large energy gap between the $1\pi_g$ and the $2\pi_u$ orbitals indicates that the first excitation will lie in the deep ultraviolet part of the spectrum. This is what is found experimentally;[1] the first excitation is to a ${}^1B_2$ state [arising from the $(4b_2) \otimes (6a_1)$ open-shell configuration], at 46,000 cm^{-1}. In NO_2, the first excitation would be $6a_1 \rightarrow 2b_1$. Attempts to analyze the absorption spectrum of this apparently simple molecule have proved to be extremely frustrating;[2] the visible absorption spectrum (which accounts for the brown color of NO_2 gas and, incidentally, the characteristic color of the atmosphere in Los Angeles and other "brown smog" cities) is found to contain a dense set of lines that fall into no discernible pattern of band structure. After a great deal of labor, including laser-excited fluorescence spectroscopy and supersonic beam cooling (see chapter 10.5), it was concluded that the levels of the upper electronic state of NO_2 are thoroughly tangled with those in high vibrational levels of the electronic ground state, so that almost every vib-rotational level is perturbed. It is no wonder, then, that a set of consistent spectroscopic constants could not be derived. The one prediction of the Walsh's rule diagram (figure 9.2) borne out is that the transition is very low energy; the electronic origin has been estimated to lie at 10,000 cm^{-1}, or perhaps even less.

1. Data on the observed spectra of polyatomic molecules can be found in tables 62–82 (pp. 583–668) of reference 1.
2. For a thorough discussion of the NO_2 spectrum, including historical perspective, see reference 3.

Let us briefly consider the selection rules for this transition in NO_2. In order to have a $\langle {}^2B_1|\mu|{}^2A_1\rangle$ matrix element be nonzero in C_{2v} symmetry, we must make use of the x component of the dipole moment. This has B_1 symmetry, and $B_1 \otimes B_1$ provides the required A_1 symmetry of the whole. The transition is thus allowed, and polarized out of the plane of the molecule. In the linear configuration, however, the equivalent transition would be $\pi'_u \leftarrow \pi_u$, which is not only dipole forbidden, but, in this case, a nonexistent transition, because the initial and final orbitals would be the same.

3 Electronic Spectroscopy of Formaldehyde

Rather than attempt to give a comprehensive treatment of the electronic spectroscopy of polyatomic molecules (for which the reader is encouraged to see reference 1), we shall just give here two examples intended to illustrate some of the principles involved in such an analysis. These two well-studied examples are formaldehyde and benzene.

The formaldehyde molecule has C_{2v} symmetry. The conventional description of the molecular orbitals shown in figure 9.3, however, ignores the hydrogen atom and refers to a system having cylindrical symmetry. Thus, we have σ, π, π^*, and σ^* orbitals; the "n" orbital, corresponding to "nonbonding" electrons on the oxygen atom, has no analogous form in the cylindrical case. The C_{2v} symmetry of the σ orbital is a_1; of the π, b_1; of the n, b_2; of the π^*, b_1; and of the σ^*, a_1. Thus, the ground state, which has a $\sigma^2\pi^2n^2$ configuration, has symmetry $(a_1)^2(b_1)^2 \times (b_2)^2 = A_1$. It should be emphasized that the orbitals shown in figure 9.3 are merely schematic, meant to illustrate the symmetry properties and nodal patterns; accurately computed molecular orbital contours for this molecule may be found in reference 4.

Several types of electronic excitation are possible in this molecule. A $\pi \rightarrow \pi^*$ transition would produce an excited state with symmetry $(a_1)^2(b_1)$ $(b_2)^2(b_1) = A_1$. An $n \rightarrow \sigma^*$ transition would produce an excited state of symmetry $(a_1)^2(b_1)^2 \times (b_2)(a_1) = B_2$. The former transition can be effected by an operator of A_1 symmetry, which is the z or principal axis component of the dipole moment, which lies along the C—O bond. The latter requires an operator of B_2 symmetry, which is the y component of the dipole moment, lying in the molecular plane. Intense transitions of these symmetries are observed in the deep ultraviolet, in the 1,500–1,800-Å region. In addition, however, there is a much weaker spectrum in the 3,000–3,500-Å region,

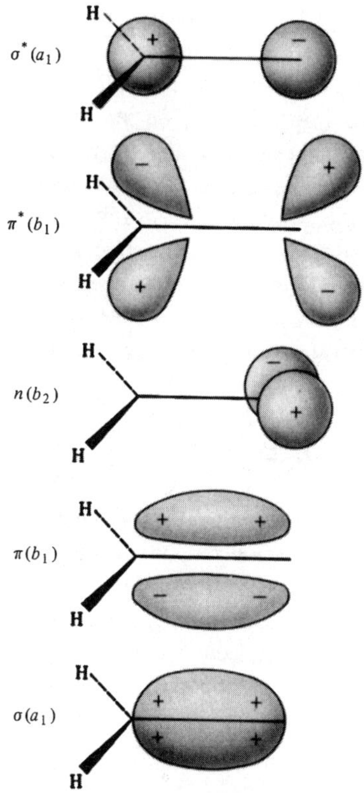

Figure 9.3
Molecular orbitals for formaldehyde for carbon and oxygen valence electrons. The "pseudodiatomic" designation is based on the nodal patterns of the orbitals; note that they are, in fact, somewhat distorted from the shapes they would have in, say, CO. The symmetry classification of each orbital in the point group C_{2v} is also shown.

which is attributed to an $n \rightarrow \pi^*$ transition. The symmetry of this excited state would be $(a_1)^2(b_1)^2(b_2)(b_1) = A_2$, so that an operator of symmetry A_2 would be required to make the transition allowed. We have noted that z has A_1 symmetry and y has B_2 symmetry; from appendix A, we see that the other component, x, has B_1 symmetry. So it appears that this transition is electric dipole forbidden; the R_z operator has A_2 symmetry, so that it could be magnetic dipole allowed. There is another way, however, in which the transition can become electric dipole allowed. This involves the effect known as *vibronic coupling*. The implicit assumption in our symmetry-based treatment of selection rules was that the electronic and vibrational wave functions were strictly separable, so that the transition matrix element we were examining could be expressed as

$$\langle ev|\boldsymbol{\mu}|e'v'\rangle = \int \psi_e(\mathbf{r}) \sum_i e\mathbf{r}_i \psi_{e'}(\mathbf{r})\, d\mathbf{r} \int \psi_v(\mathbf{Q})\psi_{v'}(\mathbf{Q})\, d\mathbf{Q}. \tag{9.2}$$

In diatomic molecules, the simple nuclear vibration preserves the symmetry of the molecule, so that the electronic wave function remains in the same symmetry species throughout the vibration, and equation (9.2) is valid to within the limits of the Born-Oppenheimer approximation. In polyatomic molecules, however, the vibration can distort the molecular shape, and thus alter the symmetry species of the electronic wave function. The quantity that must be looked at in order to determine the transition selection rules is the overall *vibronic* matrix element

$$\langle ev|\boldsymbol{\mu}|e'v'\rangle = \iint \psi_{er}(\mathbf{r};\mathbf{Q}) \sum_i e\mathbf{r}_i \psi_{e'v'}(\mathbf{r};\mathbf{Q})\, d\mathbf{r}\, d\mathbf{Q},$$

where

$$\psi_{ev} = \psi_e \psi_v$$

and

$$\Gamma(\psi_{ev}) = \Gamma(\psi_e) \otimes \Gamma(\psi_v).$$

To see how this applies in the case of formaldehyde, we must consider its normal mode vibrations, which are shown in figure 9.4. The a_1 vibrations will have no effect on the vibronic symmetry, so that they need not be considered any further in changing the selection rules for electronic transitions. Consider, however, the $^1A_2$ state with a single quantum of the $\nu_4(b_1)$ vibration. The vibronic species of the excited state is $A_2 \otimes b_1 = B_2$. Since

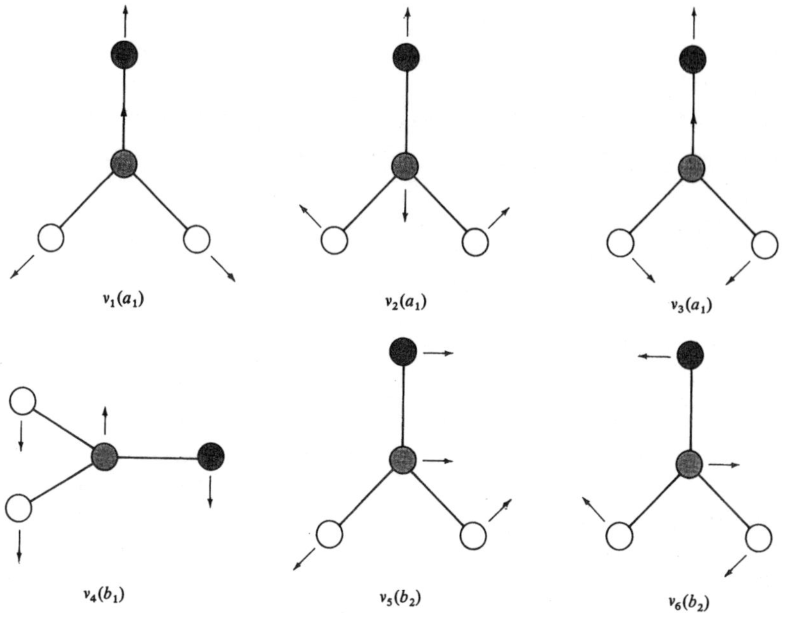

Figure 9.4
Normal coordinate vibrations for formaldehyde.

the y component of the dipole moment is of B_2 symmetry, an allowed transition is predicted from $B_2 \otimes B_2 \otimes A_1 = A_1 \otimes A_1 = A_1$. It is interesting to note that the transition, although it is polarized perpendicular to the figure axis of the molecule, is *not* directed along the direction of the distortion induced by the vibration, but rather at right angles to it, that is, in the plane of the four atoms. Similarly, since $A_2 \otimes b_2 = B_1$, transitions involving quanta of the ν_6 vibration would also be vibronically allowed, with x polarization.

The simple statement of allowedness, of course, does not quantitatively predict the intensity of any of these bands. That depends on the extent to which the electronic wave function is distorted by the nontotally symmetric vibration involved, and on the magnitude of the vibrational overlap integrals. In the case of the $n \rightarrow \pi^*$ ($^1A_2 \leftarrow {}^1A_1$) transition in formaldehyde, detailed rotational analysis has shown (reference 5) that the upper state is distorted out of plane in precisely the way caused by the ν_4 vibration; thus vibrational overlap integrals are favorable for a long progression in ν_4, which is observed.

In addition to serving as a prototype of a "small polyatomic" molecule for the analysis of high-resolution absorption spectra, formaldehyde has also been used to test various theories of the decay processes involving the electronically excited states of such molecules. These experiments are described in the last section of the present chapter.

4 The 2,600-Å System of Benzene

The so-called 2,600-Å system of benzene is shown, under low resolution, in the top panel of figure 9.5. Under higher resolution, each band is seen to possess an extensive rotational structure, as shown in the bottom panel of figure 9.5. A very crude molecular orbital description of the π electrons in benzene (see, for example, reference 6) predicts the four species of molecular orbitals shown in figure 9.6; since there are a total of six π electrons, the ground state configuration is $(a_{2u})^2(e_{1g})^4$, for a ${}^1A_{1g}$ ground electronic state. The first electronic excitation would be a "$\pi \rightarrow \pi^*$," or to a $(a_{2u})^2(e_{1g})^3(e_{2u})$ configuration. The states arising from such a configuration are given by $e_{1g} \otimes e_{2u}$, which has the irreducible representations $B_{1u} \oplus B_{2u} \oplus E_{1u}$. The first problem in analyzing the near-ultraviolet spectrum of benzene is to assign the electronic species of the upper state as belonging to one of these three symmetries.

We can immediately eliminate the E_{1u} state on the basis of the weakness of the observed spectrum. The direct product of E_{1u} and E_{1u} contains the totally symmetric species A_{1g}, and the species of the x or y components of the electric-dipole moment in D_{6h} symmetry is E_{1u} (see appendix A). Thus the transition ${}^1E_{1u} \leftarrow {}^1A_{1g}$ would be strongly allowed. Since the measured oscillator strength of the 2,600-Å system is only of the order of 0.001, the upper state in this transition cannot be E_{1u}. This assignment is reserved for the much stronger system of bands around 1,850 Å. Confirmation of this assignment is afforded by the oscillator strengths of the corresponding transitions in substituted benzenes. The 2,600-Å band is twice as intense in *ortho*- or *meta*xylene (dimethyl-substituted benzenes) than in benzene itself or toluene, and four times as intense in *para*xylene. The oscillator strength increases by an additional factor of 10 in fluorobenzenes. All this shows that the transition becomes more allowed as the D_{6h} symmetry of the molecule is broken, and that the more strongly it is broken (by the influence of the substituent on the π electron wave function), the stronger the transition becomes. This would never be the case for E_{1u}, which is already fully allowed in D_{6h} symmetry. Thus we are left with the assignment of either B_{1u} or B_{2u}

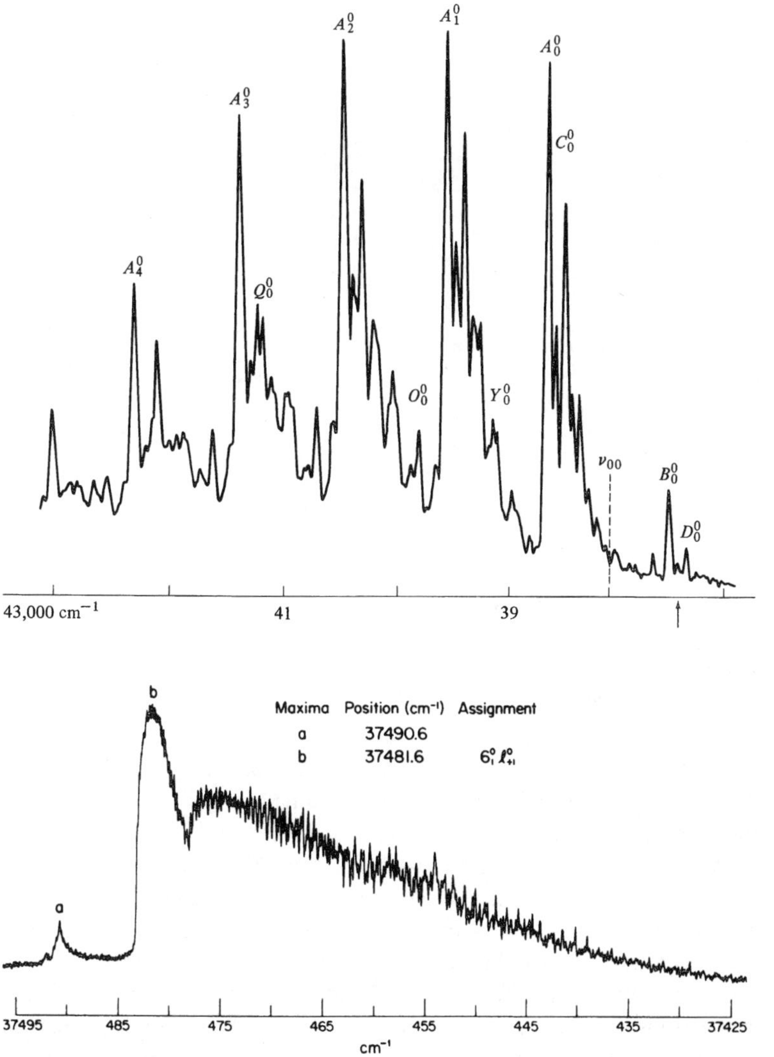

Figure 9.5
Top panel: Low-resolution absorption spectrum of the 2,600-Å system in benzene. The arrow indicates the B_0^0 band, which is shown under higher resolution in the bottom panel. [From J. H. Callomon, T. M. Dunn, and I. M. Mills, *Phil. Trans. Roy. Soc. (London)* **A259**, 499 (1966). Reproduced with permission.] Bottom panel: The $B_0^0(6_1^0)$ band of benzene under high resolution. [From G. H. Atkinson and C. S. Parmenter, *J. Mol. Spectroscopy* **73**, 52 (1978).]

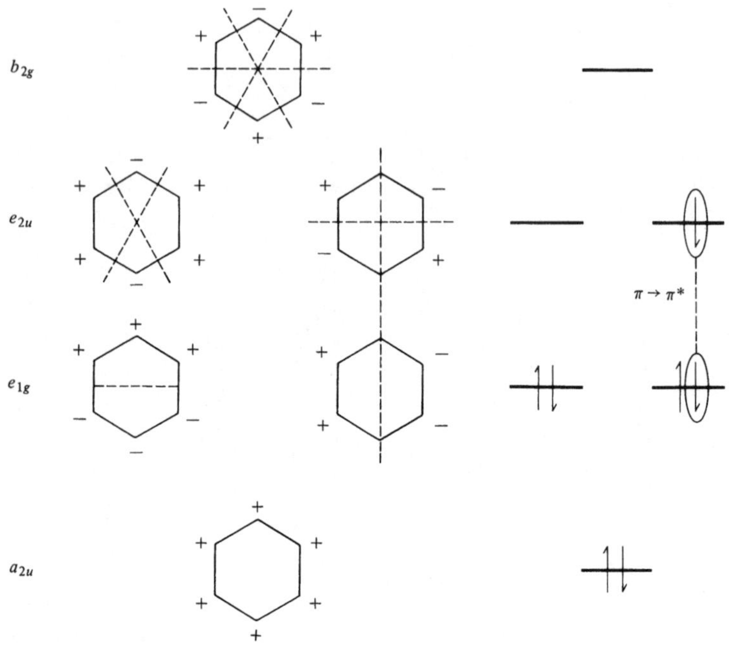

Figure 9.6
Molecular orbitals for the π electrons in benzene.

for the 2,600-Å band of benzene. Both of these transitions are forbidden, but can become vibronically allowed in the same manner as the ${}^1A_2 \leftarrow {}^1A_1$ transition in formaldehyde.

The electronic assignment must thus be decided on the basis of a vibrational analysis. A definitive analysis was provided in 1966 by Callomon and his coworkers (reference 7); more recently, this analysis has been revised and extended by Atkinson and Paramenter (reference 8). What is observed from the analysis of the spectrum shown in the top panel of figure 9.5 are two progressions in intervals of 923 cm^{-1}; the strong series of peaks, which are immediately obvious, and a weaker series, parallel to the first, which disappears at low temperatures of the vapor. It was first suggested by Sklar (reference 9) that the transition is rendered active by an e_{2g} vibration in the molecule, because either of the direct products ${}^eB_{1u} \otimes {}^ve_{2g}$ or ${}^eB_{2u} \otimes {}^ve_{2g}$ has the vibronic species ${}^{ev}E_{1u}$, which is an allowed electric dipole transition with perpendicular (that is, in-plane) polarization. In this analysis, the strong progression arises from transitions from a vibrationless ground state to an excited state with one quantum of an e_{2g} vibration designation ν_6'; the weaker progression arises from a ground electronic state molecule with one quantum of ν_6''. The 923-cm^{-1} progression corresponds to 0, 1, 2, . . . quanta of a totally symmetric a_{1g} vibration, which can be visualized as the "breathing" in and out of the benzene ring; this has a certain implication for the geometry of the excited molecule, to which we shall return shortly. This vibrational analysis is systematized in figure 9.7.

We still have not distinguished between the B_{1u} and the B_{2u} assignments, however, and this information is required in order to be able to proceed with a complete analysis. The only firm evidence available on which of the two assignments is correct is the following: A vibronic species ${}^{ev}A_{2u}$ would be allowed in absorption as a parallel transition. A careful search of the benzene spectrum has failed to reveal any parallel bands. Now, an A_{2u} vibronic state could be formed either as ${}^{e}B_{1u} \otimes {}^{v}b_{2g}$ or ${}^{e}B_{2u} \otimes {}^{v}b_{1g}$. Since benzene possesses b_{2g} vibrations, but not b_{1g} vibrations, and no parallel bands are found, the conclusion on the basis of this evidence is that the upper electronic state must be B_{2u}.

The geometrical structure of the excited state of the benzene molecule is found by an analysis of the rotational structure of the absorption bands. A rotational analysis in the conventional sense is impossible for this molecule because there are simply too many overlapping lines; what is done, instead, is a *band contour analysis*, in which the unknown rotational constants of the excited state are varied until the detailed shape of the vibrational bands is well fitted by the calculated shape. Such an analysis begins, like all others, with a statement of the energy term values and selection rules for the rotational levels. Since benzene is, at least in its ground state, an oblate symmetric top, the rotational terms are just

$$F_v(J, K) = B_v J(J + 1) + (A_v - B_v)K^2,$$

where the A axis is the sixfold symmetry axis, and the B and C axes lie in the plane of the benzene ring. Since the transition is perpendicularly polarized, we have the selection rules $\Delta K = \pm 1$, $\Delta J = 0$ and ± 1. From the usual expression $\nu = F_{v'}'(J', K') - F_{v''}''(J'', K'')$, we obtain for the three branches

$$\begin{aligned} \nu[P(J)] &= \nu_0^{\text{subband}} - (B' + B'')J + (B' - B'')J^2, \\ \nu[Q(J)] &= \nu_0^{\text{subband}} + (B' - B'')J + (B' - B'')J^2, \qquad (9.3) \\ \nu[R(J)] &= \nu_0^{\text{subband}} + (B' + B'')(J + 1) + (B' - B'')(J + 1)^2. \end{aligned}$$

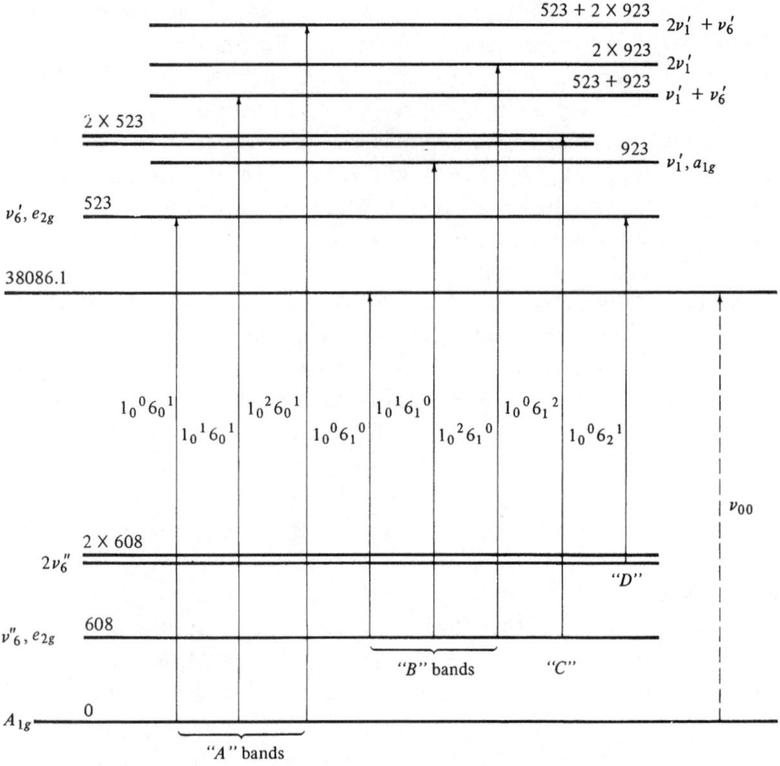

Figure 9.7
Vibrational analysis of the ultraviolet absorption spectrum of benzene. The dominant progression of "*A*" bands arises from transitions from the ground vibronic state to levels possessing $\nu_6' + n\nu_1'$, designated as $1_0^n 6_0^1$; the hot band series of "*B*" bands corresponds to transitions from ν_6'' to $n\nu_1'$, designated as $1_0^n 6_1^0$. Transitions such as 6_1^2 ("*C*" bands) and 6_2^1 ("*D*" bands) are also possible. The origin band, ν_{00}, is weakly observed in liquid benzene, where intermolecular forces break down the symmetry selection rules. [Adapted from J. H. Callomon, T. M. Dunn, and I. M. Mills, *Phil. Trans. Roy. Soc. (London)* **A259**, 499 (1966).]

The *subband origins* correspond to the two different possibilities for ΔK and are given by

$$\begin{aligned} \nu_0^{P\text{-type}} &= \nu_{\text{origin}} + (A' - B') - 2(A' - B')K \\ &\quad + [(A' - B') - (A'' - B'')]K^2 \qquad (\Delta K = -1), \\ \nu_0^{R\text{-type}} &= \nu_{\text{origin}} + (A' - B') + 2(A' - B')K \\ &\quad + [(A' - B') - (A'' - B'')]K^2 \qquad (\Delta K = +1), \end{aligned} \tag{9.4}$$

where ν_{origin} refers to the pure vibronic term difference. The complete expressions for the rotational line positions are obtained by substituting the above expressions for the subband origins into equation (9.3).

A further simplification can be effected, since the rotational constants in benzene obey the relation $B = C = 2A$, so that $A = \frac{1}{2}B$ in the previous expressions. We have shown that such a relation holds for the simpler case of a planar symmetric XY_3 molecule (see chapter 7.1). The proof that this relation also holds for a D_{6h} molecule, such as benzene, is left as an exercise (problem 3). With this simplification, the equations for the subband origins become

$$\begin{aligned} \nu_0^{P\text{-type}} &= (\nu_{\text{origin}} - \tfrac{1}{2}B') + B'K - \tfrac{1}{2}(B' - B'')K^2, \\ \nu_0^{R\text{-type}} &= (\nu_{\text{origin}} - \tfrac{1}{2}B') - B'K - \tfrac{1}{2}(B' - B'')K^2. \end{aligned} \tag{9.5}$$

From Raman work,we know that B'' has a value of 0.1896 cm^{-1} (reference 10); thus we need to find only a single parameter, B', in order to be able to describe the spectrum completely.

Since a line-by-line analysis is not possible, the analysis proceeds by means of contour fitting, using experimental data such as the spectra in the bottom panel of figure 9.5. Since the intensities that must be matched reflect the population of each absorbing level, as well as its absorption frequency, we must use the statistical expression

$$\begin{aligned} N(J, K) &= \frac{(2J + 1)\exp\{(-[BJ(J + 1) + (A - B)K^2]hc)/(kT)\}}{Q_{\text{rot}}} \\ &= \frac{(2J + 1)\exp\{(-B[J(J + 1) - \frac{1}{2}K^2]hc)/(kT)\}}{Q_{\text{rot}}}, \end{aligned}$$

using our simplifying relation $A = \frac{1}{2}B$. In order to represent the absorption band accurately, we should include contributions from levels that are populated to the extent of a few percent of the population at the maximum of the rotational distribution; a convenient rule of thumb for estimating how many levels have to be included is to require

$$\frac{BF(J_{max}, K_{max})hc}{k} \approx 3T$$

at room temperature, with $B = 0.1891$ cm^{-1}; this condition requires that we include levels with J and K up to about 75. The number of levels that must be included is reduced somewhat by realizing that we can have states only when $J > K$ because any component of the angular momentum cannot be greater than the total angular momentum. Since there are two subbands in each individual band, and three branches in each subband, this leaves us with altogether $\frac{1}{2} \times 75 \times 75 \times 2 \times 3 = 17{,}000$ lines to calculate. Since all these lines must go into a band that is no more than 30 cm^{-1} wide, we have of the order of 600 lines/cm^{-1}; it is no wonder that the band is unresolved.

The position of each of these 17,000 lines is calculated according to equations (9.3) and (9.5), weighted by its population factor, and the total intensity in each wave number resolution interval (say, 0.1 cm^{-1}) summed and plotted. The structure and overall appearance of such a calculated band is shown in figure 9.8 for the 6_0^1 band. From the result of a number of such analyses, the geometry for the excited state of benzene shown in figure 9.9 is deduced; in the excited state, the ring of carbon atoms is substantially expanded, and the C—H bond distance slightly reduced, so that the D_{6h} symmetry is preserved. This is consistent with the overall vibrational structure of the transition: The long progression in the a_{1g} breathing vibration implies that the distortion in the excited state is just that which would be produced by a number of quanta of this vibration. If the excited state had been distorted in some asymmetric manner, then we would have expected to see a long progression in the corresponding asymmetric vibration.[3] The nature of the geometry change also makes sense, in a post hoc way, on the basis of simple molecular orbital arguments: A π electron, which in some sense "holds the ring together," is being excited to a higher-energy (and thus

3. This is just a generalization of the statement in chapter 5.2 to the effect that a large change in internuclear distance of a diatomic causes a long vibrational progression, and a small change, a short progression. This was rationalized on the basis of the Franck-Condon factors for the transition. The same holds true for polyatomic molecules—a substantial change in equilibrium geometry between ground and excited states causes the vibrational overlap integral factors for the particular vibration in the direction of the distortion to have nonzero values for a wide range of v' and v''.

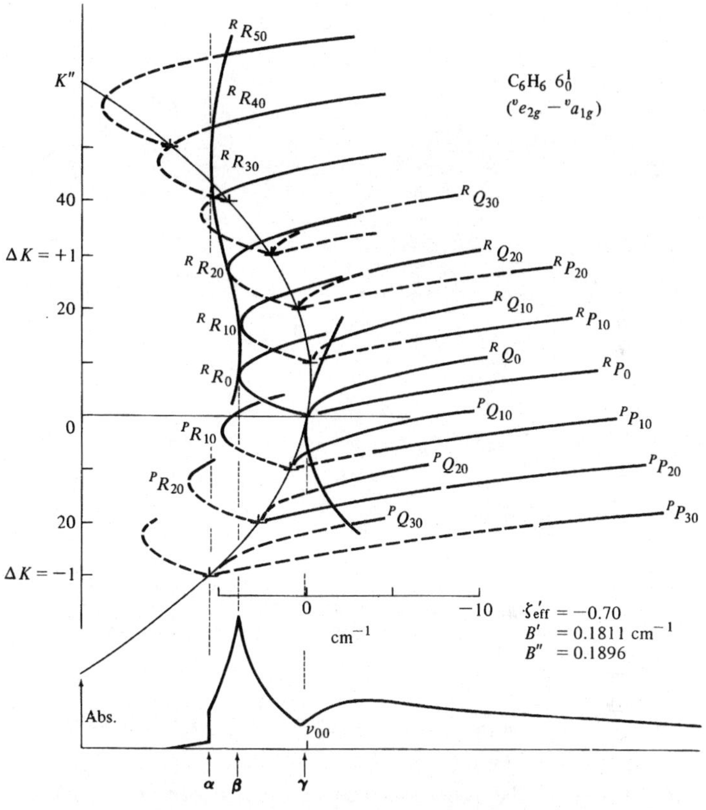

Figure 9.8
Synthesis of the 6_0^1 band of benzene. The subband origins follow the parabola opening to the left; the *P*, *Q*, and *R* branches are shown for every 10 values of *K*. A value of $B' = 0.1811$ cm^{-1} is chosen for this band. [From J. H. Callomon, T. M. Dunn, and I. M. Mills, *Phil. Trans. Roy. Soc. (London)* **A259**, 499 (1966). Reproduced with permission.]

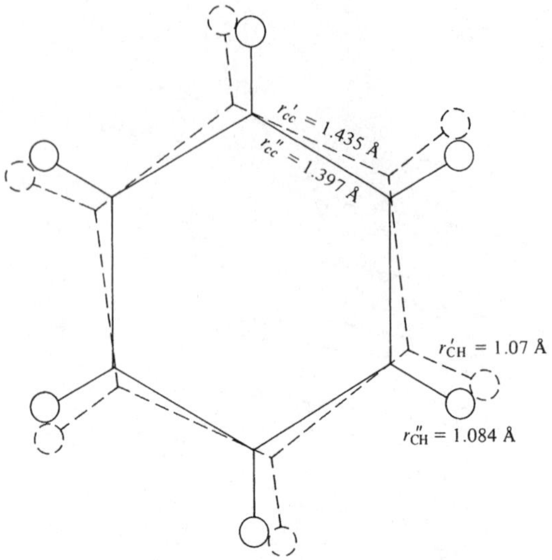

Figure 9.9
Excited state geometry of benzene, for $\Delta r(\text{C—C}) = +0.038$ Å and $\Delta r(\text{C—H}) = -0.01$ Å. The molecular frame has been slightly rotated to show the configuration changes more clearly. [From J. H. Callomon, T. M. Dunn, and I. M. Mills, *Phil. Trans. Roy. Soc. (London)* **A259**, 499 (1966). Reproduced with permission.]

"more weakly bonding") state, whereas the C—H σ electrons are, at least to a zero-order approximation, unaffected.

This excited state of benzene can also be accessed by a two-photon transition. This will be discussed, and the spectra displayed, in chapter 13.3; we note here only that the ro-vibronic selection rules for the two-photon process are found to be quite different from the ones discussed here.

5 Molecular Photoelectron Spectroscopy

If we look back at all the different types of electronic transitions discussed thus far, we see that they are actually fairly limited as to variety. They all involve the excitation of one of the electrons of a molecular system from the highest-energy occupied orbital to some one of the higher available un-occupied orbitals, at least to a reasonable degree of approximation. The molecular state produced by occupation of such a higher orbital may, of

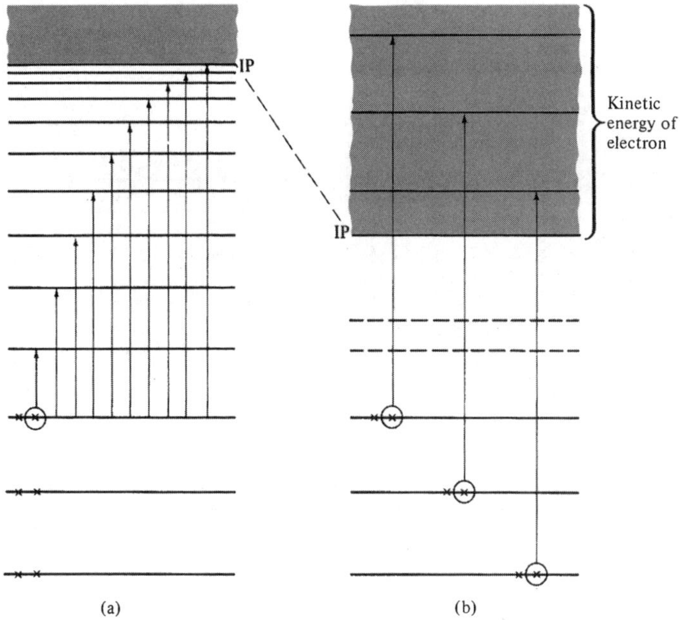

Figure 9.10
Schematic depiction of the types of spectra associated with photoionization. In (a), an electron is excited from the topmost filled orbital to a series of higher levels near the ionization limit, IP. As the limit is approached, the levels approach those of a pure Coulomb field, giving rise to a *Rydberg series*. In (b), a fixed energy photon (the arrows are all drawn to be the same length) excited electrons from a number of different orbitals into the ionization continuum. The kinetic energy of the ejected electron, which is measured in molecular photoelectron spectroscopy, is the difference between the level reached in the continuum and the limit IP.

course, be unstable with respect to dissociation; in that case, the optical absorption is a continuous spectrum, and the method of photofragment spectroscopy, discussed in Chapter 5.11, may be applicable.

As we look deeper and deeper into the ultraviolet, we usualy see a converging series of band systems, which ultimately terminate in the onset of continuous absorption. Such a series is called a *Rydberg series* and arises in the manner shown in figure 9.10a. The excited orbital for the electron, in such a series, has a higher and higher principal quantum number, and thus a larger and larger spatial extent. Ultimately, the electron finds itself in a nearly pure Coulombic field of the molecular ion, so that the orbital energies go as $-1/n^2$ up to the ionization limit corresponding to the process $M + h\nu \rightarrow$

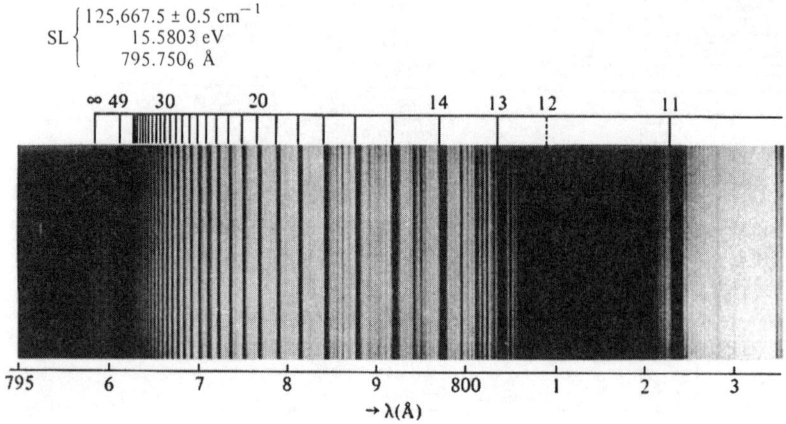

Figure 9.11
Typical Rydberg series in nitrogen in the vacuum ultraviolet. SL (upper left-hand corner) stands for "series limit." [Spectrum by Y. Tanaka, *Can. J. Phys.* **40**, 1596 (1962). Reproduced with permission.]

$M^+ + e^-$. Above this limit, the molecule can absorb light of any frequency, since an arbitrary amount of energy can be taken up as kinetic energy of the departing electron. An example of a Rydberg series converging to the ionization limit is shown in figure 9.11.

Clearly, very little molecular information is contained in the photoionization continuum alone. If, however, we could inspect the energy of the departing electron as well as of the incoming photon, then we would have some information about the state of the molecular ion left behind. Indeed, it turns out that the intensity at any given frequency in the photoionization continuum is the superposition of a number of distinct transitions involving different orbital electrons in the parent molecule, as shown in figure 9.10b. Since the ion left behind as a result of the photon absorption can be in any of a number of vibrational states governed by the Franck-Condon principle, we would expect to see vibrational, and, at least in principle, rotational structure in the electron kinetic energy spectrum as well.

Figure 9.12 shows an early version of an apparatus capable of making such measurements on molecular gases; although very much more sophisticated photoelectron spectrometers are now available, in which molecular beams and surfaces can be studied in addition to gases, this particular design clearly illustrates the principles involved. A helium resonance line at 584 Å

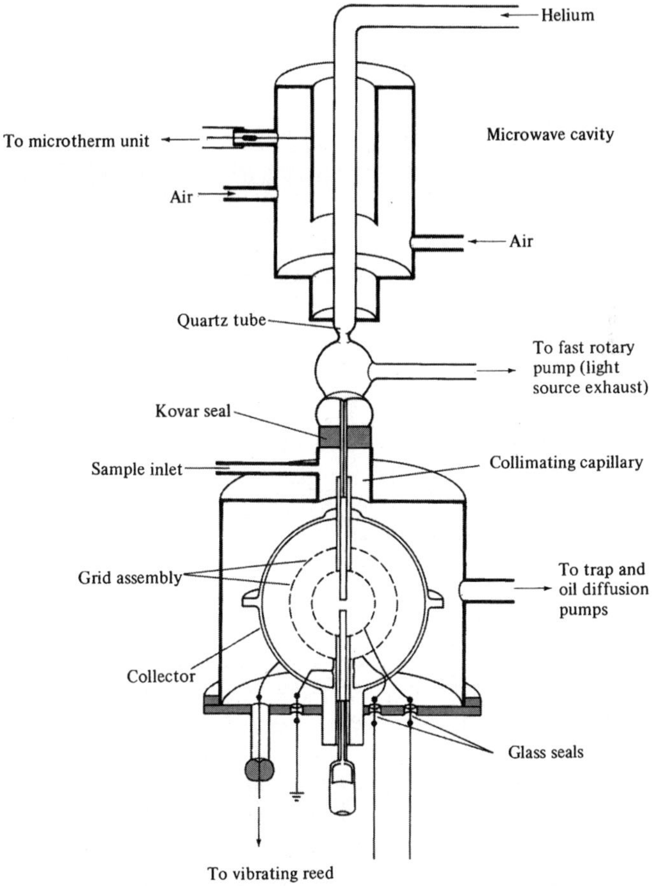

Figure 9.12
Schematic diagram of a molecular photoelectron spectrometer. Helium resonance radiation at 584 Å ($E = 21.23$ eV) is produced in the upper chamber; it is collimated and allowed to interact with the sample gas in the chamber in a well-defined geometry at the center of spherical energy-analyzing grids. [From D. C. Frost, C. A. McDowell, and D. A. Vroom, *Proc. Rov. Soc. (London)* **A296**, 568 (1966). Reproduced with permission.]

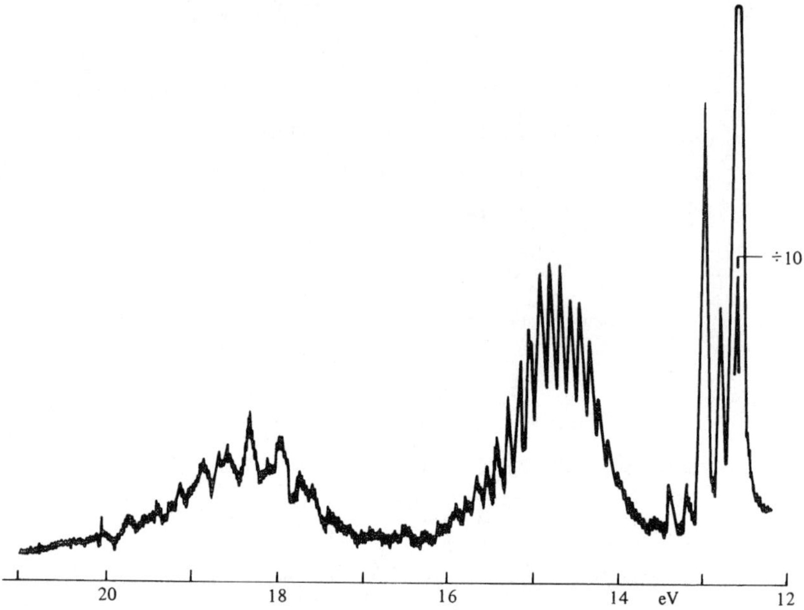

Figure 9.13
The photoelectron spectrum of H_2O. Note the three distinct regions of electron energy. [From C. R. Brundle and D. W. Turner, *Proc. Phys. Soc.* **A307**, 27 (1968). Reproduced with permission.]

is produced in the microwave cavity; this radiation imparts 21.23 eV of energy to the molecule absorbing it. The absorption takes place in a very well-defined region at the center of an electrostatic analyzing and focusing system. This system measures the energy of the emitted photoelectron; resolution of the order of 0.015 eV, or 120 cm^{-1}, can be obtained in such a device. The more recent instruments, which employ cylindrical electron energy analyzers, can resolve 1 meV, or about 10 cm^{-1}.

An example of a molecular photoelectron spectrum, that of H_2O, is shown in figure 9.13. We see that there are three distinct ranges of photoelectron energies, corresponding to three distinct spectroscopic transitions, each with superimposed vibrational structure. The ground state of the water molecule has the orbital configuration $(1a_1)^2(2a_1)^2(1b_2)^2(3a_1)^2(1b_1)^2$, for a molecular state $^1A_1$; the appearance of these orbitals is shown in figure 9.1, but the energy order for a dihydride is slightly different from that for a general XY_2

molecule in which the nuclear charges on the X and Y atoms are nearly the same, as in NO_2 or CO_2. The lowest-energy transition in figure 9.13 corresponds to the ionization of the highest-energy electron in the water molecule, to produce an H_2O^+ ion with a configuration $(1a_1)^2(2a_1)^2(1b_2)^2(3a_1)^2(1b_1)$, which is in a $^2B_1$ state. The transition is observed at 12.6 eV, and a molecular orbital calculation (reference 11) places the $1b_1$ orbital at 11.8 eV below the ionization limit. The second transition is assigned to H_2O^+ $(1a_1)^2(2a_1)^2(1b_2)^2(3a_1)(1b_1)^2$, in a $^2A_1$ state; it is observed at 13.7 eV, calculated 13.2 eV. The third transition is to H_2O^+ $(1a_1)^2(2a_1)^2(1b_2)(3a_1)^2(1b_1)^2$, in a $^2B_2$ state. This state is thought to be unstable with respect to dissociation into H atoms ($^2S_{1/2}$) and OH^+ ions ($^3\Sigma^-$); this is given as the cause for the partial diffuseness of the vibrational structure.

The photoelectron spectra of a large variety of molecules have now been obtained by this method; for further information, the reader is referred to the review (reference 12) and the book (reference 13) by Turner.

6 Radiationless Transitions

To conclude this chapter, we shall consider briefly a phenomenon connected with electronic excitation in polyatomic molecules, which has led to a good deal of theoretical speculation concerning the dynamics of highly excited molecular systems. This is the occurrence of so-called "radiationless transitions," which occur in both of the systems considered in this chapter, benzene and formaldehyde.

If benzene vapor at low pressure is excited in its 260-nm absorption band, say with a mercury lamp operating at 253.7 nm, only about 30% of the absorbed uv photons are reemitted as fluorescence, even in the zero-pressure limit (reference 14). If the molecule is in a condensed phase at low temperature, some of the energy may be reradiated in the region of the singlet-triplet absorption bands (330–430 nm); this radiation is known as *phosphorescence*. Even so, not all the energy that has been absorbed is recovered as luminescence. We know that the product of lifetime and oscillator strength is a constant [equation (1.38)]; thus behavior of this type suggests the presence of some additional, *nonradiative* process which can compete with photon emission from the excited molecule.

The existence of such a process is further suggested by lifetime data for formaldehyde. If single vibronic levels of deuterated formaldehyde (D_2CO) are selectively excited with a tunable uv laser, and radiative lifetimes

Table 9.2

Vibronic level of D_2CO, $^1A_2$	τ (μsec)	E (cm^{-1})
4^1	4.57	68
4^3	1.50	501
2^1	0.54	1,176
$2^1 4^2$	0.42	1,561
$2^1 4^3$	0.66	1,847
$1^1 4^1$	0.81	2,147
$2^2 4^1$	0.55	2,400
$4^2 5^1$	0.79	2,621
$1^1 4^3$	0.46	2,745
$2^2 4^3$	0.19	3,006
$1^1 2^1 4^1$	0.105	3,308
$2^3 4^1$	0.099	3,556
$2^1 4^2 5^1$	0.076	3,798
$1^1 2^1 4^3$	0.088	3,918
$2^3 4^3$	0.053	4,154

measured for each of these levels, the results given in table 9.2 are obtained (reference 15). In this table the notation $1^1 2^1 4^1$ indicates a D_2CO molecule with one quantum of ν_1, one quantum of ν_2, and one quantum of ν_4 in the electronically excited state; E is the excess vibrational energy above the zero-point vibrational level. The significant behavior illustrated by these data is that the lifetimes decrease nearly monotonically (except for small local fluctuations) with increasing vibrational energy. Similar behavior of the radiative lieftimes has also been observed in benzene itself (reference 16), deuterated benzene, and fluorobenzene (reference 17).

A phenomenological rate equation picture of this behavior is shown in figure 9.14. In this model, photons are absorbed in the transition from S_0, the ground singlet state, to S_1, an excited singlet state. Some fraction ϕ_F of the excited states thus produced reradiate their energy as fluorescence with a radiative rate k_{fl}. However, excited states can also go over into the triplet state T_1, lying lower in energy than S_1, with a radiationless *intersystem crossing rate* k_{ISC} which is comparable in magnitude to k_{fl}. (If the triplet state is not populated, it is customary to invoke an *internal conversion* process to high vibrational levels of S_0, having a characteristic rate k_{IC}.) The net *fluorescence yield* or *quantum yield* is given by the same relation as we had for predissociating diatomic molecules, namely, equation (5.14):

$$\phi_F = \frac{k_{fl}}{k_{fl} + k_{ISC}}.$$

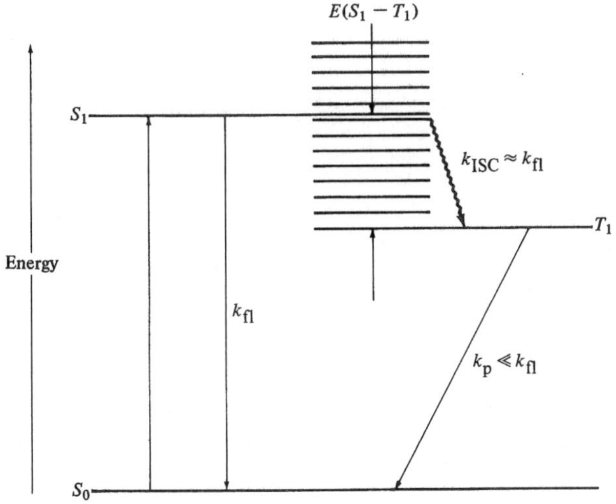

Figure 9.14
Schematic representation of intersystem crossing in large electronically excited molecules. The excited singlet S_1, the origin of which lies an energy E above the origin of the lowest triplet T_1, can either fluoresce back to S_0 with a radiative rate k_{fl} or "cross" to T_1 with an intersystem crossing rate k_{ISC}, which is comparable in magnitude to k_{fl}. The triplet state T_1 returns to S_0 via phosphorescence with a rate k_P, which is much less than k_{fl}. This scheme is often referred to as a "Jablonski diagram."

The measured lifetime of the excited state reflects both the radiative and nonradiative pathways, that is,

$$\tau^{-1} = k_{fl} + k_{ISC}.$$

Thus, as nonradiative relaxation becomes important, the quantum yield decreases, and the lifetime of the state becomes shorter, The dependence of these quantities on the internal energy of the molecules reflects the "Fermi golden rule" expression for the rates derived from first-order perturbation theory (equation (5.13); also see reference 18):

$$k_{ISC} = \frac{2\pi}{\hbar}\rho(E)|\langle S_1|\mathscr{H}'|T_1\rangle|^2. \tag{9.6}$$

Here $\rho(E)$ is the density of molecular states in the triplet manifold lying adjacent to the singlet, with E the excess energy above the origin of the triplet (see figure 9.14). $\mathscr{H}'$ is taken to be all the terms left out of the actual molecular Hamiltonian when finding the approximate states, including Born-Oppenheimer breakdown, spin-orbit coupling, and so on, as that the matrix element in equation (9.6) has the form of an electronic energy multiplied by a Franck-Condon factor for the vibrational wave functions of S_1 and T_1.

While this formulation does incorporate all of the phenomenologically observed behavior in the relaxation of large molecules, the rate equation model presents us with a serious conceptual problem. The problem is that the molecules that we are observing have never omitted any of the terms in the Hamiltonian in arriving at their own real eigenstates. The real states include the effects of all these terms, and to say that we can turn them on or off at will to effect transitions between approximate Born-Oppenheimer singlet and triplet states is just not in accord with reality. Indeed, this approach goes counter to the approach taken throughout this book, which is that the states and good quantum numbers, which we have used to describe real molecules, are, at best, the most nearly correct description for various limiting cases (see the discussion of perturbations in Chapter 5.7). For this reason, considerable effort was made in the late 1960s and early 1970s to reformulate the theory of radiationless transitions in terms of accurate molecular eigenstates. We shall state the most important results here; for further details, the reader is referred to the extensive review literature on this topic (references 19–22).

The essential point of this theory is that the vibronic level in S_1 must be regarded as embedded in a dense manifold of background states belonging to either the triplet (T_1) or the ground (S_0) electronic states. This is shown schematically in figure 9.15. The true molecular eigenstates will be a superposition of the Born-Oppenheimer states used as our basis set. When a level in S_1 is excited, its radiative properties include those of the background states, which typically possess very low oscillator strength. This has several consequences, depending on the size of the molecule and the details of the excitation process:

1. In a pure state, such as may be produced by monochromatic excitation of a diatomic molecule without predissociation, the decay of the fluorescence is a single exponential with lifetime τ^0_{rad} obeying the relations given in chapter 1.7.
2. In a small polyatomic molecule, the fluorescing state may be coupled to a small but finite number of background states. In this case, the observed

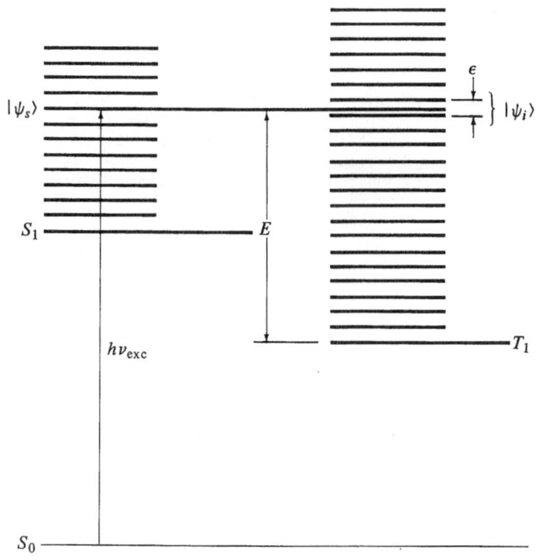

Figure 9.15
A more detailed description of the states in figure 9.14. The excited singlet state $|\psi_s\rangle$, which lies at energy $h\nu_{exc}$ above the ground state, is embedded in a dense manifold of vibrational levels belonging to T_1, separated with an average spacing ϵ. The density of states is approximately $1/\epsilon$, and increases with E.

lifetime may be *longer* than that expected from the integrated absorption strength, as a result of dilution of the oscillator strength into the background levels. This sort of behavior has been observed in the fluorescence decay of NO_2 (reference 23) and SO_2 (reference 24). The theory also predicts a modulation of the fluorescence decay, with a modulation frequency equal to the inverse spacing between the background levels. This oscillation, known as "quantum beats," has been observed in a few instances; an example is shown in figure 9.16. The number of coupled background levels, and hence the observed lifetime, fluorescence yield, and quantum beat effects, depend critically on the spectral content of the excitation source used in the experiment.

3. Finally, in a large molecule with a high density of background levels, the fluorescence decays more rapidly than τ^0_{rad}, as the excitation energy is dissipated into a large number of accessible levels. This also leads to a fluorescence yield less than unity. In principle, all the energy would be

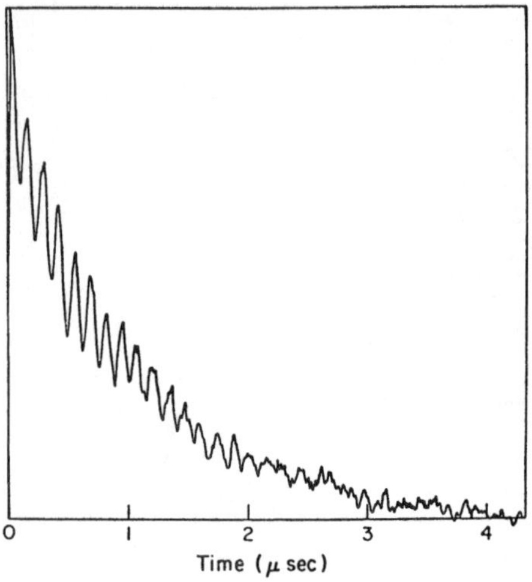

Figure 9.16
Fluorescence decay of acetylene excited at 45,297.2 cm^{-1} with a tunable ultraviolet laser. The modulation at 7.4 MHz indicates a density of background levels on the order of $(c/7.4 \times 10^6) \simeq 4{,}000$ levels/cm^{-1}. [From E. Abramson, C. Kittrell, J. L. Kinsey, and R. W. Field, *J. Chem. Phys.* **76**, 2293 (1982). Reproduced with permission.]

reemitted on a very long time scale; but this so-called recurrence time is so long that the molecule gives up a part of its energy by collisions or other processes before reradiation can occur. The nearby triplet state can act as a reservoir for the remaining energy; if the triplet is not itself quenched, the energy can appear as phosphorescence.

The radiationless transition process considered here is one example of an intramolecular relaxation process, which is characteristic of very large molecules possessing many internal states. We shall consider these processes further in chapter 14.

Problems

1. The SO_2 molecule has two equal S—O distances of 1.43 Å and a bond angle of 120° in the ground electronic state. Describe qualitatively the vibronic structure expected for absorption of ultraviolet radiation by SO_2 vapor at room temperature, if, in the upper electronic state, the S—O distances are both 1.43 Å and the bond angle is 110°.

2. Why are the absorption spectra of polyatomic molecules in the region $\lambda < 2{,}000$ Å so often continuous or diffuse?

3. Show explicitly that the rotational constants of benzene obey the relation $B = C = 2A$.

4. The methyl ion, CH_3^+, is well known from mass spectrometry, but its optical spectra have not yet been recorded. As a result of several still undiscovered advances in spectroscopic techniques, the following spectra have hypothetically been observed in a discharge and definitely attributed to CH_3^+. You are asked to deduce the structure and properties of CH_3^+ from these data and general knowledge of chemistry.

(a) Infrared absorption and Raman scattering

ν_{obs} (cm^{-1})	Raman	Infrared
1,080	Medium	Strong
1,501	Weak	Strong
3,221	Strong	—
3,421	—	Very strong
4,501	—	Weak

Deduce the point symmetry of CH_3^+ and assign the observed bands to normal vibrations. Calculate the valence force constants of CH_3^+. Predict the infrared spectrum of CD_3^+.

(b) Under higher resolution, the 1,501-cm^{-1} infrared band displays a series of peaks having a spacing of approximately 19.6 cm^{-1}. Estimate the C—H bond distance in CH_3^+, and predict the pure rotational absorption spectrum for this ion.

(c) No significant absorption is observed in the visible/uv part of the spectrum until 1,900 Å. The following diffuse bands are observed: 1,920, 1,865, 1,815, 1,767, 1,723, and 1,680 Å (all vacuum wavelengths, of course); the second and third bands possess maximum intensity.

(i) Show that the prediction from Walsh's rules is consistent with the structure for the ground state of CH_3^+ that you deduced in part (a). Predict the excited state structure from the appropriate Walsh diagram.

(ii) Assuming this structure is correct, interpret the deep-uv spectrum given above in terms of vibrational constants for the excited state of the ion.

(d) In addition to the spectra given above, an emission is observed from the discharge source in the 4,250–6,500 Å region, which shows vibrational spacing of 2,740 cm^{-1} and a very open fine structure. To what transition might this be attributed?

References

1. G. Herzberg, *Molecular Spectra and Molecular Structure*, Vol. 3: *Electronic Spectra and Electronic Structure of Polyatomic Molecules* (Van Nostrand Reinhold, New York, 1966).

2. A. D. Walsh, *J. Chem. Soc.* 2260, 2266, 2288, 2296, 2301, 2306, 2318, 2321, 2325, 2330 (1953).

3. D. K. Hsu, D. L. Monts, and R. N. Zare, *Spectral Atlas of Nitrogen Dioxide, 5530 to 6480 Å* (Academic Press, New York, 1978).

4. K. J. Miller, *J. Chem. Phys.* **62**, 1759 (1975).

5. D. C. Moule and A. D. Walsh, *Chem. Revs.* **75**, 67 (1975).

6. A. Streitweiser, *Molecular Orbital Theory for Organic Chemists* (Wiley, New York, 1961).

7. J. H. Callomon, T. M. Dunn, and I. M. Mills, *Phil. Trans. Roy. Soc. (London)* **A259**, 499 (1966).

8. G. H. Atkinson and C. S. Parmenter, *J. Mol. Spectroscopy* **73**, 20, 31, 52 (1978).

9. A. L. Sklar, *J. Chem. Phys.* **5**, 669 (1937).

10. B. P. Stoicheff, *Can. J. Phys.* **32**, 339 (1954).

11. F. O. Ellison and H. Shull, *J. Chem. Phys.* **23**, 2348 (1955).

12. D. W. Turner, *Ann. Rev. Phys. Chem.* **21**, 107 (1970).

13. D. W. Turner, C. Baker, A. D. Baker, and C. R. Brundle, *Molecular Photo-electron Spectroscopy* (*A Handbook of He 584 Å Spectra*) (Interscience/Wiley, New York, 1970).

14. G. B. Kistiakowsky and C. S. Parmenter, *J. Chem. Phys.* **42**, 2942 (1965).

15. E. S. Yeung and C. B. Moore, *J. Chem. Phys.* **58**, 3988 (1973).

16. K. G. Spears and S. A. Rice, *J. Chem. Phys.* **55**, 5561 (1971).

17. A. S. Abramson, K. G. Spears, and S. A. Rice, *J. Chem. Phys.* **56**, 2291 (1972).

18. E. Fermi, *Nuclear Physics* (University of Chicago Press, Chicago, 1950), p. 142; L. I. Schiff, *Quantum Mechanics*, 3rd ed (McGraw-Hill, New York, 1968), p. 285.

19. B. R. Henry and M. Kasha, *Ann. Rev. Phys. Chem.* **19**, 161 (1968).

20. J. Jortner, S. A. Rice, and R. M. Hochstrasser, *Advan. Photochem.* **7**, 149 (1969).

21. E. W. Schlag, S. Schneider, and S. F. Fischer, *Ann. Rev. Phys. Chem.* **22**, 465 (1971).

22. G. W. Robinson, *Excited States*, Vol. 1 (Academic Press, New York, 1974), pp. 1–34.

23. D. Neuberger and A. B. F. Duncan, *J. Chem. Phys.* **22**, 1693 (1954).

24. K. F. Greenough and A. B. F. Duncan, *J. Am. Chem. Soc.* **83**, 555 (1967).

25. E. Abramson, C. K. Kittrell, J. L. Kinsey, and R. W. Field, *J. Chem. Phys.* **76**, 2293 (1982).

10 From Molecular Beams to Masers to Lasers

The development thus far has covered the basic aspects of classical spectroscopy as developed by molecular physicists of the late nineteenth and early twentieth centuries to the point where today these methods are basic tools in every chemical laboratory in the world for the characterization of molecules. Despite the apparently closed nature of this area of spectroscopy, there still remain challenging problems in the field, such as perturbations by interacting levels (reference 1), correlating laboratory and astrophysical observations (reference 2), spectroscopy at high temperatures, for example, in flames (reference 3), and at high levels of excitation (see chapter 14). At this point, however, we would like to turn our attention to an entirely new area that within the past 20 years or so, has completely revitalized basic research in spectroscopy. This is the area of coherent optical processes, the most spectacular manifestation of which is the laser. While this area is still in the research stage, as the principles of laser spectroscopy have become more widely known, some of these methods have taken their place alongside the more classical ones as routine tools for the characterization and study of dynamic molecular processes.

The plan that we shall follow in exploring this area in the next five chapters is this. First, we shall consider the use of optical pumping to study high-resolution atomic spectra and relaxation processes. This will lead to consideration of beam spectroscopy, and thence to the first maser device. Extending these principles to shorter wavelengths gives us the optical maser, or laser. The dynamics of the two-level system are then explored rigorously, with the aid of both algebraic and geometric representations. A number of classic experiments in optical coherence are discussed, including transient nutation, photon echoes, self-induced transparency, saturation, and double-resonance spectroscopy. Nonlinear, multiphoton, and picosecond spectroscopy are surveyed, and in a concluding chapter, we attempt to project future developments in this discipline.

1 Hyperfine Structure and Line Width

Since the energy levels explored by optical-pumping experiments are frequently the fine and hyperfine structure levels, it seems appropriate to review here the energy expressions for these levels derived in chapters 2 and 7. For an atom with electronic angular momentum J and nuclear spin angular momentum I in the presence of a magnetic field B_0, the Zeeman energies are (see chapter 2.8)

$$E = E_0 + \frac{a}{2}[F(F+1) - I(I+1) - J(J+1)]$$
$$- m_F B_0 \left[g_J \mu_0 \frac{F(F+1) + J(J+1) - I(I+1)}{2F(F+1)} \right. \qquad (10.1)$$
$$\left. + g_I \mu_N \frac{F(F+1) + I(I+1) - J(J+1)}{2F(F+1)} \right],$$

in which

E_0 is the energy in the absence of magnetic interactions; the first term is the "zero-field splitting," or contact interaction, with constant

$a = (8\pi/3) g_J g_I \mu_0 \mu_N |\psi(0)|^2$;

F is the total angular momentum of the state, $\mathbf{F} = \mathbf{I} + \mathbf{J}$;
m_F is the magnetic or Zeeman quantum number;
g_J is the gyromagnetic ratio for the electron, also known as the "Landé g factor,"

$$g_J = 1 + \frac{J(J+1) + S(S+1) - L(L+1)}{2J(J+1)};$$

μ_0 is the Bohr magneton, 1.39966 MHz/G;
g_I is the gyromagnetic ratio for the nucleus; and
μ_N is the nuclear magneton, 762.3 Hz/G for hydrogen.

In a diatomic molecule, the Zeeman energy is given by (see chapter 5.8)

$$E = E_0 - \mu_0 B_0 \frac{(\Lambda + 2\Sigma)\Omega M_J}{J(J+1)}, \qquad (10.2)$$

in which Ω, the total electronic angular momentum, is the sum of Λ, the orbital electronic angular momentum, and Σ, the spin angular momentum (all referred to the internuclear axis), and J, the total angular momentum of the molecule, is the sum of Ω and R, the rotational angular momentum. In a $^1\Sigma$ state, in which $\Lambda = \Sigma = \Omega = 0$ and $J = R$, the only contribution to the magnetic moment is from rotation (see chapter 5.8), so that

$$E = E_0 - \mu_N B_0 g_{\text{rot}} \frac{M_J}{\sqrt{J(J+1)}}. \qquad (10.3)$$

The interaction of a molecule with an electrostatic field will also be useful, especially in beam spectroscopy. The Stark energies (see chapter 5.9) are given by

$$E^{(1)} = \frac{\mu E M_J \Lambda}{J(J+1)}, \tag{10.4}$$

in which μ is the electric dipole moment of the molecule, and E, the electric field strength. If $\Lambda = 0$, there is no first-order effect, but there is a second-order energy shift given by

$$E^{(2)} = \frac{\mu^2 E^2 [J(J+1) - 3M_J^2]}{2hB\,J(J+1)(2J-1)(2J+3)}, \tag{10.5}$$

in which B is the rotational constant of the molecule. Atoms, being spherically symmetric, possess only a second-order Stark shift, which is not of particularly great interest (except for the H atom, see chapter 2.8).

The splittings that arise from these electric and magnetic interactions are quite small on the scale of optical energies—in the microwave or radio-frequency region, for typical fields. This means that they will be mostly unobservable in ordinary absorption spectra because of the widths of the absorption lines themselves. These line widths arise from a number of different sources.

1. The *Doppler width* (see problem 5 in chapter 1) corresponds to frequency shifts due to the distribution of velocities of the absorbing particles with respect to the observer. It is of the order of

$$\Delta\nu_D \approx \left(7 \times 10^{-7} \sqrt{\frac{T}{M}}\right)\nu_0, \tag{10.6}$$

where T is the absolute temperature in °K, M the mass of the atom or molecule in amu, and ν_0 the frequency of the line under study. Since ν_0 in the optical region may be of the order of 10^4 cm^{-1}, the Doppler width will be about 0.01 cm^{-1}, or 300 MHz.

2. The *natural width* is a consequence of the Heisenberg uncertainty principle; if an excited state has a lifetime τ_{rad}, then a transition to that state will have a width $\delta\nu_N \sim 1/\tau_{\text{rad}}$. Since lifetimes of many atomic and molecular states are in the range of 10^{-8} sec, the line widths will be of the order of 100 MHz from this source.

3. If a grating spectrometer is used to analyze the spectrum, there is a basic limitation due to the *diffraction width* associated with the grating. The maximum theoretical resolving power of a grating is equal to the number of

ruled lines illuminated by the light times the order in which the grating is used; the best gratings now available have a resolution of the order of 10^6, so that the line width from this source $\delta\nu_G \approx 10^{-6}\nu_0$, or 300 MHz.

4. Above and beyond this theoretical limit on resolution, there is a practical one dictated by the need to admit a finite amount of light to the detector of the spectrometer. This means that a minimum spectral bandpass must be employed; for conventional, noncoherent sources, this is also typically of the order of a few hundredths of cm^{-1}.

5. Finally, in any gas phase sample, there will be a *collision broadening* whose magnitude in Hz is roughly equal to the gas kinetic collision frequency Z in the gas. This broadening is roughly of the order of 10 MHz/Torr of gas pressure.

All these sources of line broadening combine to give a line width of the order of several hundredths of cm^{-1}, or a few gigahertz (10^9 sec^{-1}) in frequency. This is more than enough to obscure most of the hyperfine splittings, which are generally in the MHz to GHz range, in any conventional absorption or emission experiment. For this reason, a number of methods have been devised for narrowing line widths. The essential principles involved in these methods are to reduce the Doppler line width by restricting the velocity distribution of the molecules under study, to reduce collision broadening by working at as low a pressure as possible, and to reduce source broadening (3 and 4) by using narrow-bandwidth, high-intensity radiation sources. The natural line width represents a fundamental lower limit to the attainable line width.

Historically, the first of these line-narrowing techniques is the optical-pumping method originally devised by Kastler (reference 4), since extended by many others; for a review, the reader is referred to the monograph by Bernheim (reference 5) and the articles by Skrotskii and Iz'iumova (reference 6) and Happer (reference 7). Molecular beam methods are discussed in sections 4 and 5. The important Lamb dip, or saturated-absorption, technique will be considered in chapter 12.6; line-narrowing methods involving double-resonance and two-photon spectroscopy will be discussed at the appropriate points in the text (chapters 12.7 and 13.3, respectively).

2 Optical Pumping

Consider the energy level scheme of an alkali atom, as shown in figure 10.1. The ground ($^2S_{1/2}$) and first excited ($^2P_{1/2}$) states are connected by one of the

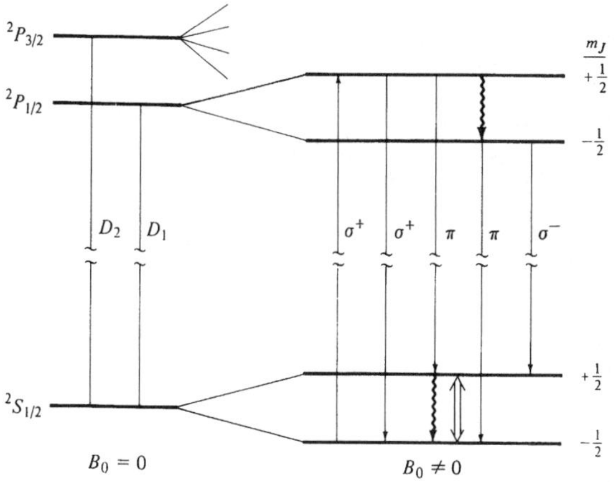

Figure 10.1
Energy levels of an alkali metal atom, at zero magnetic field (left) and in the presence of a magnetic field (right).

strong D line transitions. When a magnetic field B_0 is applied, both of these states split up into $M = +\frac{1}{2}$ and $-\frac{1}{2}$ components. But the splitting at any attainable field strength is less than the width of the D line itself, so a direct determination of, say, g_J for the transition is not possible. The lines can be resolved, however, in an optical-pumping experiment such as is sketched in figure 10.2. Here, an alkali atom resonance lamp is used to produce the desired D line, which is isolated from all the other lines by use of an interference filter. The light is circularly polarized—say right circularly polarized (RCP), to be specific—and sent through a heated bulb containing alkali atoms of the same species as is used in the resonance lamp, and on which a magnetic field in the direction of the light propagation is impressed. The light passing through the bulb is then detected.

First, let us see what happens when the lamp is turned on. A typical time development is illustrated in figure 10.3a. When the light first begins to pass through the sample, it is absorbed at a rate expected from the known absorption coefficient of the atoms. But the transmittance immediately begins to rise, and shortly levels off at some constant value corresponding to a partial transparency of the sample. If the light is then turned off for a while, and then turned back on, we find that the initial absorbance has returned to the lower value, but then rises to a plateau again as before.

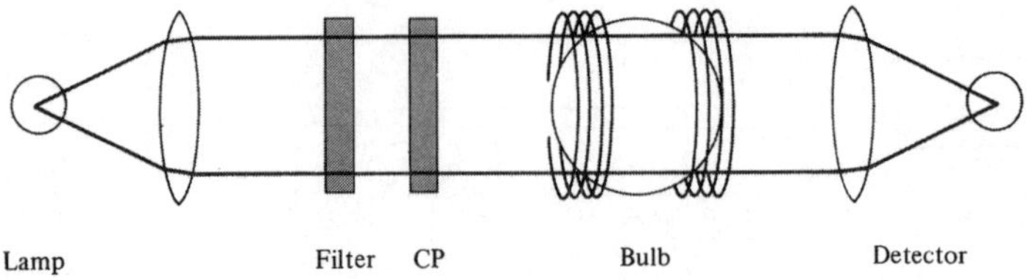

Figure 10.2
Basic apparatus for an optical-pumping experiment. The radiation from the atomic resonance lamp is filtered and right circularly polarized (most conveniently by passing through a linear polarizer and a quarter wave plate). The absorption bulb contains the same alkali metal atoms as are used in the lamp and has a magnetic field applied in the direction of the light propagation by means of a pair of Helmholz coils. Not shown is a second pair of coils for applying the radiofrequency radiation.

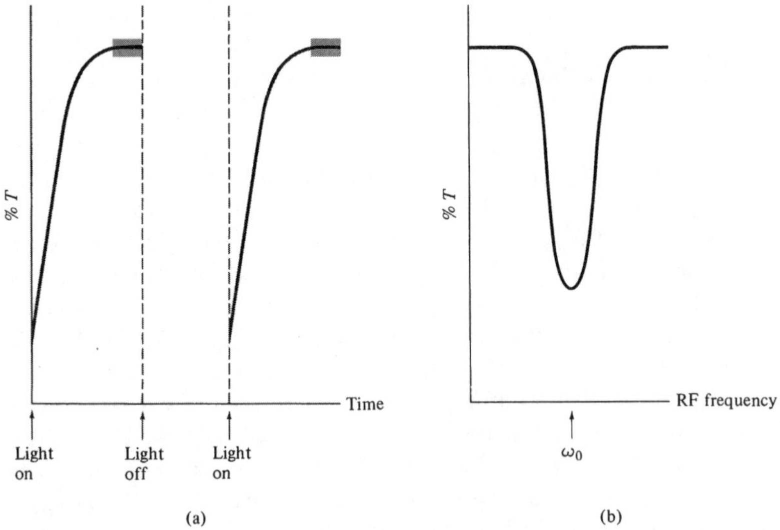

Figure 10.3
Optical-pumping signals, in terms of the percentage of radiation detected after passing through the bulb. (a) Response to turning on and off of light, showing depletion and recovery of absorbing levels. (b) Resonance signal obtained when radio frequency ω_0 is equal to the Zeeman level splitting, and when metal vapor is in the optically saturated condition.

This behavior can be understood by reference to the transitions shown in detail in figure 10.1. The right circularly polarized incident light can cause only $\Delta M = +1$ transitions, that is, from $M = -\frac{1}{2}$ of the ground state to $M = +\frac{1}{2}$ of the excited $^2P_{1/2}$ state. The excited state can reradiate this same light as fluorescence, or, with essentially equal probability, a π component ($\Delta M = 0$) that brings the atom back to the $M = +\frac{1}{2}$ state of the $^2S_{1/2}$. Once the atom is in the $^2S_{1/2}$, $M = +\frac{1}{2}$ state, it can no longer absorb the incident radiation. Thus a portion of the atoms are effectively pumped from the $-\frac{1}{2}$ to the $+\frac{1}{2}$ state, via the excited state, by the effect of the polarized resonance radiation—whence the name, "optical pumping." The partial depletion of the $-\frac{1}{2}$ state produces the partial transparency of the metal atom vapor. The only way in which the original populations can be restored is for some relaxation process to take place, either directly from the $+\frac{1}{2}$ to the $-\frac{1}{2}$ levels of the ground state or between the corresponding levels of the excited state, whereupon an atom can return to the $-\frac{1}{2}$ level by emission of a π component of the fluorescence. In either case, a collision is required to effect the relaxation (symbolized by the wavy lines in figure 10.1), and the collision rate can be made very small at sufficiently low vapor densities. This slow relaxation accounts for the recovery of the absorption when the pumping light is turned off.

From this sort of experiment, one can learn about atomic angular momentum relaxation rates; this is interesting, but still does not get us any closer to the spectroscopy of the hyperfine structure levels that we wished to investigate. This last is accomplished by applying a variable radiofrequency field to the sample bulb, through a second pair of coils (not shown in figure 10.2). When the frequency of this radiation is equal to the separation between the Zeeman components, a transition between the two magnetic sublevels is possible. This is an allowed magnetic dipole transition with $\Delta J = 0$ (but $J \neq 0$) and $\Delta M = -1$, which is symbolized by the heavy double arrow in figure 10.1. When this field is applied to the saturated system (at the time indicated by shading in figure 10.3a), the population of the $M = -\frac{1}{2}$ level is partially restored, leading to an increased absorption of light, and thus a decreased transmittance, as shown in figure 10.3b.

The advantages resulting from observing the transition in this manner become apparent by considering the various contributions to the line width. The lifetime of the ground atomic state is essentially infinite, so that the natural broadening makes a negligible contribution. The radio frequency source is a coherent oscillator, so that there is no diffraction or bandpass broadening to speak of. The atom density can be made very low (a few millitorr), so that collision broadening can be kept to 0.1 MHz or less. Finally, the Doppler broadening is applied to the radio frequency rather than the optical frequency, so that $\Delta\nu_D$ is on the order of 10^{-6} of a few hundred megahertz rather than of 10^{14} Hz. The resolution thus attainable is such that not only the fine structure components can be resolved, but all the nuclear hyperfine components as well; and the nuclear spin states of the alkali metal atoms were first determined by this technique.

3 Optical Level-Crossing Spectroscopy

In some systems, information can be obtained about the fine structure levels directly, without the use of applied radio frequency fields, by means of level-crossing spectroscopy. A typical case in which this method would be applicable is shown in figure 10.4, for a ^3P–^3S transition in an atom. As the magnetic field applied to the atom is increased, the various Zeeman sublevels may cross one another to produce accidental degeneracies at particular field strengths. Since the quantum numbers of the sublevels are different, there is no interaction or avoided crossings at these points. But if the atom is irradiated with light corresponding to the atomic transition, some of these crossings will correspond to two different states being excited simultaneously by light of the same frequency. When this occurs, the two excited states interfere with each other, and there is an observable change in the intensity and angular distribution of the fluorescence emitted from the excited atoms. From the number and magnetic field values of the intensity changes, one can work back to the quantum numbers and g value for the excited state of the atom (reference 8). The advantage of level-crossing spectroscopy, of course, is that it allows measurements to be made on excited states of atoms; optical-pumping techniques are pretty much restricted to ground state spectroscopy.

Optical pumping and level-crossing experiments are more difficult to perform on molecules than on atoms, for several reasons. First, in contrast with atoms, the oscillator strength of a molecular transition is splattered over an inhomogeneously broadened band consisting of a large number of vibration-rotation components. Second, reorienting collisions, which destroy the polarization, are much more effective on molecules than on spherically symmetric atoms. The availability of high-intensity, narrow-bandwidth lasers has made such experiments possible, however, and a wide variety of molecular systems has now been studied in this way (reference 9).

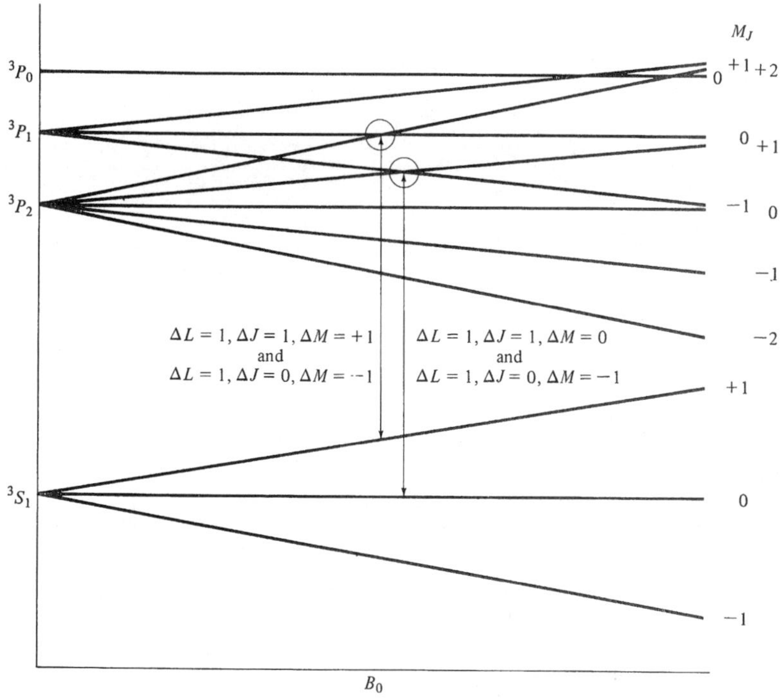

Figure 10.4
Level-crossing experiment in a ^3P–^3S transition. At the indicated field strengths, pairs of transitions (the 122–111 and 110–111, and the 121–110 and 11-1–110) are degenerate, leading to an anomalous angular distribution of fluorescence.

4 Molecular Beam Methods

While the optical-pumping and level-crossing techniques have been useful in studying a large number of atoms, as well as some molecules, there is an equally large number of systems that are not suitable for these experiments. The basic limitation is that these methods require the observation of resonance fluorescence from the system, and many systems either do not possess such fluorescence or have it in an inconvenient region of the spectrum, such as the deep vacuum ultraviolet. A particular case of this kind is the hydrogen atom, whose resonance fluorescence is the Lyman α line at 1,216 Å.

Let us look in some more detail at the Zeeman pattern to be expected from

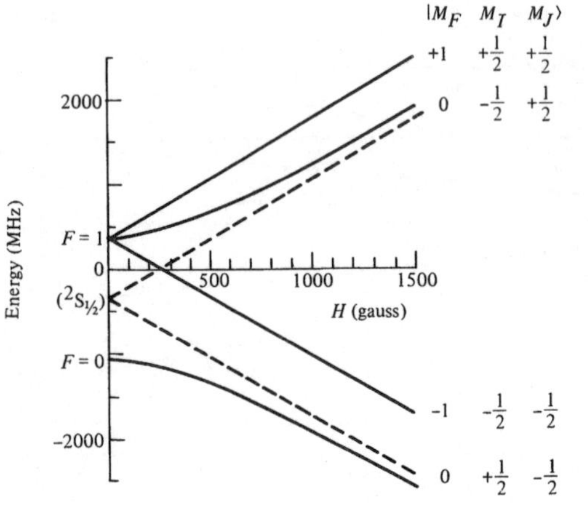

Figure 10.5
Zeeman splitting for ground state of the hydrogen atom. The dashed lines indicate the Zeeman splitting expected for a $J = \frac{1}{2}$ atom in the absence of nuclear spin. Compare figure IX.9 in Ramsey (reference 10) or figure 2.3 in Carrington and McLachlan (reference 11).

Table 10.1

F	M_F	M_J	M_I	Energy $(-E_0)$
1	+1	$+\frac{1}{2}$	$+\frac{1}{2}$	$+\frac{1}{2}(a/2) - \frac{1}{2}(g_J\mu_0 + g_I\mu_N)B_0$
1	0	$+\frac{1}{2}$	$-\frac{1}{2}$	$+\frac{1}{2}(a/2) - \frac{1}{2}(g_J\mu_0 - g_I\mu_N)B_0$
1	−1	$-\frac{1}{2}$	$-\frac{1}{2}$	$+\frac{1}{2}(a/2) + \frac{1}{2}(g_J\mu_0 + g_I\mu_N)B_0$
0	0	$-\frac{1}{2}$	$+\frac{1}{2}$	$-\frac{3}{2}(a/2) + \frac{1}{2}(g_J\mu_0 - g_I\mu_N)B_0$

the ground state of the hydrogen atom, which is a $^2S_{1/2}$ state, with a nuclear spin $I = \frac{1}{2}$, giving rise to $F = 1$ and 0 states (figure 10.5). At low fields, the $F = 1$ state will split into three components, and the $F = 0$ state will be unaffected. At intermediate fields, sufficient to uncouple **I** and **J** but not **L** and **S** (nuclear Paschen-Back limit), the energies are given by the appropriate modification of equation (10.1), namely,

$$E = E_0 + \frac{a}{2}[F(F+1) - I(I+1) - J(J+1)] - M_J B_0 g_J \mu_0 - M_I B_0 g_I \mu_N.$$

Thus the energies of the four levels in this intermediate field region are as given in table 10.1. The important point to note is that the $M_F = +1$ and

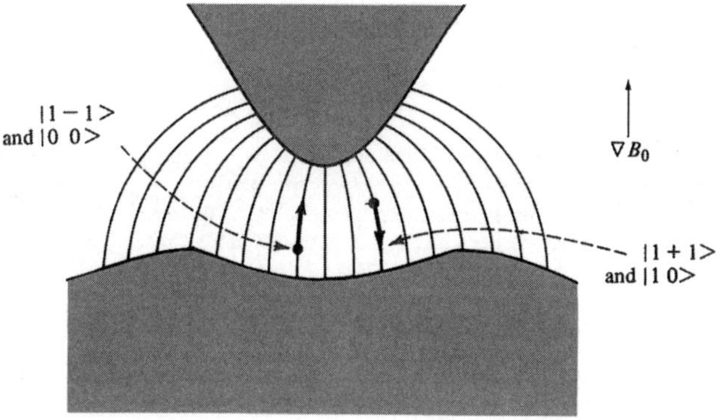

Figure 10.6
Deflection of Zeeman sublevels in an inhomogeneous magnetic field.

0 levels of the $F = 1$ state gain in potential energy as B_0 is increased, whereas the $M_F = -1$, $F = 1$, and the $F = 0$ states lose energy as B_0 is increased.

Suppose, now, that a hydrogen atom is placed in an inhomogeneous magnetic field, which possesses a field gradient in a particular direction. Then, since the force on a particle is the negative gradient of its potential energy, and the field-dependent part of the potential energy is just

$$U = -\mu_0 g_J M_J B_0 = -\gamma \hbar M_J B_0,$$

the force on the particle is

$$\mathbf{F} = -\nabla U = +\gamma \hbar M_J \nabla B_0.$$

Since an inhomogeneous magnetic field is usually set up between a pair of poles shaped as shown in figure 10.6, the effect on hydrogen atoms located between such a pair of pole pieces will be that the $|F M_F\rangle$ states $|1\ {-1}\rangle$ and $|0\,0\rangle$ will be attracted to the upper pole, and $|1\,1\rangle$ and $|1\,0\rangle$ will be attracted to the lower pole.

One consequence of this behavior is just the behavior observed in the Stern-Gerlach experiment with which we began the discussion in chapter 1, namely, the splitting up of beams of atoms possessing J units of angular momentum into $2J + 1$ subbeams in an inhomogeneous magnetic field. Beyond this phenomenon, however, is the possibility of carrying out a resonance experiment on such separated beams to measure the splitting between the various magnetic sublevels. The principle of such an experiment

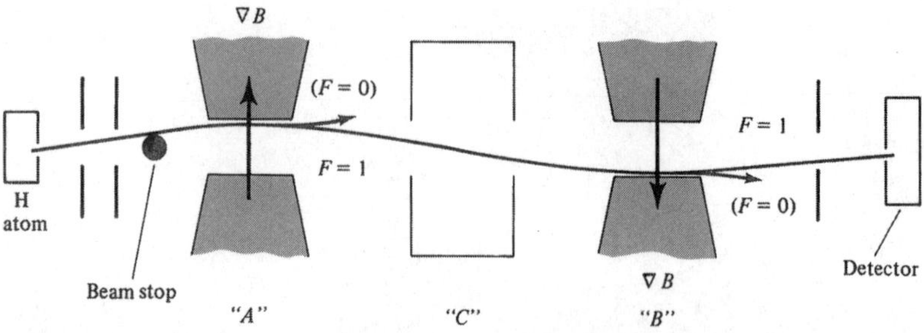

Figure 10.7
Molecular beam magnetic resonance experiment for hydrogen atoms. Atoms effuse from the source, at left, through the "A" field, where they are separated, the "C" field, where radio frequency or microwave power is applied, and the "B" field, where they are analyzed, to the detector at the right-hand end. The beam stop is placed to prevent a trajectory straight through down the axis of the beam.

is sketched in figure 10.7. In a beam resonance experiment, a beam of the atoms under study is first passed through magnetic field "A," which has the effect of deflecting the $|1\,1\rangle$ and $|1\,0\rangle$ states back toward the beam axis, if they are moving slightly off to one side; the $|1\,{-1}\rangle$ and $|0\,0\rangle$ states are deflected in the opposite direction, out of the apparatus entirely. The important point is that the beam traveling through the apparatus now contains only atoms in $F = 1$ states. This beam passes through a "C field," which we shall come back to in a moment, and passes into a "B field" region, which has a field gradient equal and opposite to that in "A." This serves to deflect the $F = 1$ atoms back toward the axis and into the detector.

The "C field" region contains a static, homogeneous magnetic field, which separates the Zeeman sublevels in energy, but does not deflect them, and provision for applying a time-varying magnetic field at radio or microwave frequencies. When the frequency of this ac field matches the energy separation between the $F = 0$ and 1 levels, with M_F equal to either 0 or ± 1 (depending on the polarization of the radiation relative to the beam), some of the $F = 1$ states in the beam make a transition to the $F = 0$ state. This leads to a loss of intensity in the beam reaching the detector, and thus a dip in measured beam current at the resonant frequency, as shown in figure 10.8. Thus the beam particles themselves are made to act as a detector of the transition. Figures 10.7 and 10.8 depict the so-called flop-out configuration, in which a decrease in beam intensity is detected; if the particle detector is located at the expected termination of the $F = 0$ trajectory out of the "B

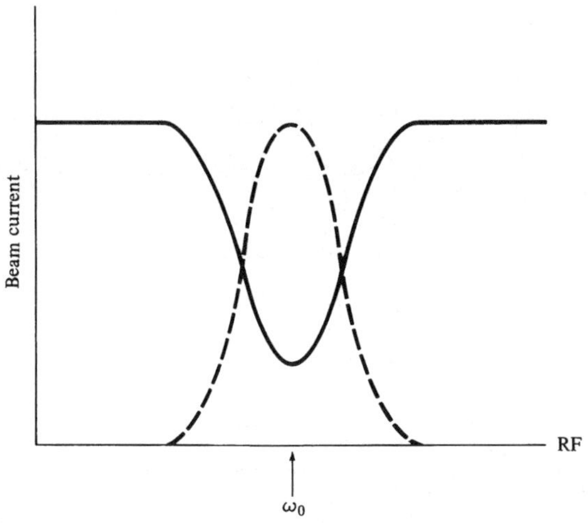

Figure 10.8
Molecular beam resonance signals through frequency ω_0. The solid curve corresponds to a "flop-out" configuration, the dashed curve, to a "flop-in."

field," then we have the "flop-in" configuration, in which an increase in beam intensity is detected.

This sort of experiment was first applied to alkali metal atoms by Kusch, Rabi, and their coworkers at Columbia University during the late 1930s (reference 12). An experiment with metastable hydrogen atoms was not actually carried out until the 1950s by Lamb and Retherford (reference 13) in their classic experiments at Yale. The reader interested in molecular beam magnetic resonance is referred to the detailed book by Ramsey (reference 10).

A similar sort of experiment can be carried out using the electrostatic Stark shift rather than the magnetic Zeeman shift; this is particularly useful for molecules (see chapter 5.9). In such a molecular beam electric resonance experiment, the "*A*" and "*B*" fields are electric dipole or quadrupole fields, which have focusing properties for electric moments similar to those of the inhomogeneous magnetic field shown in figure 10.6 for magnetic moments; the electric potential contours for a quadrupole lens are shown in figure 10.9. The "*C*" field contains a pair of Stark plates to set up a constant electric field and provision for leading in radio frequency or microwave radiation. A typical electric resonance experiment in barium oxide is shown in figure 10.10.

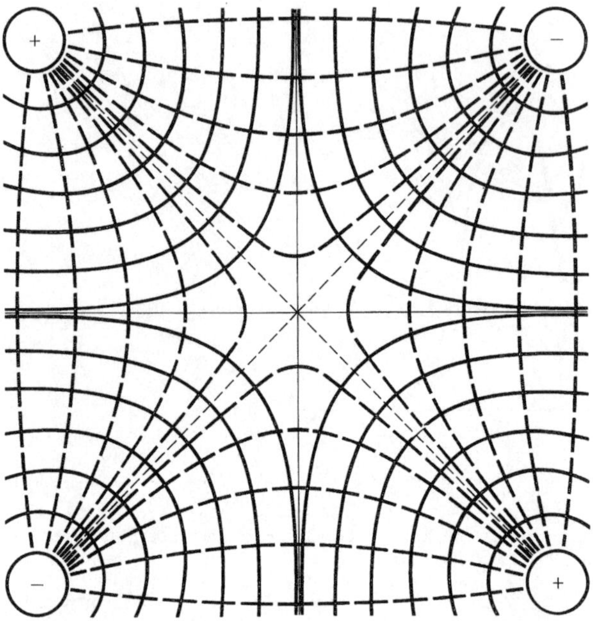

Figure 10.9
Electric potential contours (solid lines) and field strengths (dashed lines) around a quadrupole focusing lens. In order to set up a true quadrupole field, the pole pieces should actually be hyperbolic in cross section, but a very good approximation to such a field can be set up by simply using four cylindrical wires as shown (reference 10). The field is weakest at the center of the lens, so that molecules with positive Stark coefficients are focused along the axis.

One of the significant features of molecular beam spectroscopy, as distinct from spectroscopy involving measurement of the absorption or emission of radiation, which permits extremely accurate energy measurements to be made, is the very narrow line width of the transitions, even narrower than in an optical-pumping experiment. As before, there is essentially no source broadening, since coherent radio frequency or microwave oscillators are used. A molecular beam is (ideally) a collision-free medium, so there is essentially no collision broadening. Since the radiation is applied perpendicular to the beam path, there is a very small component of velocity along the observation path, so that there is hardly even any Doppler broadening. The major sources of line width arise from the finite residence time of the particles in the "*C*" field (about 1 msec, leading to a 1-kHz line width from this source) and such purely technical factors as the homogeneity of the static

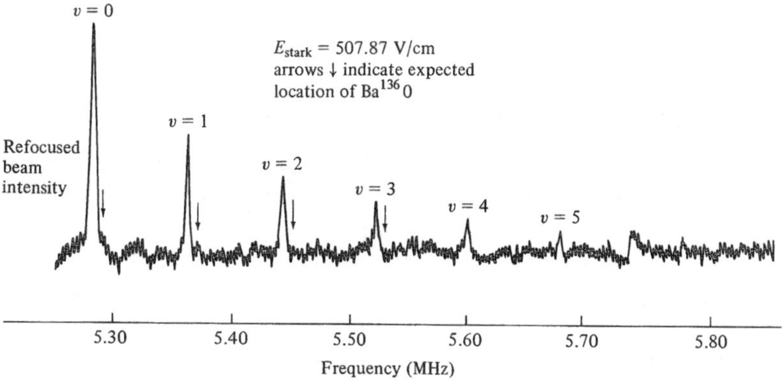

Figure 10.10
Molecular beam electric resonance spectrum of the $J = 2$, $|M_J| = 0 \rightarrow 1$ transition in barium oxide. The various components correspond to the different vibrational levels of the molecule, distributed according to a temperature of 2,200°K in the beam source. [From L. Wharton, M. Kaufman, and W. Klemperer, *J. Chem. Phys.* **37**, 621 (1962). Reproduced with permission.]

electric magnetic field. Thus, molecular constants obtained from molecular beam spectroscopy are probably the most accurately known of any that have ever been determined.

One further application of beams, not involving resonance methods at all, is the use of simple electrostatic deflection to determine molecular structure. It is plain from equations (10.4) and (10.5) that, to possess any appreciable Stark effect at all, a molecule must have a permanent electric dipole moment. Thus a polar molecule will suffer some sort of deflection in an inhomogeneous electric field, whereas a symmetric and, therefore, nonpolar molecule will be nearly unaffected. This behavior has been used to investigate the basic structural type of a number of different species (reference 14).

5 Supersonically Cooled Molecular Beam Spectroscopy

One molecular beam method that has received a great deal of attention in the past several years is the production of beams of very cold molecules by supersonic expansion. If a molecular vapor diluted in a carrier gas, such as helium or argon, is squirted through a nozzle from high pressure ($\gtrsim 1$ atm) into a low-pressure cavity, the gas will form a hydrodynamic jet. In such a jet, the forward velocity of the gas stream may be several times the sound speed in the same gas at the stagnation temperature; for this reason, such an expansion is often referred to as a *supersonic* flow. The energy for this acceleration is drawn from the other kinetic degrees of freedom of the gas; this entails two consequences that are of particular importance for spectroscopy.

1. The velocity components *transverse* to the flow are cooled to a very low temperature, typically a few °K. Thus the Doppler width observed perpendicular to the flow axis is tremendously reduced.
2. The rotational and to a lesser extent the vibrational temperatures are also considerably reduced. This concentrates the population in a few low-v and low-J, K states, so that a spectrum that is congested by reason of a high density of transitions may be expected to reduce to a relatively small number of intense lines.

The supersonic beam can be studied by a variety of spectroscopic techniques, including laser-induced fluorescence (l.i.f.), multiphoton ionization, coherent Raman scattering, and even direct absorption. Among the first molecules to be examined in this way was NO_2, using the l.i.f. technique (reference 15). It had been hoped that the remarkably dense visible NO_2 spectrum would be simplified as a result of the supersonic cooling effect. This did occur, but it was found that the resulting spectrum was still so highly irregular and perturbed that a conventional analysis of the spectrum would never be possible. Since then, a number of other molecules have been studied in this way, including I_2, *s*-tetrazine, chromyl chloride (using l.i.f.), SF_6, and UF_6 (using diode laser infrared absorption—see section 7.7). A dramatic example of the increase in resolution afforded by supersonic beam spectroscopy is demonstrated by the spectrum of phthalocyanine, shown in fig. 10.11. Phthalocyanine is a large condensed molecule with the formula $C_{32}N_8H_{26}$. The absorption spectrum of the high-temperature vapor shows absolutely no fine structure, but the l.i.f. spectrum of the cooled jet shows very sharp line structure, which does not correspond at all to the broad "bumps" in the vapor spectrum. It appears as if the vapor spectrum is dominated by transitions from vibrationally excited molecules, completely obscuring the true electronic origin.

6 Molecular Beam Masers

If we look back at figure 10.7, we should recognize that a rather unusual situation has been created. Consider just what emerges from the hydrogen

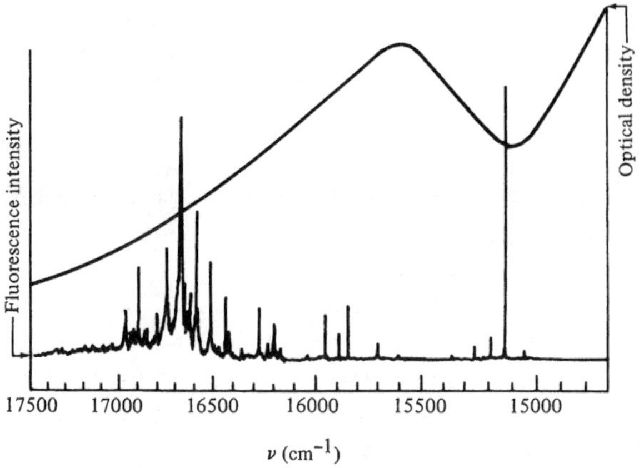

Figure 10.11
Fluorescence excitation spectrum of gas phase free base phthalocyanine cooled in a supersonic expansion. [From P. S. H. Fitch, L. Wharton, and D. H. Levy, *J. Chem. Phys.* **63**, 4977 (1975). Reproduced with permission.] The smooth curve is the absorption spectrum of a static vapor sample held at about 300°C. [From D. Eastwood, L. Edwards, M. Gouterman, and J. I. Steinfeld, *J. Mol. Spectroscopy* **20**, 381 (1966). Reproduced with permission.]

atom source and the "A" field region: This is a beam of atoms, all of which are in a higher-energy $F = 1$ state, and none of which is in the lower-energy $F = 0$ state. This is a gross departure from thermodynamic equilibrium, which would require slightly more than one-fourth of the atoms to be in the lower $F = 0$ state,[1] and thus represents a *population inversion* with respect to this particular hyperfine structure level. Now suppose we could take these $F = 1$ atoms and deposit them in a storage bulb for a while. If radiation corresponding to the $F = 0$ to $F = 1$ transition is passed through this bulb of atoms, it will be amplified, that is, increased in intensity, rather than absorbed. The basis for this is easy to see. As discussed in chapter 1.7, the processes involving atoms or molecules and the radiation field include absorption of radiation with transitions from the ground to the excited state, and two processes—spontaneous emission and induced emission—that lead to transitions from the excited to the ground state. If we retain both the

1. One-fourth because the degeneracy of the $F = 1$ level is three, compared with one for the $F = 0$ level; and only slightly more because the energy difference is so small compared with kT in the atom source.

absorption and emission terms in equation (1.35), we have for the change in light intensity I on passing through a medium containing a density n_0 of ground state molecules and n^* of excited state molecules

$$dI = Ik_{\text{abs}}(-n_0 + n^*)\,dx;$$

thus, if n^* is greater than n_0, as in the case of our state-selected hydrogen atoms, dI will be positive, and we shall get "negative absorption," or amplification of the radiation by the inverted population.

Historically, this effect was first observed (by Basov and Prokhorov, at the Lebedev Institute in Moscow, and simultaneously by Townes in 1955—reference 18) using rotational states of ammonia molecules selected by an electrostatic field. Townes called the device a maser, for *M*icrowave *A*mplification by *S*timulated *E*mission of *R*adiation.[2] Later, Ramsey and Kleppner (reference 19) developed the actual device we have been discussing, namely, a maser operating on the hyperfine transition in hydrogen atoms.

One particularly valuable feature of a stimulated emission device is its ability to amplify radiation. As is well known in electrical engineering, any amplifier can be made into an oscillator by supplying regenerative feedback. This feedback can be supplied by placing the active medium containing the population inversion into a microwave resonant cavity, or, in the case of optical masers, a Fabry-Perot interferometer cavity. Thus the maser is a source of intense, coherent optical radiation.

The coherence of maser radiation produces an extremely narrow frequency distribution in the output; in principle, the frequency spread can be as narrow as the reciprocal of the time duration of the wave train produced. The coherence of the maser output does not follow from the simple negative absorption argument we used previously, but we shall come back to this point after introducing a better treatment of the radiation field, more suitable for analysis of maser problems, in the next chapter. As an example of this frequency narrowing, the hyperfine splitting in hydrogen, which is the transition on which the hydrogen maser operates, has been measured as $1{,}420{,}405{,}751.786 \pm 0.010$ Hz; the large number of significant figures is possible because the hydrogen atom has a lifetime of a few seconds in the storage bulb, and this lifetime is the limiting factor on the width of the maser output.

2. Subsequently, Townes was known to interpret the acronym as *M*eans of *A*cquiring *S*upport for *E*xpensive *R*esearch.

The temporal coherence, which leads to this extremely narrow frequency distribution, also leads to spatial coherence, that is, the well-known property of an optical maser beam of propagating as an extremely narrow, well-collimated ray of light, which can be focused by a simple lens to a point the size of a few wavelengths of the radiation. The availability of these coherence properties in the optical region of the spectrum has had a major impact on spectroscopy, as the next few chapters will show. In the following section, we shall just indicate the principles of operation of some of the best-known types of optical masers.

7 Optical Masers

There now exist a wide variety of optical masers, or lasers (by substitution of *L*ight for *M*icrowave). We shall briefly discuss the operating principles of such devices, and then illustrate only a few representative samples, in order to indicate the availability of laser wavelengths, which may be of interest to spectroscopists.

The two requirements for laser action are, first, an *active medium* with optical gain $g = k_{abs}(n^* - n_0)l'$, and, second, some way of trapping the light spontaneously emitted by the medium and repeatedly amplifying it until the intense, coherent laser beam is produced. The optical gain is produced by having the *population inversion*, n^*/n_0, greater than unity; a number of ways of bringing this about will be discussed. The amplification is effected by placing the active medium into an *optical cavity*, as shown in figure 10.12a. Such an optical cavity is nothing more than a pair of mirrors located a distance l apart, often called a Fabry-Perot interferometer. Optical standing waves can be set up in such a cavity if the frequency fulfills the condition $\nu = nc/2l$, where n is some integral value for the order of the interferometer. Since optical frequencies are of the order of 5×10^{14} Hz, and laser cavities may typically be 100 cm long, in order to produce a large optical gain, the Fabry-Perot condition must be met with $n = 5 \times 10^6$ or so. This means that there may be many possible *cavity modes* within the gain curve $g(\nu)$ of the laser material, spaced on the order of 100 MHz or so apart, as shown in figure 10.12b.

In order to get efficient laser action, it is important to make the cavity losses as small as possible. One common method for accomplishing this is to set the ends of the active medium (gas discharge tube, crystal rod, or whatever) at *Brewster's angle* (reference 20). This is the angle for which the

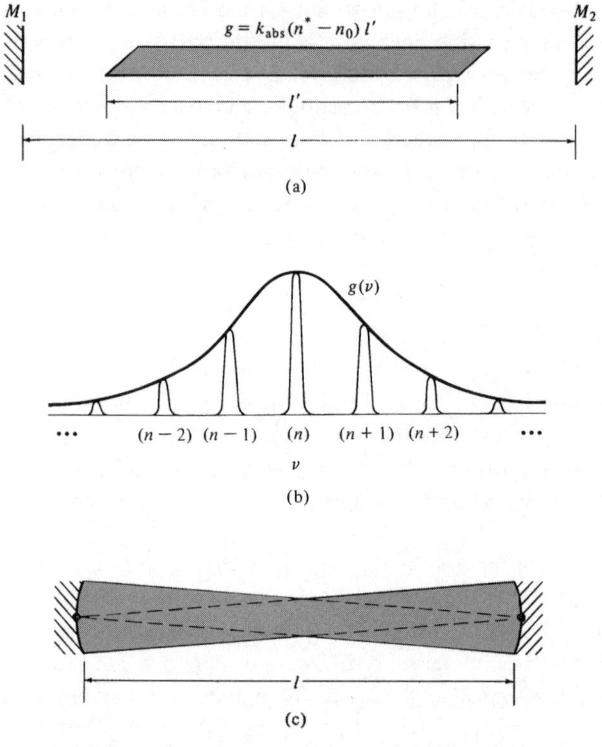

Figure 10.12
(a) Optical medium placed in a plane-parallel resonator of length l constitutes a laser. The ends of the active medium are positioned at Brewster's angle. (b) Longitudinal cavity modes of the resonator within the gain curve of the medium. Laser action may occur at the mode nearest the peak of the gain curve or at all of them. (c) Confocal resonator, with radius of curvature of each mirror equal to l.

coefficient of reflection for light polarized in the vertical direction in figure 10.12a is zero, so that this light polarization is transmitted through the ends of the laser material without loss. Another is to use, not the plane-parallel resonator shown in figure 10.12a, but the *confocal* resonator shown in figure 10.12c, in which the radius of curvature of each mirror is equal to (or somewhat greater than) the cavity length. Such a resonator is much easier to align than the plane-parallel configuration.

Sometimes, however, a judiciously arranged cavity loss can be very useful. A frequency-dependent loss inserted in the cavity can be used for selecting a particular component out of the available gain curve of the medium, that is, for *tuning* the laser within the limits of medium. A simple prism or diffraction grating is often used in this way. An *etalon*, which is simply a very short-path Fabry-Perot interferometer coupled to the laser cavity, is also very effective in eliminating unwanted cavity modes from the laser output. A time-dependent loss in the cavity can be used to produce very short, intense pulses of laser light. Since the optical quality of the cavity is called the "Q factor" (by analogy with electrical circuits), this technique is called Q switching. This can be accomplished in a variety of ways—for example, by rotating one of the cavity mirrors at high speed, by inserting an absorbing medium into the cavity which saturates at high-power densities, or by electrically or acoustically flipping the polarization of a crystal (Pockel's effect).

7.1 Optically Pumped Solid State Lasers

The first optical laser transition was found in 1960 by Maiman (reference 21), who pumped a ruby crystal with a flash lamp and obtained stimulated emission on the ${}^2E_g \rightarrow {}^4A_{2g}$ transition at 6,943 Å. The energy levels appropriate to this system are shown in figure 10.13a. The flash lamp excites a large number of the chromium atoms in the ruby to a broad manifold of levels near 20,000 cm^{-1}. These levels decay to the ${}^2E_g$ level at 14,400 cm^{-1}, which is *metastable* on account of the doublet-quartet nature of its only allowed decay route, namely, to the ${}^4A_{2g}$ ground state. This metastability permits the population of the ${}^2E_g$ levels to build up to the point at which a population inversion exists between it and the ground state. The ruby then can act as an amplifier, and, if placed in the appropriate optical cavity, laser emission results. The output is by its nature pulsed, because as the emission proceeds, the ground state is repopulated and the inversion destroyed.

An even better solid state laser is the neodymium +3 ion, whose energy levels are shown in figure 10.13b. In this case, the lower laser level is not the

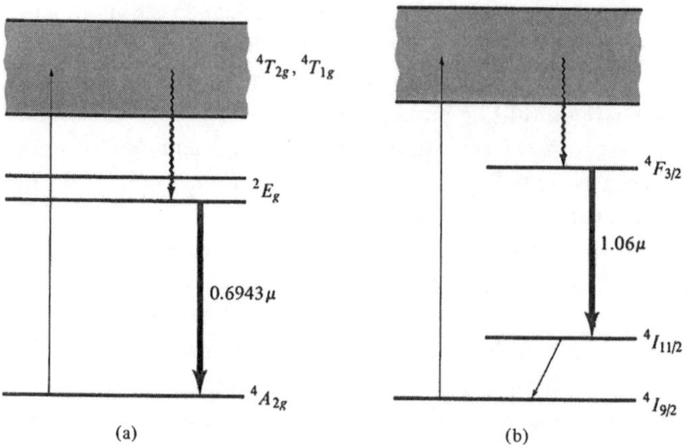

Figure 10.13
(a) Energy levels of ruby (Cr^{3+} ions in aluminum oxide crystal). (b) Energy levels of Nd^{3+} ions in yttrium aluminum garnet ("YAG").

ground state, so that it can be empited by an allowed radiative decay process, and continuous laser action at 1.06 μm is possible, and, indeed, has been observed.

One of the important features of these, as well as a number of other lasers, is that they can be "Q switched"; that is, the optical cavity can be closed by any of a number of means while the population inversion is building, and then suddenly opened to permit a large fraction of the stored energy to be dumped in a single, short pulse. Pulses of 10^{-8}-sec duration, with anywhere from 10^6 to 10^9 W peak power, can be generated in this way. The highest power achieved thus far has reached the terawatt (10^{12} W) region, using a Q-switched neodymium laser and sending the pulse through a series of neodymium amplifiers, all pumped simultaneously. It may be of interest to note that, for the few nanoseconds that the pulse was on, its power level exceeded that of all the combined electrical-generating facilities of the entire world! These lasers can also be "mode locked" to produce pulses of less than 1-psec (10^{-12} sec) duration, by a mechanism to be discussed in chapter 13.

7.2 Optically Pumped Gas Lasers

In addition to solid state glasses or crystals, a molecular gas can be used as the active gain medium in optically pumped lasers. A classic example is molecular iodine. When I_2 vapor is pumped with the 5145-Å line of an argon

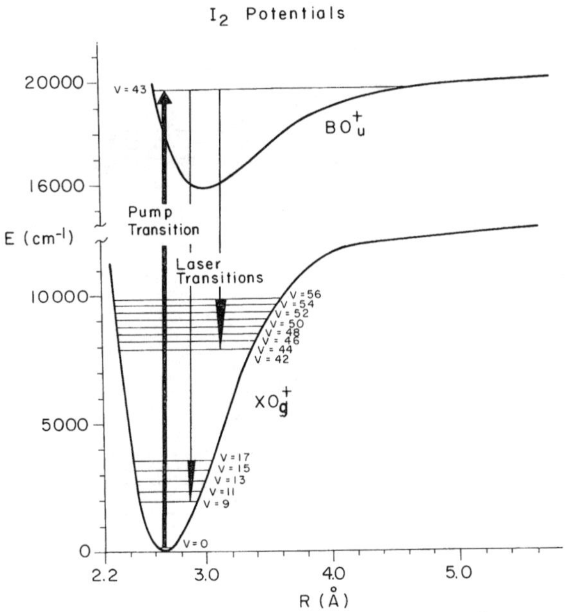

Figure 10.14
Rydberg-Klein-Rees (RKR) potential energy curves for I_2 and pump and laser transitions. [From J. B. Koffend and R. W. Field, *J. Appl. Phys.* **48**, 4468 (1977). Reproduced with permission.]

ion laser (see section 7.3), the levels $v' = 43$, $J' = 12$ and 16 are populated in the upper electronic state. These levels can be stimulated to emit to unpopulated vibrational levels of the ground electronic state [see figure 10.14]. Emission from $v' = 43$ to $v'' = 9–56$ has been observed at wavelengths extending from 5,697 to 10,274 Å (reference 22). Because of the high density of ro-vibrational levels in this molecule, a large number of different output wavelengths can be obtained, although the laser is not truly continuously tunable. Organic dye solution lasers (see section 7.6) are more practical sources of visible wavelength radiation, but the I_2 laser could be useful for down shifting into the near-infrared region, where efficient dyes are not available.

7.3 Atomic Gas Discharge Lasers

The first continuously running laser transition in a gas was the helium/neon system operating at 1.15 μm, found by Javan in 1961 (reference 23). The

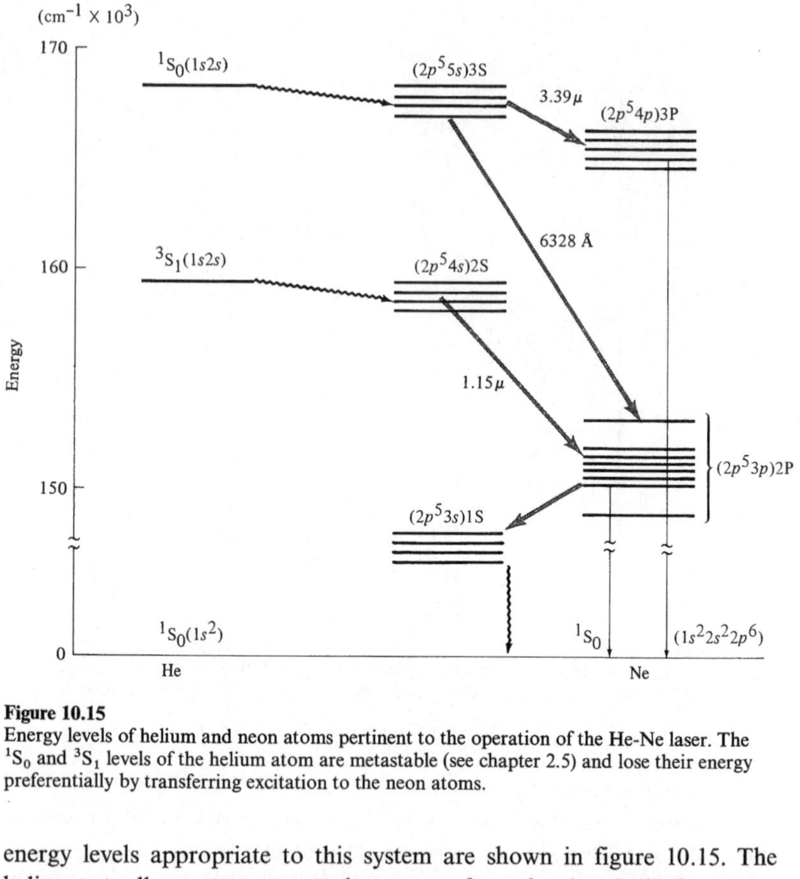

Figure 10.15
Energy levels of helium and neon atoms pertinent to the operation of the He-Ne laser. The $^1S_0$ and $^3S_1$ levels of the helium atom are metastable (see chapter 2.5) and lose their energy preferentially by transferring excitation to the neon atoms.

energy levels appropriate to this system are shown in figure 10.15. The helium actually serves to accumulate energy from the electric discharge run through the gas mixture and transfer it to several highly excited levels of the neon atom by inelastic collisions. The laser transitions are then those of the excited neon atom. In addition to the 1.15-μm output, there are a number of other possible transitions, including one at 3.39 μm and several in the red part of the visible spectrum, including the one at 6,328 Å that almost everyone has surely seen by this time. Since the lower levels of all these transitions are excited states, with decay paths of their own, these are intrinsically continuous wave (CW) transitions.

By now, hundreds of different laser wavelength arising from atomic discharges have been found. The ones most readily available and used by

Table 10.2

Laser system	Principal output wavelength (nm)
Helium/neon	632.8, 1,150, 3,390
Argon (ion)	488.0, 514.5
Krypton (ion)	520.8, 530.8, 568.2, 647.1
Cadmium ion/helium	325.0, 441.7

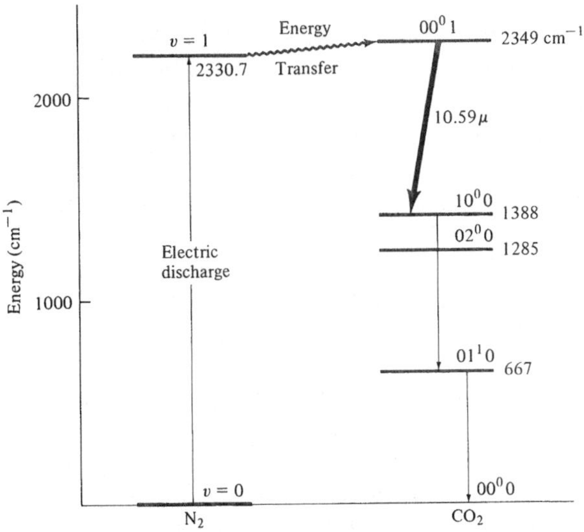

Figure 10.16
Energy levels of nitrogen and carbon dioxide molecules.

spectroscopists are given in table 10.2; the power obtainable in these transitions is in the range of 0.001 to several hundred watts.

7.4 Molecular Gas Lasers: Carbon Dioxide, Excimer

Another class of gas lasers operates on vibrational transitions in the infrared. The first of this type to be discovered was the carbon dioxide/nitrogen laser, by Patel in 1964 (reference 24). The energy levels appropriate to this system are shown in figure 10.16. Here the nitrogen acts as the energy storage and transfer agent; the metastable $v = 1$ level of the nitrogen molecule is nearly resonant in energy with the 00^01 level of the carbon dioxide, and preferentially excites that level. An inversion is produced relative to the 10^00 level, and intense laser action at 10.6 μm is produced. If the 10^00 level is deactivated to

Table 10.3

Laser system	Principal output wavelength (nm)
F_2	157
ArF	193
KrF	249
XeCl	308
XeF	351

the ground state by radiation (via the 01^10 level), or, better, by deactivation by an added gas such as helium or by extremely rapid pumping of the plasma, high-power continuous emission can result. Lasers of this type, or chemically pumped hydrogen halide systems (see section 7.5) have been reported to operate at power levels in excess of 10^6 W. It has been suggested that these high-power lasers could be used as the "death ray" of science fiction (reference 25).

Molecular lasers, such as CO_2, are also partially tunable, since output can be obtained on any one of the rotational components of the 00^01–10^00 band. A number of molecular lasers other than CO_2 have also been found, including systems based on N_2O at 11 μm and on CO at 5 to 6 μm.

A somewhat different type of electric discharge gas laser operates on *excimer* transitions. An excimer is a molecular species that is bound in an electronically excited state, but repulsive or unstable in its electronic ground state. The excimer systems used in lasers are diatomic molecules consisting of a rare gas atom and a halogen atom; these possess a stable excited state, but fall apart as soon as the excitation is given up as a photon. This feature is ideal for a laser system, since inversion can always be maintained with respect to the lower laser level. These lasers are important because they provide output in the important ultraviolet region of the spectrum (see table 10.3), and also because they are scalable to high powers.

7.5 Hydrogen Halide "Chemical" Lasers

In the laser systems considered thus far, the population inversion is produced either by optical pumping or by an electric discharge. An important class of lasers derives energy from an exothermic chemical reaction; the most important member of this class is the hydrogen fluoride chemical laser. The reaction

$$F + H_2 \rightarrow HF(v) + H$$

is exothermic by 132 kJ mole, and a large fraction of this energy goes into

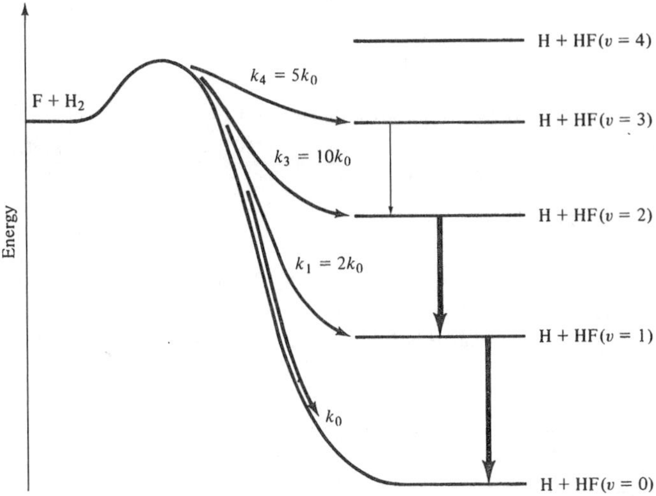

Figure 10.17
Energy versus reaction coordinate for $F + H_2 \rightarrow HF(v) + H$ reaction. Although the rate for producing $HF(v = 3)$ is less than that for $HF(v = 2)$, enough HF is produced in the $v = 3$ state to produce a partial inversion on some of the rotational components of the $3 \rightarrow 2$ band, and laser action is seen on these lines as well.

the vibration of the product HF. Indeed, twice as much HF is produced in the $v = 1$ state as in the $v = 0$ state, and five times as much in the $v = 2$ state as in the $v = 1$ state. Laser action is obtained on the $2 \rightarrow 1$ and $1 \rightarrow 0$ vibration-rotation bands at 2.5 to 3 μm (see figure 10.17). The conventional method of initiating the reaction is by an electric discharge in a mixture such as $SF_6 + H_2$, or by flash photodissociating a species such as ClF_3. It is possible, however, to operate a purely chemical laser in which the fluorine atoms are generated by decomposition of an unstable fluorine-containing compound. It is also possible to run this laser continuously, by rapidly pumping away the HF formed in the reaction. Other chemical lasers include the HCl and HBr systems and the CO system from the reaction of oxygen atoms with CS_2. An excellent source of detailed information on all of these systems is reference 26.

Considerable effort has gone into the search for a chemically pumped laser operating at visible, rather than infrared, wavelengths. To date, the only such system known is the chemically pumped oxygen-iodine system. Oxygen, produced in its metastable $^1\Delta$ state (see chapter 3.5) can transfer its energy to iodine atoms, that is,

$$O_2^*(a^1\Delta) + I(^2P_{3/2}) \rightarrow O_2(X^3\Sigma) + I^*(^2P_{1/2}).$$

The probability for this transfer is very large, since the excitation energies differ by only 280 cm^{-1}. The excited iodine atom can then lase on the weakly allowed spin-orbit transition at 1.3 μm.

7.6 Organic Dye Solution Lasers

The atomic and molecular gas lasers of the sort we have been considering have one major limitation. Although it is often possible to select one of a wide variety of discrete transitions in these lasers, they simply cannot be

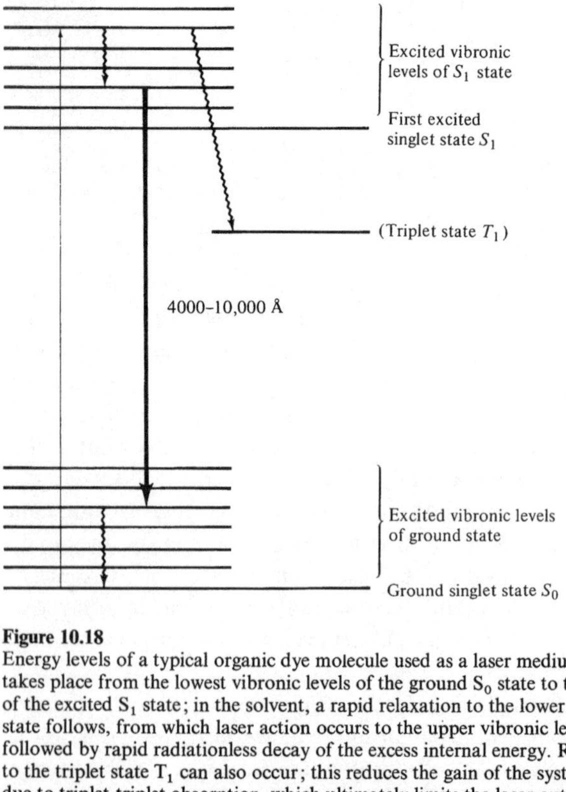

Figure 10.18
Energy levels of a typical organic dye molecule used as a laser medium. The optical pumping takes place from the lowest vibronic levels of the ground S_0 state to the upper vibronic levels of the excited S_1 state; in the solvent, a rapid relaxation to the lower vibronic levels of the S_1 state follows, from which laser action occurs to the upper vibronic levels of the S_0 state, again followed by rapid radiationless decay of the excess internal energy. Radiationless relaxation to the triplet state T_1 can also occur; this reduces the gain of the system by introducing a loss due to triplet-triplet absorption, which ultimately limits the laser output.

smoothly and continuously tuned over an appreciable range of output wavelength. One class of lasers, which are not so limited, is that of solutions of common organic dyes and scintillators. These lasers, first discovered by Sorokin in 1966 (reference 27), can be pumped by flash lamps, pulsed lasers, or by sufficiently intense (> 1 W) CW ion lasers. There are two noteworthy features of these lasers. First, they operate between two fully allowed transitions with no intervening metastable state, a feature that was thought to be impossible to realize until the accidental discovery that laser action was indeed occurring in the dyes. As figure 10.18 shows, the levels that are actually being pumped are distinct from those involved in the stimulated emission process, so that the dye system actually approximates the four-level neodymium system shown in figure 10.13b. Second, the output obtainable from these dyes is truly tunable over the entire visible and near-ultraviolet region; the fluorescence spectrum of any single dye, over which laser action can be obtained, is a band several hundred angstroms wide. In addition to the tunability of any one dye, a large number of dyes can be made to emit in this way, spanning the entire visible spectrum, as figure 10.19 indicates. It may be of some interest to note that encompassed in this class of lasers is the only known edible laser (reference 28). Stimulated emission has been observed from a dilute solution of fluorescein (a harmless food coloring) in unflavored gelatin, when the Jello-like substance is irradiated with a pulsed ultraviolet laser.

7.7 Tunable Semiconductor Lasers

An additional important class of broadly tunable lasers are the semiconductor devices, which emit in the infrared region of the spectrum. These carrier-recombination-based devices (reference 29) are commercially available, and we shall discuss them briefly here.

The nature of a recombination diode laser is sketched in figure 10.20. The current (typically 1 A) through a *P-N* junction causes electrons and holes to be raised to higher-energy states in the material; when the charge carriers recombine, infrared radiation is emitted. The optical cavity in this case is just the polished ends of the crystal itself. There are several ways of tuning such a device. First, the composition of the semiconductor determines roughly where it will emit. By varying the ratio x in compositions such as $Pb_{1-x}Sn_xTe$ and $Pb_{1-x}S_xSe$, output anywhere between 3 and 30 μm can be obtained. More usefully, as the current through the diode is varied by small amounts, the temperature of the diode (which must be mounted on a heat

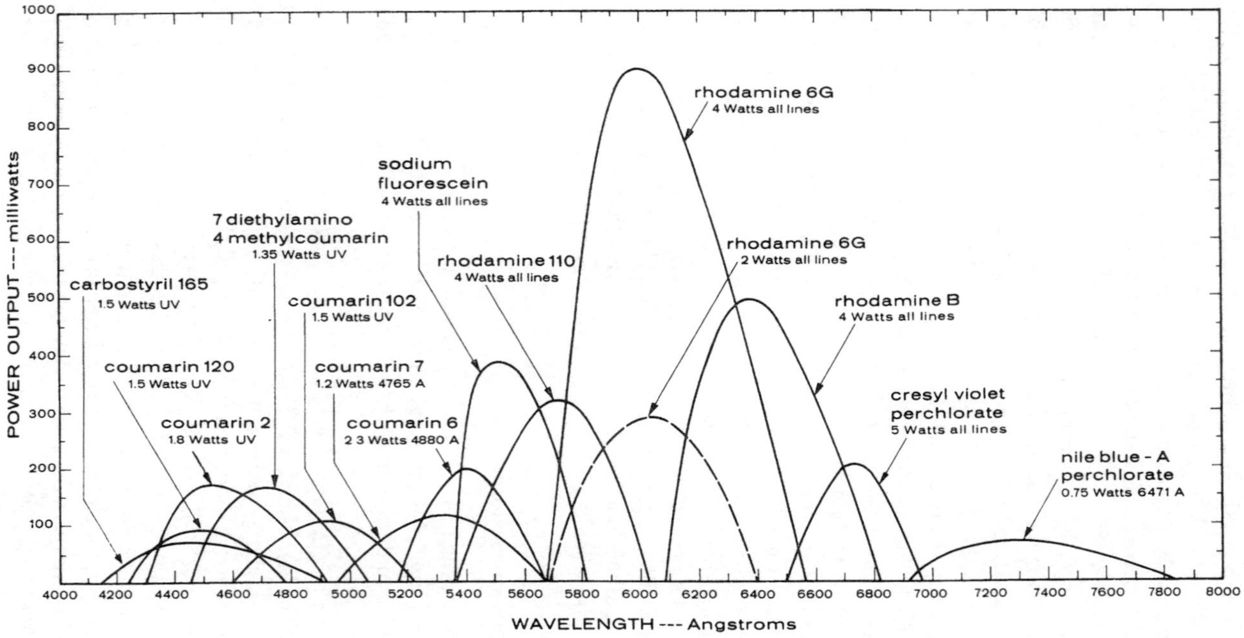

Figure 10.19
Tuning curves for an ion laser-pumped dye laser. Each curve corresponds to a different dye. Pump wavelength and power are indicated for each dye. [From Coherent, Inc., data sheet.]

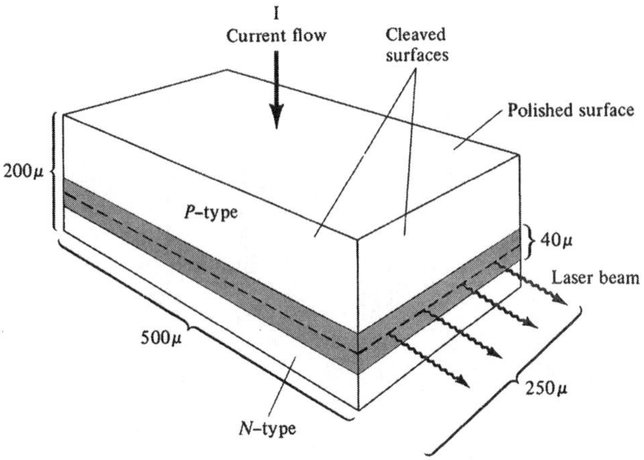

Figure 10.20
Diagram of a recombination-radiation diode laser. [From H. R. Schlossberg and P. L. Kelley, *Physics Today* **25**(7), 36 (July 1972). Reproduced with permission.]

sink) changes, thus changing the index of refraction of the material, and thus changing the exact value of the wavelength that satisfies the Fabry-Perot condition inside the cavity formed by the ends of the crystal. As the current is swept through a small range, the output wavelength is smoothly varied, until it jumps over to the next mode of the cavity. Since the crystal is usually 1 mm or less in length, there is a gap of several cm^{-1} from one mode to the next. These gaps can be bridged by physically changing the length of the crystal (that is, shaving a little off one end) to shift the cavity modes to a different wavelength. Needless to say, this is not a convenient, reversible method of tuning the laser. The output wavelength can also be varied by applying external pressure or a magnetic field to the diode.

The power obtainable from lasers of this type is small, on the order of 1 mW. Despite these drawbacks, there has been a great deal of interest in these devices because the spectral purity of the output is very sharp—of the order of 40 kHz, or about a millionth of a reciprocal centimeter. This narrow width permits one to carry out super-high-resolution absorption spectroscopy. An example is shown in figure 10.21; also refer to the spectrum of CDF_3 shown in figure 8.7. Other types of tunable lasers have been developed. These generally operate on stimulated Raman scattering, parametric amplification, or other nonlinear optical effects, which will be discussed in chapter 13.

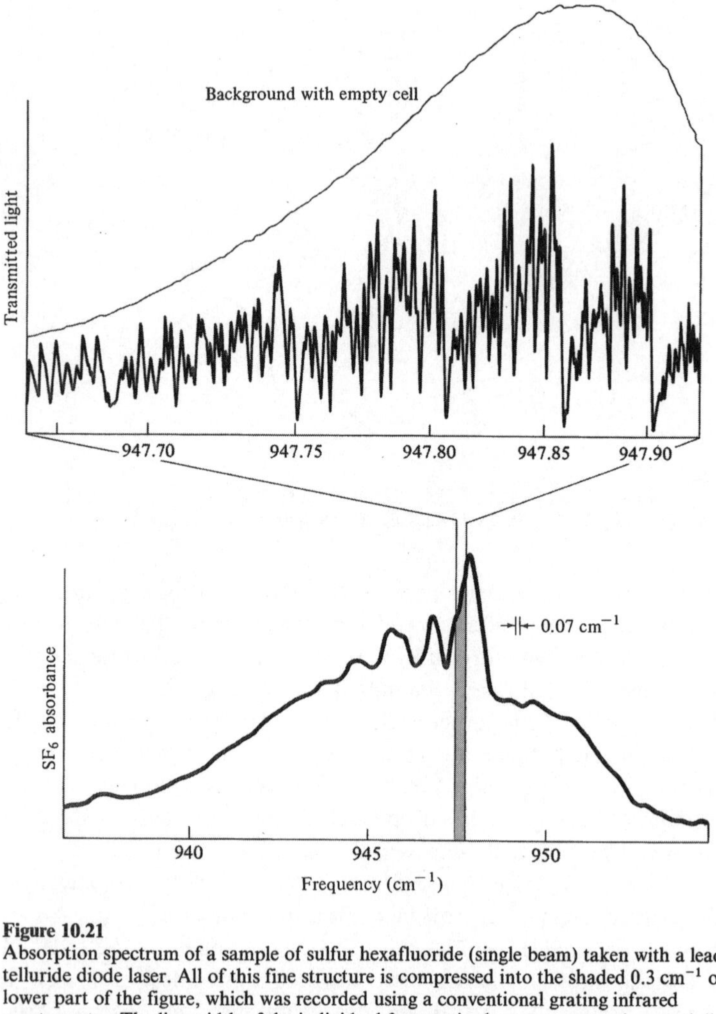

Figure 10.21
Absorption spectrum of a sample of sulfur hexafluoride (single beam) taken with a lead-tin-telluride diode laser. All of this fine structure is compressed into the shaded 0.3 cm^{-1} of the lower part of the figure, which was recorded using a conventional grating infrared spectrometer. The line width of the individual features in the top spectrum is essentially the Doppler width of the transition in the SF_6 molecule. [From H. R. Schlossberg and P. L. Kelley, *Physics Today* **25**(7), 36 (July 1972). Reproduced with permission.]

The whole field of optical masers is by now an enormous and continually developing one. What we have attempted in this section is merely to indicate the most important types of lasers that are known and have been used in the studies of coherent optical effects described in the following chapters. The reader interested in pursuing this field further is urged to begin with any of several introductory textbooks (references 31, 32), and then to go on to the advanced monographs and current literature described in appendix E.

References

1. H. Lefebvre-Brion and R. W. Field, *Perturbations in the Spectra of Diatomic Molecules* (Academic Press, New York, 1984).

2. P. G. Burke, W. B. Eissner, D. G. Hummer, and I. C. Percival, *Atoms in Astrophysics* (Plenum, New York, 1983).

3. A. G. Gaydon and H. G. Wolfhard, *Flames: Their Structure, Radiation, and Temperature*, 4th ed. (Chapman and Hall, London, 1979); D. R. Crosley, ed., *Laser Probes for Combustion Chemistry*, Symposium Series No. 134 (American Chemical Society, Washington, DC, 1980).

4. J. Brossel and A. Kastler, *Compt. Rend. Acad. Sci. (Paris)* **229**, 1213 (1949); A. Kastler, *J. Phys. Radium* **11**, 255 (1950); J. Brossel, A. Kastler, and J. Winter, *J. Phys. Radium* **13**, 668 (1952).

5. R. A. Bernheim, *Optical Pumping: An Introduction* (Benjamin, Menlo Park, CA, 1965).

6. G. V. Skrotskii and T. G. Iz'iumova, *Sov. Phys. Uspekhii* (English transl.) **4**, 177 (1961).

7. W. Happer, *Rev. Mod. Phys.* **44**, 169 (1972).

8. P. A. Franken, *Phys. Rev.* **121**, 508 (1961); R. N. Zare, *J. Chem. Phys.* **45**, 4510 (1966).

9. R. N. Zare, *Accts. Chem. Res.* **4**, 361 (1971).

10. N. F. Ramsey, *Molecular Beams* (Clarendon Press, Oxford, 1956).

11. A. Carrington and A. D. McLachlan, *Introduction to Magnetic Resonance* (Harper and Row, New York, 1967).

12. I. I. Rabi, S. Millman, P. Kusch, and J. R. Zacharias, *Phys. Rev.* **55**, 526 (1939).

13. W. E. Lamb, Jr., and R. C. Retherford, *Phys. Rev.* **79**, 549 (1950), **85**, 259 (1952), **86**, 1014 (1952); S. Triebwasser, E. S. Dayhoff, and W. E. Lamb, Jr., *Phys. Rev.* **89**, 98, 106 (1953).

14. A. Büchler, J. L. Stauffer, and W. Klemperer, *J. Am. Chem. Soc.* **86**, 4544 (1964).

15. R. E. Smalley, L. Wharton, and D. H. Levy, *J. Chem. Phys.* **63**, 4977 (1975).

16. P. S. H. Fitch, L. Wharton, and D. H. Levy, *J. Chem. Phys.* **69**, 3424 (1978).

17. D. Eastwood, L. Edwards, M. Gouterman, and J. I. Steinfeld, *J. Mol. Spectroscopy* **20**, 381 (1966).

18. J. Gordon, H. Zeiger, and C. H. Townes, *Phys. Rev.* **99**, 1264 (1955).

19. D. Kleppner, H. M. Goldenberg, and N. F. Ramsey, *Phys. Rev.* **126**, 603 (1962); N. F. Ramsey, *Am. Scientist* **56**, 420 (1968).

20. For a more detailed discussion of Brewster's angle, see G. R. Fowles, *Introduction to Modern Optics* (Holt, Rinehart and Winston, New York, 1967).

21. T. H. Maiman, *Nature* **187**, 493 (1960).

22. J. B. Koffend and R. W. Field, *J. Appl. Phys.* **48**, 4468 (1977).

23. A. Javan, W. R. Bennett, Jr., and D. R. Herriott, *Phys. Rev. Letters* **6**, 106 (1961).

24. C. K. N. Patel, *Phys. Rev.* **136A**, 1187 (1964); C. K. N. Patel, *Phys. Rev. Letters* **13**, 617 (1964).

25. "... Many think that in some way they (the Martians) are able to generate an intense heat in a chamber of practically absolute non-conductivity. This intense heat they project in a parallel beam against any object they choose, by means of a polished parabolic mirror of unknown composition .... However it is done, it is certain that a beam of heat is the essence of the matter .... Whatever is combustible flashes into flame at its touch, lead runs like water, it softens iron, cracks and melts glass, and when it falls upon water, incontinently that explodes into steam." H. G. Wells, *The War of the Worlds* (Heinemann, London, 1898), chapter 6.

26. R. W. F. Gross and J. F. Bott, *Handbook of Chemical Lasers* (Wiley-Interscience, New York, 1976).

27. P. P. Sorokin and J. R. Lankard, *IBM J. Res. Development* **10**, 162 (1966); P. P. Sorokin, J. R. Lankard, E. C. Hammond, and V. L. Moruzzi, *ibid.*, **11**, 130 (1967).

28. T. W. Hänsch, M. Permier, and A. L. Schawlow, *IEEE J. Quantum Electronics* **QE-8**, 45 (1971).

29. E. D. Hinkley, T. C. Harman, and C. Freed, *Appl. Phys. Letters* **13**, 49 (1968); E. D. Hinkley and C. Freed, *Phys. Rev. Letters* **23**, 277 (1969); E. D. Hinkley and P. L. Kelley, *Sci.* **171**, 635 (1971).

30. H. R. Schlossberg and P. L. Kelley, *Phys. Today* **25**(7), 36 (July 1972).

31. B. A. Lengyel, *Introduction to Laser Physics* (Wiley, New York, 1967); 2nd ed. (Wiley, New York), 1971.

32. C. G. B. Garrett, *Gas Lasers* (McGraw-Hill, New York, 1967).

11 Optical Resonance Spectroscopy

1 The Idealized Two-Level System

In the preceding chapter, we have surveyed some of the experimental conditions under which nonequilibrium systems may be prepared by optical pumping and coherent optical radiation may be generated. We now wish to proceed to a mathematical analysis of these phenomena. In order to carry out such an analysis, we begin with a consideration of an idealized system consisting of two nondegenerate quantum states interacting with a radiation field; this model is referred to as the "two-level system." There are a number of reasons for investigating this model in detail:

1. It is an exactly soluble model; we shall derive the general solutions later in this chapter.
2. The solutions are mathematically isomorphous with those for several important classical physics problems, such as the forced damped harmonic oscillator; the equations of motion are also the same as those encountered in nuclear magnetic resonance.
3. Finally, the two-level model is a reasonable approximation for many real atomic and even molecular systems, some of which are given in table 11.1.

In this table, only the first system (spin $\frac{1}{2}$ system in a magnetic field) can be rigorously described as a nondegenerate two-level system; this is the system typically considered in simple nuclear magnetic resonance problems. In the $F = 1 \leftarrow F = 0$ transition in the hydrogen atom (which is the basis of the H atom maser considered in the preceding chapter), the upper state is triply degenerate. The so-called inversion doublets of ammonia (see chapter 7.3.4) include the Zeeman degeneracy of the rotational levels, and the two listed optical transitions include spin degeneracy. In the Na atom, for example, the D_1 ($^2P_{1/2}$–$^2S_{1/2}$) line is actually split into a pair of doublets by the hyperfine interactions in the upper and lower states, having splittings of 1,772 and 189 MHz, respectively. Even if only a single hyperfine component were accessed, say by using a single-mode dye laser, the Zeeman degeneracy ($g = 5$ for $F = 2$, $g = 3$ for $F = 1$) would still be unresolved. Nevertheless, all of the above systems have been treated quite successfully as simple two-level atoms. We shall now proceed to show the details of this treatment, which forms the basis for the analysis of a wide variety of coherent optical phenomena.

Table 11.1

System	Quantum levels	Upper state	Lower state	Energy splitting
H atom in 14-kG magnetic field	Nuclear spin ($I = \frac{1}{2}$) projection on field axis	$m_I = +\frac{1}{2}$	$m_I = -\frac{1}{2}$	56.610 MHz
H atom in zero field	Hyperfine levels ($\mathbf{F} = \mathbf{I} + \mathbf{J}$)	$F = 1$	$F = 0$	1,420.4057518 MHz
NH_3 molecule	Inversion levels	+	−	23,232.24 MHz (in $J = 8$, $K = 7$)
Ruby crystal (Cr^{+++} ion in Al_2O_3 lattice)	Crystal field states	^2E_g	$^4A_{2g}$	14,400 cm^{-1}
Na atom	Electronic states (neglecting hyperfine structure)	$3\,^2P_{1/2}$	$3\,^2S_{1/2}$	16,960.8 cm^{-1}

2 The Rabi Solution for a Two-Level System

The semiclassical Hamiltonian for a nondegenerate two-level system interacting with an electromagnetic field $\mathscr{E}$ via an electric-dipole interaction is

$$\mathscr{H} = \mathscr{H}_0 - \mu_{ab}\mathscr{E}\cos\omega t \tag{11.1a}$$

$$= \mathscr{H}_0 - \mu_{ab}\frac{\mathscr{E}}{2}(e^{i\omega t} + e^{-i\omega t}), \tag{11.1b}$$

where μ_{ab} is the transition dipole moment and ω the angular frequency of the radiation. $E_a = \hbar\omega_a$ is the energy of the lower state and $E_b = \hbar\omega_b$, that of the upper state, so that the energy difference $\omega_0 = \omega_b - \omega_a$ is the resonant frequency for transition between the two states. The time-dependent wave function for this two-level system can be written as

$$\Psi(t) = a(t)e^{-i\omega_a t}\psi_a + b(t)e^{-i\omega_b t}\psi_b, \tag{11.2}$$

where $a(t)$ is the coefficient of the lower state with basis function ψ_a and $b(t)$ is the coefficient of the upper state with the basis function ψ_b. The equation of motion for these coefficients is easily found to be

$$\frac{d}{dt}a(t) = \frac{i}{2}\kappa\mathscr{E}b(t)\left[e^{i(\omega-\omega_0)t} + e^{-i(\omega+\omega_0)t}\right]$$

and

$$\frac{d}{dt}b(t) = \frac{i}{2}\kappa\mathscr{E}a(t)\left[e^{-i(\omega-\omega_0)t} + e^{i(\omega+\omega_0)t}\right],$$

with $\kappa = \mu_{ab}/\hbar$, so that $\kappa\mathscr{E} = V_{ab}^0/\hbar$, the matrix element introduced in chapter 1.5. We can make the rotating wave approximation (RWA) to the preceding equations by dropping out the antiresonant terms, that is, the terms oscillating with a frequency $\omega + \omega_0$; this will be a satisfactory approximation so long as $\mu_{ab}\mathscr{E}/\hbar \ll \omega_0$. With the RWA, the two first-order equations for $a(t)$ and $b(t)$ can be combined into a single second-order differential equation,

$$\frac{d^2}{dt^2}b(t) + i(\omega - \omega_0)\frac{d}{dt}b(t) + \frac{(\kappa\mathscr{E})^2}{4}b(t) = 0. \tag{11.3}$$

Note that (11.3) has the same form as that for a harmonic oscillator driven off-resonance.

The solution to (11.3) is just

$$b(t) = e^{-i\Delta t/2}(Ae^{i\Omega t/2} + Be^{-i\Omega t/2}),$$

with $\Delta = \omega - \omega_0$ and $\Omega = [\Delta^2 + (\kappa\mathscr{E})^2]^{1/2}$. This solution can then be substituted in the equation for $da(t)/dt$ to give

$$a(t) = \frac{1}{\kappa\mathscr{E}} e^{i\Delta t/2}[(\Delta - \Omega)Ae^{i\Omega t/2} + (\Delta + \Omega)Be^{-i\Omega t/2}].$$

For either two first-order or one second-order differential equation, we need to specify two independent initial conditions in order to determine the integration constants A and B. These conditions may be written as

$$a(t_0) = e^{i\theta} \qquad \text{(i.e., } 1 \times \text{a phase factor)},$$

$$b(t_0) = 0.$$

The physical content of these initial conditions is that at time t_0, all of the population is in the ground state of the system and none in the excited state. With these conditions, the solutions for $a(t)$ and $b(t)$ are

$$a(t) = e^{i\theta + i\Delta(t - t_0)/2}\left[\cos\frac{\Omega}{2}(t - t_0) - i\frac{\Delta}{\Omega}\sin\frac{\Omega}{2}(t - t_0)\right] \tag{11.4}$$

and

$$b(t) = i\frac{\kappa\mathscr{E}}{\Omega}e^{i\theta + i\Delta(t - t_0)/2}\sin\frac{\Omega}{2}(t - t_0). \tag{11.5}$$

The time-dependent populations in each level may be found by taking the squared amplitudes of each of the coefficients:

$$n_a(t) = |a(t)|^2 = \frac{\Delta^2}{\Omega^2} + \frac{(\kappa\mathscr{E})^2}{\Omega^2}\cos^2\frac{\Omega}{2}(t - t_0), \tag{11.6}$$

$$n_b(t) = |b(t)|^2 = \frac{(\kappa\mathscr{E})^2}{\Omega^2}\sin^2\frac{\Omega}{2}(t - t_0). \tag{11.7}$$

These expressions are valid for any arbitrary frequency ω of the applied radiation field and can be used to find the response of the system away from exact resonance. For the exact resonance case in which $\omega = \omega_0 = \omega_b - \omega_a$, we have $\Delta = 0$, $\Omega = \kappa\mathscr{E} = \mu_{ab}\mathscr{E}/\hbar$. If we also let $t_0 = 0$, then the preceding equations for the populations reduce to the very simple form

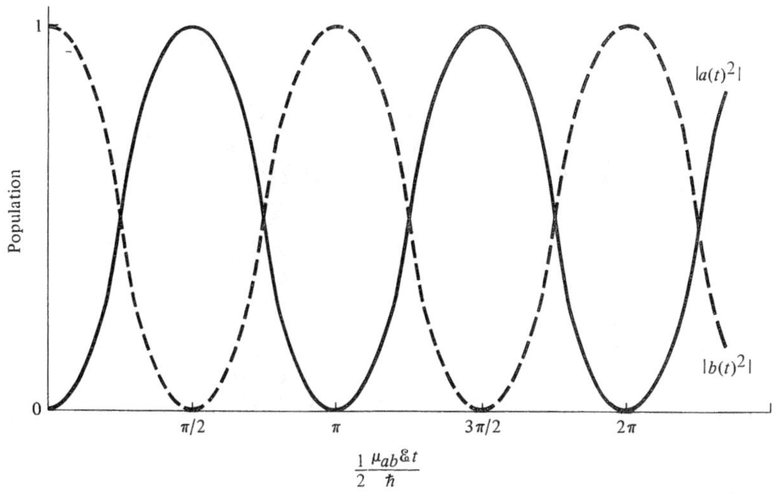

Figure 11.1
The probability that a two-level system will be in either the ground or the upper state, when driven by an oscillating field at the resonant frequency.

$$n_a(t) = \cos^2\left(\frac{\kappa\mathscr{E}}{2}t\right) = \cos^2\left(\frac{\mu_{ab}\mathscr{E}t}{2\hbar}\right),$$

$$n_b(t) = \sin^2\left(\frac{\kappa\mathscr{E}}{2}t\right) = \sin^2\left(\frac{\mu_{ab}\mathscr{E}t}{2\hbar}\right).$$

It is obvious from these equations that the normalized total population $N = n_a(t) + n_b(t)$ remains equal to one at all times. The oscillatory behavior of the ground and excited state populations is shown in figure 11.1. Note that when $\mu_{ab}\mathscr{E}t/\hbar = \pi$, the population is 100% *inverted*; that is, all of the molecules are in the upper state $|b\rangle$, and none are in the ground state $|a\rangle$. When this quantity is equal to 2π, the original configuration is restored, that is, all the population is once again in the ground state. For this reason, the quantity $\mu_{ab}\mathscr{E}/\hbar$ is termed the *Rabi frequency* ω'; roughly, this is the rate at which the system is driven between its upper and lower states by the resonant radiation field. This behavior is also seen if the field amplitude $\mathscr{E}(t)$ is slowly varying—in this case, we can define a *pulse area*

$$\theta(t) = \int_{-\infty}^{t} \kappa\mathscr{E}(t')\,dt'. \tag{11.8}$$

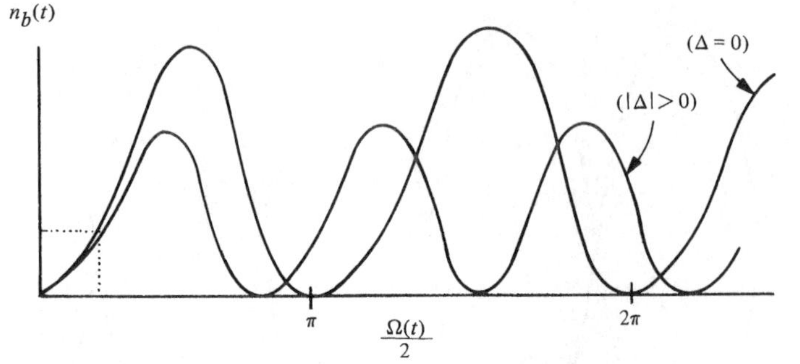

Figure 11.2
Time evolution of the population of the excited state of a two-level system subject to a coherent dipole perturbation. Note that on-resonance pumping ($\omega = 0$) results in the slowest oscillations having the greatest amplitude.

When $\theta(t) = \pi$, the system is inverted at time t, and the radiation pulse is termed a "π pulse." When $\theta(t) = 2\pi$ (a 2π pulse), the system is restored. What about nonintegral values of π, such as $\theta(t) = \pi/2$? Such a pulse leaves $n_a(t) = n_b(t) = 0.5$, but also introduces *coherence* between the two states; a proper treatment of this situation requires the use of the *pseudospin representation*, to be introduced in section 4.

In the case of detuning ($\Delta \neq 0$), we may introduce the generalized Rabi frequency, $\Omega = [\Delta^2 + (\kappa\mathscr{E})^2]^{1/2}$. The comparison between resonant and off-resonant pumping is shown in figure 11.2. Note that (i) the system never reaches 100% inversion if $\Delta \neq 0$, (ii) the Rabi oscillations are faster when $\Delta \neq 0$, and (iii) the region enclosed in the dashed box in figure 11.2 at small t corresponds to the perturbation theory result $P_{a \to b}(t) = \mu_{ab}^2 \mathscr{E}^2 t^2 / 4\hbar^2$ (chapter 1.5).

A nice example of this behavior is provided by the molecular beam resonance experiments on HCN first mentioned in chapter 1.5. There, the oscillatory behavior of the transition probability was used to demonstrate the unaveraged $\sin^2(\omega t)$ behavior predicted by the first-order time-dependent perburbation theory introduced for the weak field limit. From figure 11.3, though, we see that these oscillations are intensity dependent as well. In figure 11.3a, the radio frequency field intensity is adjusted so that $b(t) = 1$ and $a(t) = 0$ when the molecules emerge from the "C field" region, giving a strong resonance at $\omega = \omega_2 - \omega_1$. But when twice the field amplitude is applied, as in figure 11.3b, the molecules all go from state 1 to state 2 and back to state 1, so that $b = 0$ and $a = 1$ at $t = l/v_0$. It is then necessary

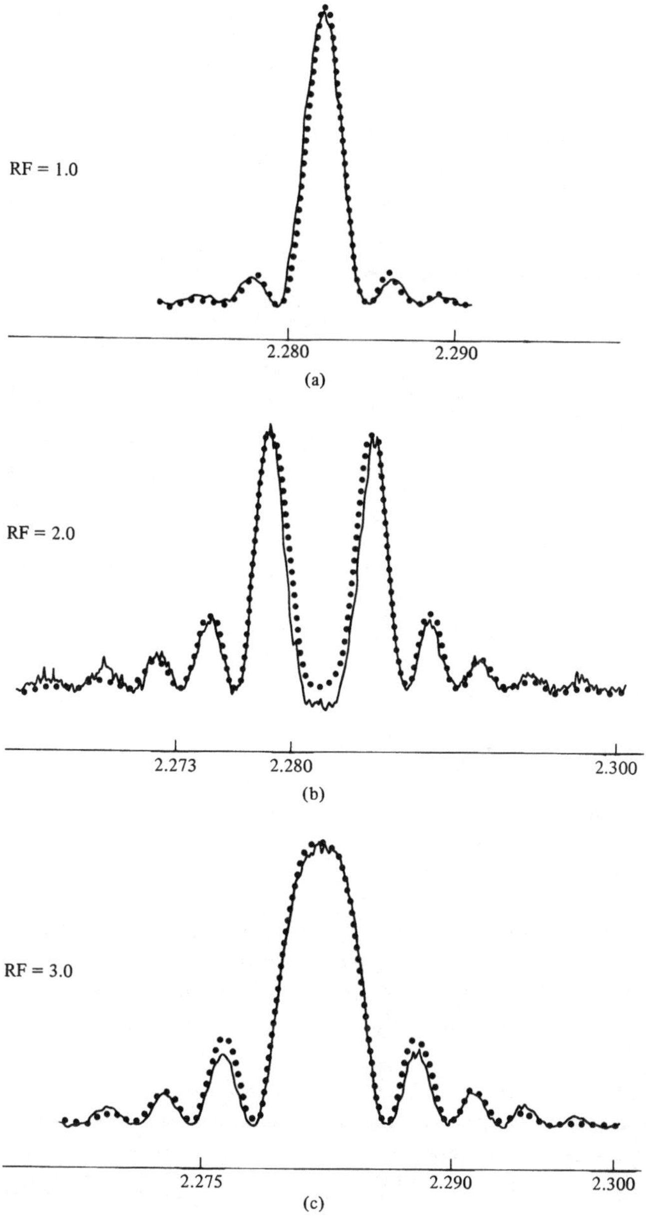

Figure 11.3
Molecular beam electric resonance curves for HCN; $J = 1$, $\Delta M_J = 1$, $\Delta M_F = 0$. Horizontal axis is in MHz. The solid lines are the experimental spectra. The dotted lines are calculated for $v_0 = (8 \times 10^4 \text{ cm sec}^{-1} \pm 10\%)$ and various values of the radio frequency field amplitude $\mathscr{E}_0$: (a) $\mathscr{E}_0 = \pi\hbar/\mu_{12}t$ (this is just the same as figure 1.3); (b) $\mathscr{E}_0 = 2\pi\hbar/\mu_{12}t$; (c) $\mathscr{E}_0 = 3\pi\hbar/\mu_{12}t$. [From T. R. Dyke, G. R. Tomasevich, W. Klemperer, and W. Falconer, *J. Chem. Phys.* **57**, 2277 (1972). Reproduced with permission.]

to go *off-resonance* by $\Delta = \pm\omega'$ to obtain a maximum in $P_{ab}(t)$, as the figure shows. The shape of the triple-amplitude resonance curve, shown in figure 11.3c, follows as a consequence of the above considerations.

3 Saturation in a Two-Level System

Thus far, we have considered a two-level system in which the only processes that can occur are radiative transitions between the upper and lower levels. But suppose that the upper state can decay with a time constant $1/\gamma = T_2'$; this decay could result from spontaneous radiation, collisional relaxation, or other processes. The number of molecules decaying per unit time is then given by a Poisson distribution,

$$dn(t) = \frac{1}{T_2'} e^{-(t-t_0)/T_2'}\, dt.$$

The time-averaged population in the upper state is thus

$$\begin{aligned} \langle |b(t)|^2 \rangle_{\text{avg}} &= \frac{1}{T_2'} \int_{-\infty}^{t} |b(t,t_0)|^2 e^{-(t-t_0)/T_2'}\, dt_0 \\ &= \frac{(\kappa\mathscr{E})^2}{\Omega^2 T_2'} \int_{-\infty}^{t} \sin^2 \frac{\Omega}{2}(t-t_0) e^{-(t-t_0)/T_2'}\, dt_0 \\ &= \frac{1}{2} \frac{(\kappa\mathscr{E})^2}{(\omega-\omega_0)^2 + (1/T_2')^2 + (\kappa\mathscr{E})^2}. \end{aligned} \tag{11.9}$$

This function is again a Lorentzian, as discussed in chapter 1.8, but with an additional width due to *power broadening*. Note that $\lim_{\mathscr{E}\to\infty} \langle |b(t)|^2 \rangle_{\text{avg}} = 0.5$; The upper- and lower-state populations are *equalized* at high incident power. This implies that the absorption at frequency ω *saturates*, as the following will show.

The rate at which power is absorbed from the radiation field can be found by multiplying the time-averaged upper-state population given by (11.9) by the number of absorbers, $n_a - n_b$, the energy per photon absorbed, $\hbar\omega$, and the rate at which energy is lost to the surroundings, $1/T_2'$. This gives

$$\Delta P = \frac{dW}{dt} = \frac{n_a - n_b}{T_2'} \frac{\hbar\omega(\kappa\mathscr{E})^2}{(\omega-\omega_0)^2 + (1/T_2')^2 + (\kappa\mathscr{E})^2}.$$

The absorption coefficient for a short optical path can be written as

$$\begin{aligned}\alpha &= \frac{\Delta P}{P_{\text{incident}}}\\ &= \frac{\Delta P}{c|\mathscr{E}|^2/4\pi} \qquad (11.10)\\ &= \frac{n_a - n_b}{T_2'} \frac{2\pi\mu_{ab}^2\omega}{c\hbar[(\omega-\omega_0)^2 + (1/T_2')^2 + (\kappa\mathscr{E})^2]}.\end{aligned}$$

Clearly, as $\mathscr{E} \to \infty$, the amount of power absorbed by the system remains constant, and the absorption coefficient α given by (11.10) tends to zero. The sample thus appears to be nearly transparent at high incident intensities. Phenomenologically, the absorption coefficient can be written as

$$\alpha = \frac{\alpha_0}{1 + I/I_{\text{sat}}},$$

where I_{sat} is the value of the intensity at which the absorption coefficient falls to half of its small-signal value, α_0. In terms of molecular quantities,

$$I_{\text{sat}} = \frac{c\hbar^2}{4\pi(T_2')^2\mu_{ab}^2} \qquad (11.11)$$

(see problem 3). Saturation is well known in radio frequency spectroscopy, where relaxation times are long, and it is easy to obtain high power levels in very narrow bandwidths from electronic oscillators. Saturation at optical frequencies generally requires the use of lasers to provide the necessary spectral brightness; some examples are shown in figure 11.4. The much higher value of I_{sat} required for the liquid solution in figure 11.4b as compared with the low-pressure gas in figure 11.4a reflects primarily the difference in relaxation times T_2'; note that in the liquid, I_{sat} is independent of concentration, while I_{sat} increases as the pressure of the gas is raised. This is because T_2' in liquids is determined primarily by molecule-solvent interactions, while in the gas, T_2' is determined by collisions between the gas molecules, and is thus inversely proportional to the pressure.

Let us also consider briefly the power *radiated* (instead of absorbed) from a coherently prepared system. In this case, the power radiated is given by

$$P_{\text{out}} = \tfrac{1}{2}(n_b - n_a)\omega\mu_{ab}\tilde{\mathscr{E}}_0,$$

where the "self-field" $\tilde{\mathscr{E}}_0$ is

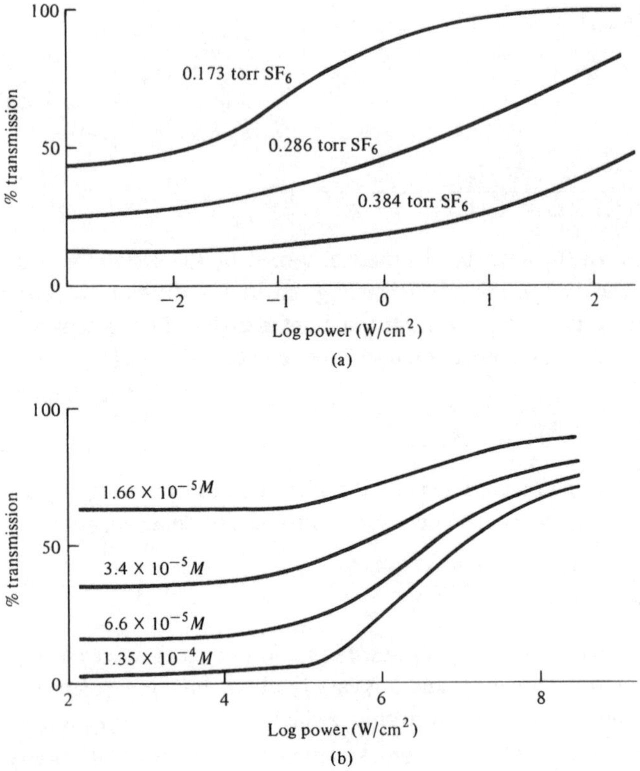

Figure 11.4
Saturation of absorption by high-power incident energy. (a) Saturation by a carbon dioxide infrared laser in sulfur hexafluoride vapor. [From I. Burak, J. I. Steinfeld, and D. G. Sutton, *J. Quant. Spectrosc. and Radiat. Transfer* **9**, 959 (1969). Reproduced with permission.] (b) Saturation by a pulsed ruby laser in cryptocyanine solutions. [From C. R. Giuliano and L. D. Hess, *I.E.E.E. J. Quantum Electronics* **QE-3**, 358 (1967). Reproduced with permission.] Note that approximately six orders of magnitude higher power is required for the onset of saturation in a liquid solution as compared with a low-pressure gas; this just reflects the difference in the relaxation time T_2' in equation (11.11).

$$\tilde{\mathscr{E}}_0 = (8\pi Q P_{out}/\omega)^{1/2}$$

(see reference 3 for details). Q is the "quality factor" of an appropriate cavity or enclosure containing the system, equal to the ratio of power stored to power radiated. If we substitute the expression for $\tilde{\mathscr{E}}_0$ into the equation for P_{out} and solve for P_{out}, we find

$$P_{out} = 2\pi Q(n_b - n_a)^2 \omega \mu_{ab}^2 \tag{11.12}$$

so long as $n_b > n_a$. This radiation from a coherently prepared ensemble, frequently termed *superradiant* emission, is characterized by being proportional to the *square* of the inversion density $n_b - n_a$; normal spontaneous emission is simply proportional to the first power of the density of excited molecules. Laser systems, such as were described in the preceding chapter, can be thought of as superradiant systems coupled to an appropriate feedback mechanism. We shall return to the topic of saturation in the following chapter.

4 Feynman-Vernon-Hellwarth Theorem

In section 2, we found the explicit Rabi solutions for a two-level system driven by an electromagnetic field. Here, we wish to develop a geometric analogue of these solutions, introduced initially by Feynman, Vernon, and Hellwarth in 1957 (reference 4). While no new physical content is introduced, this model provides additional insight helpful in understanding a variety of coherent optical effects. More extended discussions of this topic can also be found in references 5 and 6.

The Rabi solutions were found by analyzing the Schrödinger equation for a two-level system. The geometric model can be derived from either the Schrödinger or the equivalent Heisenberg equation of motion for the system (see problem 4), but follows more naturally from the latter. This equation gives the time development of the expectation value of an operator $\mathcal{O}$ as

$$\frac{d}{dt}\langle \mathcal{O} \rangle = \frac{i}{\hbar}[\mathscr{H}, \mathcal{O}] = \frac{i}{\hbar}(\mathscr{H}\mathcal{O} - \mathcal{O}\mathscr{H}) \tag{11.13}$$

using the commutator algebra introduced in (1.2). Let the operator $\mathcal{O}$ be the one-particle density matrix $\mathbf{p}$:

$$\mathbf{p} = \begin{pmatrix} bb^* & ba^* \\ ab^* & aa^* \end{pmatrix}, \tag{11.14}$$

where a and b are the coefficients of the lower and upper states, respectively. Of course, the matrix $\mathbf{p}$ contains exactly the same information as the wave function in (11.2). The Hamiltonian $\mathscr{H}$ we take to be the same as already introduced in (11.1), namely, $\mathscr{H} = \mathscr{H}_0 + \mathscr{H}'(t)$. $\mathscr{H}_0$ is diagonal, with matrix elements E_a and E_b. The time-dependent term $\mathscr{H}'(t)$ possesses only off-diagonal matrix elements V_{ab} and V_{ba}, which will turn out to be just $\mu_{ab}\mathscr{E}$. Thus the matrix representation of $\mathscr{H}$ is

$$\mathscr{H} = \begin{pmatrix} E_b & V_{ba} \\ V_{ab} & E_a \end{pmatrix}. \tag{11.15}$$

The matrices (11.14) and (11.15) are best represented in terms of the *Pauli spin matrices* (reference 7), which are the irreducible representations of a two-level system. These are

$$\sigma_1 = \frac{1}{2}\begin{pmatrix} 0 & 1 \\ 1 & 0 \end{pmatrix}, \qquad \sigma_2 = \frac{1}{2}\begin{pmatrix} 0 & -i \\ i & 0 \end{pmatrix}, \qquad \sigma_3 = \frac{1}{2}\begin{pmatrix} 1 & 0 \\ 0 & -1 \end{pmatrix}.$$

Note that the Pauli spin matrices obey simple commutation relations,

$$[\sigma_1, \sigma_2] = \frac{i}{2}\sigma_3,$$

with corresponding expressions for cyclic permutations of the subscripts 1, 2, and 3. Also,

$$\sigma_1^2 + \sigma_2^2 + \sigma_3^2 = \frac{3}{4}\begin{pmatrix} 1 & 0 \\ 0 & 1 \end{pmatrix} = \tfrac{3}{4}\mathbf{1}.$$

In terms of the Pauli spin matrices, we can write the Hamiltonian operator as

$$\hbar^{-1}\mathscr{H} = \left(\frac{V_{ab} + V_{ba}}{\hbar}\right)\sigma_1 + \left(\frac{V_{ab} - V_{ba}}{\hbar}\right)i\sigma_2 + \left(\frac{E_b - E_a}{\hbar}\right)\sigma_3 + \left(\frac{E_b + E_a}{\hbar}\right)\mathbf{1}. \tag{11.16}$$

We now let $E_b = \frac{1}{2}\hbar\omega_0$ and $E_a = -\frac{1}{2}\hbar\omega_0$, so that (11.16) simplifies to

$$\hbar^{-1}\mathscr{H} = \omega_1\sigma_1 + \omega_2\sigma_2 + \omega_0\sigma_3,$$

with $\omega_0 = (E_b - E_a)/\hbar$ and ω_1 and ω_2 simply being the coefficients of σ_1 and σ_2, respectively, in (11.16). If we further let $V_{ab} = \mu_{ab}\mathscr{E}$ and substitute this form for $\mathscr{H}$ into the Heisenberg equation of motion (11.13), we obtain separate equations of motion for the expectation values of the individual

Pauli spin matrices,[1]

$$\langle\dot{\sigma}_1\rangle = -\omega_0\sigma_2 + \frac{(\mu_{ab} - \mu_{ba})\mathscr{E}}{\hbar}\sigma_3, \tag{11.17a}$$

$$\langle\dot{\sigma}_2\rangle = +\omega_0\sigma_1 + \frac{(\mu_{ab} + \mu_{ba})\mathscr{E}}{\hbar}\sigma_3, \tag{11.17b}$$

$$\langle\dot{\sigma}_3\rangle = \frac{(\mu_{ab} + \mu_{ba})\mathscr{E}}{\hbar}\sigma_2 - \frac{(\mu_{ab} - \mu_{ba})\mathscr{E}}{\hbar}\sigma_1. \tag{11.17c}$$

We now write out explicitly the expectation values of the σ_i in the basis $|\psi\rangle = \binom{b}{a}$. These are

$$\tfrac{1}{2}r_1 = \langle\psi|\sigma_1|\psi\rangle = \tfrac{1}{2}(a^*b + ab^*), \tag{11.18a}$$

$$\tfrac{1}{2}r_2 = \langle\psi|\sigma_2|\psi\rangle = \frac{i}{2}(a^*b - ab^*), \tag{11.18b}$$

$$\tfrac{1}{2}r_3 = \langle\psi|\sigma_3|\psi\rangle = \tfrac{1}{2}(bb^* - aa^*). \tag{11.18c}$$

Note that $r_1^2 + r_2^2 + r_3^2 = (aa^* + bb^*)^2 = 1$. Further, since $\boldsymbol{\mu}$ is a symmetric operator, $\mu_{ab} = \mu_{ba}$, so that

$$\frac{(\mu_{ab} + \mu_{ba})\mathscr{E}}{2\hbar} = \kappa\mathscr{E} \qquad \text{and} \qquad \frac{(\mu_{ab} - \mu_{ba})\mathscr{E}}{2\hbar} = 0,$$

as we introduced in section 2. The equations of motion (11.17) then simplify to

$$\begin{aligned} \dot{r}_1 &= -\omega_0 r_2, \\ \dot{r}_2 &= +\omega_0 r_1 + \kappa\mathscr{E} r_3, \\ \dot{r}_3 &= -\kappa\mathscr{E} r_2 \end{aligned} \tag{11.19}$$

If we define a vector

$$\boldsymbol{\Omega} = \begin{pmatrix} -\kappa\mathscr{E} \\ 0 \\ \omega_0 \end{pmatrix},$$

1. In these equations, the electric field amplitude $\mathscr{E}$ is taken to be a physical quantity, not a quantum mechanical operator. In essence, this amounts to treating the field semiclassically.

then the preceding equations can be written compactly as a single vector equation,

$$\frac{d\mathbf{r}}{dt} = \mathbf{\Omega} \times \mathbf{r}. \tag{11.20}$$

This equation is the Feynman-Vernon-Hellwarth result. Its significance is that (11.20) has precisely the same form as the equation of motion for a spin angular momentum placed in a magnetic field, if we replace the $\mathbf{r}$ vector by the magnetic moment $\boldsymbol{\mu}$ and the $\mathbf{\Omega}$ vector by $-\gamma\mathbf{B}$, where γ is the gyromagnetic ratio and $\mathbf{B}$ is the magnetic field vector. These analogies will be considered further a little later in this chapter. Several things should be emphasized at this point, however. First, the vector equation (11.20) or the equivalent component equation (11.19) contains no new physical content, other than that already embodied in the Rabi solutions previously derived. Its usefulness lies in guiding our thinking about coherent optical processes, especially since a great deal of preexisting work in nuclear spin resonance spectroscopy can be used as the basis for predicting new coherent optical effects. Second, we must remember that, while magnetic spin moments and fields exist in real three-dimensional space, the vector quantities in (11.20) do *not* correspond to actual physical objects. The relations between $\mathbf{r}$ and $\mathbf{\Omega}$, on the one hand, and experimentally measurable quantities, on the other, will be developed as various coherent optical phenomena are discussed in the following chapter. First, however, we wish to put the set of equations represented by (11.20) in a more convenient and general form.

5 Optical Bloch Equations

The vector precession equation (11.20) describes a two-level system with energies E_a, $E_b = E_a + \hbar\omega_0$ driven by an electromagnetic field $\mathscr{E}(t) = \mathscr{E}_0 \cos\omega t = \frac{1}{2}\mathscr{E}_0(e^{i\omega t} + e^{-i\omega t})$. The frequency of the applied field, ω, is approximately equal to the resonant frequency of the two-level system, ω_0, but ω_0 is much larger than the Rabi frequency $\omega' = \kappa\mathscr{E}_0$.[2] The motion of the $\mathbf{r}$ vector in (123) space is determined by the $\mathbf{\Omega}$ vector; since $\omega_0 \gg \kappa\mathscr{E}_0$, the motion of the $\mathbf{r}$ vector would appear to be affected only slightly by the presence of the field. Since it is precisely the interaction of the system with

2. Optical frequencies are in the range 10^{13}–10^{15} sec^{-1}, while the Rabi frequency is typically on the order of 10^8 sec^{-1}, as in laser fields of 10^7–10^9 W/cm^2.

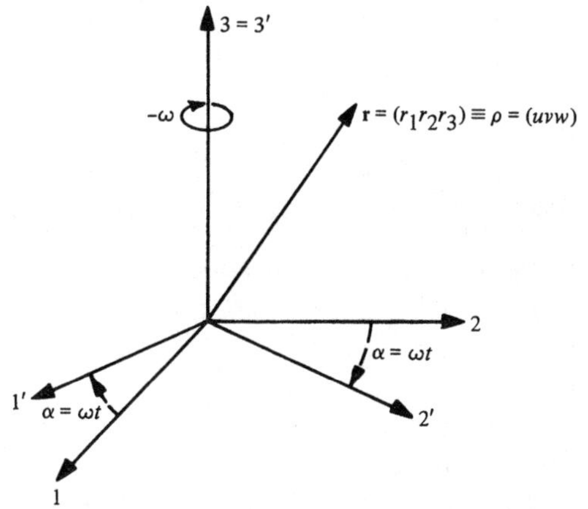

Figure 11.5
After a time t at frequency $-\omega$, coordinate system $1'2'3'$ is rotated through an angle α in the (12) plane, and the vector $\mathbf{r}$ with components $(r_1 r_2 r_3)$ is transformed into the vector $\boldsymbol{\rho}$ with components (uvw).

the electromagnetic field in which we are interested, we wish to reanalyze the system in such a way as to eliminate the motion of $\mathbf{r}$ associated solely with the field-free Hamiltonian $\mathscr{H}_0$. This can be done by transforming to a coordinate system counterrotating about the 3 axis at the frequency of the applied field, ω, as is shown in figure 11.5. As we shall see, this has the effect of eliminating the component of the motion at the frequency ω; if $\omega = \omega_0$, that is, the field is exactly on resonance, then the motion is governed solely by the Rabi frequency ω'.

The operator for rotation through an angle α is $e^{i\alpha} = e^{-i\omega t}$. The vector $\boldsymbol{\rho}$ in the rotating coordinate system (see figure 11.5) can be found from the vector $\mathbf{r}$ by matrix multiplication,

$$\boldsymbol{\rho} = \begin{pmatrix} u \\ v \\ w \end{pmatrix} = \begin{pmatrix} \cos\omega t & \sin\omega t & 0 \\ -\sin\omega t & \cos\omega t & 0 \\ 0 & 0 & 1 \end{pmatrix} \cdot \begin{pmatrix} r_1 \\ r_2 \\ r_3 \end{pmatrix}.$$

The vector equation of motion (11.20) can be written in matrix form as

$$\frac{d\mathbf{r}}{dt} = \boldsymbol{\Omega} \times \mathbf{r} = \begin{pmatrix} 0 & -\omega_0 & 0 \\ +\omega_0 & 0 & \kappa\mathscr{E}_0 \\ 0 & -\kappa\mathscr{E}_0 & 0 \end{pmatrix} \cdot \mathbf{r}.$$

If we transform $\mathbf{r}$ into $\boldsymbol{\rho}$, we can find a new equation of motion having the same form as the preceding equation, namely,

$$\frac{d\boldsymbol{\rho}}{dt} = \boldsymbol{\Omega}' \times \mathbf{r} = \begin{pmatrix} 0 & -(\omega_0 - \omega) & 0 \\ (\omega_0 - \omega) & 0 & \kappa\mathscr{E}_0 \\ 0 & -\kappa\mathscr{E}_0 & 0 \end{pmatrix} \cdot \boldsymbol{\rho}. \tag{11.21}$$

Equation (11.21) can also be derived from the vector equation for coordinate rotation,

$$\frac{d\mathbf{r}}{dt} = \left(\frac{\partial \mathbf{r}}{\partial t}\right)_{\text{rotating frame}} + (\boldsymbol{\omega} \times \mathbf{r}),$$

where the partial derivative of $\mathbf{r}$ with respect to t is taken in the rotating coordinate system. Explicitly, (11.21) can be written in component form as

$$\begin{aligned} \dot{u} &= -(\omega_0 - \omega)v, \\ \dot{v} &= +(\omega_0 - \omega)u + \kappa\mathscr{E}_0 w, \\ \dot{w} &= -\kappa\mathscr{E}_0 v. \end{aligned} \tag{11.22}$$

This is the rotating frame representation of the Feynman-Vernon-Hellwarth equations (11.19).

There may be some uncertainty at this point as to why we chose a coordinate system rotating at $-\omega_0$, when the electromagnetic field $\mathscr{E}(t)$ possesses terms in both $e^{i\omega t}$ and $e^{-i\omega t}$. The combination of the counterrotating $(-\omega)$ coordinate system and the rotating (ω) field vector gives a slowly oscillating term, while the other combination gives a rapidly time-varying term with frequency 2ω; the rotating wave approximation (RWA), introduced in section 5, says we should neglect these high-frequency terms, and we do so here. Effectively, we are replacing an oscillating field by a field rotating in the opposite sense and at nearly the same frequency as the resonance ω_0.

The physical content of the rotating coordinate transformation can be seen by looking back at figure 11.5. The "longitudinal" component $w = r_3 = bb^* - aa^*$ just measures the population difference between the upper and lower levels of the two-level system. This population difference is independent of the choice of coordinate system. Furthermore, since bb^* and aa^* are just the diagonal elements p_b and p_a, respectively, of the one-particle density matrix[3] for this system, we can write

3. The properties of the density matrix pertinent to this discussion are given in appendix C.

$$\tfrac{1}{2}\hbar\omega_0 w = \tfrac{1}{2}\hbar\omega_0(p_b - p_a) = \langle\psi|\mathscr{H}_0|\psi\rangle,$$

giving the expectation value of the two-level Hamiltonian.

The "transverse" components u and v are linear combinations of the components r_1 and r_2 introduced in (11.18), that is, involve the off-diagonal density matrix elements ab^* and a^*b. If $a(t)$ and $b(t)$ have random phase factors, then the expectation values $\langle r_1\rangle = \langle r_2\rangle = 0$; this is the proper definition of an *incoherent* system. If, however, there is a definite phase relation between $a(t)$ and $b(t)$, then r_1 and r_2 (and thus, in general, u and v) are nonzero, and we say that the system is *coherent*. The question of coherence versus incoherence is an important one, to which we shall return a number of times in the next chapter.

The polarization $\mathbf{p}$ is the expectation value of the dipole moment operator $\boldsymbol{\mu}$ and can be found as

$$\mathbf{p} = \langle\boldsymbol{\mu}\rangle = r_1\mu_{ab}\cos\omega t - r_2\mu_{ab}\sin\omega t. \tag{11.23}$$

Since we began with a field having the form $\mathscr{E}_0\cos\omega t$, the first term in (11.23) is in phase with the field, and the second term is 180° out of phase. The out-of-phase component [$v(t)$, in the rotating frame] gives the net rate of absorption of energy from the field; this can be seen from the third equation of (11.22),

$$\dot{w} = \frac{2}{\hbar\omega_0}\frac{2dE}{dt} = \kappa\mathscr{E}_0 v.$$

Therefore, we will want to solve for $v(t)$ in either (11.22) or the modified forms that we are about to derive.

For most of what we want to consider, we can let $\omega = \omega_0$ ($\Delta = 0$, or exact resonance). The $\boldsymbol{\rho}$ vector then possesses only a single component in the rotating frame, having magnitude ω' and aligned along the $\mathbf{1}'$ axis. The Rabi solution, (11.4) and (11.5), corresponds to multiple cycles of the $\boldsymbol{\rho}$ vector in the ($\mathbf{2}'\mathbf{3}'$), or (vw), plane (see figure 11.6A). For nonzero detuning ($\Delta \neq 0$), the motion is a *precession* of the $\boldsymbol{\rho}$ vector about an effective $\boldsymbol{\Omega}$ vector formed as the resultant of $(\omega_0 - \omega)\mathbf{3}'$ and $\omega'\mathbf{1}'$ in the rotating frame (see figure 11.6B).

The effect of various electromagnetic pulses can be appreciated from polar, or $\rho(\theta)$, plots in the (2′3′) plane, where the pulse area θ is defined by (11.8). The effect of a $\pi/2$ pulse, shown in figure 11.7A, is to rotate the ρ vector from its initial value of $(0, 0, -w_0)$ to a final value of $(0, +w_0, 0)$,

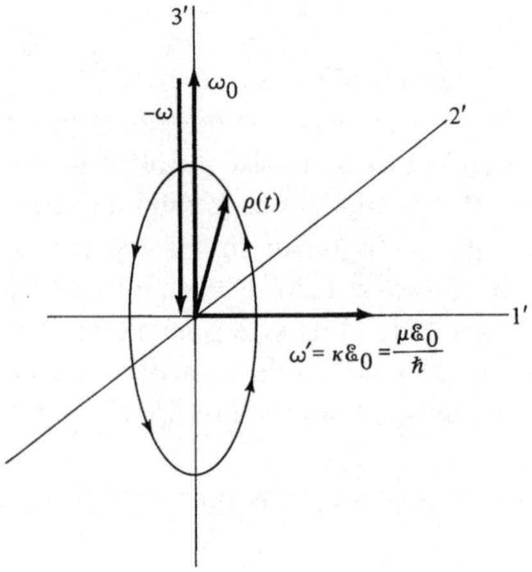

Figure 11.6A
Motion of the **ρ** vector in the (2′3′) plane when $\omega = \omega_0$, the resonant frequency.

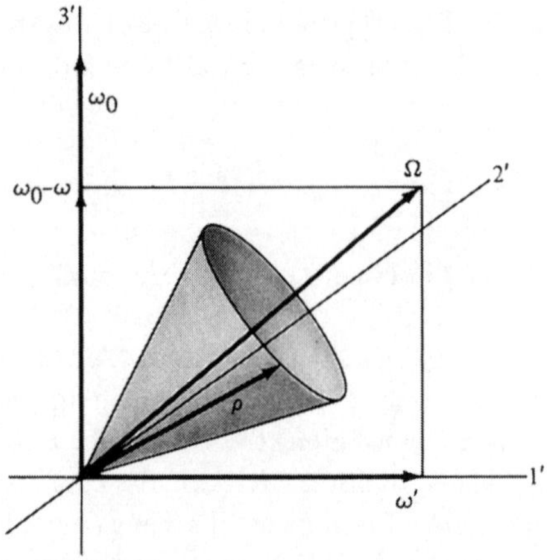

Figure 11.6B
Precession of the **ρ** vector about an effective **Ω** vector formed as the resultant of $\omega_0 - \omega$ and ω' in the rotating coordinate system.

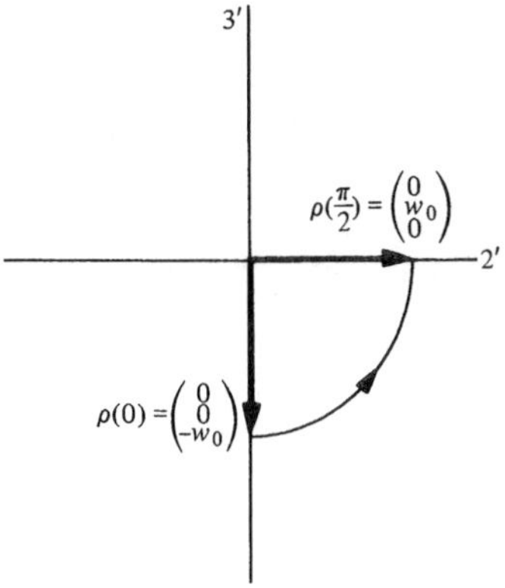

Figure 11.7A
Motion of the $\boldsymbol{\rho}$ vector in the (2′3′) plane subject to a $\pi/2$ pulse.

which lies along the positive 2′ axis. The expectation value $\langle b^*b - a^*a \rangle = 0$, meaning that the upper- and lower-state populations are equal; since there is now a nonvanishing transverse component of $\boldsymbol{\rho}$, we see that the $\pi/2$ pulse has created a *coherent* superposition state. The π pulse, shown in figure 11.7B, takes $\boldsymbol{\rho}(0) = (0, 0, -w_0)$ to $\boldsymbol{\rho}(\pi) = (0, 0, +w_0)$, that is, *inverts* the populations of the upper and lower levels. We shall make extensive use of the properties of these pulses in the following chapter, when discussing coherent transient phenomena in optical spectroscopy.

The formal development of the two-level system is now almost complete. What remains to be done is the inclusion of relaxation effects, and we do this in the same way that Bloch (reference 8) did for magnetic resonance problems, that is, introduce additional phenomenological terms describing the relaxation. The need for these terms may be inferred from figures 11.6 and 11.7; if the field is simply left on, the system will be driven from higher-energy to lower-energy to higher-energy states, repetitively, and there will be no net absorption or emission of energy, on the average. Some mechanism must be introduced to provide dissipation of the energy by the system.

A rationale for the Bloch relaxation model may be stated as follows. At equilibrium, an ensemble of two-level systems may be described by two populations, n_a in the lower state and n_b in the upper state. If the ensemble is in thermodynamic equilibrium with its surroundings at temperature T, the ratio $n_b^{\text{eq}}/n_a^{\text{eq}}$ has the value $e^{\hbar\omega_0/kT}$, where k is Boltzmann's constant

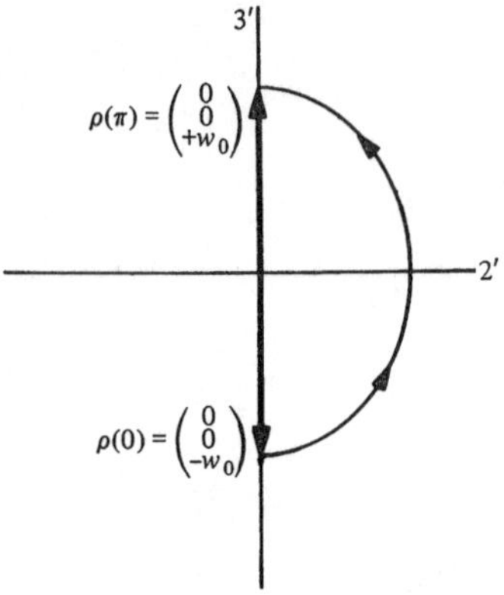

Figure 11.7B
Motion of the $\boldsymbol{\rho}$ vector in the (2′3′) plane subject to a π pulse.

(1.38×10^{-16} erg °K^{-1}). The "average" pseudospin vector describing the system thus has the form

$$\boldsymbol{\rho}^{\mathrm{eq}} = \begin{pmatrix} 0 \\ 0 \\ w^{\mathrm{eq}} \end{pmatrix} = \begin{pmatrix} 0 \\ 0 \\ b^*b^{\mathrm{eq}} - a^*a^{\mathrm{eq}} \end{pmatrix};$$

that is, the cross terms involving a^*b and ab^* are zero. For this reason, no coherence exists in the system, and since $r_1^{\mathrm{eq}} = r_2^{\mathrm{eq}} = 0$, there is zero net polarization [equation (11.24)] in either the rotating or the laboratory frame. Thus if we prepare an arbitrary initial state $|\psi\rangle = a(0)|a\rangle + b(0)|b\rangle$, the time dependence of $a(t)$ and $b(t)$ must be such that the transverse components of $\boldsymbol{\rho}$, namely u and v, will relax to zero with some decay constant. Let us call this decay constant T_2', the *dephasing* or *coherence* relaxation time. The longitudinal component $w = r_3(t)$ must relax to its equilibrium value w^{eq} with some decay constant that need not be numerically equal to T_2'. Let us call this decay constant T_1, the *energy* or *population* relaxation time. The Bloch prescription is to add these two terms to the vector precession equation (11.21) as follows:

$$\frac{d\boldsymbol{\rho}}{dt} = (\boldsymbol{\Omega}' \times \boldsymbol{\rho}) - \frac{\rho_1 \hat{\mathbf{1}}' + \rho_2 \hat{\mathbf{2}}'}{T_2'} - \frac{(\rho_3 - \rho_3^{\mathrm{eq}})\hat{\mathbf{3}}'}{T_1}, \tag{11.24a}$$

or

$$\frac{d\rho}{dt} = \begin{pmatrix} \dot{u} \\ \dot{v} \\ \dot{w} \end{pmatrix} = \begin{pmatrix} -\frac{1}{T_2'} & -\Delta & 0 \\ \Delta & -\frac{1}{T_2'} & \kappa\mathscr{E}_0 \\ 0 & -\kappa\mathscr{E}_0 & -\frac{1}{T_2'}\left(1 - \frac{w^{eq}}{w}\right) \end{pmatrix} \begin{pmatrix} u \\ v \\ w \end{pmatrix}. \tag{11.24b}$$

Written out in terms of components, this becomes

$$\begin{aligned} \dot{u} &= -\Delta v - \frac{u}{T_2'}, \\ \dot{v} &= \Delta u - \frac{v}{T_2'} + \kappa\mathscr{E}_0 w, \\ \dot{w} &= -\frac{w - w^{eq}}{T_1} - \kappa\mathscr{E}_0 v. \end{aligned} \tag{11.25}$$

Either of the preceding equivalent forms is called the *optical Bloch equation*; particular solutions for this equation can describe a wide variety of coherent transient phenomena, such as transient nutation, optical free induction decay, photon echoes, self-induced transparency, and multiphoton absorption. We shall pursue a number of these examples in the next two chapters; before doing so, however, we should pause briefly to examine the relation of this treatment of optical phenomena to that for magnetic resonance phenomena, which in fact was the area in which the theory was first developed.

6 Magnetic Resonance Analogues of Coherent Optical Spectroscopy

Magnetic resonance spectroscopy, of either nuclear or unpaired electron spin states, was developed in the late 1940s and early 1950s by Bloch, Purcell, and their coworkers. We shall not attempt to give any sort of complete treatment of this topic here (see instead references 9–12), but simply illustrate the close analogy between magnetic resonance and coherent optical phenomena. Indeed, the entire development thus far, including vector precession equations, rotating frame transformation, and Bloch equations, had all been developed for analyzing spin resonance spectroscopy well before coherent optical sources became available. As investigation proceeded in the optical regime, it soon became apparent that many of the observed phe-

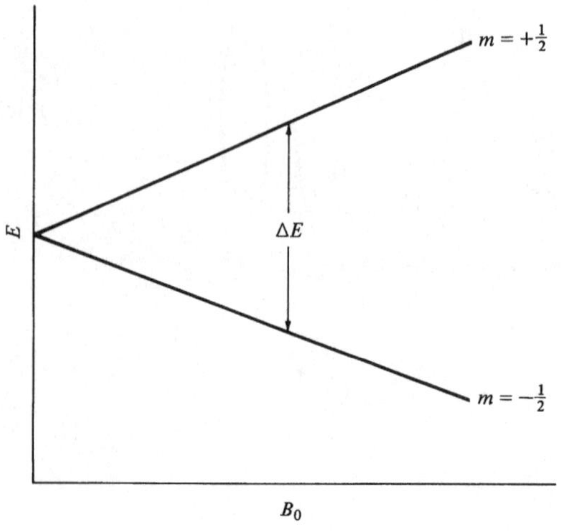

Figure 11.8
Zeeman splitting in a two-level (spin $\frac{1}{2}$) system.

nomena could simply be regarded as analogues of already understood magnetic resonance phenomena. The analogy must not be drawn too far, however, because there are also important differences between optical and magnetic resonance phenomena, particularly in the interpretation of the relaxation times T_1 and T_2'.

Let us begin by considering the simplest possible system in which a magnetic resonance can be observed: a two-level system, such as would be formed by a hypothetical atom with no electronic angular momentum ($L = S = J = 0$), and a single nucleus with $I = \frac{1}{2}$, so that $m_I = \pm\frac{1}{2}$. The energies of these two levels are given by the diagonal matrix elements of the Zeeman-Hamiltonian operator,

$$E = \langle m_I|\boldsymbol{\mu}\cdot\mathbf{B}|m_I\rangle = -m_I g_I \mu_N B_0 = -\gamma B_0 m_I,$$

so that we obtain the splitting shown in figure 11.8. A transition between these two levels is possible if we apply to the system a magnetic field B_x^0 having its field vector oriented perpendicular to B_0 and oscillating at an angular frequency $\omega = \Delta E/\hbar$. The effect of this field can be represented by an additional time-dependent term in the Hamiltonian,

$$\mathscr{H}' = -\gamma B_x^0 I_x \cos\omega t = -\tfrac{1}{2}\gamma B_x^0(I_+ + I_-)\cos\omega t,$$

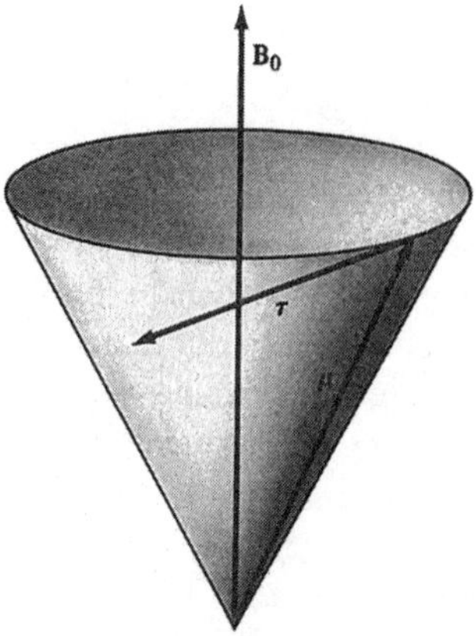

Figure 11.9
Magnetic moment in a constant external magnetic field. The torque exerted on the moment produces a precession around the field direction.

where I_x, the component of the nuclear spin angular momentum oriented parallel to the oscillating field, has been resolved into the angular momentum raising and lowering operators

$$I_+ = I_x + iI_y \qquad \text{and} \qquad I_- = I_x - iI_y,$$

which possess the nonzero matrix elements

$$\langle m|I_+|m-1\rangle \qquad \text{and} \qquad \langle m|I_-|m+1\rangle,$$

so that the selection rules for the magnetic resonance transition are $\Delta m = \pm 1$, just what is required to effect a transition between the two neighboring Zeeman levels.

The same result may be obtained from a classical analysis of the motion of the spin magnetic moment. Let us consider a single, classical magnetic moment in a static magnetic field, as shown in figure 11.9. At any instant of time, there will be a torque on the moment, exerted by the magnetic field,

$$\boldsymbol{\tau} = \boldsymbol{\mu} \times \mathbf{B}.$$

This is exactly analogous to the torque exerted on a gyroscope in a gravitational field. Since the torque is the rate of change of the angular momentum **J**, and **J** is related to the magnetic moment $\boldsymbol{\mu}$ by the gyromagnetic ratio γ, we have for the classical equation of motion of the magnetic moment

$$\frac{d\boldsymbol{\mu}}{dt} = \gamma(\boldsymbol{\mu} \times \mathbf{B}_0). \tag{11.26}$$

The solution to this classical equation is just the precession of $\boldsymbol{\mu}$ indicated in figure 11.9 in which the component of $\boldsymbol{\mu}$ along the field direction—conventionally, the z component—remains constant, and the x and y components rotate with constant magnitude in a plane perpendicular to the z axis, with frequency γB_0. This is, of course, just the analogue of the vector precession equation (11.20), with the real spin magnetic moment vector $\boldsymbol{\mu}$ replacing the pseudospin vector $\boldsymbol{\rho}$, and $-\gamma B_0$ taking the place of $\boldsymbol{\Omega}$.

Once again, it is convenient to transform this equation to a suitably rotating coordinate system. We wish to find the motion of the magnetic moment $\boldsymbol{\mu}$ in equation (11.26) at some angular frequency Ω to be specified. Since

$$\frac{d\boldsymbol{\mu}}{dt} = \frac{\partial\boldsymbol{\mu}}{\partial t} + \boldsymbol{\Omega} \times \boldsymbol{\mu} = \boldsymbol{\mu} \times \gamma\mathbf{B}_0,$$

we have

$$\begin{aligned}\frac{\partial\boldsymbol{\mu}}{\partial t} &= \boldsymbol{\mu} \times \gamma\mathbf{B}_0 - \boldsymbol{\Omega} \times \boldsymbol{\mu} \\ &= \boldsymbol{\mu} \times \gamma\mathbf{B}_0 + \boldsymbol{\mu} \times \boldsymbol{\Omega} \\ &= \boldsymbol{\mu} \times (\gamma\mathbf{B}_0 + \boldsymbol{\Omega}).\end{aligned}$$

Now if we simply choose $\boldsymbol{\Omega} = -\gamma\mathbf{B}_0$, we shall have $\partial\boldsymbol{\mu}/\partial t = 0$. That is, the magnetic moment is a *constant*, unchanging in magnitude or direction, in a coordinate system rotating with angular velocity equal in magnitude to the *Larmor frequency* $\omega_L = \gamma B_0$, and directed antiparallel to the external field direction. This means that in this coordinate system, the effect on the motion of the magnetic moment of the static field, which serves only to establish an energy difference between the two m_I states, can be ignored, and we can focus on the effects of the oscillating field, just as was done in arriving at (11.22).

To determine the effect of an oscillating magnetic field perpendicular to the static field, which can be written as

$$\mathbf{B}_1(t) = B_{10}(\hat{\mathbf{i}}\cos\omega t + \hat{\mathbf{j}}\sin\omega t),$$

we use the classical equation of motion for the magnetic moment,

$$\frac{d\boldsymbol{\mu}}{dt} = \boldsymbol{\mu} \times \gamma[\mathbf{B}_0 + \mathbf{B}_1(t)]. \tag{11.27}$$

The solution to equation (11.27) can be determined most easily by once more transforming to a coordinate system rotating at the same frequency, but opposite phase, as $B_1(t)$. Equation (11.26) becomes

$$\frac{d\boldsymbol{\mu}}{dt} = \boldsymbol{\mu} \times [\hat{\mathbf{k}}(-\omega + \gamma B_0) + \hat{\mathbf{i}}\gamma B_{10}]$$

$$= \boldsymbol{\mu} \times \gamma\left[\hat{\mathbf{k}}\left(B_0 - \frac{\omega}{\gamma}\right) + \hat{\mathbf{i}}B_{10}\right] = \boldsymbol{\mu} \times \gamma\mathbf{B}_{\text{eff}}.$$

Now, the magnetic moment precesses about a new effective field direction that is the vector sum of $B_0 - \omega/\gamma$ and B_{10}. If we choose ω equal to the Larmor frequency $\omega_L = \gamma B_0$, then $B_{\text{eff}} = B_{10}$, and the magnetic moment executes a circular motion in the (**jk**) plane in the rotating frame. This just corresponds to the resonance condition discussed earlier; the effect of the field oscillating at the resonant frequency is to cause the moment aligned parallel with the external field to be aligned antiparallel, that is, to go from $+\mathbf{k}$ to $-\mathbf{k}$, or from $m = +\frac{1}{2}$ to $m = -\frac{1}{2}$. The action of the field depends on the length of time that it is applied to the system. If B_1 is left on for a time $t = (\pi/2)/\omega_L = \pi/2\gamma B_{10}$, then the angle θ that the moment rotates in the (**jk**) plane is $\theta = \gamma B_{10}t = \pi/2$. If $t = \pi/\omega_L$, then $\theta = \pi$, and the direction of the moment is reversed; if $t = 2\pi/\omega_L$, then $\theta = 2\pi$, and the moment is returned to its original position. These three cases are summarized in figure 11.10, which is, of course, completely analogous to figure 11.6.

The Bloch equations, as originally derived for spin systems, deal with the net *magnetization* of the system, given by $\mathbf{M} = N\langle\boldsymbol{\mu}\rangle$, where N is the number of independent spins in the ensemble and $\langle\boldsymbol{\mu}\rangle$ is the average individual spin magnetic moment. The longitudinal component of the magnetization, M_z, follows the equation

$$\frac{dM_z}{dt} = \gamma(\mathbf{M} \times \mathbf{B})_z + \frac{M_0 - M_z}{T_1}.$$

The transverse components of the magnetization, M_x and M_y, follow the similar equation

$$\frac{dM_\perp}{dt} = \gamma(\mathbf{M} \times \mathbf{B})_\perp - \frac{M_\perp}{T_2}.$$

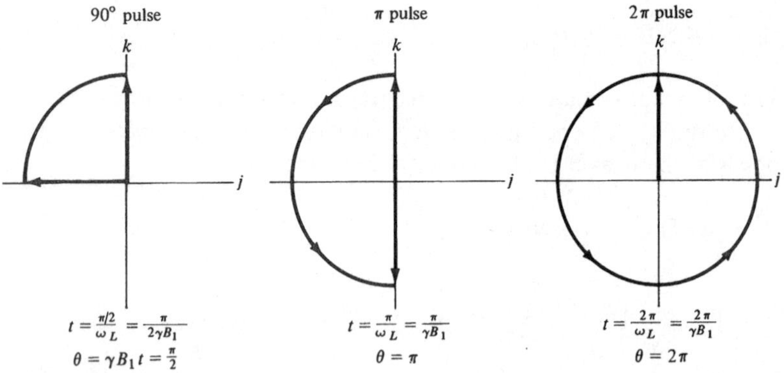

Figure 11.10
The effect of a "90° pulse," a "π pulse," and a "2π pulse" on a single magnetic moment.

The nature of the relaxation times introduced in the two preceding equations needs some comment. The T_1 process causes the z component of the magnetization, that is, the component aligned parallel with the external magnetic field, to relax from its instantaneous value to some final constant M_0, which reflects the equilibrium distribution of the individual moments in the presence of the field. T_1 processes are essentially irreversible and lead to a homogeneously broadened absorption line. In nuclear magnetic resonance (NMR), an example of a T_1 process is the "spin lattice relaxation," in which the magnetic alignment is destroyed by randomly fluctuating magnetic fields in the medium. T_2 processes, on the other hand, destroy the *coherence* among excited molecules in a system and may be reversible or irreversible. In NMR, the "spin-spin relaxation" is a typical T_2 process.

For optical systems, the interpretation of the phenomenological relaxation parameters requires particular caution. While we may certainly have *dephasing* collisions occurring with a rate $\gamma_2 = 1/T_2'$, we may also have inhomogeneous dephasing, for example, due to Doppler shifts, which are unimportant in magnetic resonance. These effects contribute an additional "width" $1/T_2^*$, so that the net T_2 to be used in the Bloch equations (11.24) is

$$\frac{1}{T_2} = \frac{1}{T_2'} + \frac{1}{T_2^*}.$$

There may also be *relaxing* or *energy-transferring* collisions, occurring with a rate $\gamma_1 = 1/T_1$. The system may also relax by spontaneous radiation, so

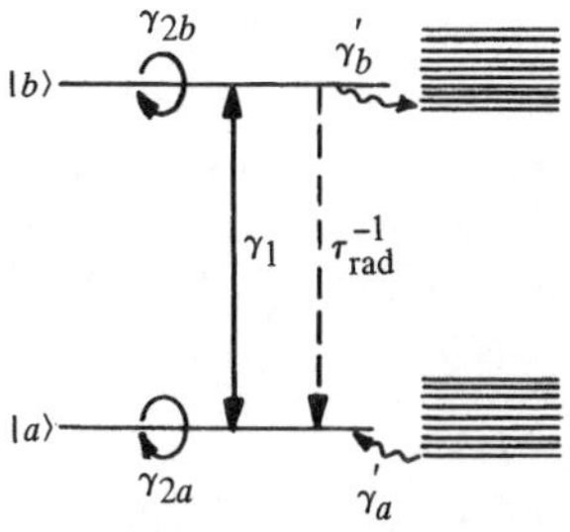

Figure 11.11
Relaxation processes in a two-level system embedded in a dense manifold of background states.

that the net T_1 is given by

$$\frac{1}{T_1^{\text{eff}}} = \gamma_1 + \frac{1}{\tau_{\text{rad}}}.$$

The situation may be complicated even further if the two-level system we are studying is actually embedded in a dense manifold of background states that are involved in the relaxation processes, as is shown in figure 11.11. This is a not at all uncommon situation for large polyatomic molecules, as we saw in chapter 9. A proper interpretation of T_1 and T_2 in this case requires a more or less detailed collision model. The simplest such model is that of *strong collisions*, in which every collision is both dephasing and energy transferring. Under these circumstance, $T_1 = T_2$. This is the usual assumption made in theories of pressure broadening. More detailed information about T_1 and T_2 may be obtained from several of the coherent transient experiments that will be discussed in the following chapter.

Problems

1. In problem 7 in chapter 2, you found the transition dipole moment matrix element between the ($n = 1, l = 0, m = 0$ [1^2S]) and the ($n = 2, l = 1, m = 1$ [2^2P]) states of atomic hydrogen to be $\langle \boldsymbol{\mu} \rangle = -(2^7/3^5)ea_0(\hat{\mathbf{i}} - i\hat{\mathbf{j}})$. Use this result in the following calculations.

(a) Calculate the Rabi frequency for a hydrogen atom placed in a field of intensity $I = 1$ W/cm^2 $= 10^7$ erg cm^{-2} sec^{-1} resonant with the $1^2\text{S} \rightarrow 2^2\text{P}$ transition (Lyman-α line). Use the relation $I = c|\mathscr{E}|^2/4\pi$.

(b) Calculate A_{21}, the spontaneous emission rate for this transition.
(c) Calculate the intensity of Lyman-α radiation that would be necessary to make the photon absorption rate equal to the spontaneous emission rate. This is approximately the intensity required to *saturate* the transition. Numerical values: e_0 (electron charge) $= 4.8 \times 10^{-10}$ esu; a_0 (Bohr radius) $= 5.3 \times 10^{-9}$ cm.

2. From the general solution for $b(t)$ derived as (11.5), recover the time-dependent perturbation theory result in the rotating wave approximation,

$$c_b(t) = -\frac{1}{2\hbar} V_{ba}^0 \left[\frac{e^{-i(\omega-\omega_{ab})t} - 1}{\omega - \omega_{ab}} \right],$$

by expanding the square root in the coefficients of integration

$$A, B = \mp \frac{\kappa \mathscr{E}}{2} [\Delta^2 + (\kappa \mathscr{E})^2]^{-1/2}$$

to lowest order.

3. Find expressions for the phenomenological constants α_0 and I_{sat} from the expression for $\alpha = \Delta P/P$ given in section 3. Let α_0 be the limiting value of α when $\mathscr{E} \to 0$ and $I = c|\mathscr{E}_0|^2/4\pi$. Estimate I_{sat} for $\mu_{ab} = 1 \times 10^{-18}$ esu cm, $\omega = \omega_0 = 2 \times 10^4$ cm^{-1}, and $T_2' = 10^{-8}$ sec.

4. In section 4, we derived the equation of motion for the pseudospin vector by using the Heisenberg operator representation. Here, you are to carry out an equivalent derivation in the Schrödinger wave function representation. The Schrödinger equation of motion is

$$i\hbar \frac{\partial \Psi}{\partial t} = \mathscr{H} \Psi,$$

with $\Psi(t) = a(t)\Psi_a + b(t)\Psi_b$. Find equations of motion for $a(t)$, $b(t)$, $a^*(t)$, and $b^*(t)$, and show that these four equations are equivalent to the equation of motion previously found,

$$\frac{d\mathbf{r}}{dt} = \mathbf{\Omega} \times \mathbf{r},$$

where $r_1 = ab^* + ba^*$, $r_2 = i(ab^* - ba^*)$, and $r_3 = aa^* - bb^*$, and $\omega_1 = (V_{ab} + V_{ba})/\hbar$, $\omega_2 = i(V_{ab} - V_{ba})/\hbar$, $\omega_3 = \omega_0$.

5. A nucleus of spin $\frac{1}{2}$ is quantized by a field B_0 in the z direction. It is in the $m = +\frac{1}{2}$ state at $t = 0$ when a rotating magnetic field of amplitude B_1 is applied for a time t_ω, producing a 90° pulse.

(a) Compute the wave function of the spin in the rotating reference system as a function of time during and after the pulse.
(b) Compute the wave function in the laboratory reference system during and after the pulse.
(c) Compute $\langle \mu_x(t) \rangle$ during and after the pulse.

References

1. I. Burak, J. I. Steinfeld, and D. G. Sutton, *J. Quant. Spectrosc. and Radiat. Transfer* **9**, 959 (1969).

2. C. R. Giuiliano and L. D. Hess, *I.E.E.E. J. Quantum Electronics* **QE-3**, 358 (1967).

3. J. D. Jackson, *Classical Electrodynamics* (Wiley, New York, 1965), chapter 9.

4. R. P. Feynman, F. L. Vernon, Jr., and R. W. Hellwarth, *J. Appl. Phys.* **28**, 49 (1957).

5. L. Allen and J. H. Eberly, *Optical Resonance and Two-Level Atoms* (Wiley, New York, 1975).

6. K. Shimoda and T. Shimizu, *Prog. Quantum Electronics* **2**, 45 (1972).

7. H. A. Bethe and E. E. Salpeter, *Quantum Mechanics of One- and Two-Electron Atoms* (Springer-Verlag, Berlin, 1957), pp. 48ff.

8. F. Bloch, *Phys. Rev.* **70**, 460 (1946).

9. C. P. Slichter, *Principles of Magnetic Resonance*, 2nd ed., Series in Solid-State Sciences No. 1 (Springer-Verlag, Berlin, 1978).

10. A. Carrington and A. D. MacLachlan, *Introduction to Magnetic Resonance* (Harper and Row, New York, 1967).

11. A. Abragam, *The Principles of Nuclear Magnetism* (Clarendon Press, Oxford, 1961).

12. J. D. Roberts, *Nuclear Magnetic Resonance* (McGraw-Hill, New York, 1959).

12 Coherent Transient Spectroscopy

Throughout this text, we have been concerned with the semiclassical interaction of an electromagnetic field with a system of charges (that is, an atom or a molecule). When the field is weak, by which we mean that the Rabi frequency $\omega' = \mu\mathscr{E}/\hbar \ll 1/T_1$, we saw in chapter 1 that perturbation theory could be used to find the behavior of the system. When the field is strong, on the other hand ($\omega' \gg 1/T_1$ or $1/T_2$), we found an exact Rabi solution for a two-level system (chapter 11.2), and also derived a vector or pseudospin representation for this solution. When both the Rabi frequency and the relaxation rates are comparable in magnitude, we must solve the optical Bloch equations derived in chapter 11.5. In this chapter, we shall be concerned with some particular solutions for these equations, by means of which a number of interesting coherent transient effects may be described. We shall first consider several examples of these effects, namely, transient optical nutation and photon echoes. We shall then see how these effects may be used in Fourier transform spectroscopy, again by analogy with magnetic resonance techniques. We shall then specify the conditions under which the optical Bloch equations reduce to a simple kinetic master equation and describe a technique of saturated absorption spectroscopy leading to extremely narrow line widths. Finally, we shall generalize the problem to include three or more levels interacting with two or more radiation fields, which will enable us to consider double-resonance and multiphoton spectroscopy.

1 Transient Nutation

The first, and simplest, solution of the optical Bloch equations that we shall consider is that for transient optical nutation. The same phenomenon was observed in magnetic resonance and analyzed by Torrey in 1949 (reference 1). This effect occurs whenever the resonant field amplitude is suddenly changed from zero to some constant value $\mathscr{E}_0$ (switched on), or from constant $\mathscr{E}_0$ to zero (switched off). For convenience, we repeat here the optical Bloch equations (11.25) in the rotating frame:

$$\dot{u} = -\frac{u}{T_2'} - \Delta v,$$

$$\dot{v} = \Delta u - \frac{v}{T_2'} + \kappa\mathscr{E}w,$$

$$\dot{w} = -\kappa\mathscr{E}v - \frac{w - w^{\text{eq}}}{T_1},$$

with $\Delta = \omega_0 - \omega$ and $\kappa\mathscr{E} = \mu\mathscr{E}_0/\hbar$ when the field is on, $\kappa\mathscr{E} = 0$ when the field is off. With this specification, we see that the preceding set of equations is a set of linear first-order differential equations with time-independent coefficients. Such systems of equations are most readily solved by using Laplace transform techniques. The Laplace transform of a time-dependent function $x(t)$ is defined as

$$\mathscr{L}(x) = \int_0^\infty e^{-pt}x(t)\,dt.$$

The properties of these transforms may be found in any advanced mathematics text (reference 2); for our purposes, we note that the transform is linear,

$$\mathscr{L}(cx) = c\mathscr{L}(x),$$

and that the transform of the derivative of $x(t)$ is

$$p\mathscr{L}(x) = x(0) + \mathscr{L}(\dot{x}),$$

where

$$x(t) = x(0) + \int_0^t \dot{x}(t')\,dt'.$$

Using these relations, the preceding optical Bloch equations transform from a set of first-order differential equations to a set of linear algebraic equations in the Laplace transforms:

$$p\mathscr{L}(u) = u_0 - \frac{1}{T_2'}\mathscr{L}(u) - \Delta\mathscr{L}(v),$$

$$p\mathscr{L}(v) = v_0 + \Delta\mathscr{L}(u) - \frac{1}{T_2'}\mathscr{L}(v) + \kappa\mathscr{E}_0\mathscr{L}(w), \tag{12.1}$$

$$p\mathscr{L}(w) = w_0 + \frac{1}{T_1}\frac{w^{\text{eq}}}{p} - \kappa\mathscr{E}_0\mathscr{L}(v) - \frac{1}{T_1}\mathscr{L}(w).$$

We shall not write out the solutions to these equations in detail, but simply outline the procedure that is employed. First, the preceding set of linear equations is solved for the Laplace transforms $\mathscr{L}(u)$, $\mathscr{L}(v)$, $\mathscr{L}(w)$. We then take the inverse transform of these functions and find that all have the same general form,

$$x(t) = Ae^{-at} + \left(B\cos st + \frac{C}{s}\sin st\right)e^{-bt} + D \qquad (x = u, v, w). \tag{12.2}$$

To find the eigenvalues a, b, and s, we take the determinant of the coefficients in (12.1), set it equal to zero, and solve:

$$\begin{vmatrix} p + \frac{1}{T_2'} & \Delta & 0 \\ -\Delta & p + \frac{1}{T_2'} & -\kappa\mathscr{E}_0 \\ 0 & \kappa\mathscr{E}_0 & p + \frac{1}{T_1} \end{vmatrix} = 0,$$

which gives an equation for p,

$$\left(p + \frac{1}{T_2'}\right)\left\{\left(p + \frac{1}{T_2'}\right)\left(p + \frac{1}{T_1}\right) + (\kappa\mathscr{E}_0)^2\right\} + \Delta^2\left(p + \frac{1}{T_1}\right) = 0. \tag{12.3}$$

Equation (12.3) possesses two real roots, a and b, which give the decay constants, and one imaginary root, is, which gives the oscillation frequency. The coefficients A, B, C, and D in (12.2) depend on the initial conditions u_0, v_0, w_0, and w^{eq}, that is, on the details of a particular experiment.

As noted earlier, we particularly want to find $v(t)$, which will give us the expected absorption signal. The general form of $v(t)$, predicted by equation (12.2), is that of a damped oscillating waveform. Explicit solutions get a bit complicated, but we can specify some simple special cases:

i. If we make the "strong-collision" assumption introduced at the end of chapter 11, namely, that $T_1 = T_2'$, we find the roots of (13.3) are

$$p_1 = \frac{1}{T_1} = \frac{1}{T_2'}, \qquad p_{2,3} = -\frac{1}{T_2'} \pm \sqrt{\Delta^2 + (\kappa\mathscr{E}_0)^2}.$$

The real roots (decay constants) are $a = b = 1/T_2' = 1/T_1$; the imaginary root (oscillation frequency) is $s = \sqrt{\Delta^2 + (\kappa\mathscr{E}_0)^2}$, which is just the generalized Rabi frequency introduced in chapter 11.2.

ii. If the field frequency is in exact resonance with the transition frequency ($\Delta = 0$), then

$$p_1 = \frac{1}{T_2'}, \qquad p_{2,3} = -\frac{1}{2}\left(\frac{1}{T_1} + \frac{1}{T_2'}\right) \pm \sqrt{(\kappa\mathscr{E}_0)^2 - \frac{1}{4}\left(\frac{1}{T_1} + \frac{1}{T_2'}\right)},$$

so that the real roots are

$$a = \frac{1}{T_2'}, \qquad b = \frac{1}{2}\left(\frac{1}{T_1} + \frac{1}{T_2'}\right),$$

and the imaginary root is the square root term in the previous expression. If, in addition to $\Delta = 0$, we have $T_1 = T_2'$, then again a and b are equal and $s = \kappa\mathscr{E}_0$. We would expect to observe an absorption signal oscillating with a period $(\Delta^2 + (\kappa\mathscr{E}_0)^2)^{-1/2}$ and decaying with a time constant T_2'.

Before turning to a discussion of specific optical nutation experiments, some explanation should be given for the use of the term "nutation" in this context. This use derives from the classical mechanics of gyroscopes and spinning tops. (See reference 3 for a lucid discussion of this subject.) If a top is spinning with angular momentum $\mathbf{J}$ in a gravitational field, the equation of motion for $\mathbf{J}$ is

$$\frac{d\mathbf{J}}{dt} = \boldsymbol{\tau} = \bar{\mathbf{r}} \times \mathbf{F},$$

where the torque $\boldsymbol{\tau}$ is provided by the cross product of the gravitational force $\mathbf{F}$ and the off-axis distance to the center of mass, $\bar{\mathbf{r}}$. This equation is, of course, exactly analogous to the equations of motion (11.21) and (11.26) for the Feynman pseudospin vector and for a magnetic moment in a static magnetic field, respectively. The top accordingly *precesses* about the gravitational field direction. But in addition, the top axis wobbles as it precesses, so that an oscillatory motion is superimposed on the precession; this wobbling motion is called *nutation*. The amplitude and period of this nutation depend on the specific impulse given to the top when it began rotating. By exact analogy, the components of $\boldsymbol{\rho}$ or $\boldsymbol{\mu}$ oscillate in the laboratory frame, reflecting the initial conditions of the interaction with the electric or magnetic field. In a physical top, the nutational motion is damped out by friction at the pivot point; in two-level systems, the $-v/T_2'$ term in the optical Bloch equations drives the oscillating component of the absorption to zero as $e^{-t/T_2'}$.

In order actually to perform a transient nutation experiment, the radiation field must be switched on and off in a time shorter than the relaxation time

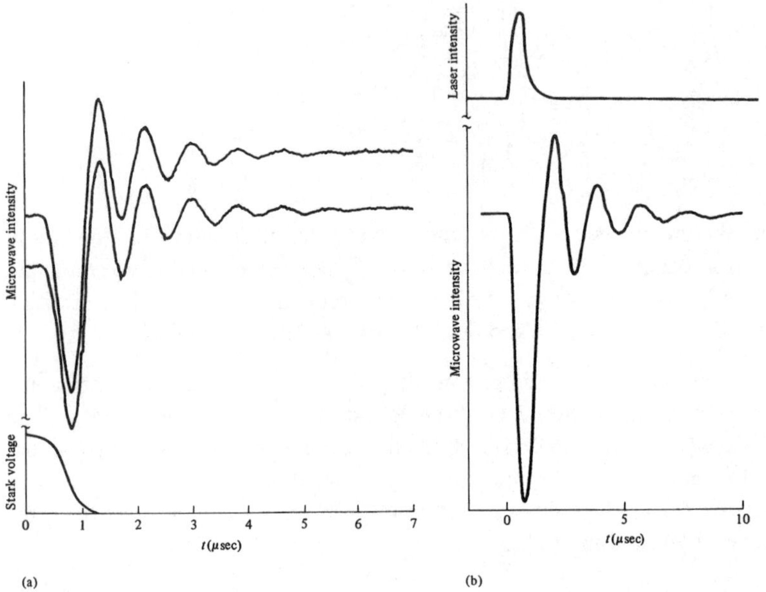

Figure 12.1
Transient nutation in the microwave absorption of the ($J = 8, K = 7$) inversion doublet of ammonia, following (a) switching off of the Stark field applied to the system and (b) pumping of the upper level by a resonant infrared laser pulse [From J. M. Levy, J. H.-S. Wang, S. G. Kukolich, and J. I. Steinfeld, *Phys. Rev. Letts.* **29**, 395 (1972). Reproduced with permission.]

T_2'; otherwise, the polarization (or magnetization) will tend simply to follow the field adiabatically. This may be done in the microwave region, for example, by using fast PIN diodes. In optical experiments in which a highly collimated laser beam is employed to excite the sample, the beam may be physically displaced out of the sample region with an acoustooptic deflector. For polar molecules, it is possible to take advantage of the Stark effect (see chapters 2.7 and 5.10) by shifting the transitions in and out of resonance with radiation at a fixed frequency and amplitude. This is frequently advantageous, since it is easier to modulate a dc electric field than the radiation source itself. This Stark-switching method has been widely applied in microwave and infrared transient experiments. An example is shown in figure 12.1a.

Another method of inducing transient nutation involves a double-resonance method (see section 7). If one level of a two-level system is coupled to a third level by a brief burst of radiation, oscillations can be induced on

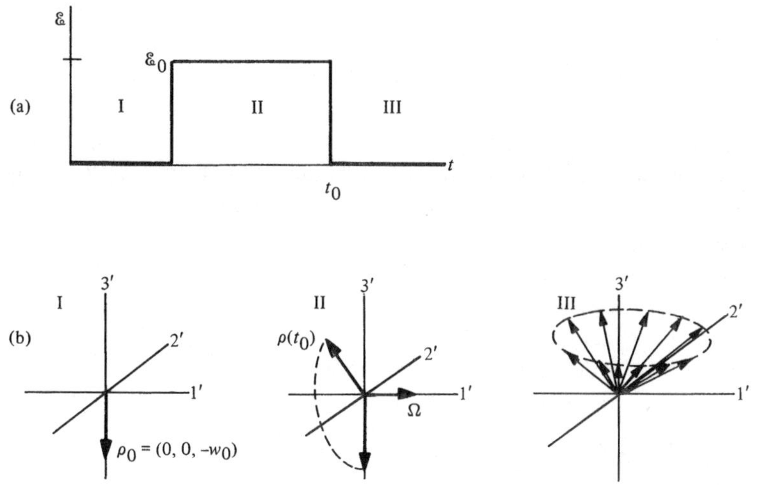

Figure 12.2
(a) Radiation field amplitude to produce free induction decay. (b) Behavior of composite pseudospin vector in rotating frame prior to pulse (I), during pulse (II), and dephasing of individual pseudospin vectors following pulse (III).

the absorption within the original two-level system. An example, shown in figure 12.1b, is the ($J = 8, K = 7$) inversion doublet of ammonia (see chapter 7.3.4). The upper level of this doublet can be coupled to the ($J' = 8, K' = 7$) level in the $v_2 = 1$ excited vibrational state by radiation from an N_2O laser at 927.739 cm^{-1}. Following such a pulse, the absorption at the inversion frequency of 23,232.24 GHz shows transient nutation. A complete analysis of this phenomenon requires consideration of the three-level system, which will be done in section 7.

A related, but somewhat different, coherent transient is coherent spontaneous emission, often called *free induction decay* by analogy with the nuclear magnetic resonance effect. This is an ensemble or cooperative effect, which requires a density matrix treatment (see appendix C) for proper analysis. Here, we shall simply sketch the concept behind this phenomenon, with the aid of figure 12.2. Suppose that we take a system initially at equilibrium, so that all of the rotating frame pseudospin vectors are aligned along the $-\hat{\mathbf{3}}'$ axis. We then apply a pulse of radiation to the system, which induces precession of $\boldsymbol{\rho}$ [rotation in the ($\mathbf{2}'\mathbf{3}'$) plane, in the rotating frame]. There will of course be transient nutation immediately after the radiation field is turned on, but that does not concern us here. After a period of time, the field is turned off, and the pseudospin vectors once again move under the influence of the field-free Hamiltonian. If the resonant frequency was precisely the same for all the individual systems, then there would (by definition) be no net torque in the rotating frame. But in a real system, individual atoms or molecules possess slightly differing resonant frequencies, for example, as a result of a Doppler width; thus the individual pseudospin vectors will begin to move away from each other, a process frequently called dephasing. After a time $T_2^* \sim 1/\Delta\nu_D$, all the microscopic moments will be nearly uniformly distributed in the (**1′2′**) plane and, even though there may still be a net inversion ($N_b - N_a > 0$), there is no net polarization, and thus no emission. A quantitative expression for the free induction decay signal is given in the article by Shoemaker (reference 5), in which a number of other coherent transient phenomena are discussed. This expression is

$$S_{\text{fid}}(t) = \frac{\mu_{ab}^2 \omega_0^2 L^2}{8\varepsilon_0 c}(N_b - N_a)^2 e^{-2(t-t_0)/T_2'} e^{-(t-t_0)^2/(T_2^*)^2}. \tag{12.4}$$

The first decay term in equation (12.4) is the ordinary homogeneous decay; the second term, which is Gaussian in $(t - t_0)$, represents the dephasing and turns out to be a *reversible* decay process. We shall explore the consequences of this in much greater detail in the following section.

In any event, we have seen that the decay envelope of the transient nutation or free induction decay signal measures the homogeneous coherence relaxation time T_2'. From the theory just outlined, it would appear that information could also be gained about the population relaxation time T_1 by carrying out the experiment with the radiation field exactly on resonance ($\Delta = 0$). In practice, this is not feasible, since small frequency instabilities produce large changes in the observed waveform. In order to measure T_1, we need to carry out still another variation of the basic transient nutation experiment, known as *delayed nutation*, or a $(\pi, \tau, \pi/2)$ pulse sequence. This experiment is outlined schematically in figure 12.3. The first pulse of radiation (or, equivalently, the Stark field) is timed to produce a pulse area $\theta = \pi$, which inverts the populations in the two-level system. After a delay time τ_D, the initially inverted population $-w_0$ has decayed to $-w_0 e^{-\tau_D/T_1}$. During interval III, a $\pi/2$ pulse is applied to generate a transverse polarization, which leads to observable transient nutation having an amplitude proportional to $(1 - 2e^{-\tau_D/T_1})$. By successively varying the delay time and measuring the signal amplitude, T_1 may be estimated directly. In the version of this experiment developed by the present author, which we have called Stark-modulated, infrared laser-enhanced transient nutation (reference 6), an

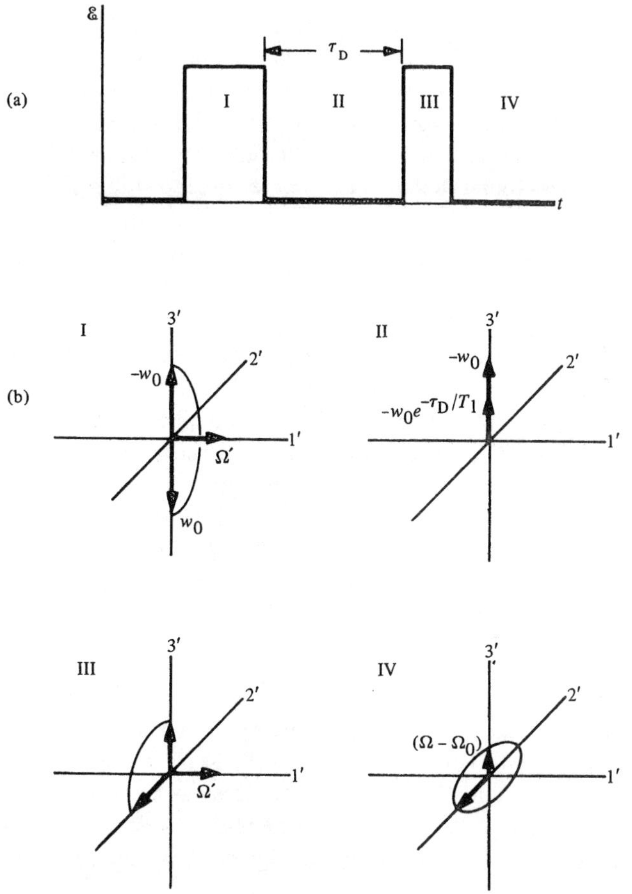

Figure 12.3
(a) Pulse sequence for $(\pi, \tau, \pi/2)$ or delayed nutation experiment. (b) Motion of pseudospin vector in rotating frame.

infrared laser pulse is used to change the populations in an inversion doublet of ammonia. Following the pulse, the populations relax back to thermal equilibrium. If the Stark field is then switched on (or off) at a time τ_D following the laser pulse, the microwave inversion signal displays nutation with an amplitude proportional to $e^{-\tau_D/T_1}$. For this particular ammonia inversion doublet, it was found that T_1 and T_2' were equal to within experimental error.

2 Photon Echo

In the discussion of free induction decay in the preceding section, we noted that the dephasing of the individual dipoles can be regarded as a reversible decay process. We now wish to see exactly what this implies, and how, in fact, the reversal can actually be accomplished. The experimental realization of this reversal produces the phenomenon of radiative echoes. These are termed "spin echoes" in nuclear magnetic resonance (NMR), and "photon echoes" in optical spectroscopy; the latter term, although widely recognized, is somewhat of a misnomer, since the photon (that is, quantized field) formalism is not at all necessary to describe the phenomenon.

The basic idea of the echo experiment is to perform the following operations on an ensemble of two-level systems:

1. Place the system into a coherent state, in the sense that the ensemble-averaged values $\bar{u}$ and $\bar{v} \neq 0$. This can be accomplished by a $\pi/2$ pulse.
2. Allow the system to dephase by free induction decay in a characteristic time T_2^* equal to the reciprocal of the inhomogeneous line width.
3. Do something that will cause the system to "rephase"; we shall see that a π pulse is able to accomplish this.
4. Allow the system to evolve spontaneously to a new coherent state.

The fact that step (4) actually occurs is the most surprising, counterintuitive aspect of the echo phenomenon; yet we shall see that it follows very straightforwardly from the mathematical description. The vector model is very helpful in gaining insight into the behavior of the system; we shall make use of this, along with the matrix formulation of the optical Bloch equations (reference 7), in the following development.

The derivation of the photon echo requires us to find the polarization of an ensemble of two-level systems subjected to the pulse sequence shown in figure 12.4. The ensemble is initially at equilibrium. The first pulse is

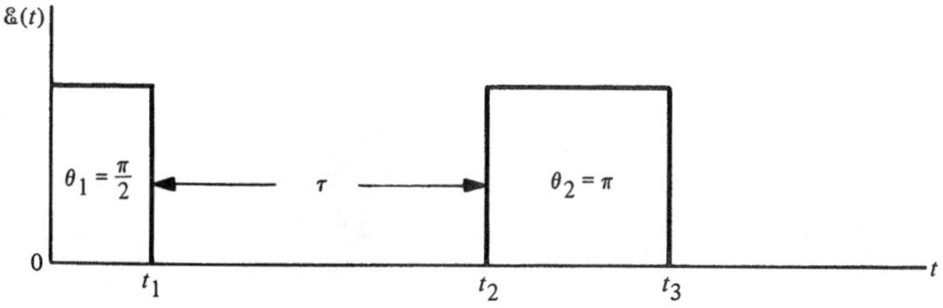

Figure 12.4
Pulse sequence for generating radiative echoes.

adjusted so that the pulse area [equation (11.8)] is

$$\theta_1 = \int_0^{t_1} \kappa\mathscr{E}\,dt = \frac{\mu\mathscr{E}_0}{\hbar}(t_1 - 0) = \frac{\pi}{2}.$$

After an interval $\tau = t_2 - t_1$, a second pulse is applied to the system, having pulse area

$$\theta_1 = \int_{t_2}^{t_3} \kappa\mathscr{E}\,dt = \frac{\mu\mathscr{E}_0}{\hbar}(t_3 - t_2) = \pi.$$

(Note that this is just the reverse of the pulse sequence used in delayed nutation.) We then inquire about the behavior of the system at times following t_3. This may be found by use of the optical Bloch equations [equation (11.21)], which are repeated here:

$$\frac{d\boldsymbol{\rho}}{dt} = \frac{d}{dt}\begin{pmatrix} u \\ v \\ w \end{pmatrix} = \begin{pmatrix} 0 & -\Delta & 0 \\ +\Delta & 0 & \kappa\mathscr{E} \\ 0 & -\kappa\mathscr{E} & 0 \end{pmatrix}\begin{pmatrix} u \\ v \\ w \end{pmatrix};$$

for the moment, we shall neglect relaxation terms.

A general, explicit solution for $\boldsymbol{\rho}(t)$ following application of a field with constant amplitude $\mathscr{E}_0$ for a time interval t may be found by using the method of successive rotations employed in reference 7. We should not forget that we are already in a coordinate system rotating at a constant frequency ω; the effect of the radiation field may be represented by an additional *fixed* rotation of this coordinate system around the $\hat{\mathbf{2}}$ axis and through an angle χ, which brings the $\boldsymbol{\Omega}$ vector into coincidence with the $\hat{\mathbf{1}}$ axis (see figure 12.5). The angle χ may be found from $\tan\chi = \Delta/\kappa\mathscr{E}_0$; we see that $\boldsymbol{\rho}$ now precesses around the $\mathbf{R}\,\hat{\mathbf{1}}$ axis in the $\mathbf{R}$ coordinate system.

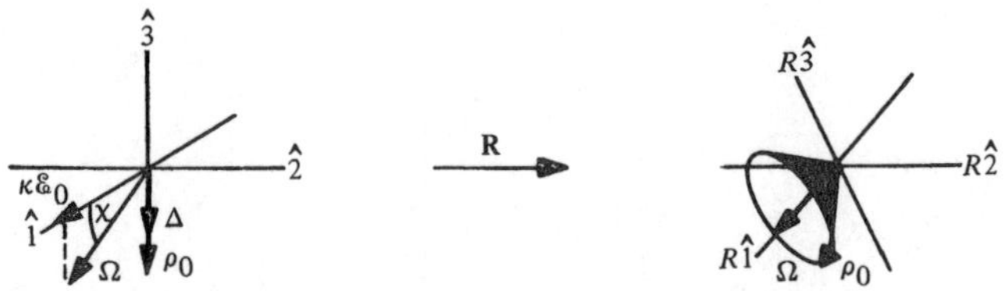

Figure 12.5
Rotation of axes in rotating coordinate system.

Algebraically, the rotation operator is given by

$$\mathbf{R} = \begin{pmatrix} \cos\chi & 0 & \sin\chi \\ 0 & 1 & 0 \\ -\sin\chi & 0 & \cos\chi \end{pmatrix}.$$

Rabi precession during the radiation pulse in the **R** coordinate system is then represented by rotation about $\mathbf{R}\hat{\mathbf{1}}$, corresponding to zero detuning and an effective Rabi frequency $\Omega' = \sqrt{\Delta^2 + (\kappa\mathscr{E}_0)^2}$:

$$(\mathbf{R}\boldsymbol{\rho}(t)) = \begin{pmatrix} 1 & 0 & 0 \\ 0 & \cos\Omega' t & -\sin\Omega' t \\ 0 & \sin\Omega' t & \cos\Omega' t \end{pmatrix} (\mathbf{R}\boldsymbol{\rho}(0)).$$

The Bloch vector $\boldsymbol{\rho}(t)$ in the canonical rotating frame may then be recovered by rotating *back* through an angle $-\chi$ about $\hat{\mathbf{2}}$:

$$\boldsymbol{\rho}(t) = \mathbf{R}^{-1}(\mathbf{R}\boldsymbol{\rho}(t)) = \begin{pmatrix} \cos\chi & 0 & -\sin\chi \\ 0 & 1 & 0 \\ +\sin\chi & 0 & \cos\chi \end{pmatrix} (\mathbf{R}\boldsymbol{\rho}(t)).$$

The explicit, and rather complicated, result for arbitrary θ may be found in reference 7, p. 58. For the specific case of a $\pi/2$ pulse, the product of the three matrices has the simple form

$$\begin{pmatrix} 1 & -\dfrac{\Delta}{\kappa\mathscr{E}_0} & -\dfrac{\Delta}{\kappa\mathscr{E}_0} \\ \dfrac{\Delta}{\kappa\mathscr{E}_0} & 0 & 1 \\ -\dfrac{\Delta}{\kappa\mathscr{E}_0} & -1 & 0 \end{pmatrix}.$$

We now have all the expressions required for computing the time evolution of $\boldsymbol{\rho}$ through the pulse sequences displayed in figure 12.4.

i. $\pi/2$ pulse, $0 \le t \le t_1$. We begin with an individual two-level system in its ground state, represented by $\boldsymbol{\rho}(0) = (0, 0, -1)$. Following the pulse,

$$\boldsymbol{\rho}(t_1) = \begin{pmatrix} 1 & -\dfrac{\Delta}{\kappa\mathscr{E}_0} & -\dfrac{\Delta}{\kappa\mathscr{E}_0} \\ \dfrac{\Delta}{\kappa\mathscr{E}_0} & 0 & 1 \\ -\dfrac{\Delta}{\kappa\mathscr{E}_0} & -1 & 0 \end{pmatrix} \begin{pmatrix} 0 \\ 0 \\ -1 \end{pmatrix} = \begin{pmatrix} +\dfrac{\Delta}{\kappa\mathscr{E}_0} \\ -1 \\ 0 \end{pmatrix}.$$

The motion of the $\boldsymbol{\rho}$ vector is shown in figure 12.6a. For $\Delta = 0$, that is, exact resonance, the vector rotates from alignment with the $-\hat{\mathbf{3}}$ axis to alignment with the $-\hat{\mathbf{2}}$ axis. For an *ensemble* of two-level systems, having a distribution of detunings $g(\Delta)$, we have the situation shown in figure 12.6b. All of the individual vectors have $w = b^*b - a^*a = 0$, and v equal to -1; there is some distribution of u over the domain $\pm 1/(\kappa\mathscr{E}_0 T_2^*)$, which is much smaller than 1 in magnitude.

ii. Free induction decay, $t_1 \le t \le t_2$. With the radiation field off, we have free precession of $\boldsymbol{\rho}$ around the residual Δ components, given by

$$\boldsymbol{\rho}(t_2) = \begin{pmatrix} \cos\Delta(t_2 - t_1) & -\sin\Delta(t_2 - t_1) & 0 \\ \sin\Delta(t_2 - t_1) & \cos\Delta(t_2 - t_1) & 0 \\ 0 & 0 & 1 \end{pmatrix} \begin{pmatrix} +\dfrac{\Delta}{\kappa\mathscr{E}_0} \\ -1 \\ 0 \end{pmatrix}$$

$$= \begin{pmatrix} \dfrac{\Delta}{\kappa\mathscr{E}_0}\cos\Delta(t_2 - t_1) + \sin\Delta(t_2 - t_1) \\ \dfrac{\Delta}{\kappa\mathscr{E}_0}\sin\Delta(t_2 - t_1) - \cos\Delta(t_2 - t_1) \\ 0 \end{pmatrix}.$$

The individual vectors begin to spread out in the $(\hat{\mathbf{1}}\hat{\mathbf{2}})$ plane, each driven by its particular value of Δ, as shown in figure 12.6c. After a time $t_2 - t_1 > T_2^*$, the individual vectors will be uniformly distributed in this plane, as in figure 12.6d, so that the ensemble average values $\overline{\overline{u}} = \overline{\overline{v}} = 0$. To an observer, the ensemble is then dephased, although as we shall see, the dephasing is not irreversible.

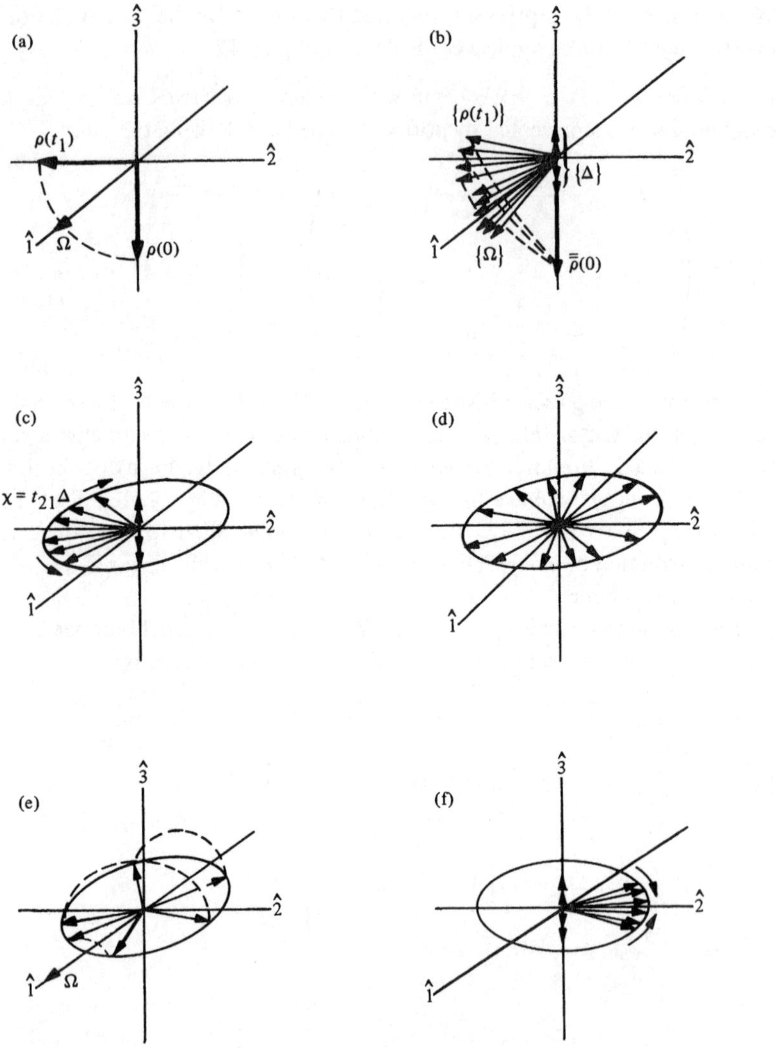

Figure 12.6
Motion of pseudospin vectors in the rotating frame during a photon echo sequence: (a) $\pi/2$ pulse for on-resonance ($\Delta = 0$); (b) motion of vector ensemble during $\pi/2$ pulse; (c) dephasing of individual vectors during free induction decay; (d) dephased ensemble for $t_{21} > T_2^*$; (e) 180° phase change during a π pulse; (f) rephasing of individual vectors following a π pulse. For the spin echo, the vectors represent the physical magnetic moments.

We can see what is happening a little more clearly by considering the polarization associated with the ensemble of two-level systems. Recall that the polarization is the expectation value of the dipole moment, or

$$\begin{aligned} p = \langle \boldsymbol{\mu} \rangle &= u\mu_{ab}\cos\omega t - v\mu_{ab}\sin\omega t \\ &= \mathrm{Re}\{(u + iv)e^{-i\omega t}\}\mu_{ab}. \end{aligned} \tag{12.5}$$

The macroscopic, ensemble average polarization is

$$P(t) = \bar{\bar{p}} = N\mu_{ab}\int \mathrm{Re}\{(u + iv)e^{-i\omega t}\}g(\Delta')\,d\Delta'. \tag{12.6}$$

At $t = t_2$,

$$\begin{aligned} u(t_2) + iv(t_2) &= -i\left(1 + i\frac{\Delta}{\kappa\mathscr{E}_0}\right)e^{i\Delta(t_2 - t_1)} \\ &= -ie^{i\Delta s}e^{i\Delta t_{21}}. \end{aligned}$$

We must average the real part of the foregoing expression over some distribution of detunings, which we take as

$$g(\Delta') = \frac{T_2^*}{\pi}e^{-(\Delta-\Delta')^2(T_2^*)^2/\pi},$$

that is, a *Gaussian* distribution with a characteristic width $2/T_2^*$. Such a distribution would arise, for example, from Doppler broadening in a gas. We obtain

$$P(t_2) = N\mu_{ab}\sin(\omega + \Delta)t_2\exp\left[-\frac{\pi}{4}\left(\frac{t_2 - t_1 + \mathscr{O}(1/\kappa\mathscr{E}_0)}{T_2^*}\right)^2\right]. \tag{12.7}$$

Examination of (12.7) shows a dipole μ_{ab} oscillating at a frequency $\omega + \Delta = \omega_0$, which is multiplied by a term decaying approximately as $e^{-(t/T_2^*)^2}$, at least for times greater than $(\kappa\mathscr{E}_0)^{-1}$; this faster than exponential decay is just the free induction decay we have encountered in the preceding section [see equation (12.4)]. In the case of nuclear magnetic resonance (that is, spin echo), the quantity in (12.7) would be just the macroscopic magnetization of the sample detected by the induction coil of the spectrometer.

iii. π pulse, $t_2 \le t \le t_3$. We now apply a π pulse to the system. Multiplying $\boldsymbol{\rho}(t_2)$, just obtained, by the rotation matrix for $\theta_{23} = \pi$ and using $t_2 - t_1 = t_{21}$ give

$$\boldsymbol{\rho}(t_3) = \begin{pmatrix} 1 & 0 & -2\left(\dfrac{\Delta}{\kappa\mathscr{E}_0}\right) \\ 0 & -1 & 0 \\ -2\left(\dfrac{\Delta}{\kappa\mathscr{E}_0}\right) & 0 & -1 \end{pmatrix} \begin{pmatrix} \dfrac{\Delta}{\kappa\mathscr{E}_0}\cos\Delta t_{21} + \sin\Delta t_{21} \\ \dfrac{\Delta}{\kappa\mathscr{E}_0}\sin\Delta t_{21} - \cos\Delta t_{21} \\ 0 \end{pmatrix}$$

$$= \begin{pmatrix} \dfrac{\Delta}{\kappa\mathscr{E}_0}\cos\Delta t_{21} + \sin\Delta t_{21} \\ -\dfrac{\Delta}{\kappa\mathscr{E}_0}\sin\Delta t_{21} + \cos\Delta t_{21} \\ -2\left(\dfrac{\Delta}{\kappa\mathscr{E}_0}\right)\sin\Delta t_{21} \end{pmatrix}.$$

We have neglected terms in $(\Delta/\kappa\mathscr{E}_0)^2$ in arriving at the above result, since we have asserted that $\Delta < \kappa\mathscr{E}_0$. Comparing $\boldsymbol{\rho}(t_3)$ with $\boldsymbol{\rho}(t_2)$, we see that $u(t_3) = u(t_2)$ but $v(t_3) = -v(t_2)$; that is, the pseudospin vectors have undergone a 180° phase change. In addition, w has picked up a nonzero longitudinal component. The motion of the representative vectors is shown in figure 12.6e.

iv. Free precession, $t \geq t_3$. Following the π pulse, the system is again free to evolve under the influence of the detunings, as in region (ii). For a later time t_4, we have

$$\boldsymbol{\rho}(t_4) = \begin{pmatrix} \cos\Delta(t_4 - t_3) & -\cos\Delta(t_4 - t_3) & 0 \\ \sin\Delta(t_4 - t_3) & \cos\Delta(t_4 - t_3) & 0 \\ 0 & 0 & 1 \end{pmatrix} \begin{pmatrix} \dfrac{\Delta}{\kappa\mathscr{E}_0}\cos\Delta t_{21} + \sin\Delta t_{21} \\ -\dfrac{\Delta}{\kappa\mathscr{E}_0}\sin\Delta t_{21} + \cos\Delta t_{21} \\ -2\left(\dfrac{\Delta}{\kappa\mathscr{E}_0}\right)\sin\Delta t_{21} \end{pmatrix}$$

$$= \begin{pmatrix} -\sin\Delta(t_{43} - t_{21}) + \dfrac{\Delta}{\kappa\mathscr{E}_0}\cos\Delta(t_{43} - t_{21}) \\ \cos\Delta(t_{43} - t_{21}) + \dfrac{\Delta}{\kappa\mathscr{E}_0}\sin\Delta(t_{43} - t_{21}) \\ -2\left(\dfrac{\Delta}{\kappa\mathscr{E}_0}\right)\sin\Delta t_{21} \end{pmatrix}.$$

Let us calculate the macroscopic polarization using (12.6). At $t = t_4$,

$$u(t_4) + iv(t_4) = +i\left(1 - i\frac{\Delta}{\kappa\mathscr{E}_0}\right)e^{i\Delta(t_{43}-t_{21})},$$

with $t_{43} = t_4 - t_3$. Thus,

$$\begin{aligned} P(t_4) &= N\mu_{ab}\int \mathrm{Re}\{u(t_4) + iv(t_4)e^{i\omega t_4}\}g(\Delta')\,d\Delta' \\ &= -N\mu_{ab}\sin(\omega + \Delta)t_4 \exp\left[-\frac{\pi}{4}\left(\frac{t_{43} - t_{21} - 1/\kappa\mathscr{E}_0}{T_2^*}\right)^2\right]. \end{aligned} \tag{12.8}$$

When $t_{43} = t_{21} + (\kappa\mathscr{E}_0)^{-1}$, the Gaussian term in (12.8) goes through a *maximum*. This polarization will be observed as a pulse of radiation of frequency $\omega + \Delta$, and lasting approximately $2T_2^*$ sec. This pulse is the photon (or spin) echo, so called because it occurs at a delay time following the π and the $\pi/2$ pulses. In the rotating frame (figure 12.6f), the microscopic vectors are very nearly *rephased* at $t_4 \simeq t_3 + t_{21}$.

At this point, it may be helpful to review the entire sequence (a)–(f) in figure 12.6, to see the time history of the system vectors. Several analogies from more familiar experience have been put forward in order to aid one's visualization of the process. Abragam (reference 8) has given a particularly felicitous model: Imagine a group of ants crawling around the edge of a pancake. They start out as one clump, but, because of their different crawling speeds, soon space themselves out all around the circumference (a T_2^* process). Now someone comes along and turns the pancake over (a π pulse). The ants all turn around and continue to crawl in the same direction as before. However, the faster ants now have a longer distance to travel to get back to their starting point than the slower ones. After an appropriate interval, the ants will all find themselves bunched together again. A similar, but more trendy, model (runners on a track) has been given by Allen and Eberly (reference 7); in this case, the π pulse is an order to the runners to about-face and run back around the track in the reverse direction. A mechanical analogue of the spin echo exists, also suggested by Hahn. This involves the reversible mixing and "unmixing" of a soluble dye in a viscous solvent such as glycerine, and it must be seen to be appreciated.[1]

The effect of irreversible decay processes (in the physical analogues:

1. An electromechanical analogue of the photon echo has been suggested as well: see the article by V. P. Chebotayev and B. Ya. Dubetskii, *Appl. Phys.* **B31**, 45 (1983).

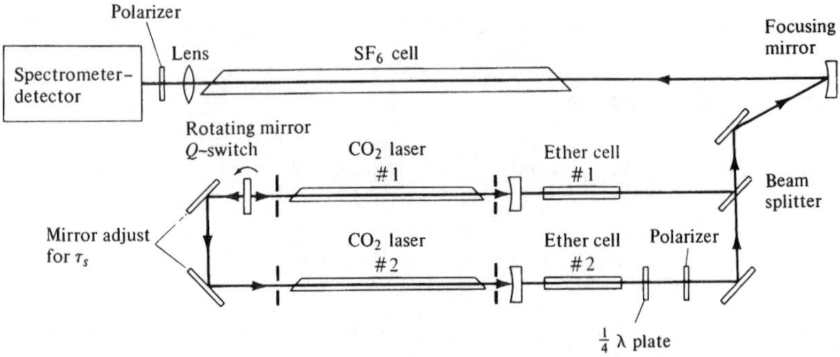

Figure 12.7
Apparatus for detecting infrared photon echoes in gases. The lasers are Q switched by the same rotating mirror, with the interval between the pulses adjusted by means of the two 45° mirrors. The ether cell acts as a neutral attenuator for the CO_2 laser radiation, so that the intensities can be adjusted to produce the proper pulse amplitudes. [From J. P. Gordon, C. H. Wang, C. K. N. Patel, R. E. Slusher, and W. J. Tomlinson, *Phys. Rev.* **179**, 294 (1969). Reproduced with permission.]

ants falling off the pancake, or runners having heart attacks) can be treated by adding the suitable relaxation terms, namely $-u/T_2'$, $-v/T_2'$, and $-(w - w^{eq})/T_1$, to the optical Bloch equations. This complicates the solutions, of course; the essential result is that $u(t) + iv(t)$ goes as $e^{i\Delta t_{21}}e^{-t_{21}/T_2'}$. The echo amplitude thus decays with a time constant T_2', furnishing another method for the measurement of this quantity by varying the delay between the π and $\pi/2$ pulses and recording the decrease in the echo amplitude.

The derivation of the photon echo amplitude places some constraints on the experimental parameters. The pulse amplitudes must be such that $(\kappa\mathscr{E}_0) > (\Delta \sim 1/T_2^*)$, but at the same time the pulse durations t_{10} and t_{32} must be less than T_2'. In the optical regime, this restricts systems showing echo phenomena to either low-temperature solids or low-pressure gases. Thus, while the spin echo was observed in 1950 (reference 9), the first optical photon echo, using ruby laser pulses and a ruby crystal at liquid helium temperature, was not observed until 1964 (reference 10). The first photon echo in a gas (SF_6, using CO_2 lasers) was seen by Patel and Slusher in 1968 (reference 11), using the apparatus shown in figure 12.7. The appearance of the polarizing pulses and the echo is shown in figure 12.8. By using the Stark-switching method to produce the π and $\pi/2$ pulses, echo experiments have been carried out in a large number of molecules. These results are

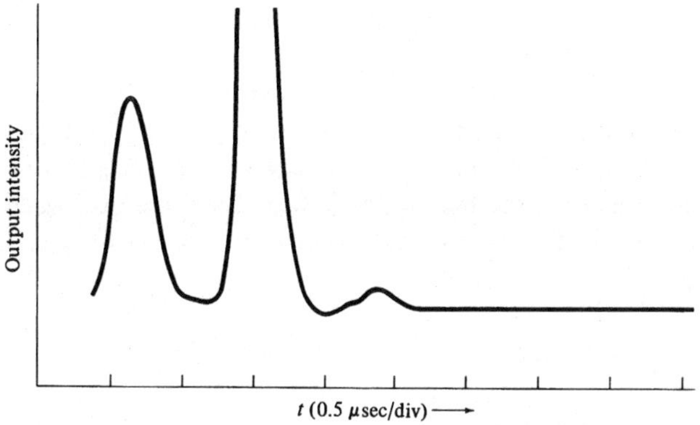

Figure 12.8
Photon echo in SF_6. The first pulse is the 90° pulse, the second (off scale) the 180° pulse, and the third small pulse is the "echo" of superradiant infrared emission. [From J. P. Gordon, C. H. Wang, C. K. N. Patel, R. E. Slusher, and W. J. Tomlinson, *Phys. Rev.* **179**, 294 (1969). Reproduced with permission.]

described in reference 5, which also discusses a variety of more complicated multiple-pulse experiments.

One additional aspect of photon echoes is their directionality, that is, that the emitted echo radiation is propagated in a particular direction in space determined by the incident beams. This is an example of a coherent propagation effect, which we shall consider in great detail in section 4. Before doing so, however, let us consider an important recent application of coherent transient techniques of spectroscopy, namely, the use of Fourier transform techniques.

3 Fourier Transform Spectroscopy

In our treatment of coherent transient phenomena, such as transient nutation or photon echoes, we have thus far considered only a single resonant transition interacting with the radiation field. What if several nearly resonant transitions are excited simultaneously? The result would be a *superposition* of the transient signals arising from each of the individual transitions. If such a time-varying signal is transformed back into the frequency domain, the frequency differences between each of the transitions and the applied field will be displayed. This can be very helpful in elucidating details of spectra in which the frequency differences are too small to be resolved by conventional means, or in which the absorption signals are too weak to be observed directly.[2]

This principle has been exploited in a series of experiments by the late W. H. Flygare. A sequence of microwave pulses is applied to a sample; the pulses, of duration t_p and separated by an interval t_0, are generated by directly switching on and off the microwave power. There are no special pulse area restrictions; all that is required is that $t_p \ll (T_2', T_1) \ll t_0$ and that the field amplitude

$$\kappa\mathscr{E}_p = \frac{\mu\mathscr{E}_p}{\hbar} \gg |\Delta\omega_j| = |\omega_j - \omega_p|.$$

Thus signal is collected from all transitions lying within a "Rabi frequency" of the pump frequency ω_p. The repetitive transient signal is averaged and then Fourier analyzed to yield the spectrum. A detailed description of the method may be found in reference 12.

This technique has found its greatest usefulness in measuring microwave spectra of van der Waals complexes in a supersonic nozzle beam, which is injected into a Fabry-Perot microwave cavity. Systems that have been studied in this way include hydrogen halide complexes with rare gases (references 13, 14), nitrogen (reference 15), hydrogen cyanide (reference 16), and hydrocarbons (references 17, 18). A typical set of spectra, for the acetylene-HF complex, is shown in figure 12.9.

4 Self-Induced Transparency

In our previous discussion of transient nutation, photon echo, and so on, we have not explicitly considered the *propagation* of the coherent radiation through an absorbing medium. We now shall turn our attention to this subject, generally following the treatment given by Allen and Eberly (reference 7) and Bullough et al. (reference 19).

Let us consider a pulse of monochromatic light for which the pulse width τ_p is much greater than ν^{-1}, so that there are many oscillations of the radiation field during the pulse. However, τ_p is at the same time much less than T_2', the relaxation time of the system we are going to investigate. This system is taken to be a simple two-level system, with $E_a - E_b = \hbar\omega_0$ and $\omega_0 = 2\pi\nu$, so that the field is in resonance with the system. Under ordinary circum-

2. This technique is widely used in NMR, where commercial instrumentation for Fourier transform spectroscopy can be obtained.

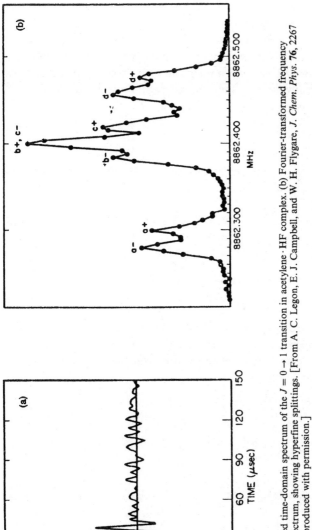

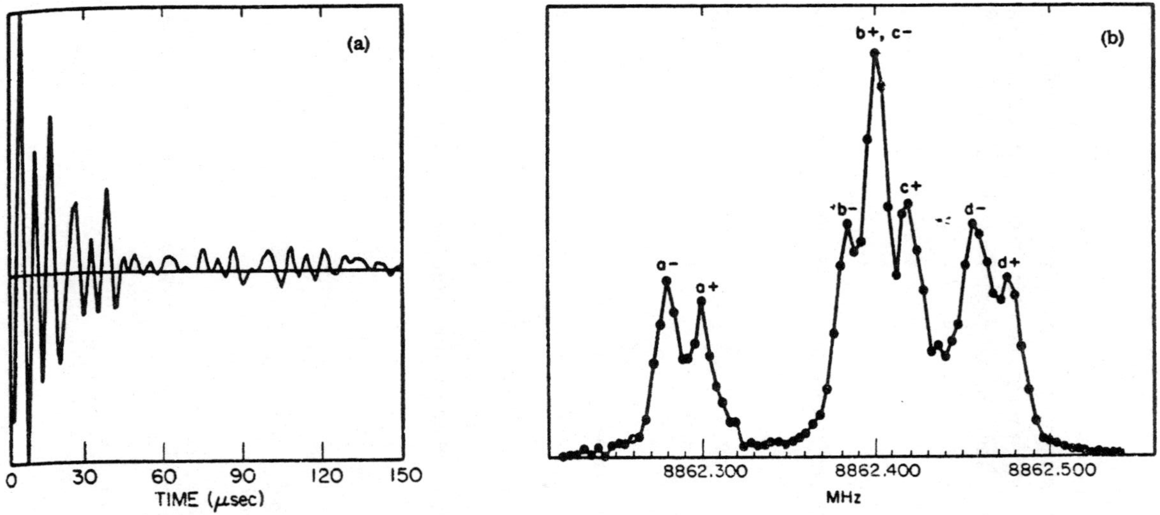

Figure 12.9
(a) Observed time-domain spectrum of the $J = 0 \rightarrow 1$ transition in acetylene · HF complex. (b) Fourier-transformed frequency domain spectrum, showing hyperfine splittings. [From A. C. Legon, E. J. Campbell, and W. H. Flygare, *J. Chem. Phys.* **76**, 2267 (1982). Reproduced with permission.]

stances, a definite fraction of the incident radiation would be absorbed on passing through the medium, independent of the incident intensity. For the right pulse conditions, however, the amount of light transmitted may be much greater than that predicted by the linear absorption coefficient; in terms of the two-level Rabi solution, the incident photons may be absorbed and reemitted many times during their passage through the medium.

In free space, our plane-polarized light pulse has the form[3]

$$\mathbf{E} = \hat{\mathbf{i}}\mathscr{E}(z, t)\cos[\nu t - kz]. \tag{12.9}$$

To see how the pulse will be modified in going through the medium, we shall need the appropriate Maxwell's equations in the presence of a nonconducting, nonmagnetic medium. These are

$$\begin{aligned} &\nabla \cdot \mathbf{D} = 0, \qquad \nabla \cdot \mathbf{B} = 0, \\ &\nabla \times \mathbf{E} = -\frac{\partial \mathbf{B}}{\partial t}, \qquad \nabla \times \mathbf{H} = -\frac{\partial \mathbf{D}}{\partial t}, \end{aligned} \tag{12.10}$$

with

$$\mathbf{D} = \epsilon_0 \mathbf{E} + \mathbf{P} \quad \text{and} \quad \mathbf{B} = \mu_0 \mathbf{H}, \tag{12.11}$$

where $\mathbf{P}$ is the macroscopic polarization of the medium, and ϵ_0 and μ_0 are the electrical permittivity and magnetic susceptibility, respectively, of free space ($\epsilon_0 \mu_0 = 1/c^2$). We obtain the wave equation for the electric field in the usual way.

$$\nabla \times \left(\frac{\mathbf{B}}{\mu_0}\right) = \epsilon_0 \frac{\partial \mathbf{E}}{\partial t} + \frac{\partial \mathbf{D}}{\partial t},$$

$$\begin{aligned} \nabla \times (\nabla \times \mathbf{E}) &= -\nabla \times \left(\frac{\partial \mathbf{B}}{\partial t}\right) = -\frac{\partial}{\partial t}(\nabla \times \mathbf{B}) \\ &= -\epsilon_0 \mu_0 \frac{\partial^2 \mathbf{E}}{\partial t^2} - \mu_0 \frac{\partial^2 \mathbf{P}}{\partial t^2}, \end{aligned}$$

or

$$\nabla^2 \mathbf{E} + \mu_0 \epsilon_0 \frac{\partial^2 \mathbf{E}}{\partial t^2} = -\mu_0 \frac{\partial^2 \mathbf{P}}{\partial t^2}.$$

3. A full description would include a phase term $\delta(z, t)$ in the argument of the cosine in (12.9), but this can be conveniently neglected in the treatment that follows.

To find $\nabla^2\mathbf{E}$ for our plane-polarized light pulse, we must first evaluate

$$\nabla \times \mathbf{E} = \begin{vmatrix} \hat{i} & \hat{j} & \hat{k} \\ \partial_x & \partial_y & \partial_z \\ \mathbf{E} & 0 & 0 \end{vmatrix} = \hat{j}\frac{\partial \mathbf{E}}{\partial z},$$

and then

$$\nabla^2\mathbf{E} = \nabla \times (\nabla \times \mathbf{E}) = \begin{vmatrix} \hat{i} & \hat{j} & \hat{k} \\ \partial_x & \partial_y & \partial_z \\ 0 & \dfrac{\partial \mathbf{E}}{\partial z} & 0 \end{vmatrix} = -\hat{i}\frac{\partial^2 \mathbf{E}}{\partial z^2}.$$

That is, all the vector quantities in the wave equation point in the x direction, so that we can write it as a single scalar equation,

$$-\frac{\partial^2 E}{\partial z^2} + \mu_0\epsilon_0\frac{\partial^2 E}{\partial t^2} = -\mu_0\frac{\partial^2 P_x}{\partial t^2}. \tag{12.12}$$

We assume that polarization $P = P_x$ will have the same time and space dependence as E, multiplied by some linear response factors:

$$P = C(z,t)\cos[\nu t - kz] + S(z,t)\sin[\nu t - kz]; \tag{12.13}$$

for this form of P, $\partial^2 P/\partial t^2 = -\nu^2 P$, so that

$$\frac{\partial^2 E}{\partial z^2} - \mu_0\epsilon_0\frac{\partial^2 E}{\partial t^2} = -\mu_0\nu^2 P.$$

If we substitute equation (12.9) for the electric field E and collect the cosine and sine terms separately, we obtain two equations, namely, cosine terms

$$\frac{\partial^2 \mathscr{E}}{\partial z^2} - \mathscr{E}k^2 - \mu_0\epsilon_0\left[\frac{\partial^2 \mathscr{E}}{\partial t^2} - \mathscr{E}\nu^2\right] = -\nu^2 C(z,t)$$

and sine terms

$$-2\frac{\partial \mathscr{E}}{\partial z}(-k) + \mu_0\epsilon_0\left[2\nu\frac{\partial \mathscr{E}}{\partial t}\right] = -\nu^2 S(z,t).$$

If we make the reasonable assumptions that the envelope of the pulse is slowly varying, $\partial^2\mathscr{E}/\partial t^2$ and $\partial^2\mathscr{E}/\partial z^2$ can be neglected relative to the first derivatives. Then the equation for $C(z,t)$ reduces to

$$-\mathscr{E}k^2 + \mu_0\epsilon_0\mathscr{E}\nu^2 = -\nu^2 C(z,t).$$

But in a transparent medium, $\mu_0\epsilon_0 = 1/c^2 = k^2/\nu^2$, so that the two terms on the left-hand side cancel, giving $C(z,t) = 0$. In other words, there is no component of the polarization **P** which is in phase with the electric field. For the sine terms,

$$-\frac{\nu^2}{2k}S(z,t) = \frac{\partial\mathscr{E}}{\partial z} + \mu_0\epsilon_0\frac{\nu}{k}\frac{\partial\mathscr{E}}{\partial t},$$

or

$$-\frac{\nu^2}{2k}S(z,t) = \frac{\partial\mathscr{E}}{\partial z} + \frac{1}{c}\frac{\partial\mathscr{E}}{\partial t}, \tag{12.14}$$

which is a wave equation for the envelope function $\mathscr{E}(z,t)$.

So far, all we have done is develop Maxwell's equations for the propagation of a pulse through a medium with some response function $S(z,t)$. In order to relate this to the microscopic properties of the systems making up the medium, we need some description in terms of energy levels. This is provided by the results we obtained in the previous section for coherently driven transitions in a two-level system. If the total wave function for our system is

$$\Psi(r,t) = a(t)\psi_a + b(t)\psi_b,$$

then the time-varying polarization will be given by

$$\begin{aligned} P(z,t) &= \langle\bar{\bar{\mu}}_{ab}\rangle \\ &= N\mu_{ab}\int[u\cos(\nu t - kz) - v\sin(\nu t - kz)]g(\Delta')\,d\Delta', \end{aligned}$$

where $u = a^*b + b^*a$, $v = a^*b - b^*a$, and $g(\Delta')$ is the distribution over detunings from exact resonance. The result of our analysis of Maxwell's equations has shown that

$$P = C\cos(\nu t - kz) + S\sin(\nu t - kz),$$

and we have seen that $C = 0$ in free space and

$$S(z,t) = -\frac{2k}{\nu^2}\left(\frac{\partial\mathscr{E}}{\partial z} + \frac{1}{c}\frac{\partial\mathscr{E}}{\partial t}\right).$$

We therefore must have

$$\frac{\partial\mathscr{E}}{\partial z} + \frac{1}{c}\frac{\partial\mathscr{E}}{\partial t} = \frac{\nu^3}{2k}N\mu_{ab}\int v(t,x,\Delta')g(\Delta')\,d\Delta'. \tag{12.15}$$

Equation (12.15) is a simple form of a *coupled Maxwell-Bloch equation*; this type of equation is very important in quantum electronics. Several methods are available for the solution; here, we shall follow the procedure outlined in reference 7.

The optical Bloch equations, including relaxation, which govern the time evolution of the system are those given in (11.25). Since we have already specified that τ_p is much shorter than the relaxation times T_1 and T_2', we can drop the relaxation terms to arrive at a simplified set of equations,

$$\dot{u} = -\Delta v,$$

$$\dot{v} = \Delta u + \kappa\mathscr{E}w,$$

$$\dot{w} = -\kappa\mathscr{E}v,$$

which corresponds to (11.22), except that $\mathscr{E}$ is not a constant, but is slowly varying in time.

The simplest form of the solution is obtained by specifying exact resonance ($\Delta = 0$), and the system initially dephased with all molecules in the lower level, so that

$$\boldsymbol{\rho}(0) = (0, 0, -1).$$

Then $du/dt = 0$ and $u(t = 0) = 0$, so that u remains equal to zero at all times. The preceding set of equations reduces to the trivially simple form

$$\dot{v} = \kappa\mathscr{E}w,$$

$$\dot{w} = -\kappa\mathscr{E}v,$$

which can be combined into a single second-order equation,

$$\ddot{w} = -(\kappa\mathscr{E})^2 w.$$

This is, of course, just the two-level Rabi solution discussed in chapter 11.2.

The solutions for the propagating radiation field can be written

$$v(t, z, \Delta = 0) = -\sin\gamma(t, z),$$

$$w(t, z, \Delta = 0) = -\cos\gamma(t, z),$$

where we define

$$\gamma(t, z) = \int_{-\infty}^{t} \kappa\mathscr{E}(t', z)\, dt'. \qquad (12.16)$$

For the full pulse, the area $\theta(z)$ [see equation (11.8)] is

$$\theta(z) = \int_{-\infty}^{\infty} \kappa \mathscr{E}(t', z)\, dt' = \gamma(\infty, z) \tag{12.17}$$

and

$$\frac{d\theta}{dz} = \kappa \int_{-\infty}^{\infty} \frac{\partial \mathscr{E}}{\partial z} dt'.$$

Note, however, that

$$\int_{-\infty}^{\infty} \left[\frac{\partial \mathscr{E}}{\partial z} + \frac{1}{c} \frac{\partial \mathscr{E}}{\partial t'} \right] dt' = \int_{-\infty}^{\infty} \frac{\partial \mathscr{E}}{\partial z} dt' + \frac{1}{c} \mathscr{E} \bigg|_{t'=-\infty}^{t'=\infty}$$

and that the last term vanishes because the pulse envelope goes to zero as t goes to $\pm\infty$. Thus

$$\begin{aligned} \frac{d\theta}{dz} &= \kappa \int_{-\infty}^{\infty} \frac{\partial \mathscr{E}}{\partial z} dt' \\ &= \kappa \int_{-\infty}^{\infty} \left[\frac{\partial \mathscr{E}}{\partial z} + \frac{1}{c} \frac{\partial \mathscr{E}}{\partial t'} \right] dt' \\ &= \frac{\mu_{ab}}{\hbar} \cdot \frac{\nu^2}{2\kappa} N \mu_{ab} \int_{-\infty}^{\infty} dt' \int v(t, z, \Delta') g(\Delta')\, d\Delta'. \end{aligned}$$

If we consider just the on-resonance case, that is, $\Delta' = 0$, the integral over the inhomogeneous lineshape reduces to a factor $g(\Delta' = 0) = g(\nu)$, and we have

$$\frac{d\theta}{dz} = -\frac{N \mu_{ab}^2 \nu^2 g(\nu)}{2\kappa\hbar} \int_{-\infty}^{\infty} \sin \gamma(t', z)\, dt'. \tag{12.18}$$

Remember that at the upper limit of the integral in (12.18), $t' = +\infty$ and $\gamma = \theta(z)$; at the lower limit, $t' = -\infty$ and $\gamma = 0$. If we define the optical absorption coefficient of the medium as

$$\alpha(\nu) = \frac{N \mu_{ab}^2 \nu^2 g(\nu)}{\kappa\hbar}$$

and note that the limit of $\sin \gamma(t', z)$ as $t' \to \infty$ is just $\sin \theta(z)$, we obtain the simple and useful result

$$\frac{d\theta}{dz} = -\frac{\alpha}{2} \sin \theta. \tag{12.19}$$

Equation (12.19) is frequently called the "Hahn-McCall area theorem"; Hahn, however, has termed this equation "Booze's law," since it is actually a stronger form of Beer's law, as we shall shortly see.

Let us see what (12.19) has to say about pulse propagation through an absorbing medium. The integral form of (12.19) is

$$\int \frac{d\theta}{\sin\theta} = -\int_0^l \frac{\alpha}{2}\,dz$$

for small θ, which can be obtained from low-power pulses, $\sin\theta \simeq \theta$, so that

$$\int \frac{d\theta}{\theta} = \ln\theta - \ln\theta_0 = -\frac{\alpha}{2}l,$$

or

$$\frac{\theta}{\theta_0} = \frac{(2\mu_{ab}/\hbar)\displaystyle\int_{-\infty}^{\infty} \mathscr{E}(l,t)\,dt}{(2\mu_{ab}/\hbar)\displaystyle\int_{-\infty}^{\infty} \mathscr{E}(0,t)\,dt} = e^{-\alpha l/2}.$$

For this to be true, for arbitrary pulse shape, we must have

$$\frac{\mathscr{E}(l)}{\mathscr{E}(0)} = e^{-\alpha l/2},$$

so that the attenuation of the optical energy is

$$\frac{I}{I_0} = \left|\frac{\mathscr{E}^2(l)}{\mathscr{E}^2(0)}\right| = e^{-\alpha l},$$

which is, of course, just Beer's law for absorption. For large θ, which is obtained from high-power pulses, direct integration of equation (12.19) yields

$$\int \frac{d\theta}{\sin\theta} = \log\tan\frac{\theta}{2}\bigg|_{\theta_0}^{\theta(l)} = -\frac{\alpha}{2}l,$$

so that

$$\tan[\tfrac{1}{2}\theta(l)] = \tan[\tfrac{1}{2}\theta(0)]\,e^{-\alpha l/2}. \tag{12.20}$$

Equation (12.19) is a transcendental equation, which must be solved for the actual transmitted pulse shape. We can see its behavior for certain cases. If $0 < \theta < \pi$, then $\sin\theta > 0$, $d\theta/dz < 0$, so that

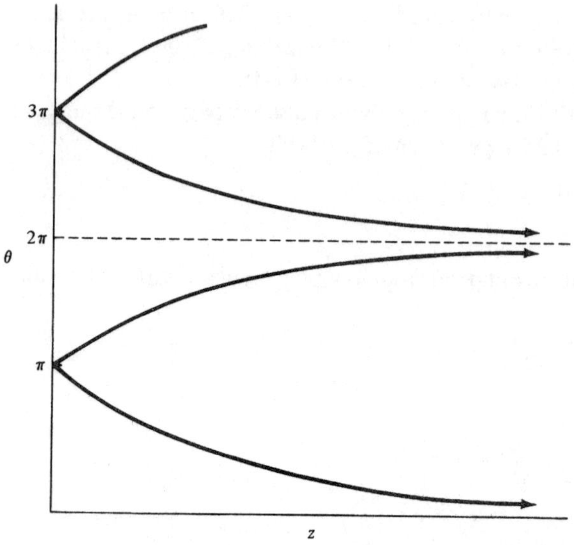

Figure 12.10
The behavior of $\theta(z)$ for various multiples of π.

$$\left(\frac{2\mu_{ab}}{\hbar}\right)\int \mathscr{E}\,dt \rightarrow 0;$$

that is, the pulse is attenuated. If $\pi < \theta < 2\pi$, however, then $d\theta/dz > 0$, and

$$\left(\frac{2\mu_{ab}}{\hbar}\right)\int \mathscr{E}\,dt$$

increases until $\theta = 2\pi$, when $d\theta/dz = 0$. Thus we may say that a pulse transmitted thorough this medium "seeks the nearest 2π pulse shape." The situation is depicted schematically in figure 12.10. Whether a particular pulse is attenuated or not depends on whether it has a value of θ slightly less than or slightly more than an odd multiple of π, as shown in figure 12.11. When a 2π pulse is reached, it is transmitted without further attenuation.

The reader may be worrying at this point about the conservation of energy, with $\theta(z)$ increasing as the pulse goes through the medium. There is no difficulty about this point—the pulse *energy* is

$$\Phi = \frac{4\pi}{c}\int \mathscr{E}^2(z,t)\,dt,$$

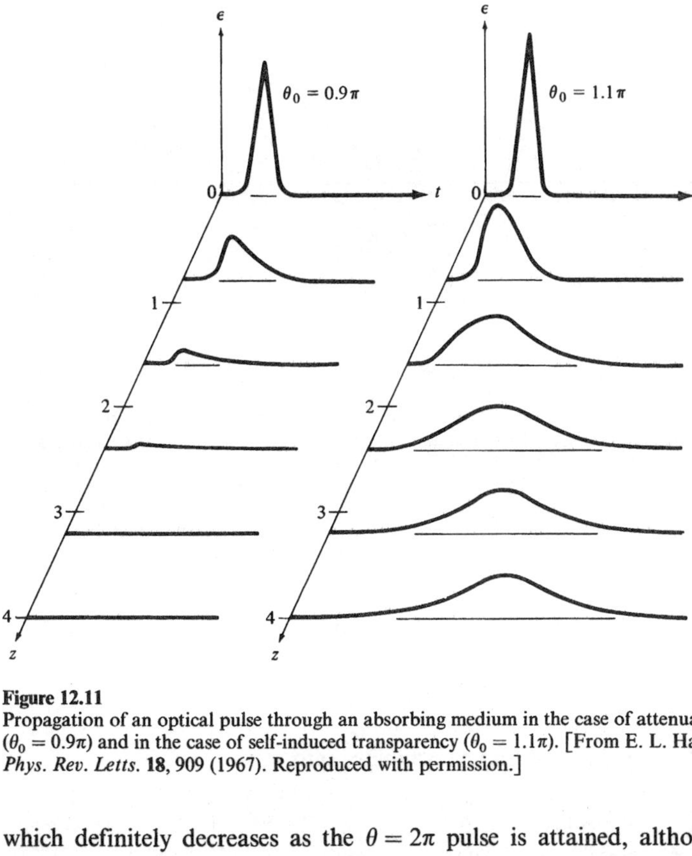

Figure 12.11
Propagation of an optical pulse through an absorbing medium in the case of attenuation ($\theta_0 = 0.9\pi$) and in the case of self-induced transparency ($\theta_0 = 1.1\pi$). [From E. L. Hahn, *Phys. Rev. Letts.* **18**, 909 (1967). Reproduced with permission.]

which definitely decreases as the $\theta = 2\pi$ pulse is attained, although not nearly as much as a simple Beer's law for absorption would predict.

Hahn has suggested an ingenious mechanical analogy for self-induced transparency, which is shown in figure 12.12. The basis for this model (which, unfortunately, has not yet actually been constructed), is an important mathematical similarity between the pulse propagation analysis which we have presented here and the equation of motion for a physical pendulum. Recall the definition of the time-dependent integral pulse area in (12.16), from which we can write

$$\mathscr{E}(z,t) = \frac{\hbar}{\mu_{ab}} \frac{\partial \gamma(z,t)}{\partial t}.$$

The one-dimensional Maxwell-Bloch equation (12.15), with the notation

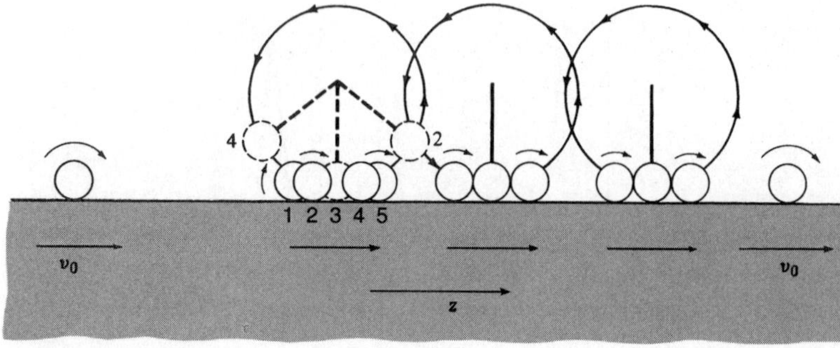

Figure 12.12
Hahn's mechanical analogy for self-induced transparency. The light pulse is the ball rolling in from the left at velocity v_0. The absorbing system is represented by the collection of freely swinging pendulums. If the momentum with which the ball enters the system is such that, when it strikes the bob of the first pendulum, the bob is not able to travel through an angle of 180°, then the ball will be stopped, or "absorbed," by the system. If the momentum is great enough, however, for the pendulum to go through π and come back down on the other side, then the pendulum smacks the ball from the rear and sends it on to the next pendulum, and so on through the entire system. This is an analogue of a "π pulse" becoming a "2π pulse." [From *Physics Today* **20**(8), 47 (August 1967). Reproduced with permission.]

we have introduced and with $\Delta = 0$ so that $w = v$, is

$$\frac{\partial \mathscr{E}}{\partial z} + \frac{1}{c}\frac{\partial \mathscr{E}}{\partial t} = \frac{v^2}{2\kappa} N\mu_{ab} g(v) \sin\gamma(z, t).$$

The preceding equations can be combined into

$$\frac{\partial^2 \gamma}{\partial z\,\partial t} + \frac{1}{c}\frac{\partial^2 \gamma}{\partial t^2} = \frac{\alpha}{2}\sin\gamma,$$

which is technically known as a "sine-Gordon equation" and appears in the theory of the pendulum (see reference 19).

In order to have a pulse that propagates at velocity v without distortion, that is, a "2π-pulse," we require that

$$\gamma(z, t) = \gamma\left(t - \frac{z}{v}\right),$$

so that

$$\frac{\partial \gamma}{\partial z} = -\frac{1}{v}\frac{\partial \gamma}{\partial t},$$

or

$$-\frac{1}{v}\frac{\partial^2\gamma}{\partial t^2}+\frac{1}{c}\frac{\partial^2\gamma}{\partial t^2}=\frac{\alpha}{2}\sin\gamma. \tag{12.21}$$

The reader will recognize (12.21) as the equation for a pendulum. The solution to (12.21) can be variously written as

$$\gamma = 2\sin^{-1}\tanh\left[\frac{1}{\tau}\left(t-\frac{z}{v}\right)\right]$$

(reference 19) or as

$$\gamma = 4\tan^{-1}\exp\left[\frac{1}{\tau}\left(t-\frac{z}{v}\right)\right],$$

where τ is a characteristic period of the system. (reference 7); the proof of the identity of these and other expressions is left as an exercise (problem 1). In any case, the pulse shape found from either of these solutions is

$$\mathscr{E}=\frac{\hbar}{\mu_{ab}}\frac{\partial\gamma}{\partial t}=\frac{\hbar}{\mu_{ab}\tau}\operatorname{sech}\left[\frac{1}{\tau}\left(t-\frac{z}{v}\right)\right], \tag{12.22}$$

which can be shown to have an integrated pulse area equal to 2π (problem 2). The propagation velocity v for this pulse through the medium is given by

$$\frac{1}{v}=\frac{1}{c}+\frac{\alpha\tau_p}{2}. \tag{12.23}$$

Note that this velocity can be very much less than the speed of light c, or even c/n (n = refractive index, which is typically between 1.0 and 5.0). For example, we can easily have $\alpha \approx 1$ cm^{-1}, so that with $\tau_p \approx 10^{-8}$ sec,

$$\frac{1}{v}=\frac{1}{3\times10^{10}}+\frac{1\times10^{-8}}{2},$$

or $v \approx 1.98\times10^{8}$ cm sec^{-1} $\approx 0.0066c$; the "effective" refractive index is on the order of 150. This slowing down of course leads to a substantial pulse delay, given by

$$\tau_d=\frac{l}{v}-\frac{l}{c}=\tfrac{1}{2}(\alpha l\tau_p).$$

In this example, with a propagation length $l = 10$ cm,

$$\tau_d = \tfrac{1}{2}(1 \times 10 \times \tau_p) = 5\tau_p,$$

which would be easily observed.

Self-induced transparency was first observed experimentally in two systems: in a low-temperature solid, ruby laser pulses passing through a ruby crystal, by Hahn and McCall in 1967 (reference 21), and in a low-pressure gas, CO_2 laser pulses passing through SF_6 gas, by Patel and Slusher later that same year (reference 22). In the optical experiment, the light source is a ruby laser operating at 77°K, and the absorber is a ruby crystal at 4.2°K. The energy levels for this system have been shown in figure 10.12a; the radiation in this case falls at 6943 Å. The temperatures of the laser and the absorber have been chosen to make the line widths and relaxation times satisfy the condition $(\Delta\nu)^{-1} \ll \tau_p \ll T_2$.

We can calculate the intensity of radiation required to produce transparency. The radiative lifetime of the ^2E level in ruby has been measured to be 0.003 sec. From the radiation relations given in chapter 1.7,

$$A = \tau_{\text{rad}}^{-1} = \frac{g_1}{g_2} B_{12} 8\pi h \left(\frac{\nu}{c}\right)^3$$

$$= \frac{g_1}{g_2} \frac{64\pi^4}{3} \frac{\mu_{ab}^2}{h} \left(\frac{\nu}{c}\right)^3,$$

so that

$$\mu_{ab}^2 = \frac{g_2 3h}{g_1 64\pi^4 (\nu/c)^3 \tau_{\text{rad}}}$$

$$= 3.5 \times 10^{-40} \text{ (esu cm)}^2;$$

hence $\mu_{ab} = 1.9 \times 10^{-20}$ esu cm $= 0.019$ D. In order to produce transparency, we need

$$\frac{\mu_{ab}}{\hbar} \int_{-\infty}^{\infty} \mathscr{E}(t)\, dt \approx \frac{\mu_{ab}}{\hbar} \bar{\mathscr{E}} \tau_p \gtrsim \pi,$$

where $\bar{\mathscr{E}}$ is, approximately, the peak electric field of the pulse. The Q-switched ruby laser used in the experiment has a pulse width of approximately 2×10^{-8} sec, so that the field required is

$$\bar{\mathscr{E}} \approx \frac{\pi\hbar}{\mu_{ab}\tau_p}$$

$$\approx 8 \text{ statV/cm}.$$

The intensity necessary to produce this field may be found from

$$I = \frac{c}{4\pi}|\bar{\mathscr{E}}|^2$$

$$\approx 1.6 \times 10^{11}\ \text{erg/cm}^2\,\text{sec} \approx 16\ \text{kW/cm}^2.$$

This intensity is easily obtainable from a Q-switched laser, in which peak powers in the MW/cm^2 range are easily obtainable; thus the experiment does indeed work.

In the analogous infrared experiment, the laser used is a CO_2 laser operating at 10.59 μm (see figure 10.14), and the absorber is a gas of SF_6 molecules, which has an exceedingly strong absorption to a ν_3 vibration at that wavelength. The absorber linewidth, in this case, is predominantly due to Doppler broadening, for which

$$(\Delta\nu_D) \approx 7 \times 10^{-7}\sqrt{\frac{T}{M}}\,\nu_0$$

$$\approx 3 \times 10^7\ \text{Hz},$$

so that $(\Delta\nu)^{-1} \approx 3 \times 10^{-8}$ sec. The T_2 relaxation time may be assumed to be equal to the gas kinetic mean free time between collisions, which is approximately 10^{-5} sec at a pressure of 10 mTorr; the restrictions on the pulse width are easily met by pulses produced by a Q-switched CO_2 laser, for which $\tau_p = 3 \times 10^{-7}$ sec, as long as the absorber pressure is kept at a few hundredths of a torr or below. The dipole strength of the transition is found, in this case from the integrated absorption, to be 3×10^{-19} esu cm, and the radiation field necessary for transparency is 0.017 statV/cm, corresponding to an intensity of 0.1 W/cm^2. In the experiment, transparency begins to appear at a power level of the order of 10 W/cm^2; this results from the fact that the SF_6, far from being an ideal two-level system, is actually excited by the laser to a high rotational level with $J = 59$ (reference 23) having extensive M_J degeneracy.

Perhaps the classic self-induced transparency experiment has been carried out by Slusher and Gibbs (reference 24) in Rb vapor. The 5^2P–5^2S transition at 7,947.64 Å was tuned into resonance with a Hg^+ laser oscillating at 7,944.66 Å by placing the Rb atoms in a 74.5-KG magnetic field. The result is shown in figure 12.13. Local maxima in the transmission are observed at $\theta = 2\pi$, 4π, and 6π; in addition, very large pulse delays are also observed.

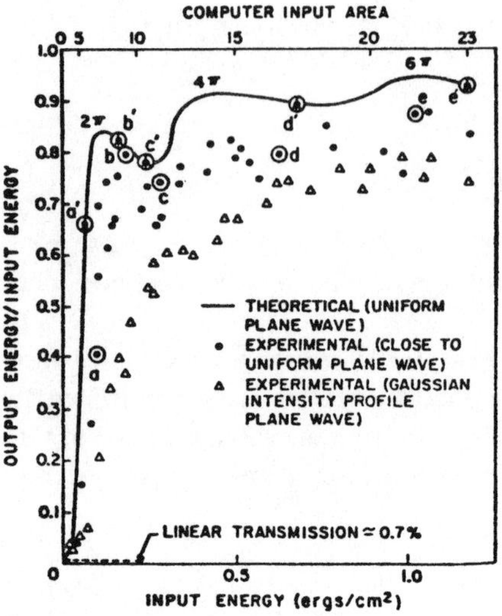

Figure 12.13
Self-induced transparency, or nonlinear transmission, in Rb vapor. Solid curve is a uniform plane wave computer solution. Solid dots are data taken with 200-μm output aperture to approximate a uniform plane wave. Triangles are data with no aperture, corresponding to a plane wave input with Gaussian intensity profile. [From reference 24.]

5 Saturation Revisited

In chapter 11.3, we considered the phenomenon of optical saturation in a two-level system. Let us return to this problem, in terms of the optical Bloch equation treatment we have been exploiting in this chapter. First, let us treat the system as a simple rate equation problem, with an optical pumping rate R coupling levels 1 and 2, and a single relaxation time $1/T_1 = k_{12} + k_{21}$. (This implies that the levels can relax only from one to the other, and not to any other possible state.) The rate equations describing this are

$$\frac{dn_2}{dt} = -Rn_2 + Rn_1 - \frac{n_2}{T_1},$$
$$\frac{dn_1}{dt} = +Rn_2 - Rn_1 + \frac{n_2}{T_1}. \tag{12.24}$$

If we let $n_2 - n_1 = Nw$, where N is the total number of particles in the ensemble and w is the inversion, corresponding to the longitudinal component of the $\boldsymbol{\rho}$ vector, (12.24) can be written as the single equation

$$\frac{d}{dt}(n_2 - n_1) = \frac{dw}{dt} = -2Rw - \frac{(w+1)}{T_1}.$$

However, the optical Bloch equations (11.25) that we have been using state that

$$\frac{dw}{dt} = -\kappa\mathscr{E}v - \frac{(w - w^{\text{eq}})}{T_1}.$$

The obvious question at this point is how can the two preceding expressions for dw/dt be reconciled? This can actually be done quite easily, and an important principle is brought out by doing so. Let us look at the full system of equations (11.25) once again:

$$\dot{u} + \frac{u}{T_2'} = -\Delta v,$$

$$\dot{v} + \frac{v}{T_2'} = \Delta u + \kappa\mathscr{E}w,$$

$$\dot{w} + \frac{w - w^{\text{eq}}}{T_1} = -\kappa\mathscr{E}v.$$

Now, let T_2' be *very small*; specifically, $u/T_2' \gg du/dt$, and also $T_2' \ll T_1$. We can then find an approximate solution by using the steady-state condition, $\dot{u} = \dot{v} = 0$. This gives

$$(u)_{\text{ss}} = -T_2'\Delta(v)_{\text{ss}}$$

and

$$\begin{aligned}(v)_{\text{ss}} &= T_2'\Delta(u)_{\text{ss}} + T_2'\kappa\mathscr{E}w \\ &= -(T_2'\Delta)^2(v)_{\text{ss}} + T_2'\kappa\mathscr{E}w,\end{aligned}$$

so that

$$(v)_{\text{ss}} = \kappa\mathscr{E}T_2'\frac{1}{1 + (\Delta T_2')^2}w$$

and

$$\dot{w} = -\frac{(\kappa\mathscr{E})^2 T_2'}{1 + (\Delta T_2')^2}w - \frac{(w - w^{\text{eq}})}{T_1}.$$

If we let the coefficient of $-w$ in the preceding equation be equal to $2R$, this becomes identical to the rate equation result. The initial rate of transitions from level 1 to level 2, when $n_1 = N$, $n_2 = 0$, and $w = w^{eq} = -1$ is given by

$$R_{1\to 2} = \frac{\mu_{12}^2 \mathscr{E}^2 T_2'}{\hbar^2[1 + (\Delta T_2')^2]},$$

which is seen to possess three factors, first a dipole strength $(\mu_{12}/\hbar)^2$, second an intensity of radiation (proportional to $\mathscr{E}^2$), and third a Lorentzian line shape, corresponding to the perturbation result discussed in chapter 1.

The simple analysis just given can be generalized as an important principle, namely, that a rate equation treatment applies whenever the dephasing rate $1/T_2'$ is much faster than either optical pumping or other (T_1) relaxation rates. The system then remains *incoherent* during its interaction with the radiation field. For a multilevel system, the resulting set of equations is known as a *master equation*; this will be important in the treatment of multiphoton processes to be discussed in the next chapter.

We can go further and recover the expression for the saturated absorption coefficient by considering the solution for w at constant $\mathscr{E}$ and as $t \to \infty$ (that is, $t \gg T_1$). We then also have $dw/dt = 0$, so that a steady-state solution is found for w:

$$(w)_{ss} = \frac{w^{eq}}{1 + \{(\kappa\mathscr{E})^2 T_1 T_2'/[1 + (\Delta T_2')^2]\}}.$$

With $w^{eq} = -1$, as in the preceding discussion, this becomes

$$(w)_{ss} = \frac{1 + (\Delta T_2')^2}{1 + (\Delta T_2')^2 + T_1 T_2'(\kappa\mathscr{E})^2}.$$

This clearly goes to zero for $(\kappa\mathscr{E})^2 \gg (T_1 T_2')^{-1}$. The significance of this is that an incoherent field can never *invert* populations (as in the photon echo or self-induced transparency), but only *saturate*, that is, equalize, them.

If we calculate the power absorbed from the field, we find

$$P_{abs} = -\frac{\hbar w^{eq}\omega}{2T_1} \frac{T_1 T_2'(\kappa\mathscr{E})^2}{1 + (\Delta T_2')^2 + T_1 T_2'(\kappa\mathscr{E})^2},$$

and the saturated absorption coefficient is

$$\alpha(I, \Delta) = \frac{P_{abs}}{P_{in}} = \frac{2\pi(n_1 - n_2)\mu_{12}^2\omega}{\hbar c T_2'}\left[\frac{1}{\Delta^2 + (1/T_2')^2 + (\kappa\mathscr{E})^2}\right],$$

identical with (11.10). It is particularly interesting to look at α for a particular value of the detuning Δ; this predicts a saturated Lorentzian (homogeneous) line shape at $\nu = \nu_0 + \Delta$ superimposed on an inhomogeneous line. This is the principle behind the method of saturated absorption spectroscopy, which we consider in the following section.

6 Saturated Absorption Spectroscopy: The "Lamb Dip"

The principle of saturated absorption spectroscopy is illustrated in figure 12.14. Essentially, the sample to be investigated is subjected to two light waves, of the same frequency, but propagating in opposite directions through the sample. This can be accomplished, for example, by enclosing the sample in a resonant cavity containing the radiation, since the standing waves in such a cavity are equivalent to two traveling waves propagating in opposite directions. If the light is sufficiently intense, it will "burn a hole" into the absorption profile of the gas by saturating the molecules lying within a homogeneous line width of ω (see chapter 1.8). If the light traveling in the $+k$ direction saturates the gas at $\omega = \omega_0 - \delta$, this saturation will affect the transmission of light in the $-k$ direction at $\omega' = \omega_0 + \delta$, by the Doppler effect (figure 12.14b). Under the special circumstances $\omega = \omega' = \omega_0$, that is, when the applied frequency is set to the exact peak of the molecular line, the same set of molecules having zero velocity along the direction of propagation will interact with both light waves simultaneously. This produces a sharp dip in the absorption of light through the sample, known as the "Lamb dip" (after W. E. Lamb, Jr., who first predicted this effect). The line width of this dip is the homogeneous line width $\Delta\nu_N$, which is generally much less than the full Doppler width $\Delta\nu_D$ of the line. Thus the net effect is to produce a very narrow line right at the center position of a molecular transition. If a number of transitions are blended into one inhomogeneously broadened line, this type of saturation spectroscopy provides a method for resolving them.

Numerous methods are available for detecting the Lamb dip. The simplest is to pass the output of a narrow-bandwidth laser through an absorption cell in two counterpropagating directions, by reflecting the beam from a mirror placed behind the cell, and to measure the power absorbed from the beam. Examples of spectra obtained in this way are shown in figure 12.15. If the sample is placed *inside* the laser cavity, the saturated (reduced) absorp-

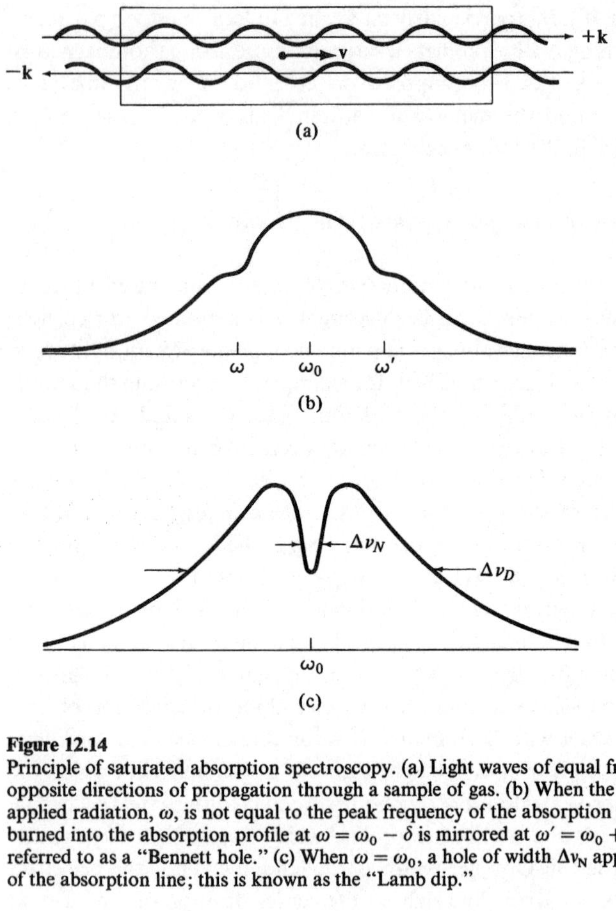

Figure 12.14
Principle of saturated absorption spectroscopy. (a) Light waves of equal frequency and opposite directions of propagation through a sample of gas. (b) When the frequency of the applied radiation, ω, is not equal to the peak frequency of the absorption line, ω_0, a "hole" burned into the absorption profile at $\omega = \omega_0 - \delta$ is mirrored at $\omega' = \omega_0 + \delta$; this is often referred to as a "Bennett hole." (c) When $\omega = \omega_0$, a hole of width $\Delta\nu_N$ appears at the center of the absorption line; this is known as the "Lamb dip."

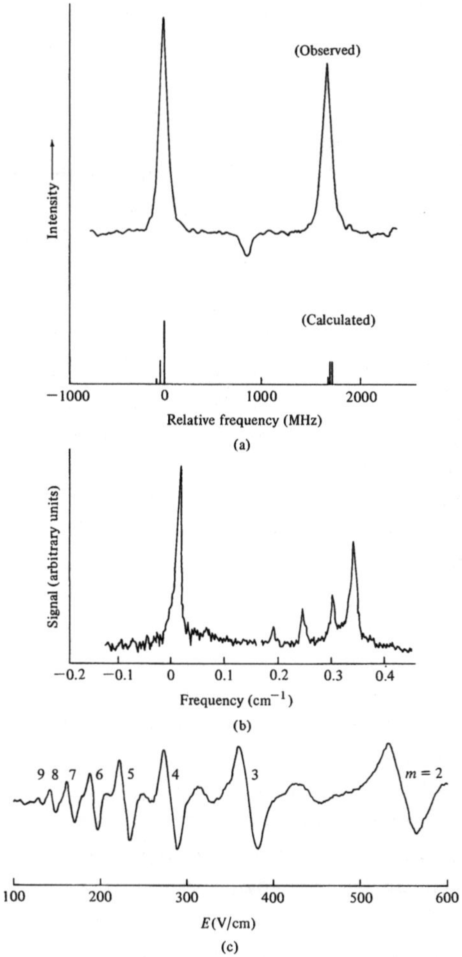

Figure 12.15
Examples of Lamb dip or saturated absorption spectroscopy. (a) Nuclear spin hyperfine structure in the D_2 line of sodium at 5,889.96 Å. This was obtained by measuring the absorption of a beam from a pulsed, narrow-band tunable dye laser through a cell of sodium atom vapor [T. W. Hänsch, I. S. Shahin, and A. L. Schawlow, *Phys. Rev. Letts.* **27**, 707 (1971)]. (b) Nuclear spin hyperfine structure in the H_α Balmer line of hydrogen [$(n = 3) \leftarrow (n = 2)$], obtained by a method similar to that used for spectrum (a) [see T. W. Hänsch, I. S. Shahin, and A. L. Schawlow, *Nature (Phys. Sci.)* **235**, 63 (1972)]. (c) Stark splittings in the $^{Q}P_6(10)$ line of the ν_1 band of methyl fluoride. This spectrum was obtained by putting a sample of CH_3F into the cavity of an He/Ne laser set to oscillate at 2,947.9 cm^{-1}. The resonant frequencies ω_0 of the successive Stark components were then pulled into coincidence with the laser frequency by varying the electric field applied across the cell containing the molecules. [From A. C. Luntz, J. D. Swalen, and R. G. Brewer, *Chem. Phys. Letts.* **14**, 4 (1972). Reproduced with permission.]

tion at the center frequency of the transition produces a sharp spike in laser output at that frequency, known as an "inverted Lamb dip." This is a particularly useful method for stabilizing the laser's output frequency, by locking to the frequency of the inverted Lamb dip; for example, the 3.39-μm transition of the He/Ne laser can be stabilized to a few parts in 10^{13} by using methane as the intracavity absorber. The Lamb dip can also be detected by recording the fluorescence emitted at right angles to the beams, for suitable species such as sodium atoms or iodine molecules (see, for example, reference 25). In the latter system, the inhomogeneous Doppler width in the visible $B \leftarrow X$ transition for room temperature vapor is about 430 MHz (0.014 cm^{-1}), while the natural lifetime (determined primarily by radiative decay, at sufficiently low pressures) is on the order of a few MHz. Thus the Lamb dip method affords an improvement in resolution by a factor of several hundred, which allows the hyperfine structure to be observed and analyzed. The Lamb dip has even been observed in microwave spectra (reference 26); the resolution advantage is much less marked here, since Doppler widths in the microwave are on the order of 10 kHz. Nevertheless, it is possible to reduce the line width to less than 1 kHz by saturated absorption techniques, making possible highly precise measurement of microwave frequencies. Other spectroscopic detection techniques, such as optogalvanic and optoacoustic spectroscopy, have also been utilized in the measurement of Lamb dips. The entire subject has been extensively reviewed in a series of monographs by the Soviet spectroscopists V. Letokhov and V. Chebotayev (references 27–29).

Several other types of Doppler-free two-photon spectroscopy can also be carried out. The optical double-resonance method will be discussed in the context of other double-resonance techniques in the following section, and the two-photon method will be considered in chapter 13.

7 Double-Resonance Spectroscopy

Thus far in this chapter, we have concentrated exclusively on the dynamics of a nondegenerate two-level system interacting with a single radiation field having a specified frequency and field amplitude. For many physical problems, however, we shall need to consider two independent radiation fields interacting with two or three levels in an atom or molecule, a situation that is denoted by the general term "double-resonance spectroscopy." In the

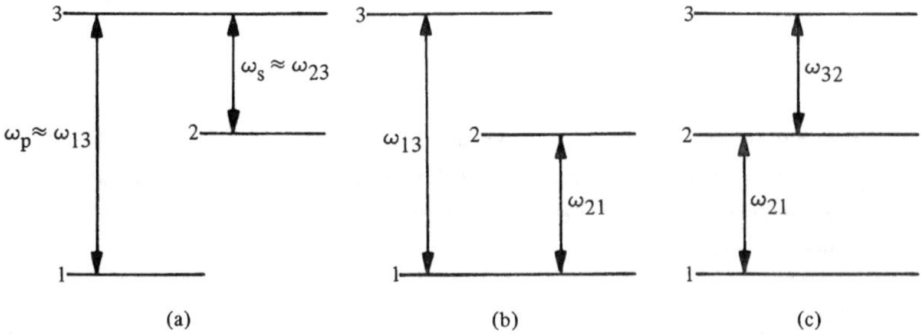

Figure 12.16
Three-level double-resonance configurations: (a) upper level in common, as discussed in the text; (b) lower level in common; (c) intermediate level in common.

theoretical treatment to be given here, we shall follow the approach of Shimoda and Shimizu (reference 30); a review of the experimental literature may be found in reference 31. In dealing with double-resonance problems, the principal change that must be made in our thinking is that we will have to give up the pseudospin vector representation introduced by Feynman; since we now have two waves of different frequencies rotating at the same time, a single rotating frame representation cannot conveniently be found, and the more rigorous but less intuitive algebraic solutions must be employed.

We begin by defining the levels and fields as shown in figure 12.16a.[4] For the purposes of this discussion, we shall consider only the situation when the upper level is the one that is common to the two transitions; the other possible double-resonance configurations are shown in figures 12.16b and 12.16c. We define the "pump" field, having amplitude $\mathscr{E}_p$, as that at frequency $\omega_p \approx \omega_{13} = (E_3 - E_1)/\hbar$; the "probe" field, with amplitude $\mathscr{E}_s$, is at the frequency $\omega_s \approx \omega_{23} = (E_3 - E_2)/\hbar$. The total radiation field experienced by the system is the sum of these two,

$$\begin{aligned}\mathscr{E}(t) &= \tfrac{1}{2}\mathscr{E}_p(e^{i\omega_p t} + e^{-i\omega_p t}) + \tfrac{1}{2}\mathscr{E}_s(e^{i\omega_s t} + e^{-i\omega_s t}) \\ &= \mathscr{E}_p \cos \omega_p t + \mathscr{E}_s \sin \omega_s t.\end{aligned} \tag{12.25}$$

4. One problem encountered in this subject is that the notation for the various energy levels and fields is not at all consistent in the literature, with almost every author choosing a different convention. The remark made by Mark Twain with regard to German grammar (reference 32) is quite appropriate to this problem as well.

To arrive at (12.25), we have chosen the convention that the pump and probe fields are 90° out of phase and have taken $\mathscr{E}_p$ and $\mathscr{E}_s$ to be real envelope functions. In an actual experiment, there may be additional phase terms that would have to be taken into account; the inclusion of these terms only complicates the analysis, without introducing any new physical content.

The wave function for the system can be written as a superposition of the three basis functions,

$$|\psi\rangle = \sum_{n=1}^{3} a_n(t)e^{-iE_nt/\hbar}|n\rangle. \tag{12.26}$$

The equations of motion for the coefficients in (12.26) are just the Schrödinger equation written in the rotating wave approximation (RWA):

$$\frac{da_1}{dt} = \frac{i}{2}(\kappa_{13}\mathscr{E}_p)a_3e^{i(\omega_p-\omega_{31})t},$$

$$\frac{da_2}{dt} = \frac{i}{2}(\kappa_{23}\mathscr{E}_s)a_3e^{i(\omega_s-\omega_{32})t},$$

$$\frac{da_3}{dt} = \frac{i}{2}\{(\kappa_{13}\mathscr{E}_p)^*a_1e^{-i(\omega_p-\omega_{31})t} + (\kappa_{23}\mathscr{E}_s)^*a_2e^{-i(\omega_s-\omega_{32})t}\}.$$

Note that, as yet, no relaxation terms have been included.

Even in the absence of relaxation terms, *no general analytic solutions for $a_n(t)$ are known.* We can find approximate solutions for the case of a "strong" pump field and a "weak" probe field (hence the nomenclature), that is, $\mathscr{E}_p \gg \mathscr{E}_s$. With this assumption, we are able to ignore the second term in the equation for $\dot{a}_3$. If we further assume that the dipole strengths for the two transitions are approximately equal, so that $\kappa_{13} \approx \kappa_{23}$, and that the initial conditions are $a_1(t_0) = e^{i\theta}$, $a_2(t_0) = a_3(t_0) = 0$, we are able to carry out the integral for $\dot{a}_3$ to obtain

$$a_3(t) = i\frac{\kappa_{13}\mathscr{E}_p}{\Omega_p}\exp\left[i\theta - \frac{i}{2}(\omega_p - \omega_{31})(t - t_0)\right]\sin\frac{\Omega_p(t - t_0)}{2},$$

with

$$\Omega_p = [(\omega_p - \omega_{31})^2 + (\kappa_{13}\mathscr{E}_p)^2]^{1/2}.$$

Note that Ω_p is just a generalized Rabi frequency for the pump transition. Substituting this expression for $a_3(t)$ in the equation for $\dot{a}_2$, we obtain

$$\frac{da_2}{dt} = -\frac{(\kappa_{23}\mathscr{E}_s)(\kappa_{13}\mathscr{E}_p)}{2\Omega_p} e^{i\theta} e^{i(\omega_s-\omega_{32}-\omega_p+\omega_{31})t_0}$$
$$\times\, e^{i[2(\omega_s-\omega_{32})-(\omega_p-\omega_{31})](t-t_0)/2} \sin\frac{\Omega_p(t-t_0)}{2}.$$

Integrating $\int_{t_0}^{t} \dot{a}_2(t')\,dt'$ gives the probability amplitude for transitions from level 1 to level 2, which is identical to the coefficient $a_2(t)$ when $a_1(t_0) = 1$. This quantity is

$$a_{1\to2}(t) = \frac{(\kappa_{23}\mathscr{E}_s)(\kappa_{13}\mathscr{E}_p)}{2\Omega_p} \exp[i\theta + i(\omega - \omega_{32} - \omega_p - \omega_{31})t_0]$$
$$\times \left\{ \frac{\exp[i(\Omega_p + \delta)(t-t_0)/2] - 1}{\Omega_p + \delta} + \frac{\exp[-i(\Omega_p - \delta)(t-t_0)/2] - 1}{\Omega_p - \delta} \right\},$$

with $\delta \equiv 2(\omega_s - \omega_{32}) - (\omega_p - \omega_{31})$. The transition probability is found as the square of this amplitude,

$$|a_{1\to2}(t)|^2 = \frac{(\kappa_{23}\mathscr{E}_s)^2(\kappa_{13}\mathscr{E}_p)^2}{\Omega_p^2} \left\{ \frac{\sin^2(\Omega_p + \delta)[(t-t_0)/4]}{(\Omega_p+\delta)^2} + \frac{\sin^2(\Omega_p - \delta)[(t-t_0)/4]}{(\Omega_p-\delta)^2} + \frac{\cos^2\Omega_p[(t-t_0)/2] - \cos\Omega_p[(t-t_0)/2]\cos\delta[(t-t_0)/2]}{\Omega_p^2 - \delta^2} \right\}. \tag{12.27}$$

Corresponding expressions for $a_{3\to2}(t)$ and $|a_{3\to2}(t)|^2$ may be found by similar manipulations.

In all this formalism, we should not lose sight of our ultimate goal, which is to calculate the power absorbed from the probe field at ω_s when the pump field at ω_p is present. In order to do this realistically, we must include relaxation terms, and calculate $\langle|a_{12}|^2\rangle_{\text{avg}}$ and $\langle|a_{32}|^2\rangle_{\text{avg}}$. The simplest relaxation model that we can choose is one in which all population transfer and dephasing rates are equal, with a value of $1/\tau$. Then the mean residence time of an individual system in a particular level is given by a Poisson function,

$$g(t - t_0) = (1/\tau)e^{-(t-t_0)/\tau}, \tag{12.28}$$

and the average transition rate is found by integrating (12.27) over g,

$$\langle|a_{12}|^2\rangle_{\text{avg}} = \int |a_{12}(t-t_0)|^2 g(t-t_0)d(t-t_0)$$

$$= \frac{(\kappa_{23}\mathscr{E}_s)^2(\kappa_{13}\mathscr{E}_p)^2\tau^2}{2\Omega_p^2}\left\{\frac{1}{4+(\Omega_p+\delta)^2\tau^2}+\frac{1}{4+(\Omega_p-\delta)^2\tau^2}\right.$$
$$\left. -\frac{8-(2+\Omega_p^2\tau^2)(\Omega_p^2-\delta^2)\tau^2}{(1+\Omega_p^2\tau^2)[4+(\Omega_p+\delta)^2\tau^2][4+(\Omega_p-\delta)^2\tau^2]}\right\}. \tag{12.29a}$$

A similar expression is found for

$$\langle|a_{32}|^2\rangle_{\text{avg}} = \frac{(\kappa_{23}\mathscr{E}_s)^2\tau^2}{2\Omega_p^2}\left\{\frac{(\Omega_p+\omega_p-\omega_{31})^2}{4+(\Omega_p+\delta)^2\tau^2}+\frac{(\Omega_p-\omega_p+\omega_{31})^2}{4+(\Omega_p-\delta)^2\tau^2}\right.$$
$$\left. +\frac{(\kappa_{13}\mathscr{E}_p)^2[8-(2+\Omega_p^2\tau^2)(\Omega_p^2-\delta^2)\tau^2]}{(1+\Omega_p^2\tau^2)[4+(\Omega_p+\delta)^2\tau^2][4+(\Omega_p-\delta)^2\tau^2]}\right\}. \tag{12.29b}$$

The net power absorbed at the probe frequency is found by using the two preceding expressions in

$$\Delta P(\omega_s) = \frac{\hbar\omega_s}{\tau}\{-(N_1-N_2)\langle|a_{12}|^2\rangle_{\text{avg}} + (N_2-N_3)\langle|a_{32}|^2\rangle_{\text{avg}}\}. \tag{12.30}$$

Note that if $N_1 > N_2 > N_3$, as would be the case for an optical system at Boltzmann equilibrium, the first term of (12.30) gives *negative* absorption, that is, gain at the probe frequency. Physically this is simply the everpopulation of level 3 by optical pumping out of level 1, coupled with stimulated emission from level 3 to level 2. The second term in (12.30) corresponds to "normal" absorption from level 2 to level 3 at the probe frequency.

The preceding expressions are considerably simplified if the pump is on-resonance, that is, $\omega_p = \omega_{31}$ exactly. They $\delta = 2(\omega_s - \omega_{32})$, or just twice the probe detuning. Since in a typical experiment, the pump frequency is held fixed and the probe is scanned across the double-resonance line shape, $\delta/2$ makes a convenient variable against which to plot the result. It is also convenient to introduce a quantity

$$\omega_{\pm} = \omega_s - \omega_{32} \pm \left(\frac{\kappa_{13}\mathscr{E}_p}{2}\right)$$

in terms of which the results of (12.29) take on the pleasingly symmetrical form

$$\langle |a_{12}|^2 \rangle_{\text{avg}} = \frac{(\kappa_{23}\mathscr{E}_s)^2\tau^2}{8} \left\{ \frac{1}{1+\omega_+^2\tau^2} + \frac{1}{1+\omega_-^2\tau^2} - \frac{2 + [2 + (\kappa_{13}\mathscr{E}_p)^2\tau^2)]\omega_+\omega_-\tau^2}{[1 + (\kappa_{13}\mathscr{E}_p)^2\tau^2](1+\omega_+^2\tau^2)(1+\omega_-^2\tau^2)} \right\}$$

and

$$\langle |a_{32}|^2 \rangle_{\text{avg}} = \frac{(\kappa_{23}\mathscr{E}_s)^2\tau^2}{8} \left\{ \frac{1}{1+\omega_+^2\tau^2} + \frac{1}{1+\omega_-^2\tau^2} + \frac{2 + [2 + (\kappa_{13}\mathscr{E}_p)^2\tau^2)]\omega_+\omega_-\tau^2}{[1 + (\kappa_{13}\mathscr{E}_p)^2\tau^2](1+\omega_+^2\tau^2)(1+\omega_-^2\tau^2)} \right\}.$$

The net result for the power absorbed when the pump is on resonance is

$$\Delta P(\omega_s; \omega_p = \omega_{31}) = \frac{\hbar\omega_s(\kappa_{23}\mathscr{E}_s)^2\tau}{8} \left\{ (2N_2 - N_1 - N_3) \times \left(\frac{1}{1+\omega_+^2\tau^2} + \frac{1}{1+\omega_-^2\tau^2} \right) + (N_1 - N_3) \frac{2 + [2 + (\kappa_{13}\mathscr{E}_p)^2\tau^2)]\omega_+\omega_-\tau^2}{[1 + (\kappa_{13}\mathscr{E}_p)^2\tau^2](1+\omega_+^2\tau^2)(1+\omega_-^2\tau^2)} \right\}. \tag{12.31}$$

The first term in (12.31) gives a net gain whenever $N_1 + N_3 > 2N_2$, with two peaks located at

$$\omega_s = \omega_{32} - \frac{\kappa_{13}\mathscr{E}_p}{2} \qquad (\text{when } \omega_+ = 0)$$

and

$$\omega_s = \omega_{32} + \frac{\kappa_{13}\mathscr{E}_p}{2} \qquad (\text{when } \omega_- = 0).$$

That is, the gain term is split into two lines separated by the pump Rabi frequency. This effect, called "resonant modulation splitting," may be seen in the line shape shown in figure 12.17. An experimental example is shown in figure 12.18; in this example, the molecule (deuterated fluoroform, CDF_3) is simultaneously pumped by the 10R(10) CO_2 laser line at 969.14 cm^{-1} probed by a tunable diode laser (reference 33).

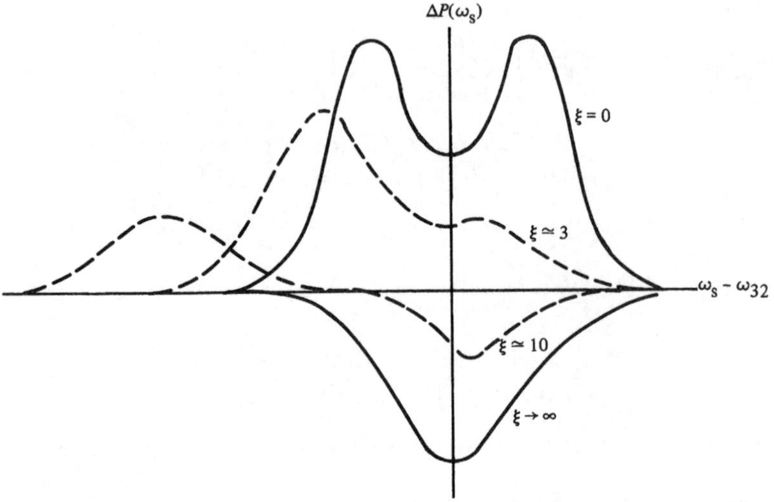

Figure 12.17
Plots of the double-resonance lineshape function for several values of the detuning parameter ξ. The Lorentzian line for $\xi \to \infty$ is the same as that obtained when $\mathscr{E}_p \to 0$, that is, when the pump field is turned off.

If the pump is not on resonance, then the full expressions of (12.29) must be employed. A more complex asymmetric line shape is obtained, which is illustrated in figure 12.17 for several values of the detuning parameter $\xi = (\omega_p - \omega_{31})\tau$. It is intructive to check the limiting form of (12.31) for pumping very far off resonance ($\xi \to \infty$) or, equivalently, $\mathscr{E}_p \to 0$, that is, no pump at all. In this case, $\omega_\pm = \omega_s - \omega_{32}$, and

$$\Delta P(\omega_s; \mathscr{E}_p = 0) = \frac{\hbar\omega(\kappa_{23}\mathscr{E}_s)^2\tau}{8}\left\{(2N_2 - N_1 - N_3)\left[\frac{2}{1 + (\omega_s - \omega_{32})^2\tau^2}\right]\right.$$

$$\left. + (N_1 - N_3)\left[\frac{2}{1 + (\omega_s - \omega_{32})^2\tau^2}\right]\right\}$$

$$= \frac{\omega\mu_{23}^2\mathscr{E}_s^2\tau}{2\hbar}(N_2 - N_3)\frac{1}{1 + (\omega_s - \omega_{32})^2\tau^2},$$

which is just the normal Lorentzian absorption line.

General expressions for the double-resonance line shape have been pub-

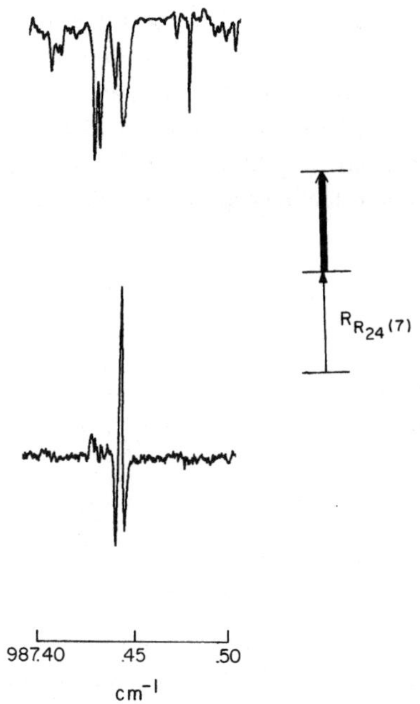

Figure 12.18
Infrared double-resonance signal in deuterated fluoroform (CDF_3), obtained by scanning the 987.4–987.5 cm^{-1} region with a tunable diode laser while simultaneously pumping with the 10R(10) line of a CO_2 laser. The situation corresponds to that shown in figure 12.16c, with $\omega_p = \omega_{32}$. The double-resonance line shape shown in the lower trace is characteristic of on-resonance pumping between the upper two levels of the three-level system (reference 36); the upper trace is the normal absorption spectrum of CDF_3 in this region. [From D. Harradine, Ph.D. thesis, MIT (1984). Reproduced with permission.]

lished by Mollow (reference 34) for the two-level double-resonance case (that is, where levels 1 and 2 are identical and $\omega_p \approx \omega_s$) and by Hänsch and Toschek (reference 35) for the three-level case. Both of these treatments deal with essentially nondegenerate systems; extension to spatially degenerate levels, such as are encountered in high-J states of molecules, has been carried out by Galbraith, Dubs, and Steinfeld (reference 36).

One last comment concerning double-resonance spectroscopy deals with the question of spectral line narrowing, considered in the previous section. Suppose the system under investigation corresponds to the configuration shown in figure 12.16c. For example, ω_{21} and ω_{32} could be visible wavelength optical frequencies, and population in level 3 could be detected by ultraviolet fluorescence, an arrangement frequently called "optical-optical double resonance," or OODR (reference 37). If the pump and probe sources are extremely narrow-bandwidth lasers, so that $\delta\omega_L \ll \delta\omega_D$, then only a very narrow portion of the total velocity distribution will be populated in level 2. When this set of molecules is probed, very narrow OODR signals will be observed, permitting resolution of sub-Doppler splittings in the $3 \leftarrow 2$ transition.

Problems

1. In the literature on self-induced transparency, the following three expressions may be found for the quantity γ introduced in (12.16):

$$\gamma = 2\sin^{-1}\tanh x,$$

$$\gamma = 2\tan^{-1}\sinh x,$$

$$\gamma = 4\tan^{-1}\exp x,$$

where $x \equiv \omega(t - z/v)$. Show that each of these expressions is a solution to the sine-Gordon equation and that $\mathscr{E}(z,t) = (\hbar/\mu_{ab})[\partial\gamma(z,t)/\partial t]$ is a unique function. Are the three expressions algebraically equivalent?

2. We have seen that a "2π pulse" can propagate without distortion or attenuation in an absorbing medium. Show that the wave form

$$\mathscr{E}(z,t) = \frac{\hbar}{\mu_{ab}\tau}\operatorname{sech}\left[\frac{1}{\tau}\left(t - \frac{z}{v}\right)\right]$$

is indeed a 2π pulse (v is the group velocity of light in the medium). Find the pulse width, that is, full width at half-maximum, of this pulse in terms of the parameter τ.

References

1. H. C. Torrey, *Phys. Rev.* **76**, 1059 (1949).

2. R. V. Churchill, *Operational Mathematics*, 3rd ed. (McGraw-Hill, New York, 1972).

3. J. Walker, "The Amateur Scientist," *Scientific American*, 182–192 (March 1981).

4. J. M. Levy, J. H.-S. Wang, S. G. Kukolich, and J. I. Steinfeld, *Phys. Rev. Letts.* **29**, 395 (1972).

5. R. L. Shoemaker, in *Laser and Coherence Spectroscopy*, J. I. Steinfeld, ed. (Plenum, New York, 1978), pp. 197–371.

6. G. M. Dobbs, R. H. Micheels, J. I. Steinfeld, J. H.-S. Wang, and J. M. Levy, *J. Chem. Phys.* **63**, 1904 (1975).

7. L. Allen and J. H. Eberly, *Optical Resonance and Two-Level Atoms* (Wiley, New York, 1975).

8. A. Abragam, *The Principles of Nuclear Magnetism* (Clarendon Press, Oxford,1961), p. 63.

9. E. L. Hahn, *Phys. Rev.* **80**, 580 (1950).

10. N. A. Kurnit, I. D. Abella, and S. R. Hartmann, *Phys. Rev. Letts.* **13**, 567 (1964).

11. C. K. N. Patel and R. E. Slusher, *Phys. Rev. Letts.* **20**, 1087 (1968).

12. T. G. Schmalz and W. H. Flygare, in *Laser and Coherence Spectroscopy*, J. I. Steinfeld, ed. (Plenum, New York, 1978), pp. 125–196.

13. T. J. Balle, E. J. Campbell, M. R. Keenan, and W. H. Flygare, *J. Chem. Phys.* **72**, 922 (1980).

14. M. R. Keenan, E. J. Campbell, T. J. Balle, L. W. Buxton, T. K. Minton, P. D. Soper, and W. H. Flygare, *J. Chem. Phys.* **72**, 3070 (1980).

15. P. D. Soper, A. C. Legon, W. G. Read, and W. H. Flygare, *J. Chem. Phys.* **76**, 292 (1982).

16. A. C. Legon, E. J. Campbell, and W. H. Flygare, *J. Chem. Phys.* **76**, 2267 (1982).

17. W. G. Read and W. H. Flygare, *J. Chem. Phys.* **76**, 2238 (1982).

18. J. A. Shea and W. H. Flygare, *J. Chem. Phys.* **76**, 4857 (1982).

19. R. K. Bullough, P. J. Caudrey, J. C. Eilbeck, and J. D. Gibbon, *Opto-Electronics* **6**, 121 (1974).

20. *Physics Today* **20**(8), 47 (August 1967).

21. S. L. McCall and E. L. Hahn, *Phys. Rev. Letts.* **18**, 908 (1967).

22. C. K. N. Patel and R. E. Slusher, *Phys. Rev. Letts.* **19**, 1019 (1967).

23. R. S. McDowell, H. W. Galbraith, B. J. Krohn, C. D. Cantrell, and E. D. Hinkley, *Opt. Commun.* **17**, 178 (1976).

24. R. E. Slusher and H. M. Gibbs, *Phys. Rev.* **A5**, 1634 (1972).

25. M. S. Sorem and A. L. Schawlow, *Opt. Commun.* **5**, 148 (1972).

26. C. Costain, *Can. J. Phys.* **47**, 2431 (1969).

27. V. P. Chebotayev and V. S. Letokhov, *Prog. Quantum Electronics* **4**(2), 111–206 (1975).

28. V. S. Letokhov and V. P. Chebotayev, *Principles of Nonlinear Laser Spectroscopy* (in Russian; Nauka, Moscow. 1975).

29. V. S. Letokhov and V. P. Chebotayev, *Nonlinear Laser Spectroscopy*, Ser. Opt. Sci. No. 4 (Springer-Verlag, Berlin, 1977).

30. K. Shimoda and T. Shimizu, "Nonlinear Spectroscopy of Molecules," *Prog. Quantum Electronics* **2**(2), 45 (1972).

31. J. I. Steinfeld and P. L. Houston, in *Laser and Coherence Spectroscopy*, J. I. Steinfeld, ed. (Plenum, New York, 1978), pp. 1–123.

32. M. Twain, *A Tramp Abroad*, appendix D (Harper and Brothers, New York, 1879).

33. D. M. Harradine, Ph.D. thesis, MIT (1984).

34. B. R. Mollow, *Phys. Rev.* **A5**, 2217 (1972), **A8**, 1949 (1973).

35. T. W. Hänsch and P. Toschek, *Z. Physik* **236**, 213 (1970).

36. H. W. Galbraith, M. Dubs, and J. I. Steinfeld, *Phys. Rev.* **A26**, 1528 (1982).

37. R. W. Field, *Disc. Faraday Soc.* **71**, 111 (1981).

13 Multiple-Photon Spectroscopy

In all of the spectroscopy that has been considered until this point, the processes have been ones in which a single photon was required to effect a transition between two energy levels. Even the nonlinear coherent optical processes treated in the preceding chapters were all described as coherently prepared ensembles of one-photon states. We had noted, in chapter 1.6, that terms that were higher than first order in the field strength could contribute to radiative transition probabilities, at sufficiently high light intensities. We shall now turn our attention to such nonlinear, multiple-photon processes. First, we shall consider the nonlinear polarizability of optical materials, and its consequences, such as frequency doubling and parametric generation. A general treatment of multiple-photon transitions will be given and two specific examples considered: infrared multiple-photon absorption and resonant multiphoton ionization.[1] Next, a simple theory of stimulated Raman processes will be presented. We close this chapter with a brief discussion of methods for production of ultrashort light pulses, which may be used to study molecular processes occurring at picosecond time scales.

1 Nonlinear Polarizability and Second-Harmonic Generation

With a high-intensity pulsed laser, we have available a very high photon flux, which is associated with a very high electric field amplitude at the peak of the pulse. This permits the atoms or molecules in the sample being investigated to interact with more than one photon at the same time, or, what amounts to the same thing classically, for the macroscopic polarization to respond to the square of the electric field rather than linearly. We shall use the latter classical model for most of the ensuing discussion; for further details, the reader is referred to the text by Yariv (reference 1).

Since we are going to be using classical radiation theory, we shall begin with the appropriate Maxwell's equation for the propagation of electric and magnetic fields in nonconducting, nonmagnetic media, namely,

$$\nabla \times \mathbf{H} = \frac{\partial \mathbf{D}}{\partial t} = \frac{\partial}{\partial t}(\epsilon_0 \mathbf{E} + \mathbf{P}). \tag{13.1}$$

We have usually put $\mathbf{P} = \epsilon \mathbf{E} = \chi \epsilon_0 \mathbf{E}$, where ϵ is the dielectric constant, and

1. The terms "multiple-photon" and "multiphoton" are often used interchangeably; here, we shall use the former to denote *sequential* absorption of photons, the latter for *simultaneous* transitions.

χ the electric susceptibility of the medium. If **E** is very large, however, we shall have to contend with additional, nonlinear terms, so that

$$\mathbf{P} = \chi\epsilon_0\mathbf{E} + \mathbf{P}_{\mathrm{NL}}, \tag{13.2}$$

where any given component of the nonlinear polarizability is related to two components of the electric field by the nonlinear susceptibility tensor **d**—for example,

$$(P_{\mathrm{NL}})_i = d_{ijk}E_jE_k. \tag{13.3}$$

The tensor **d** has 18 independent components, and the general form of equation (13.3) is

$$\begin{pmatrix} (P_{\mathrm{NL}})_x \\ (P_{\mathrm{NL}})_y \\ (P_{\mathrm{NL}})_z \end{pmatrix} = \begin{pmatrix} d_{11}d_{12}\cdots d_{16} \\ d_{21}d_{22}\cdots d_{26} \\ d_{31}d_{32}\cdots d_{36} \end{pmatrix} \begin{pmatrix} E_x^2 \\ E_y^2 \\ E_z^2 \\ 2E_zE_y \\ 2E_zE_x \\ 2E_xE_y \end{pmatrix}.$$

The d coefficients are very small in most materials. There exist a number of substances, however, for which these coefficients are sufficiently large that the linear and nonlinear parts of the polarizability have comparable magnitudes at light intensities of 10^6 W/cm^2 and above, which is in the range of Q-switched and mode-locked laser peak powers. For a fuller discussion of the choice of materials for producing nonlinear optical effects, the reader is again referred to Yariv's text (reference 1).

An especially important manifestation of this sort of nonlinearity is optical second-harmonic generation—that is, the production of light at angular frequency $2\omega_0$ by conversion of light at frequency ω_0. To the extent that this conversion occurs, the photon flux is reduced by half, but since the energy per photon is just proportional to ω, the energy flux through the medium is, of course, unchanged. We can analyze this process in terms of classical electrodynmaics. In addition to Maxwell's equation (13.1), with **P** given by equation (13.2), we shall need the relation

$$\nabla \times \mathbf{E} = -\frac{\partial}{\partial t}(\mu_0\mathbf{H}). \tag{13.4}$$

Equation (13.1) can be rewritten as

$$\nabla \times \mathbf{H} = \frac{\partial}{\partial t}(\epsilon_0 \mathbf{E} + \chi\epsilon_0 \mathbf{E} + \mathbf{P}_{\rm NL})$$

$$= \frac{\partial}{\partial t}(\epsilon' \mathbf{E} + \mathbf{P}_{\rm NL}),$$

with $\epsilon' = \epsilon_0(1 + \chi)$. Thus

$$\begin{aligned}\nabla \times \nabla \times \mathbf{E} &= -\nabla \times \frac{\partial}{\partial t}(\mu_0 \mathbf{H}) = -\mu_0 \frac{\partial}{\partial t}(\nabla \times \mathbf{H}) \\ &= -\mu_0 \frac{\partial^2}{\partial t^2}(\epsilon' \mathbf{E} + \mathbf{P}_{\rm NL}).\end{aligned} \tag{13.5}$$

Let us consider three independent light waves propagating through our sample. Each is propagating in the z direction, and each is plane polarized, so that we can neglect the x and y derivatives in the gradient operatior. These three light waves can be represented analytically by

$$E_i^{(\omega_1)}(z, t) = \tfrac{1}{2}[E_{1i}(z)e^{i(\omega_1 t - k_1 z)}],$$

$$E_j^{(\omega_2)}(z, t) = \tfrac{1}{2}[E_{2j}(z)e^{i(\omega_2 t - k_2 z)}],$$

$$E_k^{(\omega_3)}(z, t) = \tfrac{1}{2}[E_{3k}(z)e^{i(\omega_3 t - k_3 z)}],$$

where $(ijk) = x$ or y, and we have chosen $\omega_1 = \omega_3 - \omega_2$. The Laplacian operator ∇^2 applied to the first of these light waves gives

$$\nabla^2 \mathbf{E}_1 = \frac{\partial^2}{\partial z^2} E_i^{(\omega_1)}(z, t) = \frac{1}{2}\frac{\partial^2}{\partial z^2}[E_{1i}(z)e^{i(\omega_1 t - k_1 z)}]$$

$$= -\frac{1}{2}\left[k_1^2 E_{1i}(z) + 2ik_1 \frac{dE_{1i}}{dz}\right] e^{i(\omega_1 t - k_1 z)},$$

in which we have made the usual approximation that the envelope function $E_{1i}(z)$ varies so much more slowly than the oscillation frequency ω_1 that

$$\frac{d^2 E_{1i}}{dz^2} \ll k_1 \frac{dE_{1i}}{dz}$$

and can therefore be neglected. We shall obtain analogous expressions for $\nabla^2 \mathbf{E}_2$ and $\nabla^2 \mathbf{E}_3$. Substituting the expression we have just obtained into the wave equation (13.5) in gives

$$\left[\frac{k_1^2}{2} E_{1i}(z) - ik_1 \frac{dE_{1i}}{dz}\right] e^{i(\omega_1 t - k_1 z)} = \omega_1^2 \mu_0 \epsilon \frac{E_{1i}}{2} d^{i(\omega_1 t - k_1 z)} - \mu_0 \frac{\partial^2}{\partial t^2} P_{\rm NL}^{(\omega_1)},$$

with

$$P_{\mathrm{NL}}^{(\omega_1)} = \tfrac{1}{2} d_{ijk} E_{3j}(z) E_{2k}^*(z) \exp i[(\omega_3 - \omega_2)t - (k_3 - k_2)z].$$

Combining these expressions gives

$$\left[\frac{k_1^2}{2} E_{1i}(z) - ik\frac{dE_{1i}}{dz}\right] e^{i(\omega_1 t - k_1 z)} = \frac{k_1^2}{2} E_{1i} e^{i(\omega_1 t - k_1 z)}$$
$$- \tfrac{1}{2}\mu_0 \omega_1^2 d_{ijk} E_{3j} E_{2k}^* e^{i[\omega_1 t - (k_3 - k_2)z]},$$

or, after cancellation,

$$\frac{dE_{1i}}{dz} = -\frac{i}{2}\mu_0 \omega_1 \left(\frac{\omega_1}{k}\right) d_{ijk} E_{3j} E_{2k}^* e^{-i(k_3 - k_2 - k_1)z}.$$

We note that ω_1/k equals the speed of light in the medium, which, in turn, equals $(\mu_0 \epsilon)^{-1/2}$, so that finally

$$\frac{dE_{1i}}{dz} = -\frac{i\omega_1}{2}\sqrt{\frac{\mu_0}{\epsilon}}\, d_{ijk} E_{3j} E_{2k}^* e^{-i(k_3 - k_2 - k_1)z}. \tag{13.6a}$$

In a similar manner, we obtain equations for E_2 and E_3:

$$\frac{dE_{2k}^*}{dz} = +\frac{i\omega_2}{2}\sqrt{\frac{\mu_0}{\epsilon}}\, d_{kij} E_{1i} E_{3j}^* e^{-i(k_1 - k_3 + k_2)z} \tag{13.6b}$$

and

$$\frac{dE_{3j}}{dz} = -\frac{i\omega_3}{2}\sqrt{\frac{\mu_0}{\epsilon}}\, d_{jik} E_{1i} E_{2k} e^{-i(k_1 + k_2 - k_3)z}. \tag{13.6c}$$

The first example we shall take of the use of these equations is that of frequency doubling, in which $\omega_1 = \omega_2$ and $\omega_3 = 2\omega_1$. For simplicity of treatment, we shall assume that E_1 is nearly constant along the length of the doubling medium; that is, the process is of relatively low efficiency, so that only a small fraction of the fundamental power gets converted, and also that there is no absorption loss in the medium. We then have a single equation with which to deal, namely,

$$\frac{dE_{3j}}{dz} = -\frac{i\omega_3}{2}\sqrt{\frac{\mu_0}{\epsilon}}\, d_{jik} E_{1i} E_{1k} e^{-i(2k_1 - k_3)z}$$
$$= -\frac{i\omega_3}{2}\sqrt{\frac{\mu_0}{\epsilon}}\, d_{jik} E_{1i} E_{1k} e^{i\Delta k z}.$$

The solution to this equation, for the electric field amplitude at ω_3 at the end of a doubling medium of length L, can be written down immediately by inspection. It is just

$$E_{3j}(z=L) = \frac{i\omega_3}{2}\sqrt{\frac{\mu_0}{\epsilon}}\, d_{jik} E_{1i} E_{1k} \frac{e^{i\Delta k L}-1}{i\Delta k},$$

so that

$$I(\omega_3, L) = |E_{3j}|^2 = \frac{\mu_0}{\epsilon}\omega_3^2 d_{jik}^2 E_{1i}^2 E_{1k}^2 L^2 \frac{\sin^2(\Delta k\, L/2)}{(\Delta k\, L/2)^2}.$$

This result contains several immediately interesting results. First, we see that $I(\omega_3)$ is proportional to $I^2(\omega_1)$, which is of course required to be so if we are to turn two "red" photons into one "blue" photon. Then, we see that the doubling efficiency, that is, $I(\omega_3)/I^2(\omega_1)$, is proportional to the square of the length L of the doubling medium. Usually, this is not the actual physical length of the medium because if it were simply that, then infinite doubling efficiencies could be obtained by simply using a sufficiently long path. Rather, it is the optical *coherence length* in the doubling medium, usually a crystal, which is determined by the uniformity and homogeneity of the material. Finally, we see that the doubling efficiency is a very sensitive function of the quantity $(\Delta k\, L/2)$ and is sharply peaked about $(\Delta k\, L/2) = 0$. This means that we must have $\Delta k = k_3 - 2k_1 = 0$ for efficient frequency doubling to occur.

This particular requirement, known as the "phase-matching" criterion, is an important problem in practical doubling techniques. The magnitude of the wave vector $\mathbf{k}$ is given by $\omega/v = \omega n/c$, where v is the velocity of light, and n the refractive index at frequency ω, in the medium. Thus

$$\Delta k = k_3 - 2k_1 = \frac{2\omega_1 n^{(2\omega)}}{c} - \frac{2\omega_1 n^{(\omega)}}{c}$$

$$= \frac{2\omega_1}{c}[n^{(2\omega)} - n^{(\omega)}].$$

If we are to have $\Delta k = 0$, then clearly we must have $n^{(2\omega)} = n^{(\omega)}$; the phase-matching criterion becomes a refractive index-matching criterion. The difficulty is that, for most materials, the refractive index shows a *dispersion*, that is, a wavelength dependence, as illustrated in figure 13.1. The dispersion arises from the absorption terms at λ_i in Sellmeier's modification of the Drude equation (reference 2),

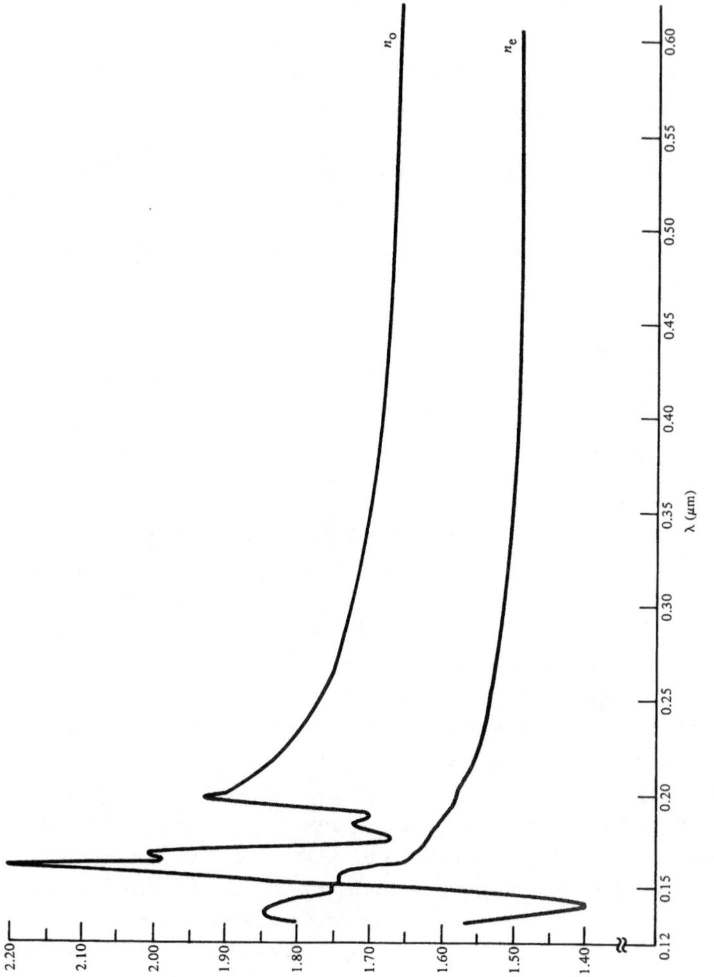

Figure 13.1
Refractive index dispersion curve for calcite. Both the ordinary (n_o) and the extraordinary (n_e) indices are displayed. [Data are from reference 3.]

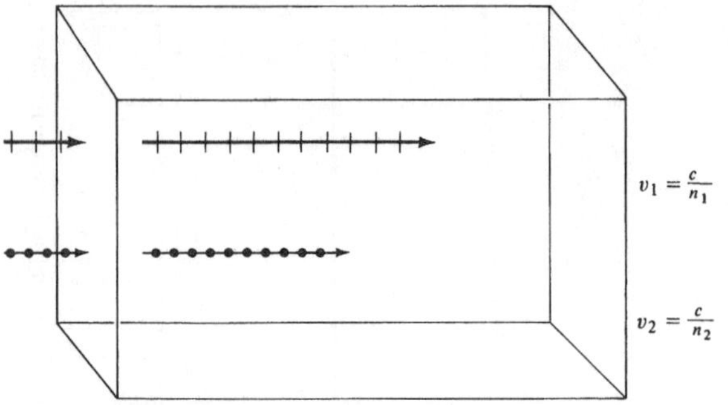

Figure 13.2
Different refractive indices for light of orthogonal polarizations in a birefringent medium.

$$n_\lambda^2 - 1 = \sum_i \frac{(D_i\lambda)^2}{(\lambda^2 - \lambda_i^2)}.$$

The structure shown at short wavelengths in figure 13.1 corresponds to these absorption resonances. Although it might seem possible to find frequencies ω and 2ω in the anomalous dispersion region for which the index-matching requirement is satisfied, this is not practical for frequency doubling, since the anomalous dispersion regions are characterized by high absorption coefficients, so that the radiation would be attenuated by the material.

The solution to this problem is to use a *birefringent* material, which possesses different refractive indices for different polarizations of the light. This can occur, for example, in a crystal of sufficiently low symmetry, so that the two mutually perpendicular polarizations of the incident light sample quite distinct optical environments, as in figure 13.2. The difference in refractive indices for different polarizations implies that, for nonnormal incidence, the angles of refractions will also differ for the two polarizations. This makes possible such effects as double refraction in calcite crystals, shown in figure 13.3, and the use of a double prism as a light polarizer, shown in figure 13.4.

A birefringent crystal makes the solution of the index-matching problem possible in the following way. A material is found in which the index of refraction for the extraordinary ray at 2ω is equal to that for the ordinary ray at ω, as shown in figure 13.5, *and* the matrix elements of **d** have the form appropriate for generating E_3 from E_1 with the correct relative polarizations.

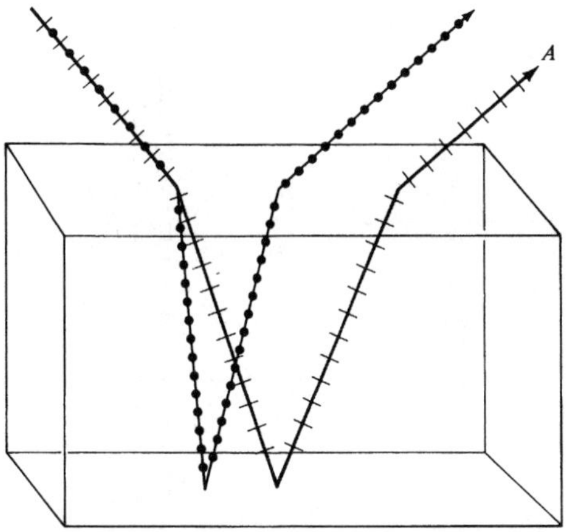

Figure 13.3
Double refraction in a birefringent crystal. The observer at A sees light coming from two different locations at the base of the crystal, thus producing a double image.

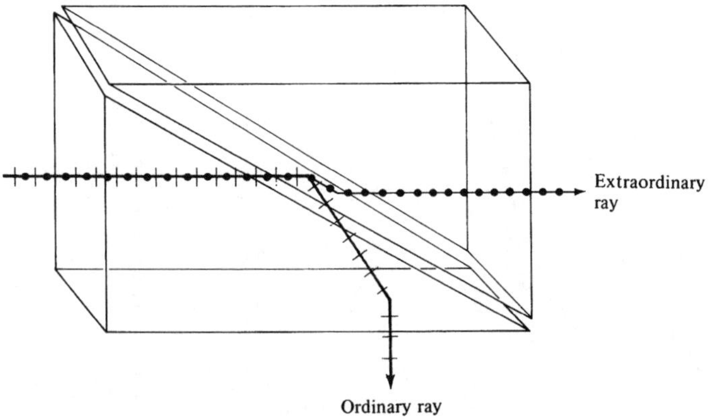

Figure 13.4
Polarization produced by a double prism of birefringent material, of the Glan-Thompson type. The prism angle is cut so that the ordinary ray is reflected from the thin layer of isotropic material between the two prisms, while the extraordinary ray is transmitted. The two orthogonal polarizations are thus separated spatially.

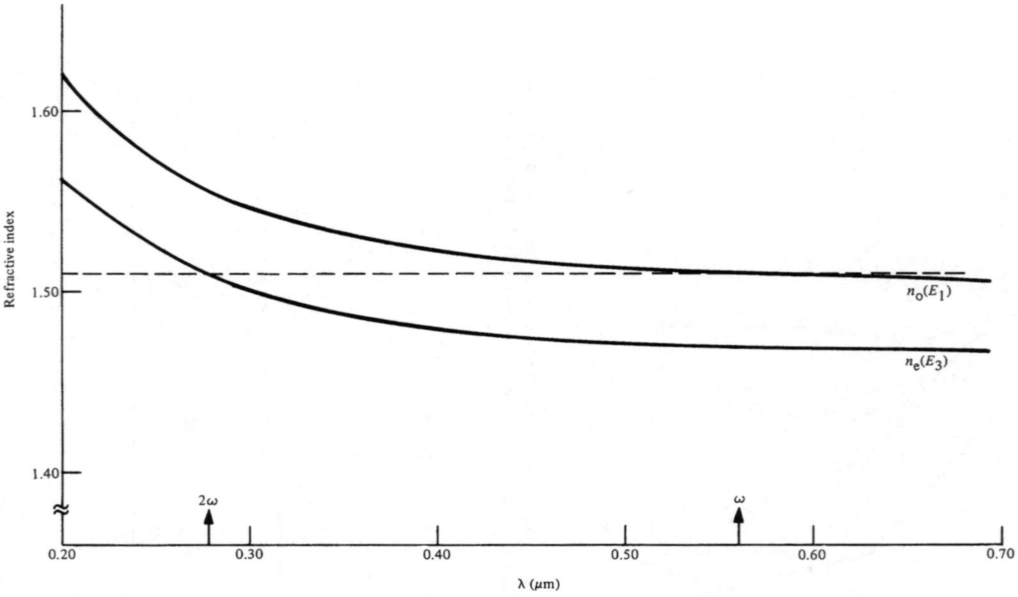

Figure 13.5
Index matching in a birefringent medium (potassium dihydrogen phosphate). E_1, at a wavelength $\lambda \simeq 0.56\ \mu m$, is arranged to have the "ordinary" polarization, while E_3, at a wavelength $\lambda/2 \simeq 0.28\ \mu m$, is arranged to have the "extraordinary" polarization. Thus $n_o(\omega) = n_e(2\omega)$, and the $\Delta k = 0$ criterion is satisfied. [Dispersion data are from reference 3.]

Further details of index matching and of the kinds of substances that can be employed are given by Yariv (reference 1), and a fuller discussion of birefringence can be found in the text by Fowles (reference 4). Materials, which possess all the required properties and have been used successfully for frequency doubling, include such substances as potassium dihydrogen phosphate (KDP), lithium niobate, barium sodium niobate ("banana"), and lithium iodate.

2 Optical Parametric Oscillation

The same principles as applied to second-harmonic generation are equally valid when the frequencies of the three light waves are all different. In this more general case, equations (13.6a) to (13.6c) for the propagation of the light waves can be written as

$$\frac{dE_1}{dz} = -i\omega_1 S E_2^* E_3 e^{-i\Delta k z},$$

$$\frac{dE_2}{dz} = +i\omega_2 S E_1 E_3^* e^{i\Delta k z},$$

and

$$\frac{dE_3}{dz} = -i\omega_3 S E_1 E_2 e^{-i\Delta k z},$$

with $S = \frac{1}{2}\sqrt{\mu_0/\epsilon}\, d_{ijk}$. We shall require that $\omega_3 = \omega_1 + \omega_2$, and also that $\Delta k = k_3 - (k_1 + k_2) = 0$. The first requirement is just the conservation of energy in the photon conversion process: If a photon of light at frequency ω_3 is converted by the parametric medium into a photon at frequency ω_2 and another at frequency ω_1, then we must have $\hbar\omega_3 = \hbar\omega_1 + \hbar\omega_2$. The second is just the corresponding conservation of momentum associated with the light waves, which is expressed by the vector equation $\mathbf{k}_3 = \mathbf{k}_1 + \mathbf{k}_2$. If the three waves are propagating collinearly down the z axis, then this becomes a simple scalar relation, $\hbar k_3 = \hbar k_1 + \hbar k_2$. This last condition allows us to simplify even further the three preceding equations. If we further define $A_1 = E_i/\sqrt{\omega_i}$, and $\kappa' = \sqrt{\omega_1\omega_2\omega_3}\, S$, these become

$$\frac{dA_1}{dz} = -i\kappa' A_2^* A_3,$$

$$\frac{dA_2}{dz} = +i\kappa' A_1 A_3^*,$$

and

$$\frac{dA_3}{dz} = -i\kappa' A_1 A_2.$$

The A_i quantities have the following physical significance: $|A_i|^2 = |E_i|^2/\omega_i = cI/\hbar\omega_i$, where I is the light intensity, so that the square of A_i is proportional to the numerical flux of photons at frequency ω_i.

In order to find a solution of these three equations, let us once more take $A_3 = \text{constant}$, so that we are assuming that the conversion efficiency is low and that absorption losses are negligible. If we define a "gain parameter" $g = 2\kappa' A_3$, we have simply

$$\frac{dA_1}{dz} = -i\frac{g}{2}A_2^* \qquad \text{and} \qquad \frac{dA_2}{dz} = +i\frac{g}{2}A_1,$$

which have the solutions

$$A_1(z) = A_1(0)\cosh\frac{gz}{2} - iA_2^*(0)\sinh\frac{gz}{2},$$

$$A_2^*(z) = A_2^*(0)\cosh\frac{gz}{2} + iA_1(0)\sinh\frac{gz}{2}.$$

If power at frequency ω_3 is applied to a suitable parametric medium, and a small signal at frequency ω_1 is introduced at one end, this signal will experience approximately exponential amplification in its passage through the medium. At the same time, an "idler" wave at frequency ω_2 is generated. The photon flux for these two amplified light waves at the end of the medium of length L is given by

$$n_1(L) = n_2(L) \approx \tfrac{1}{4}n_1(0)^2 e^{gL},$$

where the gain parameter is, by combining all our previous definitions,

$$g = \frac{1}{\epsilon\hbar}\sqrt{\frac{(\omega_1\omega_2)}{\omega_3}}\, d_{ijk} I_3.$$

One practical use of these optical parametric generators is as tunable coherent optical oscillators, using the scheme shown in figure 13.6. The initial photon flux $n_1(0)$ is supplied by random spontaneous emission noise in the medium; the material is placed in an optical cavity to provide the feedback necessary to maintain oscillation. The only frequencies that can be amplified

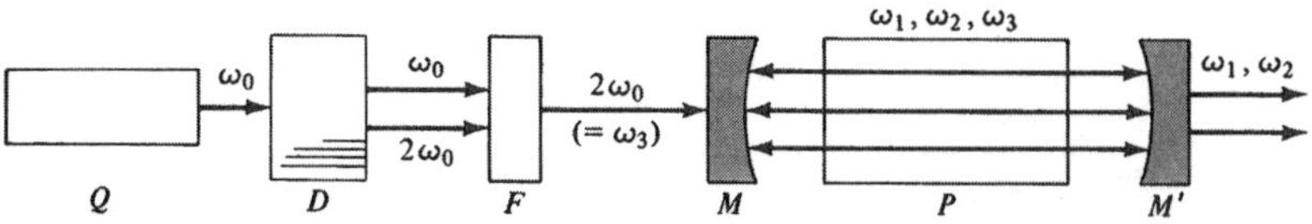

Figure 13.6
Schematic arrangement of an optical parametric oscillator system. Intense pumping light is produced by a Q-switched neodymium laser Q, at $\lambda_0 = 1.06$ μm. This is passed through a doubling crystal D to produce second-harmonic radiation at $\lambda_0/2 = 0.53$ μm. A filter F removes the fundamental and allows only the second harmonic to pass into the parametric medium P. Radiation at ω_1 and ω_2 builds up in P; mirror M allows $\omega_3 = 2\omega_0$ to pass into the medium, but reflects ω_1 and ω_2, while mirror M' reflects the unused pumping light ω_3 back into the medium, but allows a portion of ω_1 and ω_2 to emerge. The medium P is enclosed in an oven by means of which its temperature, and thereby its refractive index at ω_1 ω_2, can be varied to select different frequency outputs.

in the oscillator are those for which $k_3 - k_1 - k_2 = 0$, so that the output of the oscillator is very nearly monochromatic. The significance of this device is that the ks, and thus the frequencies, of the three light waves that meet the phase-matching conditions can be continuously varied by changing the refractive index of the medium; this, in turn, can be done by varying the temperature of the material. Typical tuning curves for lithium niobate, pumped by the second harmonic of Nd:YAG ($\lambda_0 = 0.53$ μm), are shown in figure 13.7. A summary of practical parametric oscillator systems may be found in reference 5.

3 Multiphoton Spectra of Molecules

The phenomena of second-harmonic generation and optical parametric amplification, though interesting and useful in themselves, are not really representative of the kinds of spectroscopy that we have been discussing in the rest of this book. Most of the processes we have been dealing with involve resonant transitions between two definite energy levels in the molecules under investigation, and we would now like to consider such transitions in which two or more photons are involved. An example of such a transition is two-photon fluorescence, which, as we shall see, will enable us to display picosecond laser pulses. In this process, an excited state, which is inaccessible by single-photon absorption, can be reached by the simultaneous absorption of two photons whose energies sum to the required amount. Another process of this nature is stimulated Raman scattering, which will be taken up in section

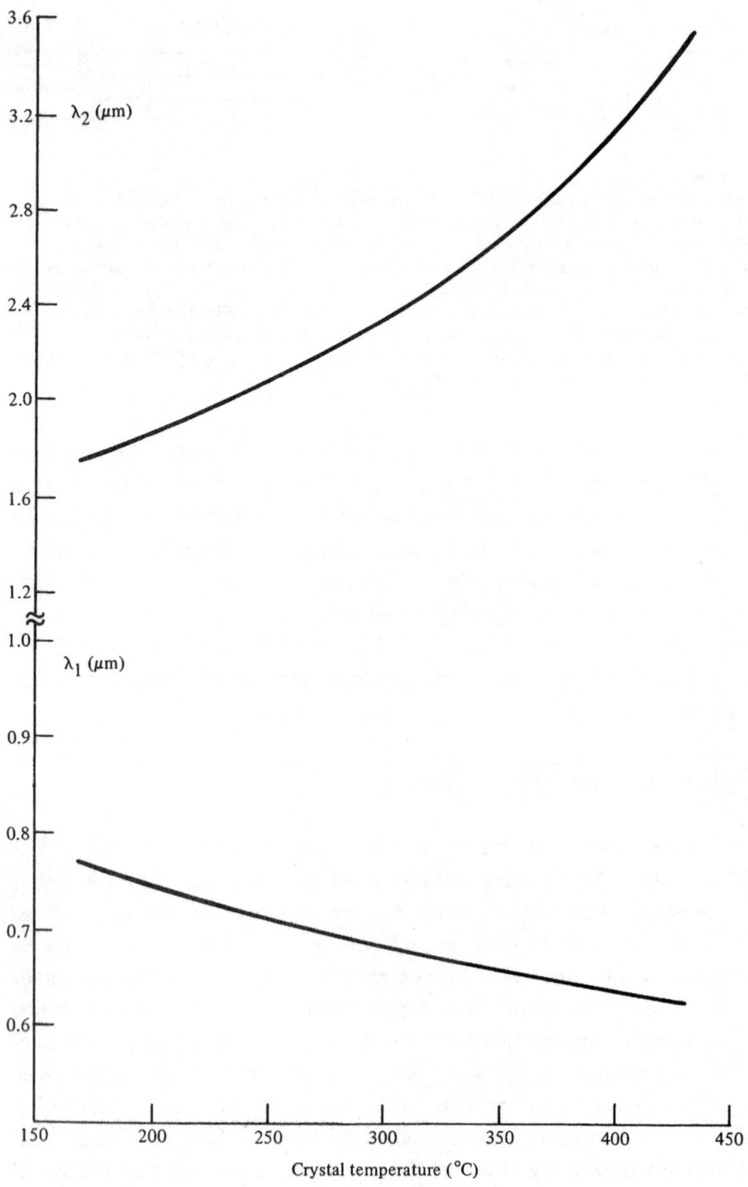

Figure 13.7
Parametric tuning curves for signal (λ_1) and idler (λ_2) waves in lithium niobate.

6. Fuller discussion of these processes can be found in the text by Yariv (reference 1) and the review article by Peticolas (reference 6). Actually, the possibility of two-photon absorption spectroscopy was first suggested by M. Goeppert-Mayer (reference 7), many years before lasers were even thought of.

In order to see just how multiphoton spectra can come about, let us go all the way back to the perturbation theory of the interaction of radiation with a molecular system, introduced in chapter 1.5. The Hamiltonian for the system, including the oscillating radiation field, was written as

$$\mathscr{H} = \frac{p^2}{2m} + V_0$$

$$= \mathscr{H}_0 + \mathscr{H}'(t),$$

where $\mathscr{H}_0$ included all the time-independent parts, and $\mathscr{H}'(t)$ included all the time-dependent parts of the energy. We wrote the momentum as

$$p = i\hbar\nabla - \frac{e}{c}\mathbf{A}(t),$$

where

$$\mathbf{A}(t) = \mathbf{A}_0 e^{\pm i(\omega t - \mathbf{k}\cdot\mathbf{r})};$$

we then obtained

$$\mathscr{H}'(t) = \frac{ie\hbar}{mc}\mathbf{A}\cdot\nabla + \frac{e^2}{2mc^2}\mathbf{A}\cdot\mathbf{A}.$$

Using first-order perturbation theory, we obtained an equation for the rate of change of the coefficient of the nth quantum state, assuming that the system began its career in the mth quantum state, as

$$\frac{dc_n}{dt} = -\frac{i}{\hbar}e^{i(E_n - E_m)t/\hbar}\langle n|\mathscr{H}'|m\rangle. \tag{13.7}$$

The approximations made at this point were, first, to discard the $\mathbf{A}\cdot\mathbf{A}$ term, and, second, to make a dipole approximation to the $\mathbf{A}\cdot\nabla$ term, to give

$$\frac{dc_n}{dt} \simeq -\frac{i}{\hbar}\exp\left[\frac{i(E_n - E_m \pm \hbar\omega)t}{\hbar}\right]\boldsymbol{\mu}_{nm}\cdot\mathbf{E}_0.$$

This was integrated to yield

$$c_n^{(1)}(t) = \mu_{nm} E_0 \left\{ \frac{1 - \exp[i(E_m - E_n \pm \hbar\omega)t/\hbar]}{i(E_m - E_n \pm \hbar\omega)} \right\}$$
$$= \frac{\mu^n{}_m E_0}{i\hbar} \left[\frac{1 - e^{-i(\omega_{mn} \pm \omega)t}}{\omega_{mn} \pm \omega} \right]. \tag{13.8}$$

We pointed out that, as the dipole approximation was relaxed, magnetic dipole and electric quadrupole selection rules began to be significant, but we still neglected the $\mathbf{A} \cdot \mathbf{A}$ term in the interaction.

We now wish to extend this treatment in a manner that will be appropriate for the very high photon fluxes found in laser beams. This involves two complications, namely, retention of the $\mathbf{A} \cdot \mathbf{A}$ term and use of second-order perturbation theory. Recall that equation (13.7) came from the complete expression

$$\frac{d}{dt} c_n^{(1)}(t) = -\frac{i}{\hbar} \sum_k c_k^{(0)} \langle n | \mathscr{H}' | k \rangle \, e^{i\omega_{nk} t},$$

in which we took $c_k(0) = \delta_{km}$; that is, all the probability was in the mth state at $t = 0$. By extension, the corresponding expression for the time dependence of the lth coefficient, correct to second order, will be

$$\frac{d}{dt} c_l^{(2)}(t) = -\frac{i}{\hbar} \sum_n c_n^{(1)}(t) \langle l | \mathscr{H}' | n \rangle \, e^{-i\omega_{nl} t},$$

but in this case, $c_n^{(1)}(t)$ is given by equation (13.8), so that

$$\frac{d}{dt} c_l^{(2)}(t) = \frac{i}{\hbar} \sum_n \left[\frac{1}{\hbar} \langle n | \mathscr{H}' | m \rangle \frac{1 - e^{-i(\omega_{mn} \pm \omega)t}}{\omega_{mn} \pm \omega} \right] \langle l | \mathscr{H}' | n \rangle \, e^{-i(\omega_{nl} \pm \omega)t}.$$

In integrating the preceding equation, we have to be careful to keep only those terms in which the denominator approaches zero; the others make no contribution to the accumulating probability. In order to keep track of this, let us take $|m\rangle$ as the ground state, and $E_l > E_n > E_m$. With this convention, $\omega_{mn} = E_m - E_n < 0$; thus we keep the term in $\omega_{mn} + \omega$ and discard the term in $\omega_{mn} - \omega$. Similarly, $\omega_{nl} = E_n - E_l < 0$, and we keep only the term in $\omega_{nl} + \omega$. This gives

$$c_l(t) = \frac{i}{\hbar^2} \sum_n \langle n | \mathscr{H}' | m \rangle \langle l | \mathscr{H}' | n \rangle \int_0^t \frac{1 - e^{-i(\omega_{mn} + \omega)t}}{\omega_{mn} + \omega} e^{-i(\omega_{nl} + \omega)t'} \, dt'.$$

Expanding the above integral over t' gives

$$\int_0^t \frac{e^{-i(\omega_{nl}+\omega)t'} - e^{-i(\omega_{nl}+\omega_{mn}+2\omega)t'}}{\omega_{mn}+\omega} dt',$$

and integrating that gives

$$i\frac{e^{-i(\omega_{nl}+\omega)t'}}{(\omega_{mn}+\omega)(\omega_{nl}+\omega)}\Bigg|_{t'=0}^{t'=t} - i\frac{e^{-i(\omega_{nl}+\omega_{mn}+2\omega)t'}}{(\omega_{mn}+\omega)(\omega_{nl}+\omega_{mn}+2\omega)}\Bigg|_{t'=0}^{t'=t}.$$

But in accordance with our accepted convention for the relative order of the energies, $\omega_{nl} + \omega_{mn} = \omega_{ml}$; so that our expression for $c_l(t)$ is, finally,

$$c_l(t) = \frac{1}{\hbar^2}\sum_n \langle l|\mathscr{H}'|n\rangle\langle n|\mathscr{H}'|m\rangle$$
$$\times\left[\frac{e^{-i(\omega_{nl}+\omega)t} - 1}{(\omega_{nl}+\omega)(\omega+\omega_{mn})} + \frac{e^{-i(\omega_{ml}+2\omega)t} - 1}{(\omega_{ml}+2\omega)(\omega+\omega_{mn})}\right].$$

The important point to note is that, in second order, a resonance now appears at $\omega_{ml} = 2\omega$, corresponding to a two-photon absorption process. The probability of the process occurring can be written as

$$P_{m\to l}(t) = |c_l(t)|^2 = \sum_n \frac{|\langle l|\mathscr{H}'|n\rangle\langle n|\mathscr{H}'|m\rangle|^2}{\hbar^4(\omega-\omega_{nm})^2}\frac{\sin^2[\frac{1}{2}(2\omega-\omega_{lm})t]}{\frac{1}{4}(2\omega-\omega_{lm})^2}.$$

The bookkeeping of all the different terms that come into these expressions can be greatly simplified by using the time-ordered Feynman diagrams, as discussed by Peticolas (reference 6). As an example, consider the diagram given by figure 13.8a for single-photon absorption, and that given by figure 13.8b for single-photon emission. These diagrams correspond to the

$$\frac{\langle m|\mathbf{A}\cdot\nabla|n\rangle}{\omega_{mn}\pm\omega}$$

terms in the perturbation expansion. In absorption, $|m\rangle$ is the initial state and $|n\rangle$ is the final state, so that $\omega_{mn} = E_m - E_n$ is negative, and the resonance occurs when $\omega_{mn} + \omega$ goes to zero. In emission (figure 13.8b), this situation is reversed, and the resonance occurs when $\omega_{mn} - \omega$ goes to zero.

The term in $\mathbf{A}\cdot\mathbf{A}$, in first order, is represented by the diagrams given by figures 13.8c and 13.8d. The diagram given by figure 13.8c represents two-photon absorption, for which the term is

$$\frac{\langle m|\mathbf{A}\cdot\mathbf{A}|n\rangle}{\omega_{mn}+(\omega_2+\omega_1)};$$

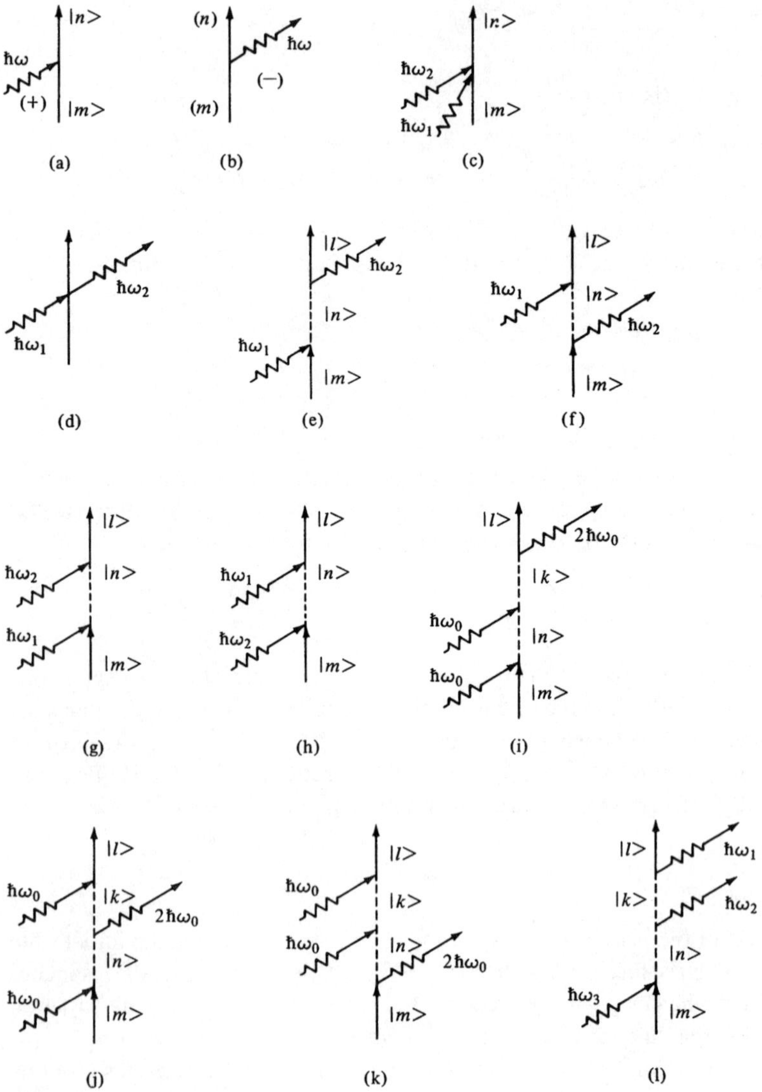

Figure 13.8
Photon absorption and emission diagrams. See text for explanation.

the sum of the energies of the two incoming photons must add to the energy separation of the initial and final states. The diagram given by figure 13.8d corresponds to light scattering, for which the term is

$$\frac{\langle m|\mathbf{A}\cdot\mathbf{A}|n\rangle}{\omega_{mn}+(\omega_1-\omega_2)}.$$

Let us look at this case in a little more detail. The matrix element $\langle m|\mathbf{A}\cdot\mathbf{A}|n\rangle$ can be written out as $|A_0|^2\langle m|e^{2i\mathbf{k}\cdot\mathbf{r}}|n\rangle$. In the dipole approximation, this is approximately equal to $|A_0|^2\langle m|n\rangle = |A_0|^2\delta_{mn}$. Thus this term can connect a given state only with itself, so that $\omega_{mn}=0$. This means that all terms of the form of the diagram given by figure 13.8c vanish, so that the $\mathbf{A}\cdot\mathbf{A}$ part of the Hamiltonian is ineffective in two-photon absorption; in the case of figure 13.8d, we must have $\omega_1=\omega_2$, which corresponds to *Rayleigh*, or purely elastic, light scattering.

The diagrams that correspond to second-order perturbation are given by figures 13.8e–13.8h, which include a sum over virtual intermediate states $|n\rangle$. The diagram given by figure 13.8e, which represents terms having the form

$$\sum_n \frac{\langle m|\mathbf{A}\cdot\nabla|n\rangle\langle n|\mathbf{A}\cdot\nabla|l\rangle}{(\omega_{mn}-\omega_1)(\omega_{nl}+\omega_2)},$$

and the diagram given by figure 13.8f, which represents terms having the form

$$\sum_n \frac{\langle m|\mathbf{A}\cdot\nabla|n\rangle\langle n|\mathbf{A}\cdot\nabla|l\rangle}{(\omega_{mn}-\omega_2)(\omega_{nl}+\omega_1)},$$

are the terms that are effective in Raman, or inelastic, light scattering. Raman scattering has previously been considered in chapters 4.6, 7.2, and 8.4 as a tool in molecular structure determination, and the stimulated version of this process will be considered in section 6. We may note, from inspecting these diagrams, that in two-photon scattering processes, terms in which emission formally precedes or follows absorption make fully equivalent contributions to the total cross section; this only points up the fact that no physical reality should be attributed to the set of intermediate states which appear in formal perturbation expansion.

Two-photon absorption processes are represented by the diagrams given by figures 13.8g and 13.8h, which correspond to terms of the form

$$\sum_n \frac{\langle m|\mathbf{A}\cdot\nabla|n\rangle\langle n|\mathbf{A}\cdot\nabla|l\rangle}{(\omega_{mn}+\omega_1)(\omega_{nl}+\omega_2)}$$

and

$$\sum_n \frac{\langle m|\mathbf{A}\cdot\nabla|n\rangle\langle n|\mathbf{A}\cdot\nabla|l\rangle}{(\omega_{mn}+\omega_2)(\omega_{nl}+\omega_1)},$$

respectively. If we make an electric dipole approximation for each of the matrix elements, we see that the two-photon absorption strength will be proportional to the *square* of the quantity

$$\sum_n |E|^2\langle m|\mu|n\rangle\langle n|\mu|l\rangle.$$

Since each dipole matrix element embodies selection rules of the type, for example, of $u \rightarrow g$ or $\Delta J = 1$, the selection rules for two-photon processes will be of the form $u \leftrightarrow u$ and $g \leftrightarrow g$, and $\Delta J = 0$ or 2. That is, the states reached by a two-photon process will be those inaccessible to ordinary single-photon electric dipole absorption.[2] This is borne out, for example, in the two-photon absorption spectrum of benzene (figure 13.9A), which was recorded by excitation of the two-photon fluorescence in the ${}^1B_{2u}$–${}^1A_{1g}$ system (reference 8). This is to be compared with the one-photon spectrum shown in the top panel of figure 9.5. The differences are striking—in the one-photon spectrum, the strongest progression is built on a symmetric e_{2g} vibration (ν_6). This mode is entirely absent in the two-photon spectrum, which instead has its greatest intensity in ν_{14} (e_{1u}) progressions. The high-resolution two-photon spectrum of this molecule has also been recorded (reference 9), and a typical band contour is shown in figure 13.9B. Note that the rotational intensity distribution is quite different from that of a one-photon transition (compare the bottom panel of figure 9.5), reflecting the different selection rules: $\Delta K = 0$ for the two-photon spectrum, while $\Delta K = \pm 1$ for the one-photon (figure 9.8). Note also that the polarization of the laser photons has a significant effect on the intensity distribution. The two-photon fluorescence technique used to record these spectra corresponds to the absorption processes of the diagrams given by figures 13.8g and 13.8h, with $\omega_1 = \omega_2$, followed by the uncorrelated emission of a fluorescence photon at ω' at some later time. Since the rate of light absorption is

2. In molecules that do not possess a center of symmetry, one-photon and two-photon operators can couple to the same set of states. The intensity distribution of the two kinds of transitions may be very different in the two cases, however.

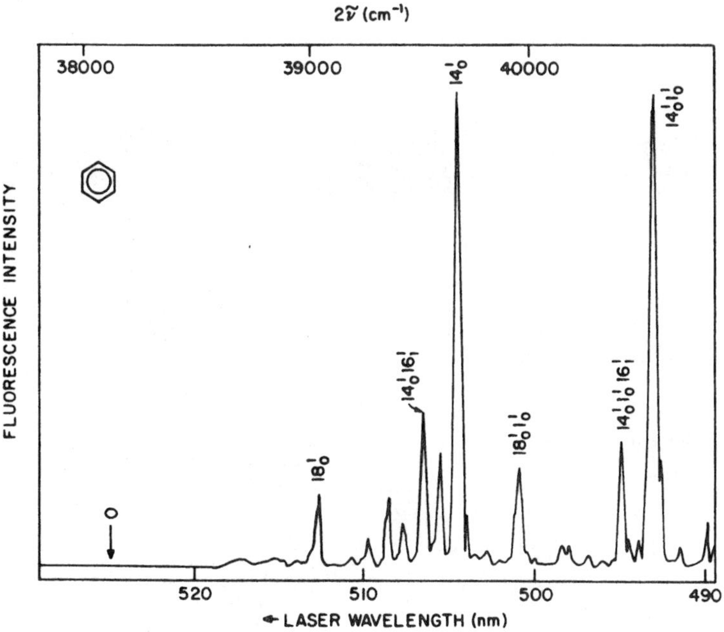

Figure 13.9A
Two-photon fluorescence excitation spectrum of benzene vapor. [From L. Goodman and R. P. Rava, *Adv. Chem. Phys.* **54**, 177 (1983). The lower scale shows the laser photon wavelength, and the upper scale the two-photon energy in wave number units. This spectrum should be compared with the corresponding region (38,000–41,000 cm^{-1}) in the top panel of figure 9.5.

proportional to $|E|^4$, it is proportional to the light intensity squared. Thus the two-photon fluorescence observed in this experiment should be proportional to the square of the exciting light intensity, which is indeed found to be the case. This behavior was also found in earlier two-photon fluorescence experiments on anthracene using ruby laser excitation (reference 10), which are shown in figure 13.10.

The processes of frequency doubling and parametric oscillation, considered in the previous sections, involve the third order of perturbation in the interaction with the radiation field. Diagrams for frequency doubling are of the types given in figures 13.8i–13.8k, with terms of the form

$$\sum_{n,k} \frac{\langle m|\mathbf{A}\cdot\nabla|n\rangle\langle n|\mathbf{A}\cdot\nabla|k\rangle\langle k|\mathbf{A}\cdot\nabla|l\rangle}{(\omega_{mn}+\omega_0)(\omega_{nk}+\omega_0)(\omega_{kl}-2\omega_0)}, \ldots$$

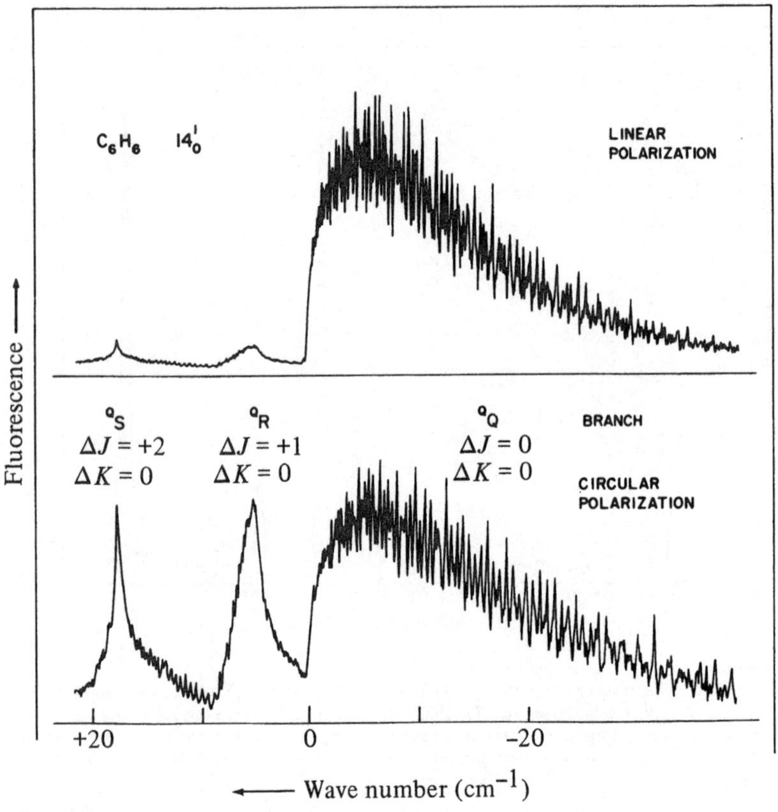

Figure 13.9B
High-resolution two-photon fluorescence excitation spectrum of the 14^1_0 band of benzene. [From J. R. Lombardi, R. Wallenstein, T. W. Hänsch, and D. M. Friedrich, *J. Chem. Phys.* **65**, 2357 (1976). Reproduced with permission.] This should be compared with the high-resolution one-photon spectrum shown in the bottom panel of figure 9.5.

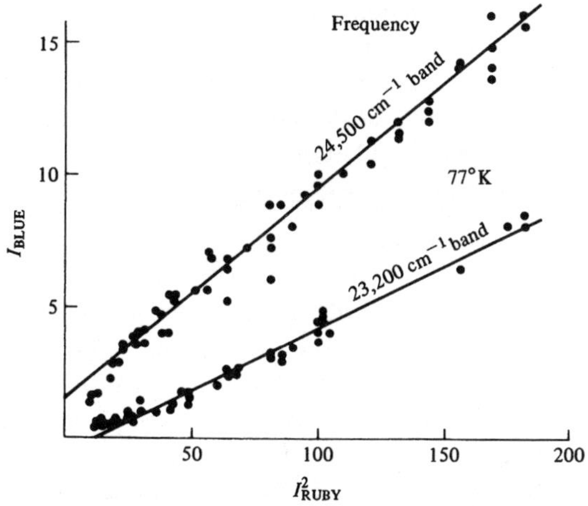

Figure 13.10
Dependence of the two-photon fluorescence intensity in anthracene on the exciting laser pulse intensity. The linearity of the plots of the fluorescence intensity (I_{blue}) against the square of the laser excitation intensity shows that the fluorescent state is excited by two-photon absorption. [From S. Singh and B. P. Stoicheff, *J. Chem. Phys.* **38**, 3032 (1963). Reproduced with permission.]

The more general case of parametric oscillation, in which frequencies 1, 2, and 3 are all different, involve diagrams of which figure 13.8l is an example.

One additional important feature of two-photon absorption spectroscopy may be noted here. If we are dealing with direct two-photon processes, such as are represented by the diagrams given by figures 13.8g and 13.8h, and the excitation beam is split into two beams counterpropagating through the sample, then the Doppler shift cancels to second order (reference 11). This may easily be seen by adding the two photon energies in the molecule's reference frame,

$$\hbar\omega_1 = \hbar\left(\omega_0 - \frac{v_z}{c}\omega_0\right),$$

$$\hbar\omega_2 = \hbar\left(\omega_0 + \frac{v_z}{c}\omega_0\right),$$

$$\overline{\Delta E = 2\hbar\omega_0}$$

exactly. The consequence of this is that *all* the molecules can absorb at $2\omega_0$, independently of their velocity. This Doppler-free two-photon resonance will appear as an intense, narrow line having only the collision- or radiation-broadened line width superimposed on the much broader Doppler-broadened line. In addition to making sub-Doppler measurements possible, the collapse of the absorption strength into a single narrow resonance facilitates observation of these frequently weak two-photon transitions.

4 Infrared Multiple-Photon Absorption

A particular nonlinear process that has attracted a good deal of attention in the past several years is infrared multiple-photon absorption, or IRMPA. This phenomenon was first observed by Isenor and coworkers in 1973 (reference 12); they subjected SiF_4 vapor to intense CO_2 laser pulses and observed luminescence from electronically excited fragments. The molecular nature of the process was spectacularly demonstrated by Ambartzumyan and coworkers in 1975 (reference 13). They found that SF_6, which strongly absorbs the CO_2 P(20) laser line at 10.6 μm, is selectively dissociated at that frequency: the ^{32}S-containing species, which absorbs at that wavelength, is extensively dissociated, but the ^{34}S-containing species, which absorbs 20 cm^{-1} to the red, is unaffected. This permitted them to obtain a 2,000-fold enrichment of the heavier isotope. This experiment was soon repeated in many laboratories around the world; since then, a wide range of molecules has been found to undergo infrared multiple-photon dissociation or isomerization (reference 14).

The energy required to break a molecular bond, or to surmount the potential-energy barrier between isomeric forms, may be 120–350 kJ/mole. Each CO_2 laser photon at 1,000 cm^{-1} carries only about 12 kJ/mole. Therefore, from 10 to 30 or more photons must be absorbed by the molecule to effect the observed chemical change. Of course, the molecule must be chosen to have a fundamental absorption at the CO_2 laser frequency; but the anharmonicity of the absorption (see chapter 7) will drive the absorption off resonance before the requisite number of photons can be absorbed. The central question in understanding this process is thus, how can an isolated molecule absorb the number of low-frequency photons necessary to overcome the barrier to reaction?

The currently accepted model for explaining this process is shown in figure 13.11. In the low-energy region (I), the molecule is described by a "ladder" of vibrational states v_p which are in resonance, or nearly so, with the applied laser frequency. In this region, the pumping may be coherent,

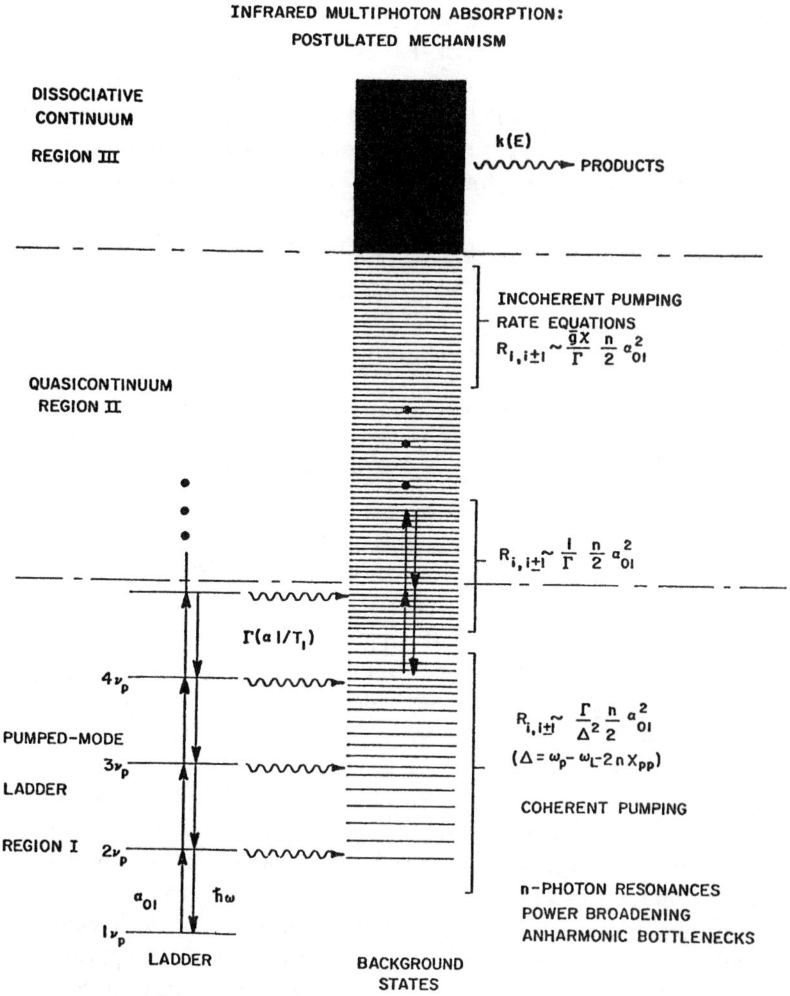

Figure 13.11
Schematic model for infrared multiphoton absorption, showing transition from coherent pumping in the pump mode ladder (region I) to incoherent pumping in the quasicontinuum (region II) to dissociation in the true continuum (region III). [Courtesy of D. M. Brenner.]

so that direct multiphoton resonances ($n\nu_p \leftarrow 0$) and power broadening may be important. Eventually, though, the absorption process must "bottleneck" as the anharmonic detuning becomes too great to be compensated.

The feature that makes the multiple-photon absorption process possible in polyatomic molecules is the presence of many other vibrational modes, which form a set of background states. The directly pumped modes may couple to these background states with a rate (better, a characteristic inverse lifetime for energy to remain in the pumped mode) given by the Fermi golden rule as

$$\Gamma_n = 2\pi g \rho(E = nh\nu_p),$$

where g is the coupling matrix element between the pump and background modes and $\rho(E)$ is the density of vibrational states. Eventually, this density function becomes sufficiently large that the background states form a *quasi-continuum*, which is defined as that region in which the average spacing between vibrational levels, $[\rho(E)]^{-1}$, becomes less than the natural width of the levels. Typically, a density of several thousand states/cm^{-1} must be reached before this criterion can be satisfied. In the quasi-continuum, incoherent single-photon absorption can occur, so that the absorption is best described as a sequential multiple-photon process, rather than as a true multiphoton absorption. We shall take another, more searching look at the concepts of intramolecular vibrational relaxation and the quasi-continuum in the following chapter.

When the molecule has absorbed a sufficient number of photons to surmount an energy barrier on its potential surface, it can then dissociate or isomerize with an energy-dependent rate $k(E)$ given by unimolecular rate theory (reference 15). At this point, the question becomes one of kinetics, rather than spectroscopy, and so we shall not pursue it here.

In order to analyze the process just outlined and obtain quantitative predictions, it is necessary to set up and solve a multilevel optical Bloch equation. Such a solution would entail manipulating matrices of the order of $((n/2)^2) \times ((n/2)^2)$. For SF_6, with $n = 33$, this is an impractical task. The usual approach is to treat absorption in the quasi-continuum as an incoherent process and reduce the optical Bloch equation to an n-level master equation, as was discussed in chapter 12.5. Numerical solution is still required, but the $(n \times n)$ problem is readily computed. The results of such analyses have validated the model presented here. For example, it is clear why no diatomics, few (if any) triatomics, and only a small number of 4-atomic molecules have been found to undergo IRMPA—the density of background states is too small (nonexistent, in the case of diatomics) for the quasi-continuum to be reached at any energy accessible by direct pumping from the ground state. Large polyatomic molecules, on the other hand, possess a high density of background levels and in general are easily dissociated.

For further references on this subject, the reader is urged to consult any of a number of review articles and monographs (references 14, 16–18).

5 Resonant Multiphoton Ionization

A particularly useful process involving multiphoton absorption is multiphoton ionization (MPI). Here, near-ultraviolet or visible wavelength photons are used to drive a molecular system through one or more intermediate excited electronic states to a final, photoionized condition. The advantages offered by this technique are the following:

1. Ionization can be brought about by near-ultraviolet or visible wavelength photons, which can be efficiently generated, instead of requiring vacuum-ultraviolet sources.
2. Typical one-photon ionization spectra are diffuse and nearly featureless, since the final state is part of an ionization continuum. Since the resonant MPI process goes through a real intermediate state, the spectra will display the fine structure associated with that state. The spectrum is thus distinctive for a particular species; conversely, an MPI wavelength can be chosen to ionize selectively a single component in a mixture.
3. Since ions and electrons can be detected with high efficiency, MPI is a much more sensitive detection technique than direct absorption or even, in many cases, laser-induced fluorescence.

An example of a resonantly enhanced MPI (or REMPI) spectrum is shown in figure 13.12. In this example, NO is excited to its $C^2\Pi$ state by two-photon absorption at 382 nm, having an aggregate energy of 6.5 eV [see figure 3.14]. A third photon at the same wavelength takes the molecule from the C state to the $X^1\Sigma^+$ state of NO^+.

6 Stimulated Raman Scattering

The Raman effect, which we have considered in previous chapters, involves inelastic scattering of photons from a molecule in initial state $|0\rangle$, with the

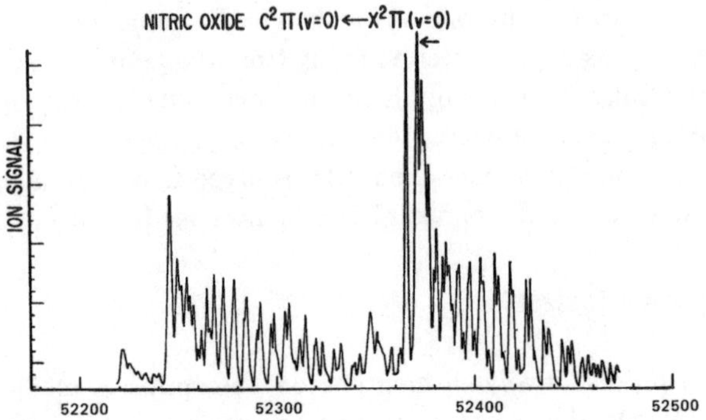

Figure 13.12
Resonantly enhanced MPI spectrum of NO; horizontal axis gives energy in cm^{-1}. [From J. S. Hayden and G. J. Diebold, *J. Chem. Phys.* 77, 4767 (1982). Reproduced with permission.] The energy scale is that corresponding to the two-photon-excited $C^2\Pi$ state. In this work, REMPI was used to detect NO molecules scattered from a cooled copper surface.

molecule ending up in final state $|f\rangle$ and the frequency of the scattered light given by $\hbar\omega' = \hbar\omega_0 - \hbar\omega_i$, where $\hbar\omega_i$ is the energy difference between molecular states $|0\rangle$ and $|f\rangle$. In the case that state $|0\rangle$ is an excited state, and final state $|f\rangle$ lies below it, the frequency of the photon is increased by the corresponding energy difference. This is termed anti-Stokes-Raman scattering, and is much weaker than ordinary Stokes scattering because the population in excited states is generally less than in the ground state. The intensity of ordinary spontaneous Raman scattering is quite small anyway; the cross section for the process is given by terms corresponding to the diagrams given by figures 13.8e and 13.8f, namely

$$\sigma_{\text{Raman}} \propto \sum_n \frac{\langle 0|\mathbf{A}\cdot\nabla|n\rangle\langle n|\mathbf{A}\cdot\nabla|f\rangle}{(\omega_{0n}+\omega_0)(\omega_{nf}-\omega')}.$$

If we make the dipole approximation for the matrix elements,

$$\sigma_{\text{Raman}} \propto E^2 \sum_n \frac{\langle 0|ex|n\rangle\langle n|ex|f\rangle}{(\omega_{0n}+\omega_0)[\omega_{nf}-(\omega_0-\omega_i)]}.$$

But $\omega_{nf}+\omega_i=\omega_{0n}$, so we can write this as

$$\sigma_{\text{Raman}} \propto E^2 \sum_n \frac{e^2\langle 0|x|n\rangle\langle n|x|f\rangle}{(\omega_{0n}+\omega_0)(\omega_{0n}-\omega_0)}$$

$$\propto E^2 \sum_n \frac{e^2\langle 0|x|n\rangle\langle n|x|f\rangle}{\omega_{0n}^2-\omega_0^2}.$$

This last expression is just that for the polarizability of the molecule, so that the Raman-scattering probability is proportional to αE^2, just as we had obtained earlier from classical considerations.

When we come to consider stimulated Raman scattering, a rather more elaborate analysis is called for. We shall follow closely the one presented in chapter 23 of Yariv (reference 1). This treatment begins with a consideration of a medium containing a density of N atoms/cm³ and of depth z. Each atom contains a Thomson-type electron, which oscillates about a fixed point with resonant frequency ω_v, amplitude X, and damping factor γ. The equation of motion for these charges is thus

$$\frac{d^2X(z,t)}{dt^2}+\gamma\frac{dX}{dt}+\omega_v^2X=\frac{F(z,t)}{m}. \tag{13.9}$$

From classical electrostatics, the total energy stored in the medium is

$$W=\tfrac{1}{2}\epsilon E^2,$$

where

$$\epsilon=\epsilon_0(1+N\alpha)=\epsilon_0\left\{1+N\left[\alpha_0+\left(\frac{\partial\alpha}{\partial X}\right)_0X\right]\right\}.$$

The total atomic polarizability α is divided into a static polarizability α_0 and a part dependent on the motion of the atomic charges. Thus the force in equation (13.9) is given by the derivative of the total energy,

$$F(z,t)=\frac{1}{N}\frac{\partial W}{\partial X}=\tfrac{1}{2}\epsilon_0\left(\frac{\partial\alpha}{\partial X}\right)_0E^2(z,t).$$

We shall take the electric field in all these equations as that of a plane wave propagating down the z axis,

$$E_z=\tfrac{1}{2}E_0e^{i(\omega_0t-k_0z)},$$

and that associated with the scattered radiation oscillating at ω_R (which, as we shall see, is not necessarily identical with $\omega_0-\omega_v$) is

$$E_z'=\tfrac{1}{2}E_1e^{i(\omega_Rt-k_Rz)},$$

so that the total field is given by

$$E(z,t)=\tfrac{1}{2}E_0e^{i(\omega_0t-k_0z)}+\tfrac{1}{2}E_1e^{i(\omega_Rt-k_Rz)}. \tag{13.10}$$

From equation (13.6), the rate of change of the field associated with the coherent Raman-scattered light wave propagating down the medium is

$$\frac{dE_1^*}{dz} e^{-i(\omega_R t - k_R z)} = -i\frac{\mu_0}{k_R}\frac{\partial^2}{\partial t^2} P_{NL}^{(\omega_R)}(z,t).$$

The nonlinear part of the polarizability in the previous equation is

$$P_{NL} = \epsilon_0 N\left(\frac{\partial\alpha}{\partial X}\right)_0 XE.$$

If we assume that $X(z,t)$ has the form $\frac{1}{2}X_0(z)e^{i\omega t}$ and substitute this and the electric field from equation (13.10) into the expression for the nonlinear polarizability, we obtain

$$P_{NL} = \tfrac{1}{4}\epsilon_0 N\left(\frac{\partial\alpha}{\partial X}\right)_0 [X_0(z)e^{i\omega t}][E_0 e^{i(\omega_0 t - k_0 z)} + E_1 e^{i(\omega_R t - k_R z)}].$$

Now, we substitute our expression for $F(z,t)$, with E as given by equation (13.10), and our assumed form for $X(z,t)$ into the equation of motion [equation (13.9)]. This gives

$$(\omega_v^2 - \omega^2 + i\omega\gamma)\frac{X_0}{2}e^{i\omega t} = \frac{\epsilon_0}{8m}\left(\frac{\partial\alpha}{\partial X}\right)_0 E_1^* E_0 \exp\{i[(\omega_0 - \omega_R)t - (k_0 - k_R)z]\}$$

From this, we see that $\omega = \omega_0 - \omega_R$ and so obtain

$$X_0(z) = \frac{\epsilon_0(\partial\alpha/\partial X)_0}{4m[\omega_v^2 - (\omega_0 - \omega_R)^2 + i(\omega_0 - \omega_R)\gamma]} E_1^* E_0 e^{-i(k_0 - k_R)z}.$$

If we carry out the indicated time derivatives and substitute the previous expression for $X_0(z)$ into the equation for the propagation of E_1, we obtain

$$\frac{dE_1^*}{dz} = \frac{i}{4}k_R\left(\frac{\epsilon_0}{\epsilon}\right)N\left(\frac{\partial\alpha}{\partial X}\right)_0 E_0^* X_0(z),$$

or, assuming that E_0 is constant along the length of the medium, that is, that the Raman conversion efficiency is relatively small,

$$\frac{dE_1^*}{dz} = \left[\frac{ik_R(\epsilon_0^2/\epsilon)N(\partial\alpha/\partial X)_0^2 E_0^2}{16m[\omega_v^2 - (\omega_0 - \omega_R)^2 + i(\omega_0 - \omega_R)\gamma]}\right]E_1^*.$$

We can see several things immediately from the form of this equation.

1. First, and most important, the Raman intensity is *amplified*.
2. The peak in the scattering spectrum occurs at $\omega_R = \omega_0 - \omega_v + \gamma^2/4\omega_v$; it is slightly shifted from the spontaneous Raman frequency.

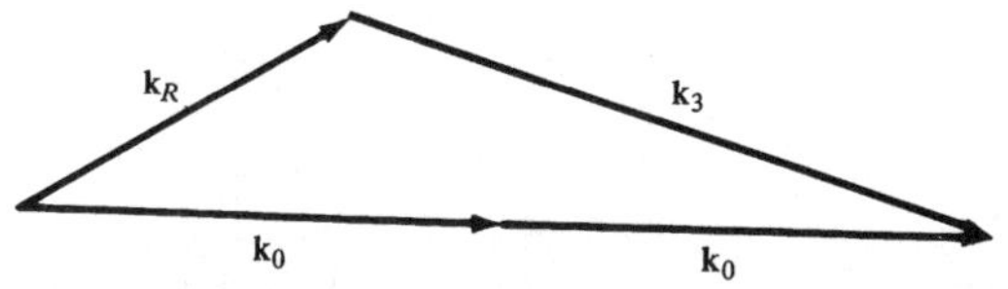

Figure 13.13
Phase-matching requirements for stimulated anti-Stokes-Raman emission. $\mathbf{k}_3$ is the wave vector of the anti-Stokes line.

3. E_1 propagates along the original direction, that is, down the z axis.
4. The Raman active medium itself oscillates coherently because

$$X(z,t) = -\frac{i\epsilon_0(\partial\alpha/\partial X)_0}{8m\omega_v\gamma} E_0 E_1^* \exp\{i[\omega_v t - (k_0 - k_R)z]\}.$$

This last fact is the source of another interesting aspect of this process, namely, that of stimulated anti-Stokes-Raman scattering. If the medium oscillates coherently, then light can scatter off these oscillations. Let us look at light propagating with a frequency $\omega_3 = \omega_0 + \omega_v$. The nonlinear polarizability at this frequency will be given by

$$P_{NL}^{(\omega_3)} = \tfrac{1}{4}\epsilon_0 N\left(\frac{\partial\alpha}{\partial X}\right)_0 X_0(z)E_0 \exp\{i[\omega_v + \omega_0)t - k_0 z]\},$$

and the equation for the propagation of E_3 is thus

$$\frac{dE_3}{dz}e^{i(\omega_3 t - k_3 z)} = \frac{i\omega_3^2\mu_0\epsilon_0(\partial\alpha/\partial X)_0}{4} X_0(z)E_0 \exp\{i[(\omega_v + \omega_0)t - k_0 z]\}.$$

Substituting our previous expression for $X_0(z)$, we have

$$\frac{dE_3}{dz} = \frac{\omega_3\epsilon_0^2\sqrt{\mu_0/\epsilon}N(\partial\alpha/\partial X)_0^2}{16m\omega_v\gamma} E_0^2 E_1^* \exp\{-i[(2k_0 - k_R) - k_3]z + \pi\}.$$

This equation represents an amplification of the electric field E_3, associated with the light wave at frequency ω_3, by radiation at frequency ω_R. Since this is a parametric type of process, we have the phase-matching requirement $k_3 = 2k_0 - k_R$. This can be satisfied geometrically in the manner shown in figure 13.13. The stimulated anti-Stokes-Raman light is propagated at an angle β to the z axis, as shown in figure 13.14. It can be shown (reference 20) that the angle β is given by the fairly simple relation

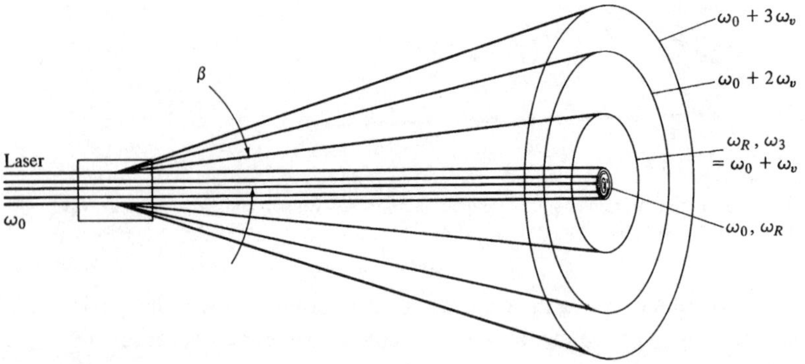

Figure 13.14
Geometric form of stimulated Raman scattering. The undisturbed laser frequency ω_0 and stimulated Raman ω_R are propagated coaxially down the active medium. Some of the ordinary Raman and the first anti-Stokes line propagate in a cone, making an angle β with this axis. Successively higher anti-Stokes lines propagate in coaxial cones of larger and larger apex angles as shown.

$$\beta = \left\{ \frac{1}{n_0} \frac{\omega_R}{\omega_3} \left[n_3 + n_R - 2n_0 + \frac{\omega_0 - \omega_R}{\omega_0} (n_3 - n_R) \right] \right\}^{1/2}, \tag{13.11}$$

where the ns are the refractive indices of the medium at the various frequencies indicated. Successively higher anti-Stokes modes can be excited, each propagating at its particular angle.

Stimulated Raman Scattering has been put to use in frequency shifting the output of fixed frequency lasers. In the example shown in figure 13.15, the ultraviolet output of a rare gas-halide excimer laser (see chapter 10.7.4) is passed through a cell containing high-pressure hydrogen gas. New frequencies are generated at both Stokes and anti-Stokes positions, shifted by up to five times the H_2 Raman frequency, with reasonably good efficiency.

7 Coherent Anti-Stokes Raman Scattering

A stimulated Raman process involving the simultaneous interaction of four waves is coherent Anti-Stokes Raman Scattering, or CARS. The transitions involved in this process are diagrammed in figure 13.16. The system under investigation is simultaneously pumped at frequencies ω_1 and ω_2. Under the right conditions, a new radiation field at frequency $\omega_3 = 2\omega_1 - \omega_2$ is generated.

The power generated at ω_3 is given by the expression (references 21, 22)

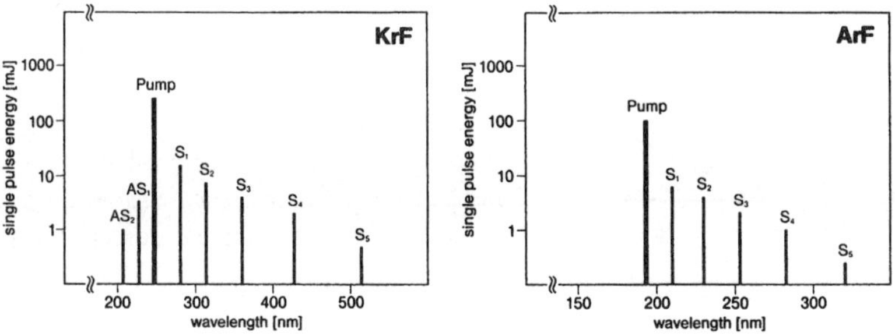

Figure 13.15
Output of a commercial Raman shifter when pumped by a KrF (left) or ArF (right) excimer laser.

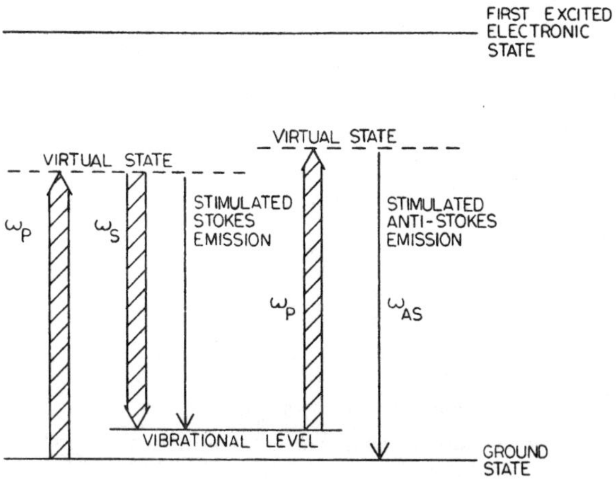

Figure 13.16
Diagram of the coherent Anti-Stokes Raman scattering process.

$$P_3 = \left(\frac{2}{\lambda}\right)^2 \left(\frac{4\pi^2\omega_3}{c^2}\right)^2 \frac{1}{n_1 n_2 n_3} |3\chi^{(3)}_{\text{CARS}}|^2 P_1^2 P_2. \tag{13.12}$$

In this equation, n_i is the refractive index of the medium at ω_i, λ is the wavelength of ω_1, and P_1 and P_2 are the powers at ω_1 and ω_2, respectively. Note that the CARS power is quadratic in P_1, corresponding to the two ω_1 photons employed in figure 13.16. $\chi^{(3)}_{\text{CARS}}$ is the third-order nonlinear susceptibility given by a classical expression,

$$\chi^{(3)}_{\text{CARS}} = \chi^{\text{NR}} + \frac{2N\Delta c^4}{\hbar\omega_2^4} \sum_{if} \left(\frac{d\sigma}{d\Omega}\right)_{if} \frac{1}{2(\omega_{if} - \omega_1 + \omega_2) - i\Gamma_{if}}. \tag{13.13}$$

Note that when the frequency difference $\omega_1 - \omega_2$ approaches that of one of the Raman active transitions, ω_{if}, there is a sharp maximum in the susceptibility function, and a CARS signal is generated at $\omega_3 = \omega_1 + \omega_{if}$. $N\Delta$ is the net population difference between the ground state and the Raman active level, $(d\sigma/d\Omega)_{if}$ is the spontaneous Raman scattering cross section for that transition, Γ_{if} is the spontaneous Raman line width, and the sum is taken over all Raman active transitions. (The major contribution will, of course, be from the most nearly resonant.) By using the relation between Raman cross section and polarizability mentioned previously in chapter 4.6, we can rewrite equation (13.13) as

$$\chi^{(3)}_{\text{CARS}} = \frac{2N\Delta}{\hbar} \sum_{if} \alpha_{if}^2 \frac{1}{2(\omega_{if} - \omega_1 + \omega_2) - i\Gamma_{if}}.$$

In the above, χ^{NR} is a real, nonresonant scattering term associated with the medium under investigation (as well as every other optical element located in the beam path); it is this nonresonant scattering that ultimately limits the sensitivity of the CARS method.

In practice, CARS spectra are produced by combining a strong, fixed-frequency pump beam at ω_1 with a variable frequency beam at ω_2, which is scanned across the region $\omega_1 - \omega_{if}$. Part of the ω_1 pump beam is often used to pump a dye laser that generates ω_2, as shown in figure 13.17. The two beams are combined in the sample, and the CARS beam at ω_3 is selected out by a prism or dichroic filter. An example of a CARS spectrum of vibrationally and rotationally excited molecular oxygen in a low-pressure gas is shown in figure 13.18. A particularly useful application of CARS spectroscopy is as a probe for species present in high-temperature combustion sources (reference 24).

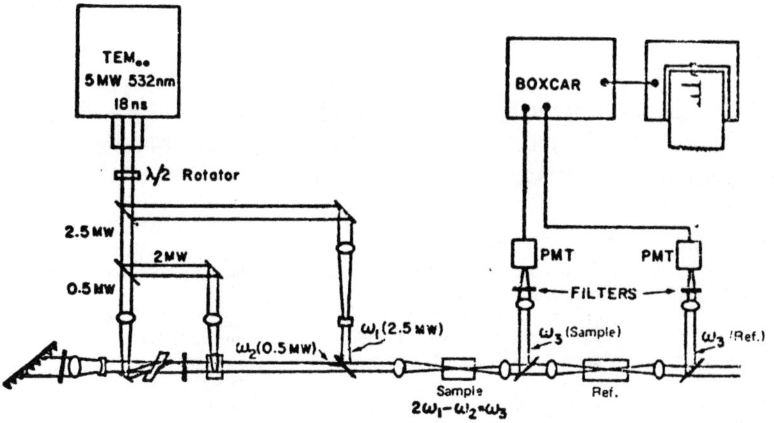

Figure 13.17
Schematic of experimental arrangement for producing and detecting CARS signals. A frequency-doubled, Q-switched Nd:YAG laser is used to produce ω_1, part of which is used to generate ω_2 in a dye laser amplifier system, and part of which is combined with ω_2 in the sample. CARS radiation at ω_3 is filtered out and detected.

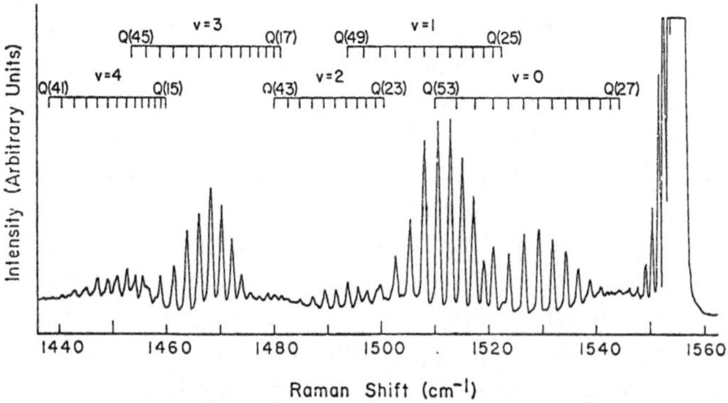

Figure 13.18
CARS spectrum of gaseous oxygen in excited vibrational and rotational states. [From J. J. Valentini, D. S. Moore, and D. S. Bomse, *Chem. Phys. Letts.* **83**, 217 (1981). Reproduced with permission.]

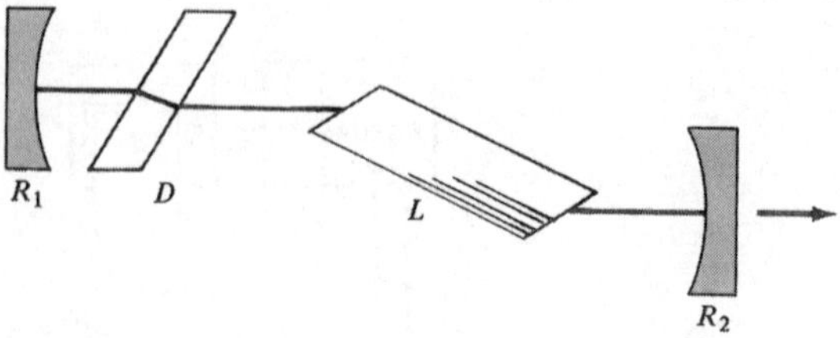

Figure 13.19
Optical arrangement for the production of mode-locked laser pulses: R_1, totally reflecting mirror; D, cell containing a saturable dye to act as a passive Q switch; L, flash-pumped ruby or neodymium-glass rod; R_2, partially reflecting mirror from which the laser beam emerges. The ends of the laser rod are cut at Brewster's angle to minimize unwanted reflections, thus accounting for the offset in the laser beam as it travels through the cavity.

8 Production and Detection of Picosecond Light Pulses

Shortly after the discovery of passively Q-switched ruby and neodymium lasers (chapter 10.7.1), it was noticed that if the laser rod and bleachable dye cell were placed in a somewhat longer optical cavity than usual—say of the order of 1 m, as in figure 13.19—then a new, unusual type of laser activity was obtained (reference 25). Instead of the customary single 10–20-nsec pulse, a long train of much shorter pulses was obtained, which were distinguished as *mode-locked* pulses. These pulse widths were observed to be less than 1 nsec, but how much less could not be determined, because none of the optical detectors in ordinary use possessed an intrinsic time constant much less than 1 nsec.

The nature of these pulses was revealed by means of a new, and extraordinarily simple, means of detection (references 26, 27). The train of pulses was simply led into a solution of a dye that exhibited the phenomenon of two-photon fluorescence, which we have previously discussed in section 3. The arrangement for observing the mode-locked pulses is shown in figure 13.20a. The pulse train is sent through a solution of the dye having two-photon fluorescence activity, and reflected back upon itself from a simple plane mirror. A picture is then taken, using an ordinary camera, of the dye cell. The appearance of the picture is illustrated in figure 13.20b. A streak of fluorescence is visible along the track of the laser beam through the dye, but there are bright spots regularly spaced along the track. These bright spots are just the regions of the dye solution where one of the reflected pulses overlaps with

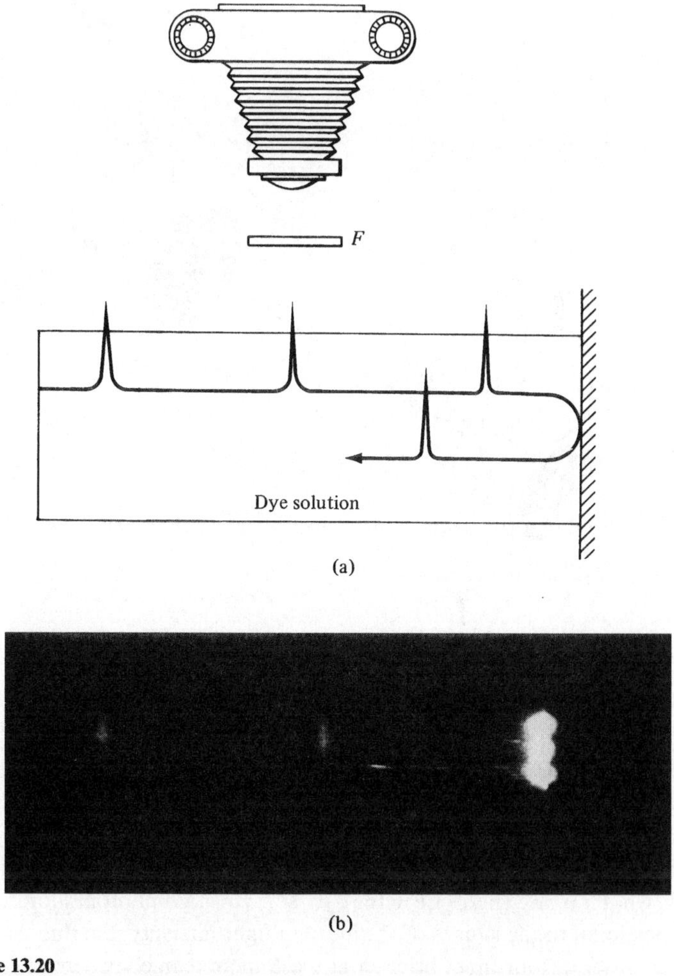

Figure 13.20
(a) Arrangement for making two-photon fluorescence measurements of mode-locked pulse widths. The laser beam enters from the left and is reflected back upon itself by the mirror at *M*. A picture of the experiment is taken by the camera *C*; the filter *F* serves to remove intense scattered light at the laser fundamental frequency. (b) Fluorescence trace produced by a train of mode-locked pulses spaced 67 psec apart, entering the cell from the left, and reflected upon itself by a mirror (not visible in the photograph) located at the right end of the track. [Reproduced with the permission of P. M. Rentzepis, Bell Telephone Laboratories.]

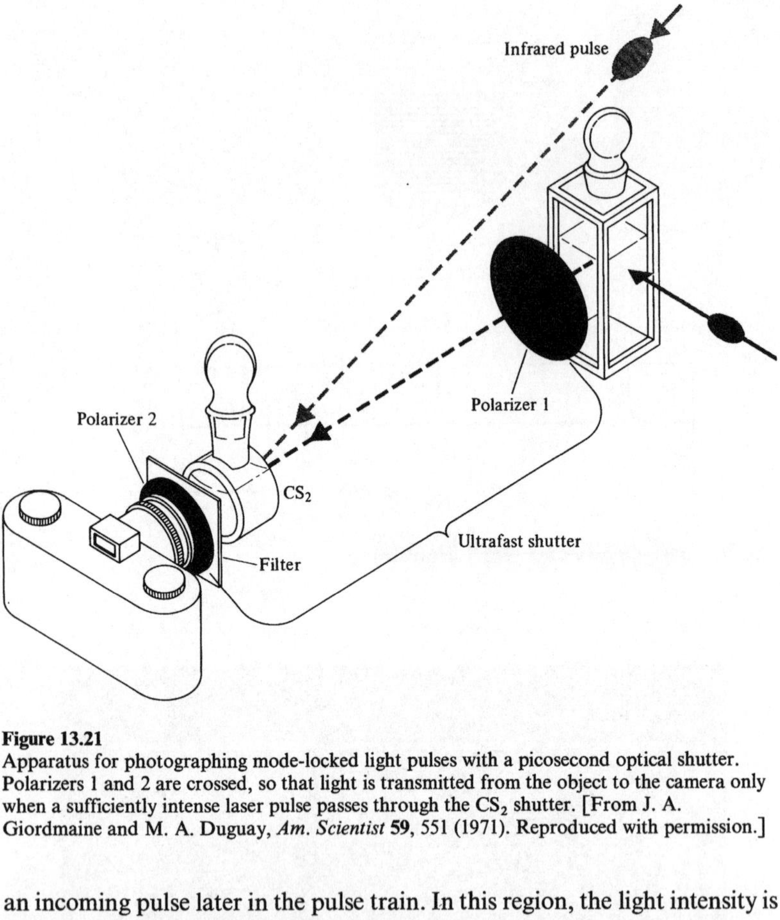

Figure 13.21
Apparatus for photographing mode-locked light pulses with a picosecond optical shutter. Polarizers 1 and 2 are crossed, so that light is transmitted from the object to the camera only when a sufficiently intense laser pulse passes through the CS_2 shutter. [From J. A. Giordmaine and M. A. Duguay, *Am. Scientist* **59**, 551 (1971). Reproduced with permission.]

an incoming pulse later in the pulse train. In this region, the light intensity is double what it is anywhere else in the dye; since the two-photon fluorescence is proportional to the square of the incident light intensity, the fluorescence is of the order of four times brighter at these spots than elsewhere along the track. Since the spots in figure 13.20b have a spatial width of less than 1 mm, and the refractive index of the dye solvent is about 1.3, the temporal width of the laser pulses, which produced the spots, must be

$$\tau_p \lesssim \frac{1.3 \times 10^{-1}\ \text{cm}}{3 \times 10^{10}\ \text{cm/sec}} = 4 \times 10^{-12}\ \text{sec};$$

that is, the individual pulses must be less than 4 psec in duration.

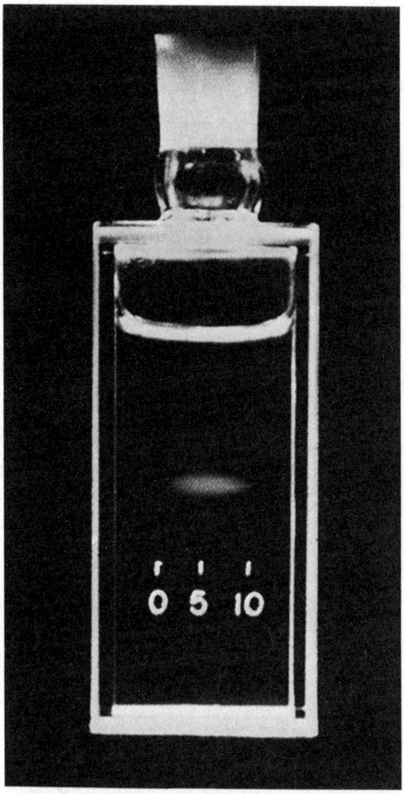

Figure 13.22
Photograph of a single mode-locked light pulse, taken with the apparatus shown in figure 13.21. The pulse is traveling (at the speed of light) from right to left through an ordinary glass cuvette filled with slightly milky water, which scatters a portion of the light. The scale marked on the front of the cuvette is in millimeters. The pulse appears slightly elongated due to the finite aperture time of the shutter. [From J. A. Giordmaine and M. A. Duguay, *Am. Scientist* **59**, 551 (1971). Reproduced with permission.]

Although historically the two-photon fluorescence technique was the method whereby the existence of picosecond pulses was first demonstrated, much more powerful techniques are now available for investigating these phenomena. The most general of these is simply photography using an ultra-fast shutter, operating on the optical Kerr effect. When an electric field is applied to a polarizable medium, the medium may become sufficiently anisotropic to rotate the plane of polarization of plane-polarized light; this is the simple Kerr effect. Thus, if a Kerr cell is placed between two crossed polarizers, light may be transmitted when the electric field is applied. In the optical Kerr effect, the electric field is supplied by an intense mode-locked laser pulse, which passes through a suitable liquid medium, generally carbon disulphide. The effect disappears as the molecules in the liquid relax after the pulse has passed through, which requires only a few picoseconds. Thus, this effect can be the basis of an optical shutter system with a several-picosecond aperture time. Such a system is diagrammed in figure 13.21. A picture of a single mode-locked light pulse, photographed with such a system, is shown in figure 13.22.

The field of picosecond pulse spectroscopy has undergone rapid development. Techniques are now available for selecting a single picosecond pulse out of a mode-locked pulse train; for breaking up this single pulse into a sequence of pulses spaced 2 or 3 psec apart, using a simple glass etalon; and for generating a 2-psec white "flash," covering the entire visible region of the spectrum at once. As of this writing, the shortest pulse that has been produced [by E. Ippen at MIT (reference 28)] has been 16 femtosec, or 1.6×10^{-14} sec. The spatial extent of this pulse is 5 μm, or the size of a single biological cell. By Fourier's theorem, the pulse must contain a frequency bandwidth of the order of several hundred cm^{-1}. With light now available in such unusual packages, a number of investigations of ultrafast molecular processes, especially in liquids, may be carried out.

Problem

1. The output wavelength of a krypton fluoride laser is 249 nm, and the fundamental vibrational frequency of hydrogen is 4,400 cm^{-1}. Calculate the wavelengths of the five Stokes and two anti-Stokes SRS lines shown in figure 13.15. Neglect the small term in $\gamma^2/4\omega_v$.

References

1. A. Yariv, *Quantum Electronics* (Wiley, New York, 1968).

2. G. S. Monk, *Light: Principles and Experiments* (Dover, New York, 1963), p. 275.

3. W. G. Driscoll and W. Vaughan, eds., *Handbook of Optics* (McGraw-Hill, New York, 1978).

4. G. R. Fowles, *Introduction to Modern Optics* (Holt, Rinehart and Winston, New York, 1967), chapter 5.

5. D. L. Weinberg, *Laser Focus* **5**(4), 35 (April 1969).

6. W. Peticolas, *Ann. Rev. Phys. Chem.* **18**, 233 (1967).

7. M. Goeppert-Mayer, *Ann. Physik* **9**, 273 (1931).

8. L. Goodman and R. P. Rava, *Adv. Chem. Phys.* **54**, 177 (1983).

9. J. R. Lombardi, R. Wallenstein, T. W. Hänsch, and D. M. Friedrich, *J. Chem. Phys.* **65**, 2357 (1976).

10. S. Singh, W. J. Jones, W. Siebrand, B. P. Stoicheff, and W. G. Schneider, *J. Chem. Phys.* **42**, 330 (1965).

11. K. Shimoda, *Appl. Phys.* **9**, 239 (1976).

12. N. Isenor, V. Merchant, R. Hallsworth, and M. Richardson, *Can. J. Phys.* **51**, 1281 (1973).

13. R. V. Ambartzumyan, Yu. A. Gorokhov, V. S. Letokhov, and G. N. Makarov, *JETP Letts.* **21**, 375 (1975).

14. J. I. Steinfeld, ed., *Laser-Induced Chemical Processes* (Plenum, New York, 1981).

15. P. J. Robinson and K. A. Holbrook, *Unimolecular Reactions* (Wiley-Interscience, New York, 1972).

16. C. D. Cantrell, ed., *Multiple-Photon Excitation and Dissociation of Polyatomic Molecules* (Springer-Verlag, Berlin, 1979).

17. M. Stitch, ed., *Laser Handbook*, Vol. 3b (North-Holland, Amsterdam, 1979), pp. 485–576.

18. M. Quack, *Adv. Chem. Phys.* **50**, 395 (1982).

19. J. S. Hayden and G. J. Diebold, *J. Chem. Phys.* **77**, 4767 (1982).

20. E. Garmire, F. Pandarese, and C. H. Townes, *Phys. Rev. Letts.* **11**, 160 (1963).

21. S. Druet and J.-P. Taran, in C. B. Moore, ed., *Chemical and Biochemical Applications of Lasers*, Vol. 4 (Academic Press, New York, 1979), pp. 187–252.

22. A. B. Harvey and J. W. Nibler, *Applied Spectrosc. Reviews* **14**, 101 (1978).

23. J. J. Valentini, D. S. Moore, and D. S. Bomse, *Chem. Phys. Letts.* **83**, 217 (1981).

24. A. C. Eckbreth, P. A. Bonczyk, and J. F. Verdieck, *Appl. Spectrosc. Reviews* **13**, 15 (1978).

25. A. J. DeMaria, D. A. Stetser, and H. Heynau, *Appl. Phys. Letts.* **8**, 174 (1966).

26. M. A. Duguay, S. L. Shapiro, and P. M. Rentzepis, *Phys. Rev. Letts.* **19**, 1014 (1967).

27. J. R. Klauder, M. A. Duguay, J. A. Giordmaine, and S. L. Shapiro, *Appl. Phys. Letts.* **13**, 174 (1968).

28. J. G. Fujimoto, A. M. Weiner, and E. P. Ippen, *Appl. Phys. Letts.* **44**, 832 (1984).

14 Spectroscopy beyond Molecular Constants

We have now nearly reached the end of this book, and it is appropriate to retrace the route over which we have traveled. We began by considering uncorrelated one-photon transitions, first in the 100-odd elemental atoms in chapter 2, and then in the $(100)^2/2$ possible diatomic molecules (neglecting valence restrictions) in chapters 3–5. The multitude of polyatomic molecules was the subject of chapters 7–9; actually, only a small fraction of these have had their spectra analyzed in detail. In chapters 10–12, we extended our view to include coherent interactions of one or more radiation fields with these systems, and in chapter 13, we considered the possibility of multiple-photon transitions between atomic or molecular energy levels.

In all of these instances, we have been dealing with transitions between two eigenstates of the system, that is,

$$|a\rangle + h\nu \rightleftarrows |b\rangle,$$

where $|a\rangle$ and $|b\rangle$ were uniquely described by a set of quantum numbers corresponding to constants of the motion, that is, conserved dynamical variables. We could thus characterize the energy states of atoms and molecules by *spectroscopic constants* that were the coefficients of these quantum numbers. We now must broaden our view and go "beyond molecular constants"[1] in order to be able to describe systems of such high complexity and/or energy content that the simple quantum number description is no longer applicable. We have already encountered instances of this, for example, in the discussion of radiationless transitions in chapter 9.6, and in the so-called intramolecular vibrational relaxation process that occurs during infrared multiphoton excitation, in chapter 13.4. In this chapter, we shall first survey the nature of the problem and several experimental approaches, and then consider what kind of theoretical model might be appropriate for the description of such systems.

1 Molecules at High Excitation Levels

We have seen that, in general, the detailed spectroscopy of molecular systems becomes progressively more complex and difficult as either the level of excitation or the dimensionality of the system increases. A useful breakdown of the various regimes has been given in reference 1, and we repeat it here.

1. I am indebted to R. W. Field of MIT for this expression.

1. Low-Excitation Regime (system in its ground or first excited vibrational state)

a. Low Dimensionality (diatomic or linear triatomic) Simple rigid rotor, harmonic oscillator models can usually be applied with confidence.
b. High Dimensionality (polyatomic molecules) The normal mode approximation is usually adequate, although vibration-rotation interaction can be important, as in the spectroscopy of SF_6. In this regime, conventional infrared, microwave, and Raman spectroscopy provide the information necessary for characterizing the structure of the molecule.

2. High-Excitation Regime (system containing 1 eV or more of vibrational energy, that is, >20–30% of a bond dissociation energy)

a. Low Dimensionality For these systems, typically diatomic molecules in high vibrational levels, Rydberg-Klein-Rees potentials and JWKB wave functions in these potentials generally provide an adequate description. Experimental techniques for investigating these states include laser-induced fluorescence and optical-optical double resonance.
b. High Dimensionality It is in this regime that the principal difficulties arise, since the nature of the interaction of a radiation field with a polyatomic molecule at high internal energy is neither simple nor intuitive. The system almost certainly does not behave as a collection of isolated two-level systems, so that the fraction of molecules interacting directly with a monochromatic field will depend on the excitation level and the intensity of the radiation. If a "hole" is "burned" into the absorption profile, relaxation processes (collisional and/or intramolecular) tending to "fill" that hole must be considered. We must consider the question of localization of energy in a single vibrational mode versus internal redistribution ("scrambling") at constant energy. Furthermore, if a coherent superposition state is produced by the exciting radiation, there remains the question of intramolecular dephasing. In other words, the entire concept of "taking a spectrum" of a polyatomic molecule at high excitation needs to be carefully examined.

As a rough but useful guide, we can say that the high-energy, high-dimensionality regime is reached whenever the density of molecular states $\rho(E)$ becomes comparable to the natural line width. This density of states is defined as

$$\rho(E) = \left.\frac{dN(E_{\text{vib}} \leq E')}{dE'}\right|_{E'=E}.$$

If the vibrational energy is given by

$$E_{\text{vib}} = \sum_i \hbar\omega_i \left(n_i + \frac{d_i}{2}\right),$$

then an approximate form for the energy density has been found to be

$$\rho(E) \simeq \frac{1}{N!} \prod_{i=1}^{N} \left(\frac{1}{\hbar\omega_i}\right)(E + \alpha E_0)^N, \tag{14.1}$$

where

$$E_0 = \sum_i \hbar\omega_i \left(\frac{d_i}{2}\right) \qquad \text{(the zero-point energy)}$$

and $N = 3N_{\text{atoms}} - 6$. In equation (14.1), α is a fitting parameter ($0 \le \alpha \le 2$), which can be adjusted to take account of anharmonicity by comparing this approximate expression with the results of an exact count of vibrational states. Other, more accurate expressions are discussed in detail in reference 2.

If we take the line width to have a nominal value of 1 MHz, the high density-of-states regime would be reached at approximately 1 state/10^6 Hz, or 3×10^4 states/cm^{-1}. Such a density is reached in typical small (5–8 atoms) polyatomic molecules at energies of around 70–100 kJ/mole, or about 10,000 cm^{-1}. This would correspond to the onset of the so-called quasi-continuum mentioned in chapter 13.4. The laws of physics tell us that certain quantities, such as total energy, total linear and angular momentum in space, and symmetry properties (if any), must have definite values, but nothing else can be specified with certainty when the molecule is in its quasi-continuum. Note also that collisions play a role, since pressure broadening increases the line width and thus lowers the threshhold for quasi-continuum behavior. In other words, the molecule cannot be regarded as an isolated system.

Several experimental methods are now available for producing and investigating molecules in these high excitation states:

1. Infrared Multiple-Photon Excitation (see chapter 13.4) IRMPE is a convenient method for producing a high degree of excitation in molecules capable of absorbing the radiation output of a CO_2 or other high-energy infrared laser. The states thus produced can be interrogated by any of several double-resonance (see chapter 12.7 and reference 3) or pump probe techniques.

a. In infrared double resonance, a low-intensity, c.w. tunable probe is used to measure the transient infrared absorption of the vibrationally excited

molecules. The probe intensity must be kept low, in order to avoid saturation and other nonlinear responses to the observing radiation.
b. Information on internal vibrational energy content and relaxation can also be obtained from measuring the changes in the optical or ultraviolet photoabsorption cross section following excitation. Even when the ultraviolet absorption is to a photodissociating state, leading to a continuous spectrum, vibrational excitation will redistribute the absorption strength in the spectrum.
c. When laser-induced fluorescence excitation and emission spectroscopy can be employed, its high sensitivity, energy resolution, and time resolution make it the method of choice for probing energy distributions following IRMPE. The limitation on this method is the relatively small number of molecules for which the simultaneous requirements of a midinfrared absorption spectrum susceptible to IRMPE and an easily accessible ultraviolet or visible transition having a high fluorescence yield can be met.
d. A way around this limitation is the use of Raman scattering as a probe for molecules in highly excited vibrational levels. The advantage of this technique is that it is applicable to all molecules, without requring specific resonant transitions, and that it is state specific; its obvious and main disadvantage is the extremely small magnitude of the Raman scattering cross section. This last limitation can be overcome, however, by using the much more intense scattering in a coherent anti-Stokes-Raman (CARS) process (see chapter 13.7).

Examples of these pump probe experiments are given in references 1 and 2.

2. Stimulated Emission Pumping SEP is a method of producing molecules in specific high vibrational levels. The principle is the same as that employed in the optically pumped laser (see chapter 10.7.2), with the addition of a second laser to direct population to a specific final state by stimulated emission from the intermediate or "transit" level at the appropriate wavelength. The presence of molecules in this final state can be detected by either laser-induced fluorescence from that state, using a third laser as the probe, or simply by monitoring the depletion of spontaneous emission from the "transit" level as a function of stimulated emission wavelength.
3. Direct Overtone Spectroscopy This method does not actually produce population in the high-excitation level, but does allow these levels to be observed with an ensemble of ground state (even expansion-cooled) molecules. Ultrasensitive modulation techniques must generally be used in order to detect the weak overtone absorptions.

Each of these methods has its advantages and disadvantages. IRMPE is a very efficient process, but it produces a distribution of states in the excited molecules. SEP and direct overtone spectroscopy access very specific energy levels, but only a restricted set of these levels can be observed experimentally. An illustration of this will be given at the end of the following section.

2 Spectra from Molecular Dynamics

In the practice of spectroscopy discussed throughout this text, one derives spectroscopic constants from measured and assigned transition frequencies and term values and uses these constants to determine structural parameters, such as equilibrium bond lengths and angles, curvatures of potential surfaces, and magnetic coupling coefficients. The implicit theoretical model is that of a well-behaved Hamiltonian that, with a properly chosen basis set, allows us to construct more or less accurate eigenstates. It is not at all obvious that such an approach can be fruitfully employed at high levels of excitation. An alternative approach that has been tried for this regime is to attempt to solve the dynamics classically, using an assumed potential surface and numerically integrating the equations of motion to find the trajectories of the constituent atoms on this surface. These trajectories can be related to a spectral density by means of the dipole correlation function derived in appendix D, namely,

$$I(\omega) = \frac{1}{2\pi}\int_{-\infty}^{\infty} e^{-i\omega t}\langle \boldsymbol{\mu}(0)\cdot\boldsymbol{\mu}(t)\rangle\, dt. \tag{14.2}$$

Taking the usual normal coordinate expansion of the dipole moment,

$$\boldsymbol{\mu}(Q) = \boldsymbol{\mu}(0) + \sum_{j=1}^{N}\left(\frac{\partial\boldsymbol{\mu}}{\partial Q_j}\right)Q_j + \cdots, \tag{14.3}$$

gives

$$I(\omega) = \frac{1}{2\pi}\sum_{j,k=1}^{N}\left(\frac{\partial\boldsymbol{\mu}}{\partial Q_j}\right)\left(\frac{\partial\boldsymbol{\mu}}{\partial Q_k}\right)\int_{-\infty}^{\infty} e^{-i\omega t}\langle Q_j(t)Q_k(0)\rangle\, dt. \tag{14.4}$$

Thus, if the classical trajectory $\{Q(t)\}$ of the molecular system can be found, the spectral density $I(\omega)$ can be calculated. The fluorescence or Raman scattering intensity $F(\omega)$ can be similarly defined.

The prescription of equations (14.2)–(14.4) appears straight-forward, but in fact is a very difficult job. The integrations can be done, but a potential surface must be specified. We simply do not know the detailed shapes of these surfaces, even for very simple molecules, and it turns out that small details of the surface can have profound effects on the calculated trajectories, and thus the predicted spectra. Then, we also need the dipole (or polarizability, for Raman spectra) derivatives in equation (14.3). These can be derived in principle from ab initio or semiempirical calculations; in practice, however, these quantities are very difficult to compute accurately, especially for configurations far from the equilibrium geometry. For this reason, most of the work on molecular dynamics calculation of spectral density has been for artificial two-dimensional model systems, although a few calculations on simple triatomic molecules, such as OCS, ozone, and HCN, have begun to appear (references 4–8).

A two-dimensional model potential that is very often used is the so-called Henon-Heiles potential,

$$\mathscr{H} = \tfrac{1}{2}(p_x^2 + p_y^2 + \omega_x^2 x^2 + \omega_y^2 y^2) + \lambda x y^2 + \kappa x^3. \qquad (14.5)$$

Some representative trajectories calculated in this potential are shown in figures 14.1a–14.1c. The trajectories given by figures 14.1a and 14.1b are called *quasi-periodic*, since the system undergoes repetitive (if complex) motion in the potential. These trajectories are essentially complicated Lissajous figures (see reference 9). A trajectory such as the one given by figure 14.1c is called *chaotic or stochastic*, since the motion of the system appears to be random; if the trajectory is arbitrarily prolonged, the path on the surface would eventually fill the entire energetically accessible portion of the (xy) plane. There seems to be a certain critical energy E^* above which the motion of the system makes a transition from quasi-periodic to chaotic, but the value of E^* depends very strongly on details of the assumed potential surface.

Power spectra calculated for these trajectories by using equation (14.4) are shown in figures 14.1d and 14.1e. The quasi-periodic trajectory gives sharp line spectra, while the chaotic or stochastic trajectory gives a broad frequency distribution.

Calculations of this sort, for artificial model Hamiltonians, are very suggestive, and indeed raise several profound mathematical questions about statistical mechanics and the theory of equations. As experimental spectroscopists, however, we must ask whether this sort of behavior is, in fact, actually seen in real molecules.

To conclude this section, we shall show the results of some measurements

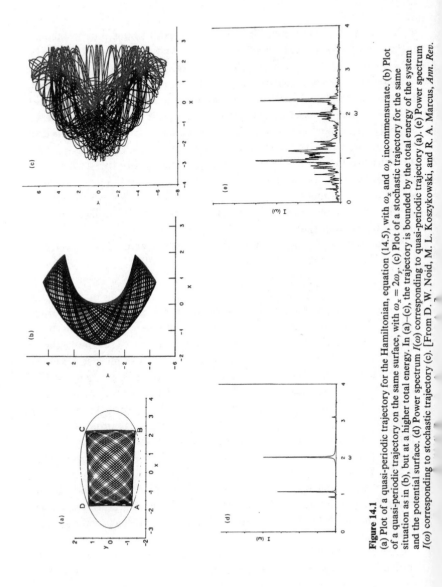

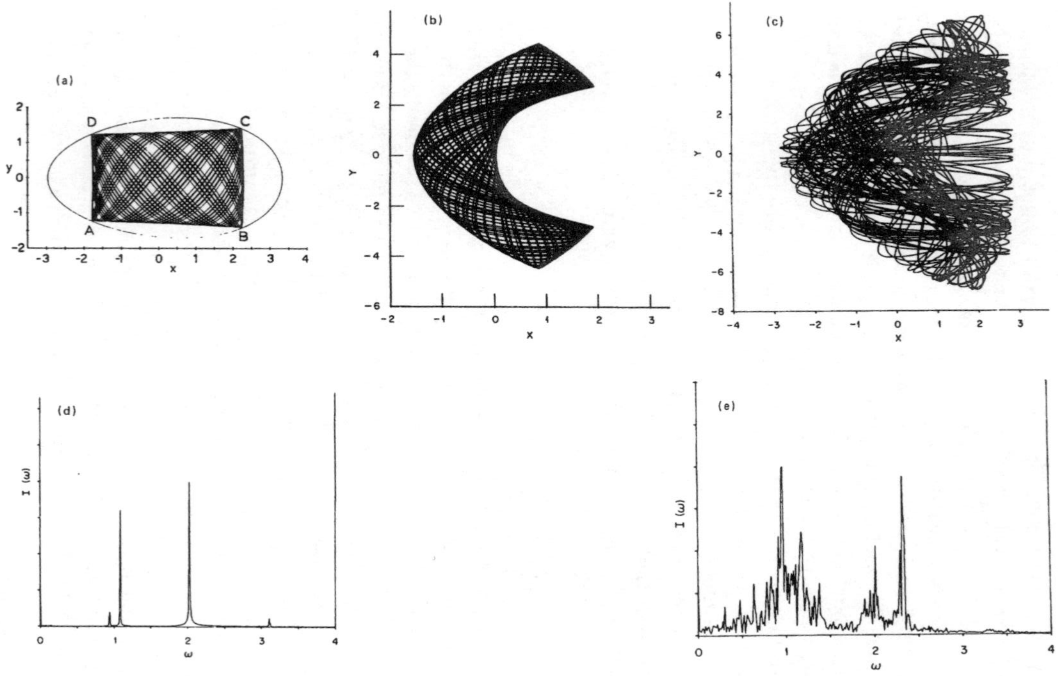

Figure 14.1
(a) Plot of a quasi-periodic trajectory for the Hamiltonian, equation (14.5), with ω_x and ω_y incommensurate. (b) Plot of a quasi-periodic trajectory on the same surface, with $\omega_x = 2\omega_y$. (c) Plot of a stochastic trajectory for the same situation as in (b), but at a higher total energy. In (a)–(c), the trajectory is bounded by the total energy of the system and the potential surface. (d) Power spectrum $I(\omega)$ corresponding to quasi-periodic trajectory (a). (e) Power spectrum $I(\omega)$ corresponding to stochastic trajectory (c). [From D. W. Noid, M. L. Koszykowski, and R. A. Marcus, *Ann. Rev.*

on highly excited vibrational levels of acetylene, using two of the techniques mentioned in the preceding section. The reader can then judge whether the behavior predicted by the molecular dynamics model is, in fact, observed.

Figure 14.2 shows measurements of transitions in acetylene at approximately 2–3 eV of internal energy. The spectrum in the top panel is a direct overtone absorption measurement, from the ground state to $v_3 = 5$, that is, the fourth overtone of a C—H stretching vibration. Only three sharp lines are seen in a 1-cm^{-1} interval. The spectrum in the bottom panel is a SEP spectrum of the same molecule, pumping through the electronically excited A state at 45,297 cm^{-1} and terminating on levels approximately 28,000 cm^{-1} above the ground state. A very congested spectrum is seen; since the "transit" state is already rotationally preselected, this must be vibrational structure in the ground electronic state (references 10 and 11).

Superficially, these results would seem to bear out the simple model discussed previously. The direct overtone ($n \leftarrow 0$) spectra consist of sharp (although severely perturbed) vibration-rotation lines, while the higher-energy states accessed in the SEP experiment have a much more irregular spectrum. However, one must be extremely cautious in such an interpretation because these different excitation methods select very different subsets of the available molecular states. Direct overtone pumping tends to have highest transition probability for "local modes," that is, oscillator configurations in which all of the vibrational energy is localized in a single C—H bond. SEP will populate a quite different set of oscillator states, namely, those having good Franck-Condon overlap with the electronically excited "transit" level. In other words, the nature of the expected spectral density function depends strongly on the nature of the excitation and probe radiation used to prepare and interrogate the system—"where you twang is where you hang" (reference 12).

The conclusion that may be drawn from this is that it is no longer sufficient to think of an isolated molecule, possessing absolute properties, interrogated by a radiation field. Instead, we are dealing with a dynamically coupled system of molecules and radiation, in which the way we ask a question has a strong influence on the answer we shall obtain. In a way, such a model is consistent with today's view of physical science: Observed phenomena are manifestations of an underlying, very complex reality, and a specific observation is produced as much by the measuring apparatus as by the object being measured. Viewed pessimistically, this would imply that it may never be possible to define a "spectrum" for a highly excited molecule, in the way

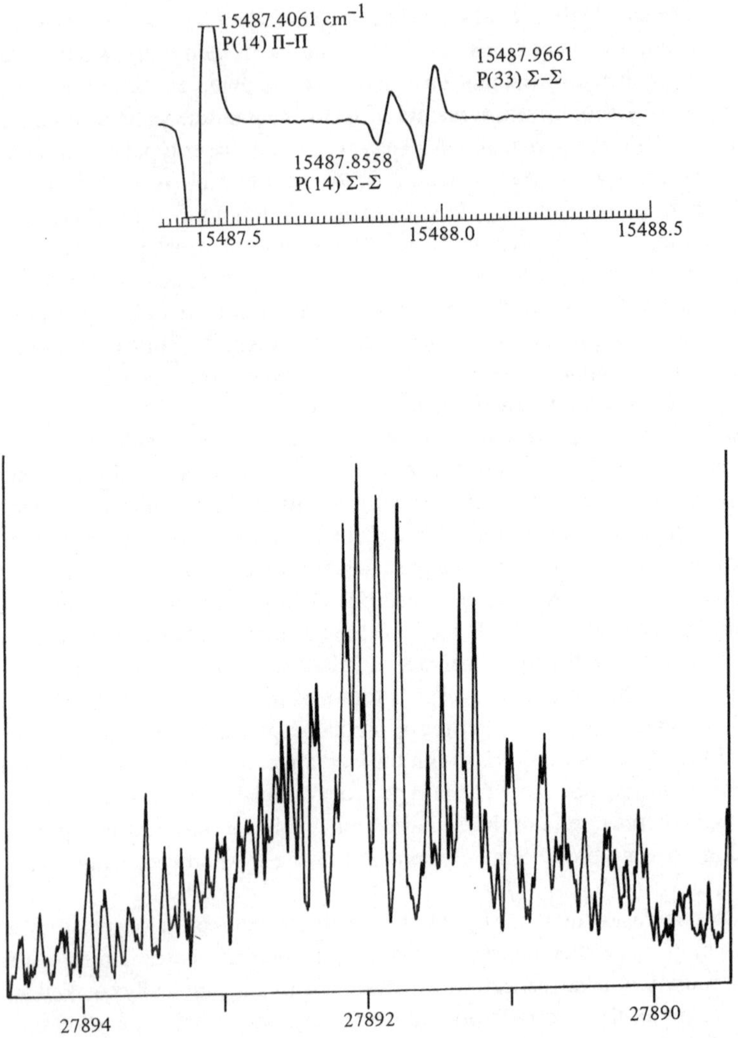

Figure 14.2
Top panel: Direct overtone spectrum of acetylene; the lines have a derivative shape because modulated photoacoustic detection is employed. [From G. J. Scherer, K. K. Lehmann and W. Klemperer, *J. Chem. Phys.* **78**, 2817 (1983). Reproduced with permission.] Bottom panel: SEP spectrum of acetylene as a function of final state vibrational energy (in cm^{-1}). [From E. Abramson, R. W. Field, D. Imre, K. K. Innes, and J. L. Kinsey, *J. Chem. Phys.* **80**, 2298 (1984). Reproduced with permission.]

that is done for a system near its equilibrium configuration, that is independent of the details of the excitation and measurement processes. To conclude on an optimistic note, however, we can take this to mean that by selection of the right excitation conditions, it may be possible to induce a molecule to behave in almost any manner we may wish it to do.

References

1. J. I. Steinfeld, in *Energy Storage and Redistribution in Molecules*, J. Hinze, ed. (Plenum, New York, 1983), pp. 1–15.

2. P. J. Robinson and K. A. Holbrook, *Unimolecular Reactions* (Wiley-Interscience, New York, 1972).

3. J. I. Steinfeld, M. Dubs, D. Harradine, S. Adler-Golden, E. Schweitzer, M. Spencer, and D. Brenner, in *Advances in Laser Spectroscopy*, Vol. 2, B. A. Garetz and J. R. Lombardi, eds. (Heyden, Philadelphia, 1983), pp. 45–71.

4. D. R. Fredkin, A. Komornicki, S. R. White, and K. R. Wilson, *J. Chem. Phys.* **78**, 7077 (1983).

5. D. W. Noid, M. L. Koszykowski, and R. A. Marcus, *Ann. Rev. Phys. Chem.* **32**, 267 (1981).

6. K. D. Hänsel, *Chem. Phys.* **33**, 35 (1978).

7. D. Carter and P. Brumer, *J. Chem. Phys.* **77**, 4208 (1982).

8. K. K. Lehmann, G. J. Scherer, and W. Klemperer, *J. Chem. Phys.* **77**, 2855 (1982).

9. D. P. Shoemaker, C. W. Garland, J. I. Steinfeld, and J. W. Nibler, *Experiments in Physical Chemistry*, 4th ed. (McGraw-Hill, New York, 1981), pp. 651–652.

10. G. J. Scherer, K. K. Lehmann, and W. Klemperer, *J. Chem. Phys.* **78**, 2817 (1983).

11. E. Abramson, R. W. Field, D. Imre, K. K. Innes, and J. L. Kinsey, *J. Chem. Phys.* **80**, 2298 (1984).

12. E. J. Heller and W. M. Gelbart, *J. Chem. Phys.* **73**, 626 (1980).

Appendix A Direct Product Tables

These tables give the symmetry species of the product of two functions with the indicated symmetry species, for each point group. Although these are all immediately derivable from the character tables for each point group (for which, see any of the references listed in chapter 6), the direct product tables should suffice for most ordinary applications, such as determining selection rules, and so on. Also tabulated are the symmetry species of the components of the electromagnetic transition operators, as follows:

electric dipole: x, y, z;
magnetic dipole: R_x, R_y, R_z;
electric quadrupole: $x^2+y^2, x^2-y^2, z^2, xy, xz, yz$.

1.

C_1	A	Operators
A	A	All

2.

C_2	A	B	Operators
A	A	B	$z, R_z, x^2, y^2, z^2, xy$
B	B	A	x, y, R_x, R_y, xz, yz

3.

C_3	A	E	Operators
A	A	E	z, R_z, x^2+y^2, z^2
E	E	$2A \oplus E$	$x, y, R_x, R_y, x^2-y^2, xy, xz, yz$

4.

C_4	A	B	E	Operators
A	A	B	E	z, R_z, x^2+y^2, z^2
B	B	A	E	x^2-y^2, xy
E	E	E	$2A \oplus 2B$	x, y, R_x, R_y, xz, yz

5.

C_{2v}	A_1	A_2	B_1	B_2	Operators
A_1	A_1	A_2	B_1	B_2	z, x^2, y^2, z^2
A_2	A_2	A_1	B_2	B_1	R_z, xy
B_1	B_1	B_2	A_1	A_2	x, R_y, xz
B_2	B_2	B_1	A_2	A_1	y, R_x, yz

6.

C_{3v}	A_1	A_2	E	Operators
A_1	A_1	A_2	E	z, x^2+y^2, z^2
A_2	A_2	A_1	E	R_z
E	E	E	$A_1 \oplus A_2 \oplus E$	$x, y, R_x, R_y, x^2-y^2, xy, xz, yz$

7.

C_{4v}	A_1	A_2	B_1	B_2	E	Operators
A_1	A_1	A_2	B_1	B_2	E	z, x^2+y^2, z^2
A_2	A_2	A_1	B_2	B_1	E	R_z
B_1	B_1	B_2	A_1	A_2	E	x^2-y^2
B_2	B_2	B_1	A_2	A_1	E	xy
E	E	E	E	E	$A_1 \oplus A_2 \oplus B_1 \oplus B_2$	x, y, R_x, R_y, xz, yz

8.

C_{5v}	A_1	A_2	E_1	E_2	Operators
A_1	A_1	A_2	E_1	E_2	z, x^2+y^2, z^2
A_2	A_2	A_1	E_1	E_2	R_z
E_1	E_1	E_1	$A_1 \oplus A_2 \oplus E_2$	$E_1 \oplus E_2$	x, y, R_x, R_y, xz, yz
E_2	E_2	E_2	$E_1 \oplus E_2$	$A_1 \oplus A_2 \oplus E_1$	x^2-y^2, xy

9.

C_{6v}	A_1	A_2	B_1	B_2	E_1	E_2	Operators
A_1	A_1	A_2	B_1	B_2	E_1	E_2	z, x^2+y^2, z^2
A_2	A_2	A_1	B_2	B_1	E_1	E_2	R_z
B_1	B_1	B_2	A_1	A_2	E_2	E_1	
B_2	B_2	B_1	A_2	A_1	E_2	E_1	
E_1	E_1	E_1	E_2	E_2	$A_1 \oplus A_2 \oplus E_2$	$B_1 \oplus B_2 \oplus E_1$	x, y, R_x, R_y, xz, yz
E_2	E_2	E_2	E_1	E_1	$B_1 \oplus B_2 \oplus E_1$	$A_1 \oplus A_2 \oplus E_2$	x^2-y^2, xy

10.

$C_s = C_{1h}$	A'	A''	Operators
A'	A'	A''	$x, y, R_z, x^2, y^2, z^2, xy$
A''	A''	A'	z, R_x, R_y, xz, yz

11.

C_{2h}	A_g	A_u	B_g	B_u	Operators
A_g	A_g	A_u	B_g	B_u	R_z, x^2, y^2, z^2, xy
A_u	A_u	A_g	B_u	B_g	z
B_g	B_g	B_u	A_g	A_u	R_x, R_y, xz, yz
B_u	B_u	B_g	A_u	A_g	x, y

12.

C_{3h}	A'	A''	E'	E''	Operators
A'	A'	A''	E'	E''	R_z, x^2+y^2, z^2
A''	A''	A'	E''	E'	z
E'	E'	E''	$2A' \oplus E'$	$2A'' \oplus E''$	x, y, x^2-y^2, xy
E''	E''	E'	$2A'' \oplus E''$	$2A' \oplus E'$	R_x, R_y, xz, yz

13.

C_{4h}	A_g	A_u	B_g	B_u	E_g	E_u	Operators
A_g	A_g	A_u	B_g	B_u	E_g	E_u	R_z, x^2+y^2, z^2
A_u	A_u	A_g	B_u	B_g	E_u	E_g	z
B_g	B_g	B_u	A_g	A_u	E_g	E_u	x^2-y^2, xy
B_u	B_u	B_g	A_u	A_g	E_u	E_g	
E_g	E_g	E_u	E_g	E_u	$2A_g \oplus 2B_g$	$2A_u \oplus 2B_u$	R_x, R_y, xz, yz
E_u	E_u	E_g	E_u	E_g	$2A_u \oplus 2B_u$	$2A_g \oplus 2B_g$	x, y

14.

D_2	A_1	B_1	B_2	B_3	Operators
A_1	A_1	B_1	B_2	B_3	x^2, y^2, z^2
B_1	B_1	A_1	B_3	B_2	z, R_z, xy
B_2	B_2	B_3	A_1	B_1	y, R_y, xz
B_3	B_3	B_2	B_1	A_1	x, R_x, yz

15.

D_3	A_1	A_2	E	Operators
A_1	A_1	A_2	E	x^2+y^2, z^2
A_2	A_2	A_1	E	z, R_z
E	E	E	$A_1 \oplus A_2 \oplus E$	$x, y, R_x, R_y, xz, yz, x^2-y^2, xy$

16.

D_4	A_1	A_2	B_1	B_2	E	Operators
A_1	A_1	A_2	B_1	B_2	E	x^2+y^2, z^2
A_2	A_2	A_1	B_2	B_1	E	z, R_z
B_1	B_1	B_2	A_1	A_2	E	x^2-y^2
B_2	B_2	B_1	A_2	A_1	E	xy
E	E	E	E	E	$A_1 \oplus A_2 \oplus B_1 \oplus B_2$	x, y, R_x, R_y, xz, yz

17.

D_5	A_1	A_2	E_1	E_2	Operators
A_1	A_1	A_2	E_1	E_2	x^2+y^2, z^2
A_2	A_2	A_1	E_1	E_2	z, R_z
E_1	E_1	E_1	$A_1 \oplus A_2 \oplus E_2$	$E_1 \oplus E_2$	x, y, R_x, R_y, xz, yz
E_2	E_2	E_2	$E_1 \oplus E_2$	$A_1 \oplus A_2 \oplus E_1$	x^2-y^2, xy

18.

D_6	A_1	A_2	B_1	B_2	E_1	E_2	Operators
A_1	A_1	A_2	B_1	B_2	E_1	E_2	x^2+y^2, z^2
A_2	A_2	A_1	B_2	B_1	E_1	E_2	z, R_z
B_1	B_1	B_2	A_1	A_2	E_2	E_1	
B_2	B_2	B_1	A_2	A_1	E_2	E_1	
E_1	E_1	E_1	E_2	E_1	$A_1 \oplus A_2 \oplus E_2$	$B_1 \oplus B_2 \oplus E_1$	x, y, R_x, R_y, xz, yz
E_2	E_2	E_2	E_2	E_1	$B_1 \oplus B_2 \oplus E_1$	$A_1 \oplus A_2 \oplus E_1$	x^2-y^2, xy

19.

D_{2d}	A_1	A_2	B_1	B_2	E	Operators
A_1	A_1	A_2	B_1	B_2	E	x^2+y^2, z^2
A_2	A_2	A_1	B_2	B_1	E	R_z
B_1	B_1	B_2	A_1	A_2	E	x^2-y^2
B_2	B_2	B_1	A_2	A_1	E	z, xy
E	E	E	E	E	$A_1 \oplus A_2 \oplus B_1 \oplus B_2$	x, y, R_x, R_y, xz, yz

20. D_{3d} has the same product rules and operator species as D_3 (table 15), with the addition of $g \otimes g = g$, $u \otimes u = g$, and $u \otimes g = u$. See table 11 for the g–u characters of the operators.

21. D_{2h} has the same product rules and operator species as D_2 (table 14), with the addition of $g \otimes g = g$, $u \otimes u =$ g, and $u \otimes g = u$. See table 11 for the g–u characters of the operators.

22.

D_{3h}	A_1'	A_2'	A_1''	A_2''	E'	E''	Operators
A_1'	A_1'	A_2'	A_1''	A_2''	E'	E''	x^2+y^2, z^2
A_2'	A_2'	A_1'	A_2''	A_1''	E'	E''	R_z
A_1''	A_1''	A_2''	A_1'	A_2'	E''	E'	
A_2''	A_2''	A_1''	A_2'	A_1'	E''	E'	z
E'	E'	E'	E''	E''	$A_1' \oplus A_2' \oplus E'$	$A_1'' \oplus A_2'' \oplus E''$	x, y, x^2-y^2, xy
E''	E''	E''	E'	E'	$A_1'' \oplus A_2'' \oplus E''$	$A_1' \oplus A_2' \oplus E'$	R_x, R_y, xz, yz

23. D_{4h} has the same product rules and operator species as D_4 (table 16), with the addition of $u \otimes u = g$, $g \otimes g = g$, $u \otimes g = u$. See table 11 for the g–u character of the operator. D_{6h} is similarly derived from D_6 (table 18).

24.

O, T_d	A_1	A_2	E	F_1	F_2	Operators
A_1	A_1	A_2	E	F_1	F_2	$x^2+y^2+z^2$
A_2	A_2	A_1	E	F_2	F_1	
E	E	E	$\begin{cases} A_1 \oplus A_2 \\ \oplus E \end{cases}$	$F_1 \oplus F_2$	$F_1 \oplus F_2$	$2z^2-x^2-y^2, \sqrt{3}(x^2-y^2)$
F_1	F_1	F_2	$F_1 \oplus F_2$	$\begin{cases} A_1 \oplus E \\ \oplus F_1 \\ \oplus F_2 \end{cases}$	$\begin{cases} A_2 \oplus E \\ \oplus F_1 \\ \oplus F_2 \end{cases}$	$\begin{cases} R_x, R_y, R_z \\ x, y, z(O) \end{cases}$
F_2	F_2	F_1	$F_1 \oplus F_2$	$\begin{cases} A_2 \oplus E \\ \oplus F_1 \\ \oplus F_2 \end{cases}$	$\begin{cases} A_1 \oplus E \\ \oplus F_1 \\ \oplus F_2 \end{cases}$	$\begin{cases} x, y, z(T_d) \\ xz, xy, yz(O) \end{cases}$

25. O_h has the same product rules and operator species as O (table 24), with the addition of $g \otimes g = g$, $u \otimes u = g$, and $u \otimes g = u$. See table 11 for the u–g character of the operators.

26.

$C_{\infty v}$	Σ^+	Σ^-	Π	Δ	$\cdots$	Operators
Σ^+	Σ^+	Σ^-	Π	Δ	$\cdots$	$x^2 + y^2, z^2, z$
Σ^-	Σ^-	Σ^+	Π	Δ	$\cdots$	R_z
Π	Π	Π	$\Sigma^+ \oplus \Sigma^- \oplus \Delta$	$\Pi \oplus \Phi$	$\cdots$	x, y, R_x, R_y, xz, yz
Δ	Δ	Δ	$\Pi \oplus \Phi$	$\Sigma^+ \oplus \Sigma^- \oplus \Gamma$	$\cdots$	$x^2 - y^2, xy$
$\vdots$	$\vdots$	$\vdots$	$\vdots$	$\vdots$		

27.

$D_{\infty h}$	Σ_g^+	Σ_u^+	Σ_g^-	Σ_u^-	Π_g	Π_u
Σ_g^+	Σ_g^+	Σ_u^+	Σ_g^-	Σ_u^-	Π_g	Π_u
Σ_u^+	Σ_u^+	Σ_g^+	Σ_u^-	Σ_g^-	Π_u	Σ_g
Σ_g^-	Σ_g^-	Σ_u^-	Σ_g^+	Σ_u^+	Π_g	Π_u
Σ_u^-	Σ_u^-	3_g^-	Σ_u^+	Σ_g^+	Π_u	Π_g
Π_g	Π_g	Π_u	Π_g	Π_u	$\Sigma_g^+ \oplus \Sigma_g^- \oplus \Delta_g$	$\Sigma_u^+ \oplus \Sigma_u^- \oplus \Delta_u$
Π_u	Π_u	Π_g	Π_u	Π_g	$\Sigma_u^+ \oplus \Sigma_u^- \oplus \Delta_u$	$\Sigma_g^+ \oplus \Sigma_g^- \oplus \Delta_g$
Δ_g	Δ_g	Δ_u	Δ_g	Δ_u	$\Pi_g \oplus \Phi_g$	$\Pi_u \oplus \Phi_u$
Δ_u	Δ_u	Δ_g	Δ_u	Δ_g	$\Pi_u \oplus \Phi_u$	$\Pi_g \oplus \Phi_g$
$\vdots$	$\vdots$	$\vdots$	$\vdots$	$\vdots$	$\vdots$	$\vdots$

$D_{\infty h}$	Δ_g	Δ_u	$\cdots$	Operators
Σ_g^+	Δ_g	Δ_u	$\cdots$	$z^2, x^2 + y^2$
Σ_u^+	Δ_u	Δ_g	$\cdots$	z
Σ_g^-	Δ_g	Δ_u	$\cdots$	R_z
Σ_u^-	Δ_u	Δ_g	$\cdots$	
Π_g	$\Pi_g \oplus \Phi_g$	$\Pi_u \oplus \Phi_u$	$\cdots$	R_x, R_y, xz, yz
Π_u	$\Pi_u \oplus \Phi_u$	$\Pi_g \oplus \Phi_g$	$\cdots$	x, y
Δ_g	$\Sigma_g^+ \oplus \Sigma_g^- \oplus \Gamma_g$	$\Sigma_u^+ \oplus \Sigma_u^- \oplus \Gamma_u$	$\cdots$	$x^2 - y^2, xy$
Δ_u	$\Sigma_u^+ \oplus \Sigma_u^- \oplus \Gamma_u$	$\Sigma_g^+ \oplus \Sigma_g^- \oplus \Gamma_g$	$\cdots$	
$\vdots$	$\vdots$	$\vdots$		

Appendix B Lagrangian Mechanics[1]

Newtonian mechanics is often not the most convenient form for solving many problems. There are two reasons for reformulating mechanics into Lagrange's equations. First, although the virtue of Newton's laws is their relation of complex motions to the much simpler forces, many forces are often simple only in non-Cartesian coordinates, for example, spherical or elliptical. It is, therefore, valuable to express Newton's law in a form that brings out this simplicity. The other reason is that we may have, in addition to simple forces determining complex motions, some complex forces determining simple motions. These are known as forces of constraint. Consider, for example, a particle confined to a circular frictionless track that stands vertical in a gravitational field. In seeking the motions of this particle, the forces exerted by the track are of no inherent interest. We know in advance that their effect is to confine the particle to the track. Lagrange's equations allow us to examine a system in terms of the most convenient coordinates that remove from the problem consideration of unallowed motions due to forces of constraint.

Let us define

$$L \equiv T - V,$$

a function we call the *Lagrangian*. We shall now show that

$$\frac{d}{dt}\left(\frac{\partial L}{\partial \dot{q}_i}\right) - \frac{\partial L}{\partial q_i} = 0,$$

where q_i are the generalized coordinates and $\dot{q}_i \equiv dq_i/dt$. These equations will be shown to be equivalent to Newton's law.

1 Some Useful Derivatives

We begin by considering certain velocity derivatives. We note that

$$\mathbf{v} = (\dot{x}, \dot{y}, \dot{z});$$

thus $\mathbf{v}$ has a certain explicit functional form. We shall consider partial derivatives with respect to explicit functional forms that depend on $\dot{x}$, $\dot{y}$, $\dot{z}$, and x, y, z. For example,

1. This appendix is taken from a set of lecture notes prepared for a graduate course in quantum mechanics by M. Gouterman of the University of Washington and R. Fulton of Florida State University, to whom thanks are due for permission to reproduce them.

$$\left.\frac{\partial \mathbf{v}}{\partial x}\right|_{y,z,\dot{x},\dot{y},\dot{z}} = 0,$$

$$\left.\frac{\partial \mathbf{v}}{\partial \dot{x}}\right|_{x,y,z,\dot{y},\dot{z}} = \hat{i},$$

and similarly for the other partial derivatives. Consider now $\mathbf{v}$ expressed in generalized coordinates:

$$\mathbf{v} = \frac{\partial \mathbf{x}}{\partial q_i}\dot{q}_1 + \frac{\partial \mathbf{x}}{\partial q_2}\dot{q}_2 + \frac{\partial \mathbf{x}}{\partial q_3}\dot{q}_3,$$

where we have equations that give $\mathbf{x} = \mathbf{x}\,(q_1, q_2, q_3)$. It may be seen that

$$\left.\frac{\partial \mathbf{v}}{\partial q_1}\right|_{q_2,q_3\ \dot{q}_1,\dot{q}_2,\dot{q}_3} = \frac{\partial^2 \mathbf{x}}{\partial q_1 \partial q_1}\dot{q}_1 + \frac{\partial^2 \mathbf{x}}{\partial q_1 \partial q_2}\dot{q}_2 + \frac{\partial^2 \mathbf{x}}{\partial q_1 \partial q_3}\dot{q}_3$$

and

$$\left.\frac{\partial \mathbf{v}}{\partial \dot{q}_1}\right|_{q_1,q_2,q_3,\dot{q}_2,\dot{q}_3} = \frac{\partial \mathbf{x}}{\partial q_1}.$$

We might also note that since

$$\frac{d}{dt}\left(\frac{\partial \mathbf{x}}{\partial q_1}\right) = \frac{\partial^2 \mathbf{x}}{\partial q_1^2}\dot{q}_1 + \frac{\partial^2 \mathbf{x}}{\partial q_1\, \partial q_2}\dot{q}_2 + \frac{\partial^2 \mathbf{x}}{\partial q_1\, \partial q_3}\dot{q}_3,$$

we have

$$\frac{d}{dt}\left(\frac{\partial \mathbf{x}}{\partial q_1}\right) = \frac{\partial \mathbf{v}}{\partial q_1}.$$

2 Derivation of Lagrange's Equations

We shall begin with d'Alembert's principle $(m\mathbf{a} - \mathbf{f})\cdot d\mathbf{x} = 0$, which clearly follows from Newton's law. We shall assume that $\mathbf{x}$ can be expressed in terms of (q_1, q_2, q_3) so that

$$d\mathbf{x} = \frac{\partial \mathbf{x}}{\partial q_1}dq_1 + \frac{\partial \mathbf{x}}{\partial q_2}dq_2 + \frac{\partial \mathbf{x}}{\partial q_3}dq_3,$$

and therefore

$$(m\mathbf{a} - \mathbf{f})\cdot\frac{\partial \mathbf{x}}{\partial q_1}dq_1 + (m\mathbf{a} - \mathbf{f})\cdot\frac{\partial \mathbf{x}}{\partial q_2}dq_2 + (m\mathbf{a} - \mathbf{f})\cdot\frac{\partial \mathbf{x}}{\partial q_{\dot{2}}}dq_3 = 0.$$

Since the dq_i are arbitrary general displacements,

$$(m\mathbf{a} - \mathbf{f}) \cdot \frac{\partial \mathbf{x}}{\partial q_1} = 0.$$

(That is, we could take $dq_2 = dq_3 = 0$ and dq_1 finite.) The kinetic energy is defined as $T = \frac{1}{2} m\mathbf{v} \cdot \mathbf{v}$, and its derivative is

$$\frac{\partial T}{\partial \dot{q}_1} = \frac{\partial}{\partial \dot{q}_1}(\tfrac{1}{2} m\mathbf{v} \cdot \mathbf{v}) = m\mathbf{v} \cdot \frac{\partial \mathbf{v}}{\partial \dot{q}_1}.$$

Using the results just proved, we see that

$$\frac{\partial T}{\partial \dot{q}_1} = m\mathbf{v} \cdot \frac{\partial \mathbf{x}}{\partial q_1}.$$

Consider now

$$\frac{d}{dt}\left(\frac{\partial T}{\partial \dot{q}_1}\right) = m\frac{d\mathbf{v}}{dt} \cdot \frac{\partial \mathbf{x}}{\partial q_1} + m\mathbf{v} \cdot \frac{d}{dt}\left(\frac{\partial \mathbf{x}}{\partial q_1}\right).$$

But

$$\frac{d}{dt}\left(\frac{\partial \mathbf{x}}{\partial q_1}\right) = \frac{\partial \mathbf{v}}{\partial q_1},$$

and therefore

$$\frac{d}{dt}\left(\frac{\partial T}{\partial \dot{q}_1}\right) = m\mathbf{a} \cdot \frac{\partial \mathbf{x}}{\partial q_1} + m\mathbf{v} \cdot \frac{\partial \mathbf{v}}{\partial q_1}.$$

However, it is clear that

$$\frac{\partial T}{\partial q_1} = m\mathbf{v} \cdot \frac{\partial \mathbf{v}}{\partial q_1},$$

with the result that

$$\frac{d}{dt}\left(\frac{\partial T}{\partial \dot{q}_1}\right) - \frac{\partial T}{\partial q_1} = m\mathbf{a} \cdot \frac{\partial \mathbf{x}}{\partial q_1}.$$

Now we also know that for the potential V,

$$\frac{\partial V}{\partial q_1} = \frac{\partial V}{\partial x}\frac{\partial x}{\partial q_1} + \frac{\partial V}{\partial y}\frac{\partial y}{\partial q_1} + \frac{\partial V}{\partial z}\frac{\partial z}{\partial q_1} = \nabla V \cdot \frac{\partial \mathbf{x}}{\partial q_1};$$

but $\nabla V = -\mathbf{f}$, so that

$$\frac{\partial V}{\partial q_1} = -\mathbf{f}\cdot\frac{\partial \mathbf{x}}{\partial q_1}.$$

It then follows from d'Alembert's principle that

$$\frac{d}{dt}\left(\frac{\partial T}{\partial \dot{q}_1}\right) - \frac{\partial T}{\partial q_1} + \frac{\partial V}{\partial q_1} = 0.$$

Now let us assume that V is a function of position only (that is, on the q_i but not on $\dot{q}_i$ or t). Then in general

$$\frac{d}{dt}\left(\frac{\partial V}{\partial \dot{q}_1}\right) = 0,$$

and we have finally

$$\frac{d}{dt}\left(\frac{\partial T}{\partial \dot{q}_1}\right) - \frac{d}{dt}\left(\frac{\partial V}{\partial \dot{q}_1}\right) + \frac{\partial V}{\partial q_1} - \frac{\partial T}{\partial q_1} = 0,$$

$$\frac{d}{dt}\left(\frac{\partial (T - V)}{\partial \dot{q}_1}\right) + \frac{\partial (V - T)}{\partial q_1} = 0,$$

$$\frac{d}{dt}\left(\frac{\partial L}{\partial \dot{q}_1}\right) - \frac{\partial L}{\partial q_1} = 0,$$

and since we have not distinguished our q_i,

$$\frac{d}{dt}\left(\frac{\partial L}{\partial \dot{q}_i}\right) - \frac{\partial L}{\partial q_i} = 0. \qquad \text{Q.E.D.}$$

Appendix C Density Matrix Methods

The density matrix provides a convenient link between microscopic expectation values and ensemble averages and is frequently employed in problems involving both coherent optical systems (for example, lasers) and magnetic resonance systems. We shall only mention here some of the most important properties of this mathematical construct; for a more thorough discussion of density matrix methods, the reader is referred to the text by Slichter (reference 1) or Vanier (reference 2).

Consider a composite system in an overall state $|\Psi\rangle$, which is describable as a linear combination of some set of basis functions

$$|\Psi\rangle = \sum_n c_n u_n.$$

Some operator M will have an expectation value in this state given by

$$\begin{aligned}\langle M\rangle &= \int \Psi^* M \Psi \, d\tau \\ &= \sum_{n,m} c_m^* c_n \langle m|M|n\rangle\end{aligned}$$

in the u basis. Let us first define a probability matrix element

$$\langle n|p|m\rangle = c_n c_m^*$$

so that we can write $\langle M\rangle$ as

$$\begin{aligned}\langle M\rangle &= \sum_{n,m} \langle n|p|m\rangle \langle m|M|n\rangle \\ &= \sum_n \langle n|pM|n\rangle = \text{Trace}(pM).\end{aligned}$$

Formally, we can write the p operator as a *projection operator*,

$$p = \sum_k |k\rangle\langle k|.$$

Then

$$\langle n|p|m\rangle = \sum_k \langle n|k\rangle\langle k|m\rangle = c_n c_m^*. \tag{C.1}$$

Equation (C.1) is especially useful when the basis set u is the set of eigenfunctions of the Hamiltonian of the system, and the operator M does not commute with the Hamiltonian. Then we can use states $|k\rangle$ in equation (C.1) that are eigenstates of M to find expectation values of this nonconserved dynamical variable.

Suppose we now have an ensemble of these systems, all of which are independent in that they can assume various eigenstates without reference to what states other systems in the ensemble are occupying. The probability that the system is in its kth state will be designated $p(k)$. Then the measured value of M, averaged over this ensemble, will be given by

$$\begin{aligned}
\langle M \rangle_{\text{avg}} &= \sum_{n,m} \langle c_m^* c_n \rangle_{\text{avg}} \langle m|M|n\rangle \\
&= \sum_{n,m} \langle n|p|m\rangle_{\text{avg}} \langle m|M|n\rangle \\
&= \sum_{n,m,k} \langle n|k\rangle p(k) \langle k|m\rangle \langle m|M|n\rangle \qquad \text{(C.2)} \\
&= \sum_{n,m} \langle n|\rho|m\rangle \langle m|M|n\rangle \\
&= \sum_{n} \langle n|\rho M|n\rangle = \text{Trace}(\rho M).
\end{aligned}$$

In equation (C.2), the density matrix is defined as

$$\rho = \sum_k |k\rangle p(k) \langle k|;$$

that is, it has the form of a probability-weighted projection operator. For instance, in an ensemble at Boltzmann equilibrium, $p(k)$ would just be given by

$$p(k) = \frac{g_k e^{-E_k/kT}}{\sum_k g_k e^{-E_k/kT}}.$$

Let us proceed to develop a few of the important properties of the density matrix. Its most important utilitarian one, of course, is that expressed by equation (C.2), that is, that the ensemble average value of any observable is given by the trace of the product of the operator for that observable with the density matrix for the ensemble. In addition, we may adduce the following additional properties.

1. The density matrix is positive definite; that is,

$$\langle j|\rho|j\rangle \geq 0.$$

The proof is simply that

$$\langle j|\rho|j\rangle = \sum_k \langle j|k\rangle p(k) \langle k|j\rangle$$
$$= \sum_k |\langle j|k\rangle|^2 p(k).$$

$p(k)$ is a statistical weight, so it can take on only real values between 0 and 1. The square of the magnitude of any quantity, complex or otherwise, is also positive, so that the positive definite property follows obviously.

2. The density matrix is Hermitean; that is,

$$\langle m|\rho|n\rangle^* = \langle n|\rho|m\rangle.$$

The proof is as follows. For the complex conjugate.

$$\langle m|\rho|n\rangle^* = \sum_k \langle m|k\rangle^* p(k)^* \langle k|n\rangle^*$$
$$= \sum_k \langle k|m\rangle p(k) \langle n|k\rangle.$$

But

$$\langle n|\rho|m\rangle = \sum_k \langle n|k\rangle p(k) \langle k|m\rangle.$$

We can move around the $\langle n|k\rangle$ and the $\langle k|m\rangle$ at will, since they are pure numbers, so that

$$\langle n|\rho|m\rangle = \sum_k \langle k|m\rangle p(k) \langle n|k\rangle = \langle m|\rho|n\rangle^*.$$

3. The density matrix has unit trace; that is,

$$\text{Trace}(\rho) = \sum_j \langle j|\rho|j\rangle = 1.$$

For a proof, we need observe only that the expectation value of the unit operator $\mathbf{1}$ is exactly 1 for all systems, so that

$$\langle \mathbf{1}\rangle = \text{Trace}(\mathbf{1}\rho) = \text{Trace}(\rho) = 1.$$

4. If the system is in a pure quantum state, then $p(k) = 1$ for that state and 0 for all other states. In such a case, the density matrix can be written $\rho = |k\rangle\langle k|$. The square of the density matrix is given by

$$\rho^2 = \sum_l |k\rangle \langle k|l\rangle \langle l|k\rangle \langle k|$$
$$= |k\rangle \langle k|k\rangle \langle k| = |k\rangle \langle k| = \rho;$$

that is, ρ is an idempotent operator, whose square equals itself, for a pure state. In a mixture, however,

$$\rho^2 = \sum_k |k\rangle [p(k)]^2 \langle k|,$$

so that

$$\text{Trace}(\rho^2) = \sum_k [p(k)]^2 \leq 1.$$

Armed with these properties for the density matrix, we can now proceed to use it in analyzing the behavior of quantum mechanical ensembles.

First, let us consider the time-dependent Schrödinger equation

$$i\hbar\frac{\partial\Psi}{\partial t} = \mathscr{H}\Psi,$$

with

$$\Psi = \sum_n c_n u_n,$$

which becomes

$$i\hbar \sum_n \dot{c}_n u_n = \sum_n c_n \mathscr{H} u_n. \tag{C.3}$$

If we take the kth matrix element of both sides of the equation, equation (C.3) becomes

$$\begin{aligned} i\hbar\int\left(\sum_n \dot{c}_n^* u_n u_k\right) d\mathbf{r} &= \sum_n c_n \int (u_k^* \mathscr{H} u_n)\, d\mathbf{r} \\ &= \sum_n c_n \langle k|\mathscr{H}|n\rangle, \end{aligned}$$

so that we obtain an equation of motion for the coefficient c_k,

$$i\hbar\dot{c}_k = \sum_n c_n \langle k|\mathscr{H}|n\rangle.$$

Now the rate of change of an element of the density matrix will be given by

$$\frac{d}{dt}\langle k|\rho|m\rangle = \frac{d}{dt}(c_m^* c_k) = \dot{c}_m^* c_k + c_m^* \dot{c}_k.$$

If we substitute

$$\dot{c}_k = \frac{1}{i\hbar}\sum_n c_n \langle k|\mathscr{H}|n\rangle = -\frac{i}{\hbar}\sum_n c_n \langle k|\mathscr{H}|n\rangle,$$

$$\dot{c}_m^* = +\frac{i}{\hbar}\sum_n c_n^* \langle m|\mathscr{H}|n\rangle = +\frac{i}{\hbar}\sum_n c_n^* \langle n|\mathscr{H}|m\rangle,$$

we obtain

$$\begin{aligned}\frac{d}{dt}\langle k|\rho|m\rangle &= \frac{i}{\hbar}\left[\sum_n c_n^* \langle n|\mathscr{H}|m\rangle c_k - c_m^* \sum_n c_n \langle k|\mathscr{H}|n\rangle\right.\\ &= \frac{i}{\hbar}\sum_n [c_n^* c_k \langle n|\mathscr{H}|m\rangle - \langle k|\mathscr{H}|n\rangle c_m^* c_n]\\ &= \frac{i}{\hbar}\sum_n [\langle k|\rho|n\rangle\langle n|\mathscr{H}|m\rangle - \langle k|\mathscr{H}|n\rangle\langle n|\rho|m\rangle]\\ &= \frac{i}{\hbar}[\langle k|\rho\mathscr{H}|m\rangle - \langle k|\mathscr{H}\rho|m\rangle],\end{aligned}$$

which we may write in the compact form

$$\frac{d\rho}{dt} = \frac{i}{\hbar}[\mathscr{H}\rho - \rho\mathscr{H}] = \frac{i}{\hbar}[\rho, \mathscr{H}] = \frac{1}{i\hbar}[\mathscr{H}, \rho].$$

This is just the same form as the Heisenberg equation of motion for the expectation value of a quantum mechanical operator, equation (11.13).

We can now extend equation (C.2) to find the time dependence of some dynamical variable of the system, with operator M. This will be given by

$$\begin{aligned}\frac{d}{dt}\langle M\rangle &= \frac{d}{dt}\text{Trace}(\rho M) = \frac{d}{dt}\sum_n \langle n|\rho M|n\rangle\\ &= \frac{d}{dt}\sum_{n,k} \langle n|\rho|k\rangle\langle k|M|n\rangle\\ &= \sum_{n,k}\left[\langle n|\frac{d\rho}{dt}|k\rangle\langle k|M|n\rangle + \langle n|\rho|k\rangle\langle k|\frac{dM}{dt}|n\rangle\right.\\ &= \frac{i}{\hbar}\sum_{n,k}[\langle n|[\rho,\mathscr{H}]|k\rangle\langle k|M|n\rangle - \langle n|\rho|k\rangle\langle k|[\mathscr{H}, M]|n\rangle].\end{aligned}$$

If M commutes with $\mathscr{H}$, we can drop out the second term in the summation, so that

$$\frac{d}{dt}\langle M \rangle = \frac{i}{\hbar}\sum_{n,k}[\langle n|[\rho, \mathscr{H}]|k\rangle\langle k|M|n\rangle].$$

Furthermore, we are free to choose a basis set in which M is diagonal, so that all we have left is

$$\frac{d}{dt}\langle M \rangle = \frac{i}{\hbar}\sum_{n}\langle n|[\rho, \mathscr{H}]|n\rangle\langle n|M|n\rangle. \tag{C.4}$$

As a simple example, let us consider a collection of spins in a static magnetic field. The density matrix for the ensemble will be given by

$$\langle n|\rho|m\rangle = \langle c_m c_n^*\rangle_{\text{avg}} = \sum_k \langle m|k\rangle p(k)\langle k|n\rangle,$$

where we assume a Boltzmann equilibrium distribution, so that

$$p(k) = \frac{1}{Q}e^{-E_k/kT},$$

with

$$Q = \sum_k e^{-E_k/kT},$$

and each of the transformation matrix elements consists of a magnitude and a phase, that is,

$$c_l = \langle l|k\rangle = |c_l|e^{i\alpha_l}.$$

Each density matrix element will have the form

$$\langle n|\rho|m\rangle = |c_m||c_n| \times (\text{probability factor}) \times \langle e^{i(\alpha_m - \alpha_n)}\rangle_{\text{avg}}.$$

In the absence of any specific knowledge about the preparation of the system, we may apply the *random phase approximation*, so that the average value of the phase factor is zero unless $m = n$. This ensures that the density matrix is diagonal; that is,

$$\langle n|\rho|m\rangle = \delta_{nm}|c_n|^2\frac{1}{Q}e^{-E_n/kT}.$$

We can write this in operator notation as

$$\rho = \frac{1}{Q}e^{-\mathscr{H}_0/kT}.$$

Let us look at the particular case of a system of spins with total spin I and compute the average magnetization in the direction of an applied field B_0, that is, the quantity $\langle M_z \rangle$. The Hamiltonian for the system is just the Zeeman-Hamiltonian,

$$\mathscr{H}_0 = \gamma\hbar B_0 I_z.$$

We have

$$\langle M_z \rangle = \text{Trace}(M_z \rho) = \frac{1}{Q}\text{Trace}(M_z e^{-\gamma\hbar B_0 I_z/kT}),$$

In this expression,

$$M_z = -\sum_k \gamma\hbar I_{zk},$$

and I_z for each particle is just the magnetic quantum number m_I. Substituting in and expanding the exponential give

$$\langle M_z \rangle = -\frac{1}{Q}\sum_m N\gamma\hbar m_I \left[1 - \frac{\gamma\hbar B_0 m_I}{kT} + \frac{1}{2}\left(\frac{\gamma\hbar B_0 m_I}{kT}\right)^2 + \cdots\right].$$

For kT much greater than the Zeeman energy $\gamma\hbar B_0 m_I$, we need to retain only the first nonvanishing term in the expansion. We have

$$\sum_m m_I = 0,$$

since every integer between $+I$ and $-I$ appears exactly once with a positive and once with a negative value. This gives us the leading term as

$$\langle M_z \rangle = \frac{N}{Q}\sum_m \frac{\gamma^2\hbar^2 B_0 m_I^2}{kT}.$$

Now the partition function

$$Q = \sum_{m=-I}^{+I} e^{-\gamma\hbar m B_0/kT}$$

remains to be evaluated. Since we have been working in the high-temperature limit, we can take $e^{-\gamma\hbar m B_0/kT} \approx 1$, so that

$$Q \approx \sum_{m=-I}^{+I} (1) = 2I + 1.$$

It is fairly easy to show that

$$\mathrm{Trace}(m_I^2) = \sum_{m=-I}^{+I} m_I^2 = \tfrac{1}{3}I(I+1)(2I+1),$$

so that our final expression for the average magnetization is

$$\langle M_z \rangle = \frac{\gamma^2 \hbar^2 I(I+1)}{3kT} N B_0, \tag{C.5}$$

which may be recognized as the Curie law for magnetization. This is an experimentally established law that states that, above a certain temperature, the magnetization of a paramagnetic substance follows the behavior C/T, where C depends on the spin quantum number I, the density N, and the magnetic field B_0 in the way predicted by equation (C.5).

References

1. C. P. Slichter, *Principles of Magnetic Resonance*, 2nd ed., Series in Solid-State Sciences No. 1 (Springer-Verlag, Berlin, 1973).

2. J. Vanier, *Basic Theory of Lasers and Masers* (Gordon and Breach, New York, 1971).

Appendix D Dipole Correlation and Spectral Density Functions

In chapter 14, we used a relation between the dipole correlation function and the spectral density function to enable us to predict spectra from classically computed molecular trajectories. Since this relation is easily derived using the treatment of reference 1, and furnishes a good example of the use of density matrices discussed in appendix C, we give this derivation here.

We begin with the expression for the transition rate in terms of electric field strength, equation (1.30). This can be written as

$$R_{fi}(\omega) = \frac{2\pi}{\hbar^2}|\langle f|\mathbf{E}_0 \cdot \boldsymbol{\mu}|i\rangle|^2 \rho_i \delta(\omega_{fi} \pm \omega), \tag{D.1}$$

where our states are labeled as initial $|i\rangle$ and final $|f\rangle$, and, since we are dealing with an ensemble of molecules, the population density of initial states ρ_i is included in the transition rate. The delta function in equation (D.1) incorporates the requirement that the radiation field have amplitude at the transition frequency ω_{fi}. Since R_{fi} is directly proportional to $|E_0|^2$, and thus to the intensity I (energy flux per unit area per unit time), the rate at which energy is removed from a light beam is given by a phenomenological equation of the form

$$-\frac{dI}{dt} = k_{\text{abs}}(\omega)I, \tag{D.2}$$

where k_{abs} is the macroscopic, frequency-dependent absorption coefficient. The microscopic picture of this same process is just

$$\begin{aligned} -\frac{dI}{dt} &= \sum_f \hbar\omega_{fi}R_{fi} \\ &= \frac{2\pi}{\hbar}\sum_f\sum_i \omega_{fi}\rho_i|\langle f|\mathbf{E}_0 \cdot \boldsymbol{\mu}|i\rangle|^2\, \delta(\omega_{fi} \pm \omega). \end{aligned}$$

In this expression, a transition will occur whenever $\omega_{fi} + \omega = 0$ or $\omega_{fi} - \omega = 0$. Since ω_{fi} is defined as $E_f - E_i/\hbar$, it will be positive whenever $E_f > E_i$, that is, when energy is absorbed and the system goes to a higher energy state. When $E_f < E_i$, radiation is emitted, and since ω_{fi} now has a negative sign, this corresponds to returning energy to, rather than absorbing it from, the light beam. Another way of expressing this same situation is to replace the density of initial states ρ_i by the population density *difference* between ground and excited states, $\rho_i - \rho_f$; we then use a single delta function corresponding to positive ω_{fi}, and the preceding equation becomes

$$-\frac{dI}{dt} = \frac{2\pi}{\hbar}\sum_f\sum_i \omega_{fi}(\rho_i - \rho_f)|\langle f|\mathbf{E}_0\cdot\boldsymbol{\mu}|i\rangle|^2\,\delta(\omega_{fi} - \omega).$$

If we assume that our molecular system is at thermal equilibrium,

$$\rho_f = \rho_i e^{-\hbar\omega_{fi}/kT},$$

and we have

$$-\frac{dI}{dt} = \frac{2\pi}{\hbar}[1 - e^{-\hbar\omega/kT}]\sum_i\sum_f \omega\rho_i|\langle f|\mathbf{E}_0\cdot\boldsymbol{\mu}|i\rangle|^2\,\delta(\omega_{fi} - \omega).$$

If we now define a normalized line shape function $I(\omega)$ as

$$I(\omega) = \frac{3\hbar(-dI/dt)}{2\pi\omega|E_0|^2(1 - e^{-\hbar\omega/kT})} \tag{D.3}$$

and substitute the previous expression for dI/dt, using the relation between light intensity and electric field strength, we obtain

$$I(\omega) = 3\sum_i\sum_f \rho_i|\langle f|\boldsymbol{\epsilon}\cdot\boldsymbol{\mu}|i\rangle|^2\,\delta(\omega_{fi} - \omega),$$

where $\boldsymbol{\epsilon}$ is a dimensionless unit vector which indicates the direction, but not the magnitude, of the electric field. Note also that the delta function permits us to replace ω_{fi} by ω.

We now introduce the well-known integral form of the delta function,

$$\delta(\omega) = \frac{1}{2\pi}\int_{-\infty}^{\infty} e^{i\omega t}\,dt.$$

Thus

$$\begin{aligned}
I(\omega) &= 3\sum_i\sum_f \rho_i|\langle f|\boldsymbol{\epsilon}\cdot\boldsymbol{\mu}|i\rangle|^2\cdot\frac{1}{2\pi}\int_{-\infty}^{\infty} e^{i(\omega_{fi}-\omega)t}\,dt \\
&= \frac{3}{2\pi}\sum_i\sum_f \rho_i\langle i|\boldsymbol{\epsilon}\cdot\boldsymbol{\mu}|f\rangle\langle f|\boldsymbol{\epsilon}\cdot\boldsymbol{\mu}|i\rangle\int_{-\infty}^{\infty}\exp i\left[\frac{E_f - E_i}{\hbar} - \omega\right]t\,dt \\
&= \frac{3}{2\pi}\int_{-\infty}^{\infty} e^{-i\omega t}\sum_i\sum_f \rho_i\langle i|\boldsymbol{\epsilon}\cdot\boldsymbol{\mu}|f\rangle\langle f|e^{iE_f t/\hbar}\boldsymbol{\epsilon}\cdot\boldsymbol{\mu}^{-iE_i t/\hbar}|i\rangle\,dt.
\end{aligned}$$

Now

$$e^{i\mathscr{H}t/\hbar}|n\rangle = e^{iE_n t/\hbar}|n\rangle$$

and

$$\langle n | e^{i\mathscr{H}t/\hbar} = \langle n | e^{iE_n t/\hbar},$$

so

$$I(\omega) = \frac{3}{2\pi} \int_{-\infty}^{\infty} e^{-i\omega t} \sum_i \sum_f \rho_i \langle i | \boldsymbol{\epsilon} \cdot \boldsymbol{\mu} | f \rangle \langle f | \boldsymbol{\epsilon} \cdot (e^{i\mathscr{H}t/\hbar} \boldsymbol{\mu} e^{-i\mathscr{H}t/\hbar}) | i \rangle \, dt$$

Recall at this point the Heisenberg form of the quantum mechanical equations of motion (reference 2), which gives the time development of the expectation value of some operator F without reference to any particular wave function representation. This equation is

$$\frac{dF}{dt} = \frac{i}{\hbar} [\mathscr{H}, F] \tag{D.4}$$

(recall the definition of the commutator in chapter 1.1). This has a solution, for $\mathscr{H}$ independent of t,[1]

$$F(t) = e^{i\mathscr{H}t/\hbar} F(0) e^{-i\mathscr{H}t/\hbar}.$$

So if the dipole moment of the system has some value μ at time $t = 0$, it has a value $e^{i\mathscr{H}t/\hbar} \mu e^{-i\mathscr{H}t/\hbar}$ at a time t later, and our expression for $I(\omega)$ can be written

$$I(\omega) = \frac{3}{2\pi} \int_{-\infty}^{\infty} e^{-i\omega t} \sum_i \sum_f \rho_i \langle i | \boldsymbol{\epsilon} \cdot \boldsymbol{\mu} | f \rangle \langle f | \boldsymbol{\epsilon} \cdot \boldsymbol{\mu}(t) | i \rangle \, dt.$$

Making use of the completeness relation [equation (1.4)] gives

$$\begin{aligned} I(\omega) &= \frac{3}{2\pi} \int_{-\infty}^{\infty} e^{-i\omega t} \sum_i \rho_i \langle i | [\boldsymbol{\epsilon} \cdot \boldsymbol{\mu}(0)][\boldsymbol{\epsilon} \cdot \boldsymbol{\mu}(t)] | i \rangle \, dt \\ &= \frac{3}{2\pi} \int_{-\infty}^{\infty} e^{-i\omega t} \langle (\boldsymbol{\epsilon} \cdot \boldsymbol{\mu}(0))(\boldsymbol{\epsilon} \cdot \boldsymbol{\mu}(t)) \rangle_{\text{ensemble average}} \, dt \\ &= \frac{1}{2\pi} \int_{-\infty}^{\infty} e^{-i\omega t} \langle \boldsymbol{\mu}(0) \cdot \boldsymbol{\mu}(t) \rangle_{\text{avg}} \, dt. \end{aligned}$$

1. *Proof* If $F(t) = e^{i\mathscr{H}t/\hbar} F(0) e^{-i\mathscr{H}t/\hbar}$, then

$$\begin{aligned} \frac{dF}{dt} &= \frac{i\mathscr{H}}{\hbar} F(t) + e^{i\mathscr{H}t/\hbar} \frac{\partial F(0)}{\partial t} e^{-i\mathscr{H}t/\hbar} + F(t) \left(-\frac{i\mathscr{H}}{\hbar} \right) \\ &= \frac{i}{\hbar} [\mathscr{H}F - F\mathscr{H}] = \frac{i}{\hbar} [\mathscr{H}, F]. \qquad \text{Q.E.D.} \end{aligned}$$

The factor of one-third is introduced by averaging the electric field direction over an isotropic absorbing medium. The quantity, $\langle \boldsymbol{\mu}(0) \cdot \boldsymbol{\mu}(t) \rangle_{\text{avg}}$ is called the *dipole correlation function* $G(t)$. We have the general result that the line shape or frequency response of a system interacting with radiation is just the Fourier transform of its dipole correlation function,

$$I(\omega) = \frac{1}{2\pi} \int_{-\infty}^{\infty} e^{-i\omega t} G(t)\, dt. \tag{D.5}$$

As an example of the use of equation (D.5), let us rederive the homogeneous or Lorentzian line shape discussed in chapter 1.8. If the excited system is relaxing due to spontaneous radiation, collisions, or some other process, the dipole includes a decay term in addition to the oscillating term, so that

$$\mu(t) = \mu_0 e^{-i\omega_0 t} e^{-\gamma t/2},$$

where γ is the decay time constant. Then

$$I(\omega) = \frac{\mu_0^2}{2\pi} \int_{-\infty}^{\infty} e^{i(\omega - \omega_0)t} e^{-\gamma t/2}\, dt$$

$$= \frac{\mu_0^2}{\sqrt{2\pi}} \frac{1}{i(\omega - \omega_0) + \gamma/2} = \frac{\mu_0^2}{\sqrt{8\pi}} \frac{\gamma/2 - i(\omega - \omega_0)}{(\omega - \omega_0)^2 + \gamma^2/4}.$$

The absorption coefficient is the real part of $I(\omega)$, so that

$$k_{\text{abs}}(\omega) = k_0 \frac{\gamma}{(\omega - \omega_0)^2 + \gamma^2/4},$$

which is identical with equation (1.39).

References

1. R. Gordon, *Advances in Magnetic Resonance* **3**, 1 (1968).

2. J. C. Slater, *Quantum Theory of Atomic Structure*, Vol. 1 (McGraw-Hill, New York, 1960), p. 101.

Appendix E The Literature of Spectroscopy

As the field of spectroscopy has developed, an extensive literature has accumulated. In this appendix, we list, with annotations, some of the most important sources in the field. Not even considered in this bibliography is the even more extensive literature of applied spectroscopy, primarily directed to analytical and organic chemists. This typically consists of simplified expanations of ultraviolet (uv), infrared (ir), and nuclear magnetic resonance (NMR) spectroscopy, empirical structural correlations, and an enormous search literature, and as such is not really pertinent to the subject of this book. The literature we do consider has been divided for convenience, if somewhat arbitrarily, into three sections: basic reference works, specialized manuals and monographs, and data compilations. A list of the abbreviations for journals cited in this book is also included.

1 Basic Reference Works

A large number of "Introduction to Spectroscopy" texts have appeared at one time or another. Some of the better ones are

G. M. Barrow, *Introduction to Molecular Spectroscopy* (McGraw-Hill, New York, 1962).

R. N. Dixon, *Spectroscopy and Structure* (Methuen, London, 1965).

G. W. King, *Spectroscopy and Molecular Structure* (Holt, Rinehart and Winston, New York, 1964).

D. H. Whiffen, *Spectroscopy*, 2nd ed. (Longman, London, 1971).

W. H. Flygare, *Molecular Structure and Dynamics* (Prentice-Hall, Englewood Cliffs, NJ, 1978) is a textbook on quantum mechanics that includes a great deal of material on spectroscopy.

A compact but surprisingly comprehensive pedagogically oriented review is

J. L. Hollenberg, "Energy States of Molecules" (Advisory Council on College Chemistry Resource Paper VIII, *J. Chem. Educ.* **47**, pp. 2–14 (1970)).

At the next level are texts for the working spectroscopist that give detailed formulas and references to the original literature. Foremost among these, by a wide margin, are the classic texts of Gerhard Herzberg:

G. Herzberg, *Atomic Spectra and Atomic Structure* (Prentice-Hall, Englewood Cliffs, N.J., 1937; Dover, New York, 1944).

G. Herzberg, *Molecular Spectra and Molecular Structure. I. Spectra of Diatomic Molecules* (Van Nostrand Reinhold, New York, 1950).

G. Herzberg, *Molecular Spectra and Molecular Structure. II. Infrared and Raman Spectra of Polyatomic Molecules* (Van Nostrand Reinhold, New York, 1945).

G. Herzberg, *Molecular Spectra and Molecular Structure. III. Electronic Spectra and Electronic Structure of Polyatomic Molecules* (Van Nostrand Reinhold, New York, 1966).

G. Herzberg, *The Spectra and Structures of Simple Free Radicals/An Introduction to Molecular Spectroscopy* (Cornell University Press, Ithaca, N. Y., 1971).

The last named is also designed to serve as an introduction to the principles of molecular spectroscopy, and does so very well.

A selection of other important reference works follows.

1.1 General Topics

A. C. G. Mitchell and M. W. Zemansky, *Resonance Radiation and Excited Atoms* (Cambridge University Press, 1961).

T. L. Cottrell, *Dynamic Aspects of Molecular Energy States* (Wiley, New York, 1965). This little volume contains some very useful and hard-to-come-by results, particularly the definition of absorption cross sections in Chapter 3.

W. G. Laidlaw, *Introduction to Quantum Concepts in Spectroscopy* (McGraw-Hill, New York, 1970).

S. Walker and H. Straw, *Spectroscopy* (Macmillan, New York, 1962). This is a two-volume survey. Volume I covers atomic, microwave, electron-spin resonance, and nuclear magnetic resonance spectroscopy. Volume II covers infrared and Raman, molecular electronic (optical), and fluorescence spectroscopy, as well as such miscellaneous topics as instrumentation and thermodynamic and astrophysical applications.

There is, of course, a steady stream of review articles and conference proceedings concerning various aspects of spectroscopy. Any attempt to list these would rapidly become out of date; instead, we direct the reader's attention to several series which include current titles of interest. Most notable among these are *Topics in Applied Physics*, *Series in Optical Sciences*, and *Series in Chemical Physics*, all published by Springer-Verlag (Berlin/Heidelberg/New York); *Chemical and Biochemical Applications of Lasers*, Vol. 1–, edited by C. B. Moore (Academic Press, New York); and *Advances in Laser Spectroscopy*, Vol. 1–, edited by B. A. Garetz and J. R. Lombardi (Heyden, London/Philadelphia/Rheine).

1.2 Atomic Spectra

E. U. Condon and G. H. Shortley, *The Theory of Atomic Spectra* (Cambridge University Press, 1963).

H. A. Bethe and E. E. Salpeter, *Quantum Mechanics of One- and Two-Electron Atoms* (Springer-Verlag, Berlin, 1957).

1.3 Molecular Spectra

H. B. Dunford, *Elements of Diatomic Molecular Spectra* (Addison-Wesley, Reading, Mass., 1968).

C. H. Townes and A. L. Schawlow, *Microwave Spectroscopy* (McGraw-Hill, New York, 1955).

H. C. Allen, Jr., and P. C. Cross, *Molecular Vib-Rotors* (Wiley, New York 1963).

E. B. Wilson, Jr., J. C. Decius, and P. C. Cross, *Molecular Vibrations* (McGraw-Hill, New York, 1954).

S. P. McGlynn, T. Azumi, and M. Kinoshita, *Molecular Spectroscopy of the Triplet State* (Prentice-Hall, Englewood Cliffs, N. J., 1969).

1.4 Molecular Beams

N. F. Ramsey, *Molecular Beams* (Clarendon Press, Oxford, 1956). This is still the classic work in the field; Chapters 5, 8, 9, and 10, in particular, deal with molecular-beam resonance spectroscopy.

1.5 Magnetic Resonance

C. P. Slichter, *Principles of Magnetic Resonance*, 2nd ed. (Springer-Verlag, Berlin, 1980). This is a physically oriented treatment, and includes an excellent introduction to density-matrix techniques.

A. Carrington and A. D. MacLachlan, *Introduction to Magnetic Resonance, with Applications to Chemistry and Chemical Physics* (Harper & Row, New York, 1967). A more chemically oriented treatment of nuclear magnetic and electron-spin resonance.

1.6 Optical Pumping, Quantum Electronics, and Lasers

R. A. Bernheim, *Optical Pumping: An Introduction* (Benjamin, Menlo Park, Calif., 1965).

A. Yariv, *Quantum Electronics* (Wiley, New York, 1968).

R. H. Pantell and H. E. Puthoff, *Fundamentals of Quantum Electronics* (Wiley, New York, 1969).

J. Vanier, *Basic Theory of Masers and Lasers*, (Gordon and Breach, New York, 1971).

M. Sargent III, M. O. Scully, and W. E. Lamb, Jr., *Laser Physics* (Addison-Wesley, Reading, MA, 1974).

Introductory texts about masers and lasers abound. Some of the better ones include

B. A. Lengyel, *Introduction to Laser Physics* (Wiley, New York, 1967); *Lasers*, 2nd ed. (Wiley, New York, 1971).

C. G. B. Garrett, *Gas Lasers* (McGraw-Hill, New York, 1967).

J. S. Thorp, *Masers and Lasers* (Macmillan, London, 1967).

O. Svelto, *Principles of Lasers*, 2nd ed. (Plenum, New York, 1982).

D. Röss, *Lasers, Light Amplifiers, and Oscillators* (Academic Press, London, 1969). Among other things, this book includes a bibliography of 4310 references on this subject.

The subject of chemical lasers is thoroughly covered in

R. W. F. Gross and J. F. Bott, eds., *Handbook of Chemical Lasers* (Wiley-Interscience, New York, 1976).

An unclassifiable, but totally delightful work describing all sorts of optical phenomena is the classic

R. W. Wood, *Physical Optics* (Macmillan, New York, 1934; reprinted by Dover, New York, 1967).

2 Manuals and Monographs

The experimental aspects of spectroscopy, which have not been covered in detail in this book, are discussed in

R. A. Sawyer, *Experimental Spectroscopy* (Dover, New York, 1963).

G. R. Harrison, R. C. Lord, and J. R. Loofbourow, *Practical Spectroscopy* (Prentice-Hall, Englewood Cliffs, N. J., 1948).

Summaries of calculations of line strength factors may be found in J. B. Tatum, "The Interpretation of intensities in Diatomic Molecular Spectra," *Astrophys. J.*, Supplement Series, **14**, pp. 21–55.

J. T. Hougen, "The Calculation of Rotational Energy Levels and Rotational Line Intensities in Diatomic Molecules," National Bureau of Standards Monograph 115 (U. S. Government Printing Office, Washington, D. C., 1970).

Useful atlases of molecular line spectra include

R. W. B. Pearse and A. G. Gaydon, *The Idenification of Molecular Spectra, 4 th Ed.* (Chapman and Hall, London, 1976).

D. C. Tyte, G. R. Hebert, S. H. Innanen, R. W. Nicholls, *et al.*, *Identification Atlas of Molecular Spectra* (Published by the Molecular Excitation Group, Department of Physics, University of Western Ontario, London, Ontario, 1964–1972). Vol. I: AlO Blue-Green System. Vol. II: N_2 Second Positive System. Vol. III: N_2 First Negative System. Vol. IV: O_2 Schumann-Runge System. Vol. V: C_2 Swan System. Vol. VI: The O_2 $A\,^3\Sigma_u{+} - X^3\Sigma_g{-}$ Herzberg I System. Vol. VII: The VO $C^4\Sigma^- - X^4\Sigma^-$ Yellow-Green and $B^4\Pi - X^4\Sigma^-$ Red Systems. Vol. VIII: The CN $A^2\Pi - X^2\Sigma^+$ Red System. Vol. IX: The CN $B^2\Sigma^+ - X^2\Sigma^+$ Violet System.

The visible absorption system of NO_2 is catalogued in

D. K. Hsu, D. L. Monts, and R. N. Zare, *Spectral Atlas of Nitrogen Dioxide, 5530* to *6480 Å* (Academic Press, New York, 1978).

High-resolution infrared spectra for a number of gas-phase species absorbing between 5 and 15 μm are given in

D. G. Murcray and A. Goldman, *CRC Handbook of High Resolution Infrared Laboratory Spectra of Atmospheric Interest* (CRC Press, Boca Raton, FL, 1981).

Also of interest is

Multilingual Dictionary of Important Terms in Molecular Spectroscopy, in English, French, German, Japanese, and Russian (published by the National Research Council of Canada, Ottawa, Ontario, 1966).

3 Data Compilations; Running Bibliographies

A large fraction of spectroscopic research results consists of specific molecular data, and keeping these data organized and available is no small task. A comprehensive bibliography through 1950 of diatomic molecular constants is given in Vol. I of Herzberg's *Diatomic Molecules* (cited earlier). The most recent authoritative update is

K. P. Huber and G. Herzberg, *Molecular Spectra and Molecular Structure, IV. Constants* of *Diatomic Molecules* (Van Nostrand Reinhold, New York, 1979).

Another compilation of data on diatomic molecules, not critically evaluated as is Huber and Herzberg, but that includes additional information on band systems, Franck-Condon factors, and potential curves is

S. Suchard, *Spectroscopic Data, Vol. I: Heteronuclear Diatomic Molecules* (IFI/Plenum, New York, 1975).

S. Suchard and J. E. Melzer, *Spectroscopic Data, Vol. II: Homonuclear Diatomic Molecules* (IFI/Plenum, New York, 1976).

A bibliography of molecular constants for electronic states of triatomic through twelve-atomic (that is, substituted benzene) systems appears in Vol. III of Herzberg's series (cited earlier). Current literature citations for diatomic and small polyatomic molecules appear in

Analysis of Molecular Spectra (bimonthly newsletter) (John G. Phillips, Dept. of Astronomy/ Sumner P. Davis, Dept. of Physics, University of California, Berkeley, California 94720).

The best compilations of atomic data are the tables

Charlotte E. Moore, *Atomic Energy Levels*, National Bureau of Standards Circular 467 (U. S. Government Printing Office, Washington, D. C., 1949–1958). Vol. I: ^{1}H through ^{23}V; Vol. II: ^{24}Cr through ^{41}Nb; Vol. III: ^{42}Mo through ^{57}La and ^{72}Hf through ^{89}Ac.

Energy-level and transition diagrams for lighter atoms and ions are given in

S. Bashkin and J. O. Stoner, *Atomic Energy Level and Grotrian Diagrams* (North-Holland, Amsterdam, 1976–1978). Vol. I: *Hydrogen I through Phosphorus XV*; Vol. II: *Sulphur I through Titanium XXII*.

A large amount of useful data, primarily concerned with optical masers and nonlinear optics, is collected in

C.R.C. Handbook of Laser Science and Technology, M. J. Weber, ed. (CRC press, Boca Raton, FL, 1982).

By the same publishers is

C.R.C. Handbook of Spectroscopy, J. W. Robinson, ed. (CRC Press, Boca Raton, FL, 1974–1981). This work has sections on NMR, microwave, infrared, uv, x-ray, XPS, EPR, mass, flame, and atomic emission spectroscopy.

A good source of optical data is

Handbook of Optics, W. G. Driscoll and W. Vaughan, eds. (McGraw-Hill, New York, 1978).

A number of useful compilations are included in the National Bureau of Standards–National Standard Reference Data System series, including

P. H. Krupenie, *The Band System of Carbon Monoxide* (NSRDS-NBS-5).

T. Shimanouchi, *Tables of Molecular Vibrational Frequencies*, Part I (NSRDS-NBS-6); Part II (NSRDS-NBS-11); Part III (NSRDS-NBS-17). These are continued (as Parts V–VII) in *J. Phys. Chem. Ref. Data.*

R. H. Schwendeman, *Tables for the Rigid Asymmetric Rotor* (NSRDS-NBS-12).

P. F. Wacker, M. Mizushima, J. D. Petersen, and J. R. Ballard, *Microwave Spectral Tables, Vol. I: Diatomic Molecules* (NBS Monograph 70).

P. F. Wacker and M. R. Pratto, *Microwave Spectral Tables, Vol. II: Line Strengths of Asymmetric Rotors* (NBS Monograph 70).

P. F. Wacker, M. S. Cord, D. J. Burkhard, J. D. Petersen, and R. F. Kukol, *Microwave Spectral Tables, Vol. III: Polyatomic Molecules with Internal Rotation* (NBS Monograph 70).

M. S. Cord, J. D. Petersen, M. S. Lojko, and R. H. Haas, *Microwave Spectral Tables, Vol. IV: Polyatomic Molecules Without Internal Rotation* (NBS Monograph 70).

M. S. Cord, M. S. Lojko, and J. D. Petersen, *Microwave Spectral Tables, Vol. V: Spectral Line Listing* (NBS Monograph 70).

Finally, we should mention some tables designed especially for experimental spectroscopic work. These are

G. R. Harrison, *M.I.T. Wavelength Tables* (M.I.T. Press, Cambridge, Mass., no date), which lists by air wavelength over 100,000 atomic lines normally found in the arc, spark, and discharge sources commonly used for calibration and/or spectroanalytical applications.

C. DeW. Coleman, W. R. Bozeman, and W. F. Meggers, *Table of Wavenumbers* (National Bureau of Standards Monograph 3, U. S. Government Printing Office, Washington, D. C., 1960), which is a conversion table from wavelengths in air to wavenumbers in vacuum. Vol. I: 2000 Å to 7000 Å; Vol. II: 7000 Å to 100 μ.

Tables especially designed for the calibration of spectrographs in the visible and near-uv regions are

A. Gatterer, *Grating Spectrum of Iron* (Specola Vaticana, Vatican City, 1951).

J. Junkes and E. W. Salpeter, *Spectrum of Thorium from 9400 to 2000* Å (Specola Vaticana, Vatican City, 1964).

For ultrahigh resolution work in the visible, the following gives the fluorescence excitation spectrum of I_2 with an accuracy of $\pm 0.001\ cm^{-1}$:

S. Gerstenkorn and P. Luc, *Atlas du Spectre d'Absorption de la Molecule d'Iode* (Editions du CNRS, Paris, 1978).

4 List of Journals Cited

Spectroscopy is a continuously developing field, and many results have been and continue to be reported in the research literature. In this section, we provide full identification of the journals cited by their common abbreviations in the chapter references. A number of these journals are published by professional societies, indicated as follows:

ACS: American Chemical Society,
APS: American Physical Society,
I.E.E.E.: Institute of Electrical and Electronics Engineers.

In other cases, the publisher of the journal is indicated.

Accts. Chem. Res.	*Accounts of Chemical Research* (ACS)
Adv. Chem. Phys.	*Advances in Chemical Physics* (Wiley-Interscience)
Adv. Photochem.	*Advances in Photochemistry* (Wiley-Interscience)
Am. J. Phys.	*American Journal of Physics* (APS)
Am. Scientist	*American Scientist* (Sigma Xi)
Ann. Phys.	*Annalen der Physik* (J. A. Barth, Leipzig)
Ann. Rev. Phys. Chem.	*Annual Review of Physical Chemistry* (Annual Reviews, Inc.)
Appl. Phys.	*Applied Physics* (Springer-Verlag)
Appl. Phys. Letts.	*Applied Physics Letters* (APS)
Appl. Spectrosc. Revs.	*Applied Spectroscopy Reviews* (Dekker)
Ark. Fys.	*Arkiv för Fysik* (Almqvist & Wiksell, Stockholm)
Can. J. Phys.	*Canadian Journal of Physics*
Chem. Phys.	*Chemical Physics* (North-Holland)
Chem. Phys. Letts.	*Chemical Physics Letters* (North-Holland)
Chem. Revs.	*Chemical Reviews* (ACS)
Compt. Rend. Acad. Sci.	*Comptes Rendus de l'Academie de Science* (Paris)

IBM J. Res. Development	*I.B.M. Journal of Research and Development* (I.B.M.)
IEEE J. Quantum Electronics	*I.E.E.E. Journal of Quantum Electronics* (I.E.E.E.)
J. Am. Chem. Soc.	*Journal of the American Chemical Society* (ACS)
J. Appl. Phys.	*Journal of Applied Physics* (APS)
J. Chem. Phys.	*Journal of Chemical Physics* (APS)
J. Chem. Soc.	*Journal of the Chemical Society* (London)
JETP Letts. [*Sov. Phys. JETP*]	English translation of *Pis'ma v Zhurnal Eksperimental'noi i Teoreticheskoi Fiziki* (USSR Academy of Sciences)
J. Mol. Spectroscopy	*Journal of Molecular Spectroscopy* (Academic)
J. Phys. Chem. Ref. Data	*Journal of Physical and Chemical Reference Data* (ACS & APS)
J. Quant. Spectrosc. Rad. Transfer	*Journal of Quantitative Spectroscopy and Radiative Transfer* (Pergamon)
Kgl. Dansk. Videnskab. Selskab.	*Danske Videnskabernes Selskab, Mathematisk-Fysiske Meddelelser* (Copenhagen)
Nature	*Nature*
Opt. Commun.	*Optical Communications* (North-Holland)
Phil. Trans. Roy. Soc.	*Philosophical Transactions of the Royal Society* (London)
Phys. Rev.	*Physical Review* (APS)
Phys. Rev. Letts.	*Physical Review Letters* (APS)
Physik. Z.	*Physikalisches Zeitschrift* (S. Hirzel, Leipzig)
Proc. Roy. Soc.	*Proceedings of the Royal Society* (London)
Prog. Quantum Electronics	*Progress in Quantum Electronics* (Pergamon)
Rev. Sci. Instr.	*Review of Scientific Instruments* (APS)
Sov. Phys. Uspekhii	English translation of *Uspekhii Fizicheskikh Nauk* (USSR Academy of Sciences)
Spectrochim. Acta	*Spectrochimica Acta* (Pergamon) Part A: molecular spectroscopy Part B: atomic spectroscopy
Trans. Faraday Soc.	*Transactions of the Faraday Society*
Z. Physik	*Zeitschrift für Physik* (Springer-Verlag)

Appendix F IUPAC/IUPAP Energy Conversion Constants (Revised 1973)

Unit	cm^{-1}	erg/molecule	J/mole	eV	°K	Hz (sec^{-1})
1 cm^{-1}	1	1.9863×10^{-16}	1.1962×10^{1}	1.2398×10^{-4}	1.4388	2.997924×10^{10}
1 erg/molecule	5.0345×10^{15}	1	6.0224×10^{16}	6.2418×10^{11}	7.2436×10^{15}	1.50929×10^{26}
1 cal/mole	0.34976	6.9473×10^{-17}	4.1840	4.3363×10^{-5}	0.50325	1.04854×10^{10}
1 eV	8.06573×10^{3}	1.6021×10^{-12}	9.6487×10^{4}	1	1.16049×10^{4}	2.41804×10^{14}
1°K	0.69503	1.3805×10^{-16}	8.3140	8.6170×10^{-5}	1	2.0836×10^{10}
1 Hz (sec^{-1})	3.3356×10^{-11}	6.6256×10^{-27}	3.9903×10^{-10}	4.1356×10^{-13}	4.7993×10^{-9}	1

1 Å = 10^{-8} cm = 0.1 nm = 10^{-4} μm = $10^{8}/(cm^{-1})$. Speed of light in vacuum $c = 2.99792458 \times 10^{8}$ m sec^{-1}; in October 1983 the meter was redefined as the distance traveled by light in vacuum in $1/c$ sec, so that c is now a defined constant and the second (as measured by atomic clocks) is the fundamental unit.

Index

Abelian group, 203–204
Absorption coefficient, 18, 27–29, 471, 474
saturated, 335, 390
Absorption of light
negative, 310
rates of, 23, 25–31, 471
Acetylene. *See* C_2H_2 molecule
Action integral, quantization of, 133
Alkali atoms
energy levels of, 54
optical pumping of, 296–300
Alkalis, diatomic, 91–93
Ammonia. *See* NH_3 molecule
Amplification
in masers and lasers, 309–310
in parametric systems, 414
Raman, 432
Angular momentum, 2*ff*., 39, 114, 161*ff*., 217, 294
raising and lowering operators, 349
Angular-momentum coupling, in molecules, 79–88, 161–167. *See also* Hund's coupling cases
Anharmonicity, 128–134, 250, 426
Anthracene
two-photon fluorescence in, 423–425
Antibonding orbitals, 97
Anti-Stokes. *See* Stokes and anti-Stokes scattering
Antisymmetric stretch, 239–240, 249
Association rule, 203
Asymmetric top, 216, 256
energy level expression for, 219
Asymmetry parameter κ, 219
Atomic orbitals, 40, 92, 263–264
Atomic spectra, 39–76
Aufbau principle, in molecules, 98
Axial groups. *See* $C_{\infty v}$ group; $D_{\infty h}$ group

Band contour analysis, 275–278
Band head, 159
Band origin, 158
Basis sets, 12, 463
Beer's law, 27, 381
Bend-stretch interaction, 237
Bennett hole, 392
Benzene. *See* C_6H_6 molecule
BF_3 molecule
moments of inertia, 214–215
normal modes, 251, 258
Birefringence, 410–412
Birge-Sponer extrapolation, 131–132, 155
Blackbody radiation, 25, 30
Bleaching, 335–336
Bloch equations, 351
for optical system, 340–347, 356*ff*., 428
Bohr-Einstein law, 1, 18
Bohr magneton μ_0, 70, 294
Bohr radius a_0, 40
Boltzmann statistics, 26, 116, 277–278, 464, 468
Born-Oppenheimer approximation, 77–79, 113, 146–147, 270
breakdown of, 79, 173, 288
Bose-Einstein statistics, 117
Bras. *See* State vectors
Breathing mode, 275, 278
Brewster's angle, 311–312
Brillouin-Wigner theorem, 11*n*

c. *See* Speed of light
$C_{\infty v}$ group, 83, 200, 263, 458
Calcite, 409–410
CARS. *See* Raman scattering, coherent anti-Stokes
CD_4, infrared spectrum, 255
CDF_3, infrared spectrum, 254, 401
Center of symmetry
and infrared-Raman selection rules, 250
relation to optical rotation, 184
Centrifugal distortion, 127–128
CH molecule, 102, 166, 184
C_2H_2 molecule
quantum beats in, 290
stimulated-emission pumping in, 451–452
vibrational normal modes, 245*ff*.
C_6H_6 molecule
radiationless transitions in, 285
symmetry of, 201
two-photon absorption in, 422–424
ultraviolet spectrum of, 272–280
Chaos, 449
Characters of group representations, 206–207
Cl_2 molecule, photofragment spectrum, 191
Classes, in group theory, 203–204
Closed shells, 99
Closure rule, 202
CN molecule, 164
Deslandres table, 152–153
perturbations in, 176–178
CO molecule, 122
CO_2 molecule
electronic structure, 265–267
laser action in, 250, 317–318
vibrational normal modes, 240, 248–250

Coherence
of laser radiation, 310–311
in two-level systems, 332, 343
T_2 decay of, 346, 352
Coherent transient spectroscopy, 356–388
Collision broadening, 32, 296, 446
Combination bands, 252
Combination differences, 158
Commutation rule, 203
Commutator, 3, 473
Complete set, of basis states, 4, 146
Completeness relationship, 3, 5, 473
Configurations
atomic, 44
molecular, 99–104, 265, 267, 268, 272, 284–285
Continuum spectroscopy, 187–191
Coriolis interaction, 256
Correlation diagram, 52, 263–265
Coulomb gauge, 21
Coulomb integral, 95
Coulomb interaction, in atoms, 42
Coulomb potential, 9, 105
Cross section
for optical absorption, 29
for Raman transitions, 430, 436
Crystals, for nonlinear optical effects, 413
CS_2, Zeeman effect in, 182
Curie law for magnetization, 470

$D_{\infty h}$ group, 84, 104, 201, 263, 458
D'Alembert's principle, 460–462
Degeneracy, 26, 262
of one-electron atomic states, 39*n*
related to symmetry, 80
removal by perturbation, 13–14
of rotational energy levels, 115
of rotational levels in polyatomic molecules, 218
Degrees of freedom, 113, 230, 308
Delayed nutation, 362–364
Delta function, 4, 471–472
Density matrix, 337, 342, 463–470
Density of states, 176, 287, 289, 428, 445–446
Dephasing, 346, 352, 362, 364–368, 445
Deslandres tables, 151–153, 193–194
Diatomic molecules
electronic spectra of, 145–197
electronic states of, 77–112
rotation and vibration of, 113–144
Dichroism, circular, 185
Difference bands, 250
Diffuse line, 178
Dipole approximation, 21–22, 417, 430
Dipole correlation function, 448, 471–474
Dipole moment, electric
operator, 19–20, 115
symmetry of, in axial groups, 89–90, 169
symmetry of, in point groups, 209–210, 454–458
Dipole moment, magnetic
operator, 24
symmetry of, in axial groups, 90
symmetry of, in point groups, 210, 454–458
Dipole moment derivative, 122, 248, 449
Dipole velocity operator, 22
Direct product, 207–209, 248*ff.*
tables, 208, 454–458
Direct sum, 206
Dispersion, of operators, 7
of refractive index, 186, 408–409
Dissociation energy, 129*ff.*, 155–156
of H_2^+, 97
Dissociation limit, 148
of H_2, 155–156
DNA, 200
Doppler broadening. *See* Lineshape, Doppler
Doppler effect, 38, 308, 391
cancellation in two-photon absorption, 425–426
Doppler width, formulas for, 35, 295
Double-resonance spectroscopy, 360–361, 394–402, 446–447
Drude equation, 410

Eckart conditions, 233
Eigenfunction, 4*ff.*
Eigenstate, 3*ff.*, 79, 444*ff.*
Eigenvalue, 7, 236, 358
Eigenvector, 239
Einstein *A* and *B* coefficients, 25
Electrical susceptibility, nonlinear, 405, 436
Electric-dipole transition, for rigid rotor, 115, 221
Electric field
effect on atomic spectra, 64–67, 295
effect on molecular spectra, 183–184, 294–295
of light propagating in matter, 406–408, 413, 431–433
Electrochromic effect. *See* Stark effect
Electromagnetic field, interaction of atoms and molecules with, 19–24, 417–418
Electronic transition moment, 104, 146

Emission spectra, of diatomic molecules, 150
Energy, as eigenvalue of Hamiltonian, 7–9
Energy conversion factors, 483
Energy levels
of anharmonic oscillator, 129, 133
of harmonic oscillator, 122
of hydrogen atom, 41
of rigid rotor, 115
Etalon, 313
use in picosecond spectroscopy, 442
Exchange operation, 117, 168–169
Excimer laser, 318, 434
Exclusion principle. *See* Pauli exclusion principle

Fabry-Perot cavity, 310–312, 374
Faraday effect. *See* Magnetooptic rotation
Fermi contact term, 294
Fermi-Dirac statistics, 117, 211
Fermi golden rule, 176, 287, 428
Fermi resonance, 260
Feynman diagrams, 419–420
Feynman-Vernon-Hellwarth representation, 337–340
FG method, 241–248
Field, R. W., 177, 179, 444*n*
Figure axis, 221
Fine structure, 157–161
Fine-structure constant, 50
Fluorescence, two-photon, 422–425, 438–440
Fluorescence yield, 178, 286, 289
Fluoroform. *See* CDF_3
Force constant
for harmonic oscillator, 121
for molecular vibrations, 235–238, 241
Forces of constraint, 459
Formaldehyde. *See* H_2CO molecule
Fourier's series, 4
Fourier transform spectroscopy, 373–375
Franck-Condon factors, 147, 152–154, 173, 278*n*, 288
Franck-Condon principle, 145–150, 282
Free induction decay, 361–362, 364–368
Free particle, 9
Frequency
of electromagnetic radiation, 1
vibrational, 122, 235–236, 244
Frequency doubling. *See* Second-harmonic generation
Fundamental band, 122

g-Factor. *See* Gyromagnetic ratio
Gas constant *R*, 31
Gauge transformation, 21
Gilmore diagrams, for N_2, NO, O_2, H_2, 105–109
Glan-Thompson prism, 411
Grotrian diagram, 40*n*, 41
Group frequencies, 248
Group theory, 198–213, 241, 249, 454–458
Gyromagnetic ratio, 349
anomalous, of electrons, 69*n*
determination of, by optical pumping, 196–300
Landé, 69, 294
nuclear, 70–71, 294
rotational, 180, 224, 294
Gyroscope, 349, 359

h, *ħ*. *See* Planck's constant
H atom, 9, 39–40, 295, 301–303, 327–328
hyperfine structure in, by Lamb dip spectroscopy, 393
maser, 308–310
H_2^+ molecule, 92–97, 109, 164
H_2 molecule, 99
dissociation limit, 155–156
intensity alternation, 119
potential curves, 109
Raman shifting in, 434–435
rotational *g*-factor, 180
Halogens, diatomic, 91–92, 94
Hamiltonian operator, xv, 7–9
free-particle, 9
gauge invariant, 21
harmonic oscillator, 9, 121
Hénon-Heiles, 449
hydrogen atom, 9, 39
hydrogen molecule ion, 92
hyperfine, 60
for interaction of radiation and matter, 19–23
many-electron atoms, 42
molecular, 77, 165–166, 171, 232, 288, 448
one-electron atom (*see* Hydrogen)
relativistically invariant, 20, 417
rigid-rotor, 9
spin-orbit, 49
Stark, 64
symmetry of, 80, 199
two-level system, 329, 338
Watson, 256
Zeeman, 67, 348, 469
Harmonic oscillator, xv, 4, 9, 36, 120–124, 230, 327, 445

Hartree-Fock calculations, 48
Hartree unit, 39
HC≡CCHO molecule, rotational energy levels and spectra, 220, 222
HCl molecule, infrared spectrum of, 125
HCN molecule, 234
beam resonance spectrum of, 19, 332–333
infrared spectrum of, 253
microwave spectrum of, 225
H_2CO molecule, 200
molecular orbitals, 268–269
moments of inertia, 217
normal vibrations, 271
radiationless processes in, 178, 285–286
Heisenberg equation of motion, 337, 467, 473
Heisenberg representation, 4–6
Heisenberg uncertainty principle, 33, 295
Helium, superfluidity, 118
He-Ne laser, 315–317
stabilization of, by Lamb dip, 394
Hermite polynomials, 4
Herzberg, G., xv, 155–156, 475–476, 479
HF molecule, laser action in, 318–319
HF · acetylene complex, microwave spectrum, 375
Hg atom
Grotian diagram, 59
hyperfine structure in, 62–65
H_2O molecule, photoelectron spectrum of, 284
Hole burning, 391–392, 445
Hole formalism, 102
Homogeneous broadening, 31–36
Homonuclear molecule, 83, 87, 103, 119, 122, 168, 201
Hot bands, 130, 250
Hougen, J. T., 161
Hund's coupling cases, 161–167
Hund's rule, 56
in molecules, 102
Hydrogen. *See* H atom; H_2 molecule
Hydrogenlike orbitals, 4
Hyperfine coupling constant, 61
Hyperfine structure, 62–65, 70–72, 223–224, 293–295, 393

I atom, 88
laser action in, 104, 319–320
I_2 molecule
absorption spectrum of, xx, 149, 187
Birge-Sponer extrapolation for, 131–132
isotope shift in absorption, 152–155
laser action in, 314–315
laser-induced fluorescence, 308
predissociation of, 178, 192
rotational fine structure, 120, 159–160
ICl molecule, magnetic rotation spectrum of, 187–188
Identity operator, 84, 199, 203, 207*n*
IF molecule, emission spectrum of, 175
Index matching, 408–410
Inhomogeneous broadening, 28, 31–36, 380
in spin echo, 369
Intermediate coupling cases, 167
Internal conversion, 286
Internal coordinates, 232–235, 241–245
Intersystem crossing, 286
Inverse operation, 203
Inversion doubling, 225–256
Inversion operation, 80–83, 89, 199*ff.*, 265
equivalent to exchange of nuclei, 168
Inverted population, 26, 309–321, 345
Iodine. *See* I_2 molecule
IRMPA. *See* Multiple infrared photon excitation
Irreducible representation, 205, 272
Isotope shift
Rydberg, in atoms, 60–61
in vibrational spectra, 152–155, 238, 426

J-uncoupling, 166
Jablonski diagram, 287
jj coupling, 51–52
in molecules, 85, 87
JWKB method, 133, 445

KDP, 412–413
Kerr effect, 440–442
Kets. *See* State vectors

l-doubling, 224–225
Λ-doubling, 166
Lagrange's equations, 230–231, 235, 459–462
Laguerre polynomial, 40
Lamb dip spectroscopy, 391–394
Lamb-Retherford experiment, 305
Laplace transforms, 357
Larmor frequency, 350–351
Laser-Raman method, 140
Lasers, 311–323, 438
carbon dioxide, 317–318, 372, 387, 401, 426
chemical, 318–320
dye solution, 320–322, 436–437
edible, 321

excimer, 318, 434
gas discharge, 315–318
helium-neon, 315–317, 394
iodine, 314–315
neodymium, 313–314, 415, 437
ruby, 313–314, 372, 386
semiconductor, 321–323, 401
LCAO approximation, 94
Legendre polynomial, 40, 56, 119
LeRoy-Bernstein plot, 131–132
Level-crossing spectroscopy, 300–301
Lifetime, radiative, 25–31, 287, 295
Light, speed of, 1, 27, 483
Light scattering, 135, 421
LiH molecule, 105, 110, 129, 173
$LiNbO_3$, 415–416
Linear response method, 377
Lineshape, 31–36
Doppler, 33–35, 295
homogeneous, 31–36
inhomogeneous, 31–36
Lorentzian, 33–34, 390, 400, 474
normalized, 472
Voigt, 36
Linewidths, 31–36, 178, 295, 446
diffraction-limited, 295
in molecular beam experiments, 306–307
natural, 32, 295
Lissajous figures, 449
Literature, spectroscopic, xv, 475–482
Lithium hydride. *See* LiH molecule
Lithium niobate. *See* $LiNbO_3$
Local modes, 451
Lorentzian. *See* Lineshape
Lo Surdo photograph, 66
$\mathbf{L}\cdot\mathbf{S}$ coupling, 51–52

Magnetic dipole operator, 24
selection rules, for diatomic molecules, 194
Magnetic field
effect on atomic spectra, 67–73, 294, 300, 302
effect on molecular spectra, 179–182, 184*ff.*, 224, 294
effective, in rotating coordinate system, 351
Magnetic interactions, in atoms, 44, 48–50, 61–62, 67–73
Magnetic resonance, 347–352, 374*n*
Magnetization, 351
calculation of, with density matrix, 469–470
Magnetooptic rotation (Faraday effect), 166, 184–188
Masers, 308–311. *See also* Lasers
Master equation, 390, 428
Matrix element, 1*ff.*
of density operator, 463–464
of dipole operator in atoms, 56
of dipole operator in molecules, 115, 122, 145–146, 248
of $\mathbf{I}\cdot\mathbf{J}$, 61
of $\mathbf{L}\cdot\mathbf{S}$, 50, 166
of $1/r_{12}$, 42, 47
of rotational perturbations, 173, 176
symmetry of, 198
of vibrational perturbations, 173
vibronic, 270
Matrix mechanics, 4–7
Matrix representations, 205
Maxwell-Bloch equation, 379, 383–384
Maxwell-Boltzmann law, 35
Maxwell's equations, xvi, 376, 404–406
Measurement algebra, 1*ff.*
Mercury. *See* Hg atom
Metastable states
atomic, 57
of Cr^{+++} ion, 313
of molecules, 103–104
Methane-d_4. *See* CD_4
Microwave spectroscopy, 116, 219–226, 374–375
Mixing of states, in perturbation theory, 12, 176
Mode locking, 438, 442
Modes, of optical cavities, 311–312
Molecular beams, 2, 18, 301–309
electric deflection of, 307
electric resonance spectroscopy, 183, 305–307
magnetic resonance spectroscopy, 180, 304–305
photon recoil effects, 61
spectroscopy, 188, 304–309
supersonically cooled, 307–309, 374–375
Molecular orbital theory, 92–111
for polyatomic molecules, 263–269, 272–274, 284–285
Molecular potential curves, 96–97, 105–110, 131, 134, 189
Molecular terms
from molecular-orbital configuration, 99–103
from separated-atom states, 85–88
Molecule-fixed axes, 167, 173
Moments of inertia, 214–215, 263
Monochromatic source (definition), 18

Morse function, 133
Mulliken, R. S., 82*n*, 173
Multiphoton ionization, 192, 429–430
Multiphoton spectroscopy, 404–429
Multiple infrared photon excitation, 192, 426–429, 446–448
Multiplets, from atomic configurations, 44–46
Mutual exclusion, of Raman and infrared bands, 250

$n \to \pi^*$ transition, in formaldehyde, 268–270
$n \to \sigma^*$ transition, in formaldehyde, 268
N_2 molecule
  emission spectra, 171–172, 174
  potential curves, 106
  Rydberg series in, 282
Na atom
  Grotrian diagram, 58
  hyperfine structure in, by Lamb dip spectroscopy, 363
  as two-level system, 327–328
Ne atom, laser action in. *See* He-Ne laser
Newton's laws of motion, 459–460
NH molecule, 102, 164, 184
NH_3 molecule
  inversion doubling in, 225–226
  maser action in, 310
  optical nutation in, 360–361, 364
  symmetry of, 200
  as two-level system, 327–328
NO molecule
  multiphoton ionization of, 429–430
  potential curves, 107
NO_2 molecule
  anomalous lifetime of, 289
  laser-induced fluorescence, 308
  orbital excitation in, 267–268
Nonbonding orbital, 268
Noncrossing rule, 111
Normal coordinates, 235–248
Normalization, 94
Normal-mode vibrations
  of H_2CO, 271
  of linear XY_2 molecules, 238–240, 242–244
  of linear XYYX molecules, 244–247
  of planar XY_3 molecules, 251
Nuclear spin statistics, 117–120, 160, 211, 218

O_2 molecule, 103, 164, 319
  missing *J* levels, 119
  potential curves, 108
  Raman spectrum of, 141, 436–437
Oblate top. *See* Symmetric top
O-branch, 139
Observable quantities, 1
Operator, in quantum mechanics, 3*ff*., 198, 463–468
Optical nutation, 356–364
Optical pumping, 293, 296–300
Orbital
  atomic, 40, 92, 263–264
  molecular, 95–101, 263–269, 272–274, 284–285
Order, of group, 203
Orthohydrogen, 120
Oscillator
  anharmonic, 128–134
  harmonic, xv, 9, 120–124, 445
Oscillator strength, 29–31, 272, 288–289
Overlap integral, 96, 146, 278*n*
Overtone bands, 122, 129, 250, 447, 451
Oxygen. *See* O_2 molecule

P-branch, 124, 157, 252–256, 262
π component, 72, 181, 299
π orbitals, 97–98, 268, 272, 274
$\pi/2$ pulse, 345, 352, 362, 365–368
π pulse, 332, 346, 352, 362
  in spin echo, 365–370
2π pulse, 352
  in self-induced transparency, 382–398
$\pi \to \pi^*$ transition
  in benzene, 272
  in formaldehyde, 268–270
Parahydrogen, 120
Parallel bands, 249
  rotational structure of, 252
Parametric oscillator, 413–416, 425
Parity, 104, 119, 169, 173
Particle in a box, xv, 5. *See also* Square-well potential
Paschen-Back limit, 69, 71, 302
Pauli exclusion principle, 44
  for molecules, 86–87, 99, 102, 211
Pauli spin matrices, 339
Pendulum, equations for, 383–385
Perpendicular bands, 249
  rotational structure of, 252
Perturbations, in molecular spectra, 171–179, 267, 294
Perturbation theory, 8–19,
  for degenerate states, 13–14, 67–94
  first-order, 10–11, 287
  first-order in time, 16–17, 417

second-order, 11–12, 137, 418
time-dependent, 14–19, 417–419
Phase-matching criterion, 408, 433
Phosphorescence, 285, 290
Photodissociation, 176, 188
Photoelectron spectroscopy, 280–285
Photofragment spectroscopy, 187–191, 281
Photoionization, 281–282, 429
Photon echo, 364–373
Photon recoil, 61
Phthalocyanine, 308–309
Picosecond light pulses, 314, 438–441
Placzek notation. *See* Spectroscopic notation, for polyatomic molecules
Planck's blackbody law. *See* Blackbody radiation
Planck's constant (h), 1
Point groups, 199–212
representations and product tables, 454–458
Poisson distribution, 334
Polarizability, 135–138, 249, 431
nonlinear, 404–405, 432
Polarizability ellipsoid, 136
Polarization, 343, 369, 376
Polarized light, 71–72, 124, 184–185, 297–299, 376–377, 406
and birefringence, 410–412
rotation by asymmetric molecules, 184
Polyatomic molecules
electronic spectra of, 262–292
rotational spectra of, 214–229
vibrational spectra of, 230–261
Population inversion. *See* Inverted population
Potential energy function
for diatomic molecules, 77, 96–97, 105–110, 113, 131–134, 265
for vibration, 121
Precession
of magnetic moment in magnetic field, 349–350
in rotating frame, 343–344, 359
Predissociation, 176–178
Pressure broadening. *See* Collision broadening
Principal axes, of rotation 214
Progression, 149, 271, 275–276, 422
Projection operator, 5, 463
Prolate top. *See* Symmetric top
Propynal. *See* HC≡CCHO molecule
Pseudospin representation, 332, 337–340, 356*ff.*
Pulse area theorem, 331, 380
Q-branch, 124, 142, 157, 170, 252, 262
Q-switching, 313, 314, 438
Quadrupole lens, 306
Quadrupole moment, electric, 223–224
operator, 24, 250
selection rules, for diatomic molecules, 91, 194
symmetry of, in axial groups, 91
symmetry of, in point groups, 454–458
Quantum beats, 289–290
Quantum yield. *See* Fluorescence yield
Quasicontinuum, 427–428, 446

R-branch, 124, 157, 252–256, 262
Rabi frequency, 331*ff.*, 340, 356*ff.*
Rabi solutions, 329–334, 356, 379
Radiationless transitions, 285–290
Raman scattering, 134–138, 421, 429–437, 447
coherent anti-Stokes, 434–437
stimulated, 429–434
Raman spectra
rotational, 222
vibration-rotation, 116, 134–141
vibrational, in polyatomic molecules, 249–251
Random-phase approximation, 468
Rayleigh scattering, 137, 421
Rb atom, self-induced transparency in, 387–388
Reduced mass
of electron in atom, 39
of rigid rotor, 114
for vibrations, 122*n*, 243
Reducibility, of representations, 205
Reflection operation, 80–83, 89, 199*ff.*, 265
equivalence to coordinate inversion, 168
Refractive index, 185, 323, 385, 408–412, 436
Representations, 205–209
Resonance condition, 17
Resonance integral, 96
RhC molecule, emission spectrum of, 175
Rigid rotor, 4, 9, 214–219, 445
RKR method. *See* Rydberg-Klein-Rees method
R·L coupling, 166, 173
Rotating coordinate system, 341–342, 344, 350, 356*ff.*
Rotating wave approximation, 17, 329, 342, 396
Rotation
of diatomic molecules, 113–120
of polyatomic molecules, 214–229

Rotation operation, 80–82, 199*ff.*
Rotational constant *B*, 115, 126, 218, 277
Rotational constants, for C_6H_6, 277
Rotational structure, of vibrational bands, 124–128, 157–161, 170–172, 252–256, 272–280, 422–424
Ruby
laser action in, 313–314
photon echo in, 372
self-induced transparency in, 386–387
as two-level system, 327–328
Russell-Saunders coupling. *See also* **L·S** coupling
in atoms, 51
in molecules, 85, 98
RWA. *See* Rotating wave approximation
Rydberg-Klein-Rees method, 133, 445
Rydberg series, 281–282
Rydberg states, 165, 167
Rydberg unit, 39

S-branch, 142
σ component, 72, 181
σ orbitals, 95, 268
Saturation, 334–337, 388–391
broadening, 334
in Lamb dip spectroscopy, 391–394
Scalar potential, 20
Schrödinger equation, xv, 7, 77, 466
for harmonic vibration, 121
for molecular rotation, 114
for molecular vibrations, 232, 240
for three-level system, 396
time-dependent, 15, 21
for two-level system, 337
Schrödinger representation, 4–6
Schumann-Runge bands, 104
Second-harmonic generation, 405–408, 423
Secular determinant, 14, 95, 236, 244, 247
Selection rules, 56–57, 167–171, 194, 198, 268
for ΔM transitions, 71–72
electric quadrupole, 91, 194
for electronic transition in diatomic molecules, 87–92, 157, 167–171
for hyperfine structure transitions, 62
magnetic-dipole, 91, 194
for magnetic resonance transitions, 349
for Raman transitions, 138–139, 222
for rigid rotor, 115
for symmetric tops, 221
for two-photon absorption, 422
use of group theory to determine, 80, 209–210
for vibrational transitions in polyatomic molecules, 248–252
Self-induced transparency, 374–388
Separated-atom states, 85–87
SF_6
high-resolution absorption spectrum, 324
IRMPA in, 426
photon echo in, 372–373
saturation in, 336
self-induced transparency in, 386–387
symmetry of, 201
Sine-Gordon equation, 384
Singlet states, 46*ff.*, 86*ff.*, 286–288, 320
Smog, 267
S·N coupling, 166
SO_2 molecule, anomalous lifetime of, 389
Sodium. *See* Na atom
Space-fixed axes, 113–115, 168, 173, 179, 214
Spectroscopic notation
for atoms, 39
for diatomic molecules, 82*n*, 124
for polyatomic molecules, 206–207
Speed of light (*c*), 1, 27, 483
Spherical harmonics, 4, 6
Spherical polar coordinates, 6
Spherical top, 216, 256
Spin
electron, 40*ff.*, 86, 98, 161
nuclear, 61–63, 118, 223
Spin echo. *See* Photon echo
Spin-lattice relaxation, 352
Spin-1/2 system, 327–328, 348–349
Spin-orbit coupling, 48–50, 69, 85, 288
in case (a) molecules, 161–162
Spin-rotation coupling, 164, 223
Spin-rotation interaction, 166
Spin-spin coupling, 164
Spin-spin relaxation, 352
Splitting
of degenerate levels, 14, 79, 111, 173, 224, 226
due to Stark effect, 66–67, 295
due to Zeeman effect, 67, 295, 348
zero-field, 71, 294
Spontaneous emission, 25, 309, 352
Square-well potential, 129
S·R coupling, 166, 173
Stark effect
in atoms, 64–67
internal, in diatomic molecules, 79
in molecules, 183–184
in optical nutation, 360
Stark modulation, in microwave

spectroscopy, 183, 222–223
State vectors, 2*ff.*
Stern-Gerlach experiment, 2, 303
Stimulated emission, 26, 118, 309
Stimulated emission pumping, 447, 451
Stokes and anti-Stokes scattering, 138, 430
Strong-collision model, 353, 358
Subgroups, 203
Sulfur hexafluoride. *See* SF_6
Superfluid. *See* Helium
Superradiance, 337
Supersonic cooling, 307–308
Symmetric stretch, 239–240
Symmetric top, 216–218, 252
Symmetry
of Coulomb potential, 39*n*
molecular, 198–213
of molecular vibrations, 241–248
of molecular wave functions, 168, 198–199
in one-electron integrals, 42–43
of operators, 88–91, 209–210, 454–458
Symmetry coordinates, 245
Symmetry number, 120
Symmetry orbitals, 263

T_1 relaxation, 346, 352, 379, 388
T_2 relaxation, 346, 352, 372, 379
Taylor's series, 21, 121, 122, 230
Te_2 molecule, fluorescence spectrum of, 150
Term value, 145, 262
Thomas-Kuhn sum rule, 37
Trace invariance, 47–49, 63
Transformation matrix, 4
Transition moment, 104, 146, 329
Transition probability, 17, 23
Triplet states, 86*ff.*, 186, 286–288, 320
Two-level systems, 327–328, 348, 445
coherently driven transitions in, 329–347
self-induced transparency in, 378–386
Two-photon absorption, 419–424
Two-photon fluorescence, 422–425, 438–440

Unitary transformation, 48, 246

Valence force field, 237–238, 246
van der Waals forces, 129
van der Waals molecules, 374–375
van Vleck pure precession, 166
Variational calculation, 155
Vector coupling model, 51. *See also* Angular-momentum coupling, in molecules
Vector potential, 20
Vertical transitions, 147
Vibration-rotation interaction, 126, 128 130, 157, 224–226, 256, 445
Vibrational overlap integral, 146
Vibrational progressions, 149–153, 271, 275, 276
Vibrational spectra, of polyatomic molecules, 230–261
Vibrational structure, of electronic bands, 149–155, 273–276
Vibrations
bending, 224, 233*ff.*, 244–247
of diatomic molecules, 120–124, 150–155
stretching, 232*ff.*, 243–247
Vibronic coupling
in benzene, 274–275
in formaldehyde, 270–271
Virial theorem, 240
Virtual orbital approximation, 57
Virtual states, 137

Walsh's rules, 263–268
Water. *See* H_2O molecule
Wave equation, 376–377
Wave function, 4–7, 198
harmonic oscillator, 122–123
hydrogen atom, 40
molecular, 77, 79, 113, 146
rigid-rotor, 221
for two-level system, 329
Wavelength, 1
units, 438
Wave number, 1
Wigner-Witmer rules, 87
WKB method. *See* JWKB method
Wood, R. W., 135, 478

Zeeman effect, 348, 469
in atoms, 67–73, 294, 297, 300
in H atom, 301–303
in molecules, 166, 179–182, 224, 294
Zero-point energy, 123, 188, 262, 446

A CATALOG OF SELECTED

DOVER BOOKS

IN SCIENCE AND MATHEMATICS

Astronomy

BURNHAM'S CELESTIAL HANDBOOK, Robert Burnham, Jr. Thorough guide to the stars beyond our solar system. Exhaustive treatment. Alphabetical by constellation: Andromeda to Cetus in Vol. 1; Chamaeleon to Orion in Vol. 2; and Pavo to Vulpecula in Vol. 3. Hundreds of illustrations. Index in Vol. 3. 2,000pp. 6⅛ x 9¼.
Vol. I: 23567-X
Vol. II: 23568-8
Vol. III: 23673-0

EXPLORING THE MOON THROUGH BINOCULARS AND SMALL TELESCOPES, Ernest H. Cherrington, Jr. Informative, profusely illustrated guide to locating and identifying craters, rills, seas, mountains, other lunar features. Newly revised and updated with special section of new photos. Over 100 photos and diagrams. 240pp. 8¼ x 11. 24491-1

THE EXTRATERRESTRIAL LIFE DEBATE, 1750–1900, Michael J. Crowe. First detailed, scholarly study in English of the many ideas that developed from 1750 to 1900 regarding the existence of intelligent extraterrestrial life. Examines ideas of Kant, Herschel, Voltaire, Percival Lowell, many other scientists and thinkers. 16 illustrations. 704pp. 5⅜ x 8½. 40675-X

THEORIES OF THE WORLD FROM ANTIQUITY TO THE COPERNICAN REVOLUTION, Michael J. Crowe. Newly revised edition of an accessible, enlightening book recreates the change from an earth-centered to a sun-centered conception of the solar system. 242pp. 5⅜ x 8½. 41444-2

A HISTORY OF ASTRONOMY, A. Pannekoek. Well-balanced, carefully reasoned study covers such topics as Ptolemaic theory, work of Copernicus, Kepler, Newton, Eddington's work on stars, much more. Illustrated. References. 521pp. 5⅜ x 8½.
65994-1

A COMPLETE MANUAL OF AMATEUR ASTRONOMY: Tools and Techniques for Astronomical Observations, P. Clay Sherrod with Thomas L. Koed. Concise, highly readable book discusses: selecting, setting up and maintaining a telescope; amateur studies of the sun; lunar topography and occultations; observations of Mars, Jupiter, Saturn, the minor planets and the stars; an introduction to photoelectric photometry; more. 1981 ed. 124 figures. 26 halftones. 37 tables. 335pp. 6½ x 9¼.
42820-6

AMATEUR ASTRONOMER'S HANDBOOK, J. B. Sidgwick. Timeless, comprehensive coverage of telescopes, mirrors, lenses, mountings, telescope drives, micrometers, spectroscopes, more. 189 illustrations. 576pp. 5⅝ x 8¼. (Available in U.S. only.)
24034-7

STARS AND RELATIVITY, Ya. B. Zel'dovich and I. D. Novikov. Vol. 1 of *Relativistic Astrophysics* by famed Russian scientists. General relativity, properties of matter under astrophysical conditions, stars, and stellar systems. Deep physical insights, clear presentation. 1971 edition. References. 544pp. 5⅝ x 8¼. 69424-0

Chemistry

THE SCEPTICAL CHYMIST: The Classic 1661 Text, Robert Boyle. Boyle defines the term "element," asserting that all natural phenomena can be explained by the motion and organization of primary particles. 1911 ed. viii+232pp. 5⅜ x 8½. 42825-7

RADIOACTIVE SUBSTANCES, Marie Curie. Here is the celebrated scientist's doctoral thesis, the prelude to her receipt of the 1903 Nobel Prize. Curie discusses establishing atomic character of radioactivity found in compounds of uranium and thorium; extraction from pitchblende of polonium and radium; isolation of pure radium chloride; determination of atomic weight of radium; plus electric, photographic, luminous, heat, color effects of radioactivity. ii+94pp. 5⅜ x 8½. 42550-9

CHEMICAL MAGIC, Leonard A. Ford. Second Edition, Revised by E. Winston Grundmeier. Over 100 unusual stunts demonstrating cold fire, dust explosions, much more. Text explains scientific principles and stresses safety precautions. 128pp. 5⅜ x 8½. 67628-5

THE DEVELOPMENT OF MODERN CHEMISTRY, Aaron J. Ihde. Authoritative history of chemistry from ancient Greek theory to 20th-century innovation. Covers major chemists and their discoveries. 209 illustrations. 14 tables. Bibliographies. Indices. Appendices. 851pp. 5⅜ x 8½. 64235-6

CATALYSIS IN CHEMISTRY AND ENZYMOLOGY, William P. Jencks. Exceptionally clear coverage of mechanisms for catalysis, forces in aqueous solution, carbonyl- and acyl-group reactions, practical kinetics, more. 864pp. 5⅜ x 8½. 65460-5

ELEMENTS OF CHEMISTRY, Antoine Lavoisier. Monumental classic by founder of modern chemistry in remarkable reprint of rare 1790 Kerr translation. A must for every student of chemistry or the history of science. 539pp. 5⅜ x 8½. 64624-6

THE HISTORICAL BACKGROUND OF CHEMISTRY, Henry M. Leicester. Evolution of ideas, not individual biography. Concentrates on formulation of a coherent set of chemical laws. 260pp. 5⅜ x 8½. 61053-5

A SHORT HISTORY OF CHEMISTRY, J. R. Partington. Classic exposition explores origins of chemistry, alchemy, early medical chemistry, nature of atmosphere, theory of valency, laws and structure of atomic theory, much more. 428pp. 5⅜ x 8½. (Available in U.S. only.) 65977-1

GENERAL CHEMISTRY, Linus Pauling. Revised 3rd edition of classic first-year text by Nobel laureate. Atomic and molecular structure, quantum mechanics, statistical mechanics, thermodynamics correlated with descriptive chemistry. Problems. 992pp. 5⅜ x 8½. 65622-5

FROM ALCHEMY TO CHEMISTRY, John Read. Broad, humanistic treatment focuses on great figures of chemistry and ideas that revolutionized the science. 50 illustrations. 240pp. 5⅜ x 8½. 28690-8

Engineering

DE RE METALLICA, Georgius Agricola. The famous Hoover translation of greatest treatise on technological chemistry, engineering, geology, mining of early modern times (1556). All 289 original woodcuts. 638pp. 6¾ x 11. 60006-8

FUNDAMENTALS OF ASTRODYNAMICS, Roger Bate et al. Modern approach developed by U.S. Air Force Academy. Designed as a first course. Problems, exercises. Numerous illustrations. 455pp. 5⅜ x 8½. 60061-0

DYNAMICS OF FLUIDS IN POROUS MEDIA, Jacob Bear. For advanced students of ground water hydrology, soil mechanics and physics, drainage and irrigation engineering, and more. 335 illustrations. Exercises, with answers. 784pp. 6⅛ x 9¼. 65675-6

THEORY OF VISCOELASTICITY (Second Edition), Richard M. Christensen. Complete, consistent description of the linear theory of the viscoelastic behavior of materials. Problem-solving techniques discussed. 1982 edition. 29 figures. xiv+364pp. 6⅛ x 9¼. 42880-X

MECHANICS, J. P. Den Hartog. A classic introductory text or refresher. Hundreds of applications and design problems illuminate fundamentals of trusses, loaded beams and cables, etc. 334 answered problems. 462pp. 5⅜ x 8½. 60754-2

MECHANICAL VIBRATIONS, J. P. Den Hartog. Classic textbook offers lucid explanations and illustrative models, applying theories of vibrations to a variety of practical industrial engineering problems. Numerous figures. 233 problems, solutions. Appendix. Index. Preface. 436pp. 5⅜ x 8½. 64785-4

STRENGTH OF MATERIALS, J. P. Den Hartog. Full, clear treatment of basic material (tension, torsion, bending, etc.) plus advanced material on engineering methods, applications. 350 answered problems. 323pp. 5⅜ x 8½. 60755-0

A HISTORY OF MECHANICS, René Dugas. Monumental study of mechanical principles from antiquity to quantum mechanics. Contributions of ancient Greeks, Galileo, Leonardo, Kepler, Lagrange, many others. 671pp. 5⅜ x 8½. 65632-2

STABILITY THEORY AND ITS APPLICATIONS TO STRUCTURAL MECHANICS, Clive L. Dym. Self-contained text focuses on Koiter postbuckling analyses, with mathematical notions of stability of motion. Basing minimum energy principles for static stability upon dynamic concepts of stability of motion, it develops asymptotic buckling and postbuckling analyses from potential energy considerations, with applications to columns, plates, and arches. 1974 ed. 208pp. 5⅜ x 8½. 42541-X

METAL FATIGUE, N. E. Frost, K. J. Marsh, and L. P. Pook. Definitive, clearly written, and well-illustrated volume addresses all aspects of the subject, from the historical development of understanding metal fatigue to vital concepts of the cyclic stress that causes a crack to grow. Includes 7 appendixes. 544pp. 5⅜ x 8½. 40927-9

Mathematics

FUNCTIONAL ANALYSIS (Second Corrected Edition), George Bachman and Lawrence Narici. Excellent treatment of subject geared toward students with background in linear algebra, advanced calculus, physics, and engineering. Text covers introduction to inner-product spaces, normed, metric spaces, and topological spaces; complete orthonormal sets, the Hahn-Banach Theorem and its consequences, and many other related subjects. 1966 ed. 544pp. 6⅛ x 9¼. 40251-7

ASYMPTOTIC EXPANSIONS OF INTEGRALS, Norman Bleistein & Richard A. Handelsman. Best introduction to important field with applications in a variety of scientific disciplines. New preface. Problems. Diagrams. Tables. Bibliography. Index. 448pp. 5⅜ x 8½. 65082-0

VECTOR AND TENSOR ANALYSIS WITH APPLICATIONS, A. I. Borisenko and I. E. Tarapov. Concise introduction. Worked-out problems, solutions, exercises. 257pp. 5⅝ x 8¼. 63833-2

THE ABSOLUTE DIFFERENTIAL CALCULUS (CALCULUS OF TENSORS), Tullio Levi-Civita. Great 20th-century mathematician's classic work on material necessary for mathematical grasp of theory of relativity. 452pp. 5⅝ x 8¼. 63401-9

AN INTRODUCTION TO ORDINARY DIFFERENTIAL EQUATIONS, Earl A. Coddington. A thorough and systematic first course in elementary differential equations for undergraduates in mathematics and science, with many exercises and problems (with answers). Index. 304pp. 5⅜ x 8½. 65942-9

FOURIER SERIES AND ORTHOGONAL FUNCTIONS, Harry F. Davis. An incisive text combining theory and practical example to introduce Fourier series, orthogonal functions and applications of the Fourier method to boundary-value problems. 570 exercises. Answers and notes. 416pp. 5⅜ x 8½. 65973-9

COMPUTABILITY AND UNSOLVABILITY, Martin Davis. Classic graduate-level introduction to theory of computability, usually referred to as theory of recurrent functions. New preface and appendix. 288pp. 5⅜ x 8½. 61471-9

ASYMPTOTIC METHODS IN ANALYSIS, N. G. de Bruijn. An inexpensive, comprehensive guide to asymptotic methods–the pioneering work that teaches by explaining worked examples in detail. Index. 224pp. 5⅜ x 8½ 64221-6

APPLIED COMPLEX VARIABLES, John W. Dettman. Step-by-step coverage of fundamentals of analytic function theory–plus lucid exposition of five important applications: Potential Theory; Ordinary Differential Equations; Fourier Transforms; Laplace Transforms; Asymptotic Expansions. 66 figures. Exercises at chapter ends. 512pp. 5⅜ x 8½. 64670-X

INTRODUCTION TO LINEAR ALGEBRA AND DIFFERENTIAL EQUATIONS, John W. Dettman. Excellent text covers complex numbers, determinants, orthonormal bases, Laplace transforms, much more. Exercises with solutions. Undergraduate level. 416pp. 5⅜ x 8½. 65191-6

TENSOR CALCULUS, J.L. Synge and A. Schild. Widely used introductory text covers spaces and tensors, basic operations in Riemannian space, non-Riemannian spaces, etc. 324pp. 5⅜ x 8¼. 63612-7

ORDINARY DIFFERENTIAL EQUATIONS, Morris Tenenbaum and Harry Pollard. Exhaustive survey of ordinary differential equations for undergraduates in mathematics, engineering, science. Thorough analysis of theorems. Diagrams. Bibliography. Index. 818pp. 5⅜ x 8½. 64940-7

INTEGRAL EQUATIONS, F. G. Tricomi. Authoritative, well-written treatment of extremely useful mathematical tool with wide applications. Volterra Equations, Fredholm Equations, much more. Advanced undergraduate to graduate level. Exercises. Bibliography. 238pp. 5⅜ x 8½. 64828-1

FOURIER SERIES, Georgi P. Tolstov. Translated by Richard A. Silverman. A valuable addition to the literature on the subject, moving clearly from subject to subject and theorem to theorem. 107 problems, answers. 336pp. 5⅜ x 8½. 63317-9

INTRODUCTION TO MATHEMATICAL THINKING, Friedrich Waismann. Examinations of arithmetic, geometry, and theory of integers; rational and natural numbers; complete induction; limit and point of accumulation; remarkable curves; complex and hypercomplex numbers, more. 1959 ed. 27 figures. xii+260pp. 5⅜ x 8½. 42804-4

POPULAR LECTURES ON MATHEMATICAL LOGIC, Hao Wang. Noted logician's lucid treatment of historical developments, set theory, model theory, recursion theory and constructivism, proof theory, more. 3 appendixes. Bibliography. 1981 ed. ix+283pp. 5⅜ x 8½. 67632-3

CALCULUS OF VARIATIONS, Robert Weinstock. Basic introduction covering isoperimetric problems, theory of elasticity, quantum mechanics, electrostatics, etc. Exercises throughout. 326pp. 5⅜ x 8½. 63069-2

THE CONTINUUM: A Critical Examination of the Foundation of Analysis, Hermann Weyl. Classic of 20th-century foundational research deals with the conceptual problem posed by the continuum. 156pp. 5⅜ x 8½. 67982-9

CHALLENGING MATHEMATICAL PROBLEMS WITH ELEMENTARY SOLUTIONS, A. M. Yaglom and I. M. Yaglom. Over 170 challenging problems on probability theory, combinatorial analysis, points and lines, topology, convex polygons, many other topics. Solutions. Total of 445pp. 5⅜ x 8½. Two-vol. set.
Vol. I: 65536-9 Vol. II: 65537-7

INTRODUCTION TO PARTIAL DIFFERENTIAL EQUATIONS WITH APPLICATIONS, E. C. Zachmanoglou and Dale W. Thoe. Essentials of partial differential equations applied to common problems in engineering and the physical sciences. Problems and answers. 416pp. 5⅜ x 8½. 65251-3

THE THEORY OF GROUPS, Hans J. Zassenhaus. Well-written graduate-level text acquaints reader with group-theoretic methods and demonstrates their usefulness in mathematics. Axioms, the calculus of complexes, homomorphic mapping, *p*-group theory, more. 276pp. 5⅜ x 8½. 40922-8

Math–Decision Theory, Statistics, Probability

ELEMENTARY DECISION THEORY, Herman Chernoff and Lincoln E. Moses. Clear introduction to statistics and statistical theory covers data processing, probability and random variables, testing hypotheses, much more. Exercises. 364pp. 5⅜ x 8½. 65218-1

STATISTICS MANUAL, Edwin L. Crow et al. Comprehensive, practical collection of classical and modern methods prepared by U.S. Naval Ordnance Test Station. Stress on use. Basics of statistics assumed. 288pp. 5⅜ x 8½. 60599-X

SOME THEORY OF SAMPLING, William Edwards Deming. Analysis of the problems, theory, and design of sampling techniques for social scientists, industrial managers, and others who find statistics important at work. 61 tables. 90 figures. xvii +602pp. 5⅜ x 8½. 64684-X

LINEAR PROGRAMMING AND ECONOMIC ANALYSIS, Robert Dorfman, Paul A. Samuelson and Robert M. Solow. First comprehensive treatment of linear programming in standard economic analysis. Game theory, modern welfare economics, Leontief input-output, more. 525pp. 5⅜ x 8½. 65491-5

PROBABILITY: An Introduction, Samuel Goldberg. Excellent basic text covers set theory, probability theory for finite sample spaces, binomial theorem, much more. 360 problems. Bibliographies. 322pp. 5⅜ x 8½. 65252-1

GAMES AND DECISIONS: Introduction and Critical Survey, R. Duncan Luce and Howard Raiffa. Superb nontechnical introduction to game theory, primarily applied to social sciences. Utility theory, zero-sum games, n-person games, decision-making, much more. Bibliography. 509pp. 5⅜ x 8½. 65943-7

INTRODUCTION TO THE THEORY OF GAMES, J. C. C. McKinsey. This comprehensive overview of the mathematical theory of games illustrates applications to situations involving conflicts of interest, including economic, social, political, and military contexts. Appropriate for advanced undergraduate and graduate courses; advanced calculus a prerequisite. 1952 ed. x+372pp. 5⅜ x 8½. 42811-7

FIFTY CHALLENGING PROBLEMS IN PROBABILITY WITH SOLUTIONS, Frederick Mosteller. Remarkable puzzlers, graded in difficulty, illustrate elementary and advanced aspects of probability. Detailed solutions. 88pp. 5⅜ x 8½. 65355-2

PROBABILITY THEORY: A Concise Course, Y. A. Rozanov. Highly readable, self-contained introduction covers combination of events, dependent events, Bernoulli trials, etc. 148pp. 5⅜ x 8¼. 63544-9

STATISTICAL METHOD FROM THE VIEWPOINT OF QUALITY CONTROL, Walter A. Shewhart. Important text explains regulation of variables, uses of statistical control to achieve quality control in industry, agriculture, other areas. 192pp. 5⅜ x 8½. 65232-7

Math–Geometry and Topology

ELEMENTARY CONCEPTS OF TOPOLOGY, Paul Alexandroff. Elegant, intuitive approach to topology from set-theoretic topology to Betti groups; how concepts of topology are useful in math and physics. 25 figures. 57pp. 5⅜ x 8½. 60747-X

COMBINATORIAL TOPOLOGY, P. S. Alexandrov. Clearly written, well-organized, three-part text begins by dealing with certain classic problems without using the formal techniques of homology theory and advances to the central concept, the Betti groups. Numerous detailed examples. 654pp. 5⅜ x 8½. 40179-0

EXPERIMENTS IN TOPOLOGY, Stephen Barr. Classic, lively explanation of one of the byways of mathematics. Klein bottles, Moebius strips, projective planes, map coloring, problem of the Koenigsberg bridges, much more, described with clarity and wit. 43 figures. 210pp. 5⅜ x 8½. 25933-1

CONFORMAL MAPPING ON RIEMANN SURFACES, Harvey Cohn. Lucid, insightful book presents ideal coverage of subject. 334 exercises make book perfect for self-study. 55 figures. 352pp. 5⅝ x 8¼. 64025-6

THE GEOMETRY OF RENÉ DESCARTES, René Descartes. The great work founded analytical geometry. Original French text, Descartes's own diagrams, together with definitive Smith-Latham translation. 244pp. 5⅜ x 8½. 60068-8

PRACTICAL CONIC SECTIONS: The Geometric Properties of Ellipses, Parabolas and Hyperbolas, J. W. Downs. This text shows how to create ellipses, parabolas, and hyperbolas. It also presents historical background on their ancient origins and describes the reflective properties and roles of curves in design applications. 1993 ed. 98 figures. xii+100pp. 6½ x 9¼. 42876-1

THE THIRTEEN BOOKS OF EUCLID'S ELEMENTS, translated with introduction and commentary by Thomas L. Heath. Definitive edition. Textual and linguistic notes, mathematical analysis. 2,500 years of critical commentary. Unabridged. 1,414pp. 5⅜ x 8½. Three-vol. set. Vol. I: 60088-2 Vol. II: 60089-0 Vol. III: 60090-4

GEOMETRY OF COMPLEX NUMBERS, Hans Schwerdtfeger. Illuminating, widely praised book on analytic geometry of circles, the Moebius transformation, and two-dimensional non-Euclidean geometries. 200pp. 5⅝ x 8¼. 63830-8

DIFFERENTIAL GEOMETRY, Heinrich W. Guggenheimer. Local differential geometry as an application of advanced calculus and linear algebra. Curvature, transformation groups, surfaces, more. Exercises. 62 figures. 378pp. 5⅜ x 8½. 63433-7

CURVATURE AND HOMOLOGY: Enlarged Edition, Samuel I. Goldberg. Revised edition examines topology of differentiable manifolds; curvature, homology of Riemannian manifolds; compact Lie groups; complex manifolds; curvature, homology of Kaehler manifolds. New Preface. Four new appendixes. 416pp. 5⅜ x 8½. 40207-X

History of Math

THE WORKS OF ARCHIMEDES, Archimedes (T. L. Heath, ed.). Topics include the famous problems of the ratio of the areas of a cylinder and an inscribed sphere; the measurement of a circle; the properties of conoids, spheroids, and spirals; and the quadrature of the parabola. Informative introduction. clxxxvi+326pp; supplement, 52pp. 5⅜ x 8½. 42084-1

A SHORT ACCOUNT OF THE HISTORY OF MATHEMATICS, W. W. Rouse Ball. One of clearest, most authoritative surveys from the Egyptians and Phoenicians through 19th-century figures such as Grassman, Galois, Riemann. Fourth edition. 522pp. 5⅜ x 8½. 20630-0

THE HISTORY OF THE CALCULUS AND ITS CONCEPTUAL DEVELOPMENT, Carl B. Boyer. Origins in antiquity, medieval contributions, work of Newton, Leibniz, rigorous formulation. Treatment is verbal. 346pp. 5⅜ x 8½. 60509-4

THE HISTORICAL ROOTS OF ELEMENTARY MATHEMATICS, Lucas N. H. Bunt, Phillip S. Jones, and Jack D. Bedient. Fundamental underpinnings of modern arithmetic, algebra, geometry, and number systems derived from ancient civilizations. 320pp. 5⅜ x 8½. 25563-8

A HISTORY OF MATHEMATICAL NOTATIONS, Florian Cajori. This classic study notes the first appearance of a mathematical symbol and its origin, the competition it encountered, its spread among writers in different countries, its rise to popularity, its eventual decline or ultimate survival. Original 1929 two-volume edition presented here in one volume. xxviii+820pp. 5⅜ x 8½. 67766-4

GAMES, GODS & GAMBLING: A History of Probability and Statistical Ideas, F. N. David. Episodes from the lives of Galileo, Fermat, Pascal, and others illustrate this fascinating account of the roots of mathematics. Features thought-provoking references to classics, archaeology, biography, poetry. 1962 edition. 304pp. 5⅜ x 8½. (Available in U.S. only.) 40023-9

OF MEN AND NUMBERS: The Story of the Great Mathematicians, Jane Muir. Fascinating accounts of the lives and accomplishments of history's greatest mathematical minds–Pythagoras, Descartes, Euler, Pascal, Cantor, many more. Anecdotal, illuminating. 30 diagrams. Bibliography. 256pp. 5⅜ x 8½. 28973-7

HISTORY OF MATHEMATICS, David E. Smith. Nontechnical survey from ancient Greece and Orient to late 19th century; evolution of arithmetic, geometry, trigonometry, calculating devices, algebra, the calculus. 362 illustrations. 1,355pp. 5⅜ x 8½. Two-vol. set. Vol. I: 20429-4 Vol. II: 20430-8

A CONCISE HISTORY OF MATHEMATICS, Dirk J. Struik. The best brief history of mathematics. Stresses origins and covers every major figure from ancient Near East to 19th century. 41 illustrations. 195pp. 5⅜ x 8½. 60255-9